Lecture Notes in Control and Information Sciences

Edited by M. Thoma and A. Wyner

161

L. Gerencsér,
P. E. Caines (Eds.)

Topics in Stochastic Systems: Modelling, Estimation and Adaptive Control

Springer-Verlag
Berlin Heidelberg New York
London Paris Tokyo
Hong Kong Barcelona Budapest

Editors
Professor L. Gerencsér
Computer and Automation Institute
of the Hungarian Academy of Sciences
Budapest, H-1111
Kende u 13-17
Hungary

Professor P. E. Caines
Dept. of Electrical Engineering
McGill University
3480 University Street
Montréal, Qué.
Canada H3A 2A7

ISBN 3-540-54133-0 Springer-Verlag Berlin Heidelberg New York
ISBN 0-387-54133-0 Springer-Verlag New York Berlin Heidelberg

Printed in Germany

Printing: Mercedes-Druck, Berlin
Binding: B. Helm, Berlin
61/3020-543210 Printed on acid-free paper.

PREFACE

The objective of this volume is to present a set of papers which serves to introduce the reader to the work of a group of currently active researchers in the areas of modelling, estimation and adaptive control. We asked our authors to be selective in presenting their results, choosing those which they thought would be both of topical interest to the research community and which, in addition, would involve methods which they believe will most probably lead to future advances. We asked them to do this in a rigourous but compact manner and at the same time to provide the reader with an interpretation of their results.

It is hoped that this approach of indicating the methods which are thought to be of value in the future will be helpful both as an introduction for those new to the area of stochastic systems, and of utility to those already involved in research and applications in the area.

Such surveys are plentiful in pure mathematics and serve the scientific community very well. The need for such a volume is even greater for a discipline such as stochastic systems, in which mathematical and engineering deliberations are inseparably intertwined. The rules and objectives of the engineering community must be acknowledged to provide those working in systems and control theory with a tremendous stimulus. However, we should also keep in mind the fact that the elaboration of the mathematical structures of the subject is vital for the further development of the purely engineering activity of the analysis and synthesis of control systems.

The mathematization of engineering science reached a degree of considerable maturity and mathematics became the language of a significant portion of engineering sciences. Thus quite a few reasonable questions asked by engineers quickly translate into mathematical problems. As an example we may mention the resurgence of interest in continuous time stochastic systems which is partly due to the need for high accuracy controllers in robotics.

In some other areas such as white-noise modelling or numerical solution of the nonlinear filtering equations a critical return to fundamentals seems to be necessary, rather than interpreting existing experimental results. In these instances mathematics and rigorous speculation take their appropriate position to provide direction for further experimentation.

We express the hope that our volume will clarify certain fundamental concepts and results and hence lead to a better understanding of the central problems of systems and control engineering.

This project was partially supported by the Natural Sciences and Engineering Research Council under the Grants 01329 and A1329. Finally we express our warm thanks to Jennifer Quinn, our "honorary co-editor" for the tremendous work she did in processing and performing many of the tasks that are inevitably involved in endeavours such as this.

Montreal, March 1991

L. Gerencsér
P.E. Caines

TOPICS IN STOCHASTIC SYSTEMS: MODELLING, ESTIMATION AND ADAPTIVE CONTROL

MODELLING

ESTIMATION

ADAPTIVE CONTROL

DIRECT MODELING OF WHITE NOISE IN STOCHASTIC SYSTEMS

Arunabha Bagchi
Dept. of Applied Mathematics
University of Twente
P.O. Box 217, 7500 AE Enschede
The Netherlands

Ravi Mazumdar
INRS - Telecommunications
3, Place du Commerce
Verdun, PQ-H3E 1H6
Canada

Abstract. In this paper we explain the line of development that led to a direct modeling of white noise in stochastic systems. This is followed by the mathematical theory of finitely additive white noise and some recent results on modeling the state process with white noise input. Likelihood ratio plays a central role throughout this paper.

1. WHITE NOISE IN COMMUNICATION AND CONTROL SYSTEMS.

We begin with the simple problem of estimating unknown parameters of a signal in a Gaussian white noise. Suppose that a deterministic signal $s(t;\theta)$ is observed in a time interval $[0,T]$ and the observation is corrupted by an additive Gaussian white noise $\{n_t, t \in [0,T]\}$. The observation is then one realization of a stochastic process

$$y_t = s(t;\theta) + n_t \tag{1.1}$$

Here θ denotes a vector of unknown parameters in the signal. Our problem is to obtain the maximum likelihood estimate of the unknown parameter vector θ based on the observation of y_t, $t \in [0,T]$.

In the case of discrete observations at time points $t_0, t_1, \ldots, t_N$, it is quite straightforward to calculate the probability density function of $(y_{t_0}, y_{t_1}, \ldots, y_{t_N})$ when we assume the additive noise to be a Gaussian white noise sequence (independent and identically distributed). This is then the likelihood function for the problem. The first difficulty in handling continuous data y_t, $t \in [0,T]$, is an appropriate notion of the probability density functional (we use the term functional since we are interested in obtaining a function of the continuous data). Now a realization of n_t, and therefore of y_t, is a generalized function and there is no analogue of Lebesgue measure on the "space of generalized functions", so that an extension of the idea of density from the finite-dimensional set-up is not possible). We can look to this difficulty in another way. The obvious approach for going over from a discrete to a continuous model is to perform limiting operation on the results obtained in the sampled case. Because of Gaussian assumption, the likelihood function in the discrete-time situation consists of the ratio of an exponential and a determinant term. The limit of this likelihood function is meaningless as both the determinant and the exponential term go to zero as $N \to \infty$, where N denotes the number of sampling moments. The way around this difficulty is to work with the likelihood ratio instead. This is because the like- lihood ratio eliminates the determinant term appearing in the likeli- hood function.

This sampling and limiting procedure has been justified and successfully carried out by Parzen [34] to calculate the likelihood ratio when the covariance function of the additive Gaussian noise term is reasonably smooth. When the noise term is white, a heuristic approach may be followed. Consider a partition $T_N = \{t_k^N;\ k = 0,1,\ldots,\nu(N)\}$ such that

(i) $\quad 0 = t_0^N < t_1^N < \ldots < t_{\nu(N)}^N = T$

(ii) $\quad T_{N+1} \supset T_N$

(iii) $\quad \max_k |t_{k+1}^N - t_k^N| \to 0 \qquad \text{as } N \to \infty$

We note that, for N large enough, $n_{t_k^N}(t_{k+1}^N - t_k^N)$ is Gaussian, with zero mean and approximate variance $(t_{k+1}^N - t_k^N)$. From (1.1), we may write

$$y_{t_k^N}(t_{k+1}^N - t_k^N) = s(t_k^N;\theta)(t_{k+1}^N - t_k^N) + n_{t_k^N}(t_{k+1}^N - t_k^N) \tag{1.2}$$

and, furthermore, $n_{t_j^N}(t_{j+1}^N - t_j^N)$ and $n_{t_k^N}(t_{k+1}^N - t_k^N)$ are independent for $j \neq k$.

Dividing the probability density function of $\{y_{t_k^N}(t_{k+1}^N - t_k^N);$ $k = 0,1,\ldots,\nu(N)-1\}$ by that of $\{n_{t_k^N}(t_{k+1}^N - t_k^N);\ k = 0,1,\ldots,\nu(N)-1\}$ and evaluating it for $\{y_{t_k^N}(t_{k+1}^N - t_k^N)\}$, we get the sampled likelihood ratio to be

$$L_{T_N}(y_t \mid t \in T_N)$$

$$= \exp. - \frac{1}{2}[\sum_{k=0}^{\nu(N)-1} s(t_k^N;\theta)^2 (t_{k+1}^N - t_k^N) - 2 \sum_{k=0}^{\nu(N)-1} s(t_k^N;\theta) y_{t_k^N} (t_{k+1}^N - t_k^N)] .$$

If $s(t;\theta)$ is square integrable, we can take the limit as $N \to \infty$ and obtain heuristically the likelihood ratio

$$L_T(y_. ;\theta) = \exp. - \frac{1}{2}[\int_0^T s(t;\theta)^2 dt - 2 \int_0^T s(t;\theta) y_t \, dt] \tag{1.3}$$

This formula has been used extensively from the fifties by communication engineers for signal detection and for signal parameter estimation problems [20].

The expression on the right hand side of (1.3) is not as obvious as it looks. The interpretation of the second integral requires some care. We can write the second integral as

$$\int_0^T s(t;\theta) y_t \, dt = \int_0^T s(t;\theta)^2 \, dt + \int_0^T s(t;\theta) n_t \, dt \tag{1.4}$$

where the first term is an ordinary integral, but the second term has to be defined. This is by no means obvious, as the sample paths of $\{n_t\}$ are generalized functions. As long as $s(t;\theta)$ is a deterministic function, this "so-called" white noise integral can, however, be defined as the limit of a sequence of well-defined stochastic integrals (see Wong and Hajek [36], for details). The situation changes completely when the signal is a stochastic process $s_t(\theta)$. We cannot interpret $\int_0^T s_t(\theta) n_t \, dt$ anymore as directly as in the case of a deterministic signal. The likelihood ratio formula cannot also be derived readily by following the sampling and limiting procedure discussed above.

The stochastic model that attracted particular attention in the sixties is the linear dynamical model used in Kalman filtering. In the original model of Kalman the white noise integral as defined in [36] provides adequate mathematical framework to study the filtering problem. It was soon recognized that this framework is inadequate for studying the non-linear filtering problem. Even for linear model, the likelihood ratio connot be determined with the help of white noise integral alone. These led control theorists to go to the integral version of the observation model (1.1) in the stochastic signal case. The integral of white noise has the properties of a Wiener process and this integrated form will be referred to as the "Wiener process model" as opposed to the original "white noise model".

Wiener process is mathematically a better defined object than white noise and a complete theory of stochastic calculus involving Wiener process has been developed by K. Itô in the early fifties. This helped in successfully extending Kalman filter to the nonlinear case, (see the survey paper [15]). Parallel to this development, there was intensive activity to obtain the likelihood ratio formula for models of the form (1.1) with the signal process being stochastic. This culminated in the famous formula of Kailath [23] which is remarkable in its generality, as well as its simplicity. Suppose that the observation model is given by

$$y_t = s_t + n_t, \quad 0 \le t \le T \tag{1.5}$$

The integrated form of this "white noise model" is

$$Y_t = \int_0^t y_\sigma d\sigma = \int_0^t s_\sigma d\sigma + w_t, \qquad 0 \le t \le T \tag{1.6}$$

where $\{w_t\}$ is a Wiener process. Let μ_Y denote the measure induced on $C[0,T]$, the space of continuous functions on $[0,T]$, by $\{Y_t, \ 0 \le t \le T\}$ and μ_w denote the Wiener measure thereon. Kailath [23] showed that, for the "Wiener process model" (1.6),

$$\frac{d\mu_Y}{d\mu_w}(Y) = \exp. -\frac{1}{2}\Big[\int_0^T \hat{s}_t^2 \, dt - 2\int_0^T \hat{s}_t \, dY_t\Big] \tag{1.7}$$

where $\hat{s}_t \triangleq E[s_t|\mathcal{B}_Y(t)]$, with $\mathcal{B}_Y(t)$ denoting the smallest σ-algebra generated by $\{Y_\sigma, \ 0 \le \sigma \le t\}$, and the second integral in (1.7) is an Itô integral [36]. With the phenomenal success of working with the "Wiener process model" and the associated mathematical rigor, researchers all but forgot the "white noise model" which was so prevalent just a few years ago.

Immediately after the publication of [23] attempt was made to apply this result to the aircraft parameter identification problem using real flight record data [22]. The problem considered in [22] may be formulated as follows. Given

$$\dot{x}_t = Ax_t + Bu(t) + Fn_t, \ x_0 = 0 \tag{1.8a}$$
$$y_t = Cx_t + Du(t) + G\, n_{0t} \tag{1.8b}$$

where $n_{\cdot}$ and $n_{0\cdot}$ are uncorrelated Gaussian white noise processes in appropriate dimensions, $u(\cdot)$ is a given input and $y_{\cdot}$ is the observed output of dimension m, it is desired to identify some or all of the parameters in the matrices A,B,C,D and F. The matrix G is known and we assume for simplicity that $GG^* = I$. The "Wiener process model" of (1.8a-b) is, of course,

$$x_t = \int_0^t Ax_s ds + \int_0^t Bu(s)ds + \int_0^t Fdw_s \tag{1.9a}$$

$$Y_t = \int_0^t Cx_s ds + \int_0^t Du(s)ds + Gw_{0t} \tag{1.9b}$$

with $\{w_t, \ t \geq 0\}$ and $\{w_{0t}, \ t \geq 0\}$ independent Wiener processes. The multidimensional form of (1.7) gives the likelihood ratio for the model (1.9a-b) to be

$$L_T \triangleq \frac{d\mu_Y}{d\mu_w}(Y) = \exp. -\frac{1}{2}\left\{ \int_0^T \|C\hat{x}_t\|^2 \, dt - 2\int_0^T [C\hat{x}_t, dY_t] \right\} \tag{1.10}$$

where $\|\cdot\|$ and $[\cdot,\cdot]$ denote norm and inner product, respectively, in a Euclidean space and $\hat{x}_t$ is the Kalman filter.

It was soon realized that the formula (1.10), however elegant and remarkable mathematically, cannot be directly used in the "white noise model" (1.8a-b). There is a fundamental discrepancy here between model and reality. The solution of (1.8a) may be readily interpreted by means of white noise integrals because of the linearity of the equation. In fact, the solution process is equivalent to that of the linear Itô equation (1.9a). The observation model (1.8b), even in the linear case, is more involved. The model (1.8b) is physically plausible. A stationary noise n_0 is considered whose power spectral density is approximately constant upto frequencies which are high relative to the bandwidth of the signal process. In the absence of any specific knowledge of the bandwidth of the spectrum of $\{n_{0t}\}$, we may consider an idealized situation when the spectrum is constant throughout. This is the justification for assuming $\{n_{0t}\}$ to be a white noise. The mathematical interpretation is more complicated. Suppose that we have a sequence of stationary Gaussian processes $\{n^k_{0t}\}$ which "converges" in some suitable sense to a Gaussian white noise, and yet, for each k, n^k_{0t} has well-behaved sample functions. We can think of the actual observation to be of the form

$$y_t^k = C\, x_t + Du(t) + n_{0\,t}^k \tag{1.11}$$

for some k. This is, indeed, realistic as the actual observation is always a well-behaved function, while the sample-path of the process $y_.$ defined by (1.8b) must be a generalized function. If we substitute $\int_0^t y_\sigma^k\, d\sigma$ instead of Y_t in the formula (1.10), we get the approximate likelihood ratio

$$L_T^k = \exp.\ -\frac{1}{2}\left\{ \int_0^T \|C\, \hat{x}_t^k\|^2 dt - 2\int_0^T \left[C\, \hat{x}_t^k, y_t^k\right] dt\right\} \tag{1.10'}$$

where $\hat{x}_t^k$ is given by the Kalman filter with dY_t replaced by $y_t^k dt$. If the model (1.9a-b) were to match with the actual situation, L_T^k should converge to L_T as $k \to \infty$. Unfortunately, however, this is not the case as has been observed in [7].

We can also interpret this discrepancy probabilistically. The actual observation $y_.$ has well-behaved sample paths, so that its integrated from $Y_.$ is absolutely continuous. On the other hand, the likelihood ratio formula (1.10) holds only almost surely with respect to p_W-measure and the set of absolutely continuous functions has p_W-measure zero. Therefore, the formula (1.10) has no validity when applied to operational data (in integrated form) which are always absolutely continuous. One encounters identical problem when nonlinear filtering equations derived from the "Wiener process model" are applied to the actual data. The question is whether it is possible to develop a theory of white noise which yields results that one would obtain as the limits of corresponding results for band-limited processes. Despite apparent difficulties of mathematically working with white noise can one still develop a direct theory with white noise, instead of working with the integrated version of the model? In a series of papers, Balakrishnan ([9],[10],[11]) showed that this is indeed possible, although at the cost of considerable additional mathematical complications. The basic idea is to define white noise in [0,T], directly as the "coordinate process" in $L_2[0,T]$ with canonical Gaussian measure thereon, just as a Wiener process in [0,T] may be defined as the "coordinate process" in $\mathbb{C}[0,T]$ with the Wiener measure thereon. The problem is that canonical Gaussian measure on $L_2[0,T]$ is only finitely additive on the Borel sets there and cannot be extended to be a countably additive measure. The process obtained in this manner is also not the same process as the derivative of a Wiener process, as the sample paths of the derivative would be almost surely generalized functions. These two processes, however, will determine the same probability distributions on all finite-dimensional projections. It should be mentioned that Hida [21] also gave a direct definition of white noise by introducing canonical Gaussian measure on a space of generalized functions. Since actual available observation can never be a generalized function, this theory does not help resolve the discrepancy between the model and reality for our problem.

The organization of our paper is as follows. In section 2 we give the main line of development for a direct theory of finitely-additive white noise. In section 3 we apply this to the likelihood ratio problem and discuss the connection with the Itô correction term. In section 4 some recent results on modeling the state process with white noise as input are considered. Finally, in section 5 we discuss a number of unsolved problems arising in this "white noise model" and some new questions in stochastic system modeling where the finitely- additive white noise theory may be fruitfully applied. This paper is not intended to be a survey of the whole field. The theory of finitely- additive white noise has been put into a rigorous mathematical frame- work in a series of papers by Kallianpur and Karandikar which culminat- ed in their excellent tretise on the subject ([24],[25],[26],[27]). Our approach is less formal and focusses only on those aspects of the subject not considered in sufficient detail in [27].

2. FINITELY ADDITIVE WHITE NOISE THEORY - AN OVERVIEW.

Let $\mathcal{H}$ be a real, separable Hilbert space. Let $\mathcal{P}$ be the class of all finite dimensional projections on $\mathcal{H}$. For any $P \in \mathcal{P}$, a set of the form $P^{-1}(B)$, B Borel set in range P, is called a cylinder set with B as its base. For fixed P, we denote the collection of all such sets by $\mathcal{C}_P$ which is clearly a σ-algebra. The collection $\mathcal{C} = \{C \in \mathcal{C}_P | P \in \mathcal{P}\}$ of all cylinder sets is an algebra.

A finitely additive measure μ on $(\mathcal{H},\mathcal{C})$ is called a *cylindrical measure* if the restriction of μ on $\mathcal{C}_P$ is a countably additive measure. If $\mu(\mathcal{H}) = 1$, we call $(\mathcal{H},\mathcal{C},\mu)$ a cylindrical probability space. For any finite number of elements h_i in $\mathcal{H}$, consider the map $\phi: \mathcal{H} \to \mathbb{R}^n$ defined by $\phi(h) = ([h,h_1],\dots,[h,h_n])$. The mapping induces a countably additive probability measure ν on $\mathbb{R}^n$ defined by $\nu(B) = \mu(\{h \in \mathcal{H} | \phi(h) \in B\})$, B

Borel in $\mathbb{R}^n$. That ν is countably additive follows from the definition of a cylindrical measure. In particular, for n = 1 we can define the characteristic functional corresponding to a cylindrical measure by

$$C_\mu(h') = \int_{\mathcal{H}} e^{i[h,h']} \, d\mu(h) \stackrel{\Delta}{=} \int_{-\infty}^{\infty} e^{iy} \, d\nu_{h'}(y) \tag{2.1}$$

where $\nu_{h'}$ is the measure on $\mathbb{R}$ defined by

$$\nu_{h'}\big((-\infty,y]\big) = \mu(\{h \in H | [h,h'] \le y\}).$$

Let $\mathcal{B}$ denote the Borel σ-algebra of $\mathcal{H}$; it is the smallest σ-algebra containing $\mathcal{C}$. Since μ cannot, in general, be extended to be countably additive on $\mathcal{B}$, most results of measure theory cannot be directly used in this framework. Let $f:\mathcal{H} \to \mathcal{H}'$ (another Hilbert space) be a Borel measurable map. This does not qualify f to be a random variable, since inverse images $f^{-1}(B')$, B' Borel set in $\mathcal{H}'$, are not necessarily in $\mathcal{C}$ and probabilities cannot be assigned to these "events". Note, however, that $\{f^{-1}(B') | B' \text{ Borel set in } \mathcal{H}'\}$, for fixed f, is a σ-algebra and it may well be possible to extend μ by limiting procedure to this sub-σ-algebra of $\mathcal{B}$. Those Borel measurable maps f for which this is possible will qualify to be random variables. This is formalized in the following definition:

Definition 2.1. A measurable map of the form $f \circ P$, P a finite-dimensional projection operator on $\mathcal{H}$ and $f:P\mathcal{H} \to \mathcal{H}'$, is called a *tame function*. Inverse images of tame functions can always be assigned probabilities. Let $\mathcal{L}(\mathcal{H},\mathcal{C},\mu;\mathcal{H}')$ be the class of Borel measurable functions $f:\mathcal{H} \to \mathcal{H}'$ such that for all $\epsilon > 0$, $\delta > 0$, $\exists\, P_0 \in \mathcal{P}$ such that $P_1, P_2 \in \mathcal{P}$, $P_0 \le P_i$, $i = 1,2$, ($P_0 \le P_i$ if Range $P_0 \subseteq$ Range P_i) implies that

$$\mu\big(\{h \in \mathcal{H} | \; |f \circ P_1(h) - f \circ P_2(h)| > \delta\}\big) < \epsilon \tag{2.1}$$

An element in $\mathcal{L}(\mathcal{H},\mathcal{C},\mu;\mathcal{H}')$ is called $\mathcal{H}'$-valued μ-random variable. When $\mathcal{H}' = \mathbb{R}$, we denote the space simply by $\mathcal{L}(\mathcal{H},\mathcal{C},\mu)$.

Definition 2.2. A mapping $f:(\mathcal{H},\mathcal{C},\mu) \to (\mathcal{H}',\mathcal{C}')$ is called a $\mathcal{H}'$-valued μ-weak random variable if, for all $h' \in \mathcal{H}'$,

$$(h',f)' \in \mathcal{L}(\mathcal{H},\mathcal{C},\mu)$$

where $(\cdot,\cdot)'$ is the inner product in $\mathcal{H}'$.

Remark. A $\mathcal{H}'$-valued μ-weak random variable f induces a cylindrical measure μ' on $\mathcal{H}'$ which we denote by $\mu \circ f^{-1}$.

The most important class of random variables is that of physical random variables (p.r.v.'s for short).

Definition 2.3. Let $f: \mathcal{H} \to \mathcal{H}'$ be Borel measurable and $\{P_n\}$ be a sequence of monotonically increasing finite dimensional projections converging strongly to the identity operator; that is, $P_n \overset{s}{\uparrow} I$. Suppose that $\{f(P_n h)\}$ is Cauchy in probability and

$$C(h') = \lim_{n\to\infty} \int_{\mathcal{H}} e^{i[f\circ P_n(h),h']} d\mu(h)$$

is independent of the sequence $\{P_n\}$ chosen. Then we call $f(\cdot)$ a μ-physical random variable. We can then always extend μ to events of the form $f^{-1}(B')$, B' Borel in $\mathcal{H}'$, by

$$\mu(\{f^{-1}(B'),\ B'\text{Borel in } \mathcal{H}'\})$$

$$= \lim_{n\to\infty} \mu(\{h \in \mathcal{H} | f\circ P_n(h) \in B',\ B' \text{ Borel in } \mathcal{H}'\}) \tag{2.2}$$

where the limit always exists, by definition. Furthermore, f always induces a countably additive measure on $\mathcal{B}'$, the Borel σ-algebra of $\mathcal{H}'$. We denote the class of $\mathcal{H}'$-valued physical random variables by $\mathcal{L}^1(\mathcal{H},\mathcal{C},\mu;\mathcal{H}')$. When $\mathcal{H}' \equiv \mathbb{R}$, we denote this class by $\mathcal{L}^1(\mathcal{H},\mathcal{C},\mu)$. The definition of a *$\mu$-weak physical random variable* is similar.

Definition 2.4. The cylindrical measure μ_G corresponding to the characteristic functional

$$C(h) = \exp.\ \{-\tfrac{1}{2}\ \|h\|^2\} \tag{2.3}$$

is called the *Gauss measure* on $\mathcal{H}$. μ_G can also be defined directly by

$$\mu_G(C) = \int_{\tilde{B}} G(x)dx \tag{2.4}$$

where $\tilde{B}$ is the Borel set in $\mathbb{R}^n$ isomorphic to B, base of C, and $G(\cdot)$ denotes the n-dimensional Gaussian density with zero mean and identity covariance. When we do not specifically mention the cylindrical measure, we always mean it to be the Gauss measure μ_G. We refer to μ_G-(physical) random variable simply as (physical) random variable.

Definition 2.5. The identity map $I:(\mathcal{H},\mathcal{C},\mu_G) \to (\mathcal{H},\mathcal{C})$ is called a *Gaussian white noise* on $\mathcal{H}$.

Examples.

1. Let $L: \mathcal{H} \to \mathcal{H}'$ be linear and bounded. Then L is a physical random variable if and only if L is Hilbert-Schmidt.
2. $f: \mathcal{H} \to \mathbb{R}$ is defined by $f(h) = [Lh,h]$, where $L: \mathcal{H} \to \mathcal{H}$ is linear and bounded. Then f is a physical random variable if and only if $(L + L^*)$ is nuclear.
3. Gaussian white noise is a weak physical random variable, since μ_G is countably additive on finite dimensional subspaces. It is never a physical random variable (example 1).
4. From example 2 it follows that $f: \mathcal{H} \to \mathbb{R}$ defined by $f(h) = \|h\|^2$ is never a physical random variable.

Let us now relate our abstract definition of white noise with the physical notion of it. Let $\mathcal{D}$ be any bounded domain in $\mathbb{R}^n$ and define $\mathcal{H} = L_2(\mathcal{D};H)$, the space of square-integrable functions from $\mathcal{D}$ into a Hilbert space H. We denote the Gaussian white noise by n, which is the identity map on $L_2(\mathcal{D};H)$. We may intuitively define the process n_t, $t \in \mathcal{D}$, with values in H as

$$n_t(h) = h(t); \; t \in \mathcal{D}, \; h \in L_2(\mathcal{D};H) \tag{2.5}$$

It is easy to see that h denoting a sample point in $L_2(\mathcal{D};H)$, and for $g_1, g_2 \in L_2(\mathcal{D};H)$, $[g_1,h]$ and $[g_2,h]$ are jointly Gaussian with zero mean and covariance $[g_1,g_2]$. When $g_1(\cdot)$ and $g_2(\cdot)$ have nonintersecting support in $\mathcal{D}$, this covariance is zero and, in this sense, the process n_t at two distinct points in $\mathcal{D}$ are independent (whiteness).

Examples.

5. Let $\mathcal{H} = L_2([0,T];H)$. Then the identity map on $(\mathcal{H},\mathcal{C},\mu_G)$ is a white noise on [0,T] with values in H. This formulation enables us to model white noise on [0,T] with values in an infinite dimensional space with intensity being an identity operator. In particular, if H is a Hilbert space of spatial variables, the process defined in this way will be white both in time and space. In the usual formulation of Wiener

process with values in an infinite dimensional space one has to make the restrictive assumption that the intensity associated with the covariance must be a nuclear operator.

6. Let $\mathcal{D} = [0,T_1] \times [0,T_2]$ be a rectangle in the plane. Then the identity map on $(\mathcal{H},\mathcal{C},\mu_G)$ is a white noise on the plane with values in H. We can easily extend this to any finite dimension. We may replace the rectangle by any bounded domain in $\mathbb{R}^2$ (in general, in $\mathbb{R}^n$). Furthermore, we can use the same formulation to define white noise on the boundary $\partial\mathcal{D}$ of a domain $\mathcal{D}$ with sufficiently smooth boundary (see [6] for an example).

The next question that we have to look into is the absolute continuity of two cylindrical measures. This is, of course, crucial for studying the likelihood ratio.

Definition 2.6. A cylindrical measure λ is absolutely continuous with respect to another cylindrical measure μ if, for any $\epsilon > 0$, there exists a $\delta > 0$ such that for any cylinder set $C \in \mathcal{C}$, $\mu(C) < \delta \Rightarrow \lambda(C) < \epsilon$.

Definition 2.7. Suppose that λ and μ are cylindrical measures on $(\mathcal{H},\mathcal{C})$ and λ is absolutely continuous with respect to μ. We say that f is the Radon-Nikodym derivative of λ with respect to μ if f is a μ-physical random variable and, for any increasing sequence of projections $P_n \in \mathcal{P}$ converging strongly to the idendity,

$$\lambda(C) = \lim_{n\to\infty} \int_C f(P_n h)d\mu(h)$$

uniformly in $C \in \mathcal{C}$.

Another important concept we need to define in the finitely additive white noise framework is that of conditional expectation. This is, of course, essential for studying nonlinear estimation problems. Although nonlinear estimation results will not be explicitly considered in this paper, they will appear in other problems considered here and the definition is included here for completeness. To define conditional expectation properly, we need a number of preparatory notions.

It has been shown by Gross [19] that cylindrical measures and weak distributions are identical notions. Weak distribution has been introduced by Segal [35]. The point of interest to us is that, corresponding to any cylindrical measure μ, there exist always a probability space $(\Omega_1, \mathcal{B}_1, \Pi_1)$ and a linear map L from H^* (identified with H) to the linear space $\mathcal{L}(\Omega_1, \mathcal{B}_1, \Pi_1)$ of random variables on $(\Omega_1, \mathcal{B}_1, \Pi_1)$ such that

$$\mu(\{h \in H | ([h,h_1], \ldots, [h,h_n]) \in B\})$$

$$= \Pi_1(\{\omega_1 \in \Omega_1 | (L(h_1)(\omega_1), \ldots, L(h_n)(\omega_1)) \in B\}) \tag{2.6}$$

for all Borel sets B in $\mathbb{R}^n$; $h_1, \ldots, h_n \in \mathcal{H}$ and $n \geq 1$. Such a representation is not unique.

Definition 2.8. Let f be a tame function of the form

$$f(h) = \phi([h,h_1], \ldots, [h,h_n]) \tag{2.7}$$

for some $k \geq 1$; $h_1, \ldots, h_n \in \mathcal{H}$ and $\phi: \mathbb{R}^n \to \mathbb{R}$ is a Borel function. The μ-lifting of f is defined by

$$R_\mu(f) \triangleq \phi\,(L(h_1), \ldots, L(h_n)) \tag{2.8}$$

For any arbitrary $f \in \mathcal{L}(\mathcal{H}, \mathcal{C}, \mu)$ the μ-lifting $R_\mu(f)$ is defined by

$$R_\mu(f) = \text{l.i. } \Pi_1 \text{ probability } R_\mu(f \circ P) \tag{2.9}$$

It has been shown in [27] that such a limit always exists.

Definition 2.9. Let μ' be the measure induced on $(\mathcal{H}', \mathcal{C}')$ by a weak random variable $\phi: (\mathcal{H}, \mathcal{C}, \mu) \to (\mathcal{H}', \mathcal{C}')$. We denote this by $\mu' \triangleq \mu \circ \phi^{-1}$.

We define the class of functions $\mathcal{U}(\mathcal{H}, \mathcal{C}, \mu; \phi)$ by

$$\mathcal{U}(\mathcal{H}, \mathcal{C}, \mu; \phi) \triangleq \{g \in \mathcal{L}(\mathcal{H}', \mathcal{C}', \mu') | g \circ \phi \in \mathcal{L}(\mathcal{H}, \mathcal{C}, \mu) \text{ and } R_{\mu'}(g) = R_\mu(g \circ \phi)\}.$$

We also write it often for short as $\mathcal{U}(\phi)$.
Now we are in a position to define conditional expectation.

Definition 2.10. Let $\phi: (\mathcal{H}, \mathcal{C}, \mu) \to (\mathcal{H}', \mathcal{C}')$ be a weak random variable and $\mu' = \mu \circ \phi^{-1}$. Let $g \in \mathcal{L}^1(\mathcal{H}, \mathcal{C}, \mu)$. If there exists $g' \in \mathcal{U}(\mathcal{H}, \mathcal{C}, \mu; \phi)$ with $g' \in \mathcal{L}^1(\mathcal{H}', \mathcal{C}', \mu')$ such that for all bounded continuous real-valued tame functions f' on $\mathcal{H}'$,

$$\int_{\mathcal{H}} g(h) f'(\phi(h)) d\mu(h) = \int_{\mathcal{H}'} g'(h') f'(h') d\mu'(h') \tag{2.10}$$

then $g' \circ \phi$ is defined to be the conditional expectation of g given ϕ and is expressed as

$$E_{\mu}(g|\phi) = g' \circ \phi \tag{2.11}$$

This definition of conditional expectation is due to Kallianpur and Karandikar [27]. This definition is of fundamental importance for studying nonlinear estimation problems in the finitely additive white noise framework in a rigorous manner. We do not discuss them here. We would only like to mention that the original nonlinear filtering equations derived in the "Wiener process model" hold only a.s. and, therefore, cannot be applied to real data which belongs to a class of measure zero. To circumvent this difficulty, Davis [14] developed the theory of robust nonlinear filtering which can be meaningfully applied to real data. Robust nonlinear filtering equation differs from the usual nonlinear filtering equation by the presence of the so-called "correction term". In the "white noise model" one gets already the non-linear filtering equation in the "robust form" and this clearly shows one advantage of this approach. For nonlinear smoothing with the "white noise model", see [4]. We now consider the likelihood ratio for signal-plus-noise model discussed in Section 1 and show that the "white noise model" gives the proper expression for the likelihood ratio which can be directly applied to the real data.

3. LIKELIHOOD RATIO FOR SIGNAL-PLUS NOISE MODEL

We now go back to the original model (1.5) used by communication engineers in the fifties:

$$y_t = s_t + n_t, \qquad 0 \le t \le T \tag{3.1}$$

We have to carefully specify the probability space on which the above processes are defined. n is the Gaussian white noise on $(L_2[0,T], \mathcal{C}, \mu_G)$, following definition 2.5 with $\mathcal{H}$ replaced by $L_2[0,T]$. n being the identity map on $L_2[0,T]$, we may intuitively define the process n_t as

$$n_t(h) = h(t); \; t \in [0,T], \; h \in L_2[0,T] \tag{3.2}$$

The signal process s_t is defined on some probability space $(\Omega, \mathcal{B}, \Pi)$ and is smoother than the noise process in the sense that the sample path is in $L_2[0,T]$ and

$$E\int_0^T |s_t(\omega)|^2\, dt < \infty \tag{3.3}$$

Condition (3.3) implies that the covariance operator $\mathcal{R}: L_2[0,T] \to L_2[0,T]$ defined by ($s(\omega)$ denoting $s_.(\omega)$):

$$E\left([s(\omega),g_1][s(\omega),g_2]\right) = \left[\mathcal{R}\, g_1, g_2\right]; \; g_1, g_2 \in L_2[0,T],$$

is nuclear. It is then well known [12] that the mapping

$$s_.(\omega): \Omega \to L_2[0,T]$$

induces a countably additive measure μ_s on $L_2[0,T]$. Now s and n are defined on different probability spaces and to make (3.1) precise, we have to define s and n on the same probability space. Intuitively, it is obvious that we have to consider processes on the product space $\Gamma = L_2[0,T] \times \Omega$.

We follow [27] for formal development. Let

$$\mathcal{F} = \bigcup_{P \in \mathcal{P}} \mathcal{C}_P \otimes \mathcal{B},$$

where $\mathcal{C}_P \otimes \mathcal{B}$ is the usual product σ-algebra.

For $P \in \mathcal{P}$, $\mu_G|_P$ is a countably additive measure and we define γ_P to be the product measure of $\mu_G|_P$ and Π. Thus γ_P is a countably additive measure on $(\Gamma, \mathcal{C}_P \otimes \mathcal{B})$ and γ_P's determine a cylindrical measure γ on $\mathcal{F}$. Let s,n and y be $L_2[0,T]$-valued maps on Γ defined by

$$s(h,\omega)(t) = s_t(\omega)$$
$$n(h,\omega)(t) = h(t) = n_t(h)$$
$$y(h,\omega)(t) = s_t(\omega) + n_t(h).$$

Then the observation equation (3.1) may be reexpressed as

$$y = s + n \tag{3.4}$$

where the $L_2[0,T]$-valued functions s,n and y are defined on $(\Gamma, \mathcal{F}, \gamma)$. By our very construction, s and n are independent. The process y induces a cylindrical measure on $(L_2[0,T], \mathcal{C})$ as shown by:

Lemma 3.1. [27] The map y: $(\Gamma, \mathcal{F}, \gamma) \to (L_2[0,T], \mathcal{C})$ is measurable in the sense that $B = \{(h,\omega) \mid y(h,\omega) \in C\} \in \mathcal{F}$ for all $C \in \mathcal{C}$ and

$$\mu_y(C) = \gamma(B) = \int \mu_G(C-s(\omega))\, d\Pi(\omega), \quad C \in \mathcal{C},$$

is a cylindrical measure on $(L_2[0,T];\mathcal{C})$.

The basic result in nonlinear filtering in the "white noise model" is the white noise version of the Kallianpur-Striebel formula. It is an extension of Bayes rule to infinite dimension and is given by the following:

Proposition 3.2. [27] Let g be any integrable function on $(\Omega,\mathcal{B},\Pi)$ and Q_t be the orthogonal projection on $L_2[0,T]$ with

$$\text{Range } Q_t = \{\phi \in L_2[0,T] \,|\, \int_t^T |\phi_s|^2\, ds = 0\}.$$

Then $E_\gamma\,(g|Q_t y)$ exists and is given by

$$E_\gamma(g|Q_t y) = \frac{\int g(\omega)\exp.\{([y,s(\omega)]_t - \frac{1}{2}\,\|s(\omega)\|_t^2)\}d\Pi(\omega)}{\int \exp.\{([y,s(\omega)]_t - \frac{1}{2}\,\|s(\omega)\|_t^2)\}d\Pi(\omega)} \tag{3.5}$$

Let us now consider the likelihood ratio. The fundamental result we need for this is the following:

Proposition 3.3. ([9],[3]) μ_Y is absolutely continuous with respect to μ_G and

$$f(h) = \int_{L_2[0,T]} \exp.\left\{-\frac{1}{2}\,(\|x\|^2 - 2[x,h])\right\} d\,\mu_s(x) \tag{3.6}$$

is the Radon-Nikodym derivative of μ_Y with respect to μ_G.

Proposition 3.3 plays the central role in obtaining an explicit formula for the likelihood ratio in our signal-plus-noise model (3.1) [9]. Let $[\cdot,\cdot]_t$ denote the inner product in $L_2[0,t]$, $t \in [0,T]$. Analogous to the way μ_Y was defined in the lemma above, we can define cylindrical measure $\mu_{Y,t}$ on $L_2[0,T]$ by considering $\{y_\sigma\}$ restricted to $[0,t]$ only. From proposition 3.3, $\mu_{Y,t}$ is absolutely continuous with respect to $\mu_{G,t}$, the Gauss measure on $L_2[0,t]$ and the corresponding Radon-Nikodym derivative is

$$f(t;h) = \int_{L_2[0,t]} \exp.\left\{-\frac{1}{2}\,(\|x\|_t^2 - 2[x,h]_t)\right\} d\mu_s(x)$$

$$= \int_\Omega \exp.\left\{-\frac{1}{2}\,(\int_0^t s_\sigma(\omega)^2\, d\sigma - 2\int_0^t s_\sigma(\omega)h(\sigma)d\sigma)\right\}d\Pi \tag{3.7}$$

It has been shown in [9] that f(t;h) is absolutely continuous with respect to t and, by direct calculation,

$$\frac{d}{dt}\log f(t;h) = \frac{f'(t;h)}{f(t,h)} \qquad \text{a.e.t.}$$

$$= \frac{\int_\Omega \{-\frac{1}{2}s_t(\omega)^2 + s_t(\omega)h(t)\}\exp.\{-\frac{1}{2}\left(\int_0^t s_\sigma(\omega)^2 d\sigma - 2\int_0^t s_\sigma(\omega)h(\sigma)d\sigma\right)\}d\Pi}{\int_\Omega \exp.\{-\frac{1}{2}\left[\int_0^t s_\sigma(\omega)^2 d\sigma - 2\int_0^t s_\sigma(\omega)h(\sigma)d\sigma\right] d\Pi}.$$

From (3.5) it follows that

$$\frac{d}{dt}\log f(t;y) = -\frac{1}{2}\widehat{s_t^2} + \hat{s}_t y_t$$

$$= -\frac{1}{2}\hat{s}_t^2 + \hat{s}_t y_t - \frac{1}{2}(\widehat{s_t^2} - \hat{s}_t^2).$$

where $\widehat{s_t^2} = E_\gamma(s_t^2|Q_t y)$ and $\hat{s}_t = E_\gamma(s_t|Q_t y)$.

Let us define

$$P(t;y) \triangleq E\left[(s_t - \hat{s}_t)^2|Q_t y\right] = \widehat{s_t^2} - \hat{s}_t^2, \tag{3.8}$$

the conditional error covariance. Then we get

$$\log f(t;y) = -\frac{1}{2}\left\{\int_0^t \hat{s}_\sigma^2\, d\sigma - 2\int_0^t \hat{s}_\sigma y_\sigma d\sigma + \int_0^t P(\sigma;y)d\sigma\right\}.$$

Finally, the likelihood ratio may be written as

$$\frac{d\mu_Y}{d\mu_G}(y_.) = f(T;y) = \exp. -\frac{1}{2}\left(\int_0^T \hat{s}_t^2 dt - 2\int_0^T \hat{s}_t y_t\, dt + \int_0^T P(t;y)dt\right) \tag{3.9}$$

We are now in a position to compare the likelihood ratio formula (3.9) in the "white noise model" with the formula (1.7) in the "Wiener process model". We see that in (3.9) we replace the dY_t term in (1.7) by the $y_t dt$ term as expected, but we have to use an additional term $\int_0^T P(t;y)dt$ which does not appear in (1.7). Formula (3.9) is applicable to any real data $y_. \in L_2[0,T]$, while (1.7) has no meaning when applied to a data $Y_. \in C^1[0,T]$ which corresponds to $y_. \in L_2[0,T]$. To understand the difference more clearly, let us specialize to the linear dynamical model (1.8a-b) where, for simplicity, we take $u(t) \equiv 0$. Now the model (1.8a-b) is multidimensional, while our preceding analysis was for scalar signal and noise. The extension of our preceding analysis to the

vector case is, however, straightforward and we have to only work with $\mathcal{H} = L_2([0,T];\mathbb{R}^m)$, instead of $L_2[0,T]$, where $\mathcal{H}$ now means the space of square integrable functions from $[0,T]$ into $\mathbb{R}^m$.
The error covariance is now a matrix

$$P(t;y) = E\left[(s_t - \hat{s}_t)(s_t - \hat{s}_t)' \,|\, Q_t y\right] \tag{3.8'}$$

and the likelihood ratio changes to

$$\frac{d\mu_Y}{d\mu_G}(y_.) = \exp. -\frac{1}{2}\left(\int_0^T \|\hat{s}_t\|^2 dt - 2\int_0^T [\hat{s}_t, y_t]dt + \int_0^T \mathrm{Tr}.P(t;y)dt\right) \tag{3.9'}$$

Let us now specialize to the linear model (1.8a-b) with $u(t) \equiv 0$.

In the linear case, the error covariance matrix $P(t;y)$ is independent of the data y. The signal s_t being of the form Cx_t, we have, in fact,

$$P(t;y) = C\,P(t)C' \tag{3.10a}$$

where $P(t)$ is the state filter error covariance matrix well known in Kalman filtering and satisfies the Riccati equation

$$\dot{P}(t) = A\,P(t) + P(t)A' + FF' - P(t)C'CP(t) \tag{3.10b}$$

The likelihood ratio formula (3.9') specializes to

$$\frac{d\mu_Y}{d\mu_G}(y_.) = \exp. -\frac{1}{2}\left(\int_0^T \|C\hat{x}_t\|^2 dt - 2\int_0^T [C\hat{x}_t, y_t]dt + \int_0^T \mathrm{Tr}.(CP(t)C')dt\right) \tag{3.11}$$

where $\hat{x}_t$ is the Kalman filter.

The contrast with formula (1.10) has already been discussed above. Let us now look into the possible underlying connection closely. (1.10) was based on the observation model (1.9b) which is not realizable, since any real data must be smooth. The need, therefore, arises of approximating the observation (1.9b) by smooth processes; for example, by band-limited processes with no frequency component outside a finite band. It has been shown in [7] that, under a sufficient condition such an approximation is possible provided we add suitable correction terms. As we shall see, the correction term is precisely the trace term in (3.11). If we consider (1.10) we see that we need to evaluate the Itô-type integral

$$\int_0^T [C\hat{x}_t, dY_t]$$

Using Kalman filter equation, this integral may be rewritten as

$$\int_0^T [C\hat{x}_t, dY_t] = \int_0^T [\int_0^t CL(t,s)dY_s, dY_t] \tag{3.11}$$

where $L(t,s) = \Phi(t)\Phi(s)^{-1}P(s)C'$, with $P(t)$ the nonnegative definite solution of (3.10b) when $P(0) = 0$ and $\Phi(t)$ is the fundamental matrix solution of

$$\dot{\Phi}(t) = (A-P(t)C'C)\Phi(t), \quad \Phi(0) = I \tag{3.12}$$

To evaluate (3.11) one would use partial sums of the form:

$$\sum_i [\int_0^{t_i} C\, L(t_i,s)dY_s, Y_{t_{i+1}} - Y_{t_i}].$$

In practice, however, there is a serious difficulty with this procedure. What is observed in practice is not (1.9b) but a band-limited version, obviously of large enough bandwidth to allow the use of (1.9b) in theoretical analysis. The real data is, therefore, a band-limited approximation of $Y_.$. A convenient approximation would be

$$y_t^k(\omega) = \int_{-\infty}^{\infty} M(t-s)dY_s(\omega) \tag{3.13}$$

$$M(s) = I_m \int_{-k}^{k} e^{i2\pi fs}df = I_m \, {}^{(\sin 2\pi ks)}/_{\pi s} \tag{3.14}$$

with I_m denoting $m \times m$-matrix and the limit may be taken in the sense of quadratic mean. Then the Itô-type integral (3.11) can be approximated as:

$$\int_0^T [C\hat{x}_t, dY_t] = \lim_{k\to\infty} \int_0^T [\int_0^t CL(t,s)y_s^k ds,\ y_t^k]dt - \int_0^T \mathrm{Tr.}(CP(t)C')dt \tag{3.15}$$

In the "white noise model" the real data $y_.$ is directly modelled as the limit of band-limited approximations and that is why the correction term appears there naturally in the formula for likelihood ratio. In the "Wiener process model", we have to incorporate this correction term every time we evaluate the likelihood ratio based on the real data. We see again that "white noise model" gives results already in the "robust form". The preceding analysis has been extended in [2] to linear time-delayed systems. It has been shown that the limit of band-limited approximation in the "Wiener process model" matches exactly with the likelihood ratio formula in the "white noise model".

In the "white noise model" it is straightforward to study the multiparameter case. For illustrative purpose let us consider the two-parameter signal-plus-noise model. By t we shall now mean the pair (t_1, t_2) and let $R = [0,T_1] \times [0,T_2]$. Suppose we consider the observation equation

$$y_{t_1 t_2} = s_{t_1,t_2} + n_{t_1,t_2} \tag{3.16}$$

The likelihood ratio for this model can be studied in an analogous manner (see [3]). The explicit expression for the likelihood ratio is considerably more complicated than in the one-parameter case and involves, in addition to integrals over R, integrals over $R \times R$ as well. They correspond to the 2-integrals appearing in Wong and Zakai [38] where they study this problem in the "Wiener process model". As expected, these two models give different expressions for the likelihood ratio and the difference must have to do with the approximation of Itô integrals by band-limited processes. For filtering problem of two-parameter processes using "white-noise model" see [29].

4. DIFFERENTIAL SYSTEMS WITH WHITE NOISE INPUTS

One unsatisfactory aspect of the modelling in the previous section was that the signal and the noise were defined on different probability spaces and it was necessary to introduce a product space in which both could be defined. It is, of course, desirable to define the state process of a dynamical system on the same probability space as the observation noise process. This would necessitate studying stochastic differential equations with white noise input directly where the white noise is modelled as "coordinate map" on a Hilbert space, and not use the integrated version of Itô stochastic differential equation. This would also relax the restriction of the independence of the state and observation noises which was inherent in our approach of Section 3. For linear systems, this direct modelling of state and observation equations, (1.8a)-(1.8b), with finitely additive white noises $n_.$ and $n_{0.}$ is, indeed, possible. In fact, a complete theory can be developed for filtering and stochastic control problems even for infinite dimensional systems (see [12]). The theory enables one to include white noises which are white both in time and space, as opposed to the conventional theory with infinite dimensional Wiener processes. The problem, however, becomes considerably more difficult for nonlinear evolution of the state. For simplicity, we consider only the scalar case and take $\mathcal{H} = L_2([0,T])$. Extension to the case when $\mathcal{H} = L_2([0,T]; \mathbb{R}^n)$ is straightforward.

Consider now the following differential equation:

$$\dot{x}_t = f(x_t) + n_t, \text{ a.e.t}, \quad 0 \le t \le T \tag{4.1}$$

with x_0 given and n_t is a finitely additive white noise on $[0,T]$, as in definition 2.5. Then we have the following result:

Proposition 4.1 [12] Suppose $f:\mathbb{R} \to \mathbb{R}$ is uniformly Lipschitz; that is,

(A1) $\qquad |f(x_2) - f(x_1)| \le K \, |x_2 - x_1|, \; K < \infty.$

Then, for each x_0 and $n \in \mathcal{H}$, the solution of (4.1) is unique and continuous in $0 \le t \le T$. Furthermore, the solution $x_t = x_t(n)$ is a physical random variable and

$$\sup_{0 \le t \le T} . \; E \, |x_t(n)|^2 < \infty.$$

In addition, the solution x as a mapping from $\mathcal{H}$ into itself is also a physical random variable.

From the preceding proposition, we know that $x_t(n)$ is a physical random variable and, therefore, we can take expectation with respect to μ_G, which we denote by $E(\cdot)$. In addition to the condition imposed on f in proposition 4.1, we assume that f satisfies

(A2) $\qquad |f(x)|^2 \le K_1 \, (1 + |x|^2), \qquad K_1 < \infty$

Consider equation (4.1) in its integrated form:

$$x_s(t,x) = x + \int_t^s f\big(x_\sigma(t,x)\big)d\sigma + \int_t^s n_\sigma d\sigma \tag{4.2}$$

for $s \in [t,T]$. Define, for fixed s,

$$u(t,x) \triangleq E\big[g\big(x_s(t,x)\big)\big] \tag{4.3}$$

for $0 \le t \le s$, $x \in (-\infty,\infty)$ and $g(\cdot)$ satisfying

<u>A3</u> $\quad g(\cdot)$ is a twice continuously differentiable function vanishing outside some finite interval.

Lemma 4.2. With f, g satisfying (A1-A3),

$$\lim_{\Delta\to 0} \frac{1}{\Delta} \left(E\left[g\left(x_t(t-\Delta,x)\right)\right] - g(x)\right) = f(x)g'(x) + \tfrac{1}{2}g''(x) \tag{4.4}$$

Proof

$$g\left(x_t(t-\Delta,x)\right) - g(x)$$

$$= g\left(x_{t-\Delta}(t-\Delta,x) + \int_{t-\Delta}^{t} \dot{x}_\sigma(t-\Delta,x)d\sigma\right) - g(x)$$

$$= g\left(x + \int_{t-\Delta}^{t} f\left(x_\sigma(t-\Delta,x)\right)d\sigma + \int_{t-\Delta}^{t} n_\sigma d\sigma\right) - g(x)$$

$$= g'(x) \left[\int_{t-\Delta}^{t} f\left(x_\sigma(t-\Delta,x)\right)d\sigma + \int_{t-\Delta}^{t} n_\sigma d\sigma\right]$$

$$+ \frac{1}{2} g''(x) \left[\int_{t-\Delta}^{t} f\left(x_\sigma(t-\Delta,x)\right)d\sigma + \int_{t-\Delta}^{t} n_\sigma d\sigma\right]^2$$

$$+ \epsilon \left(\left|\int_{t-\Delta}^{t} f\left(x_\sigma(t-\Delta,x)\right)d\sigma + \int_{t-\Delta}^{t} n_\sigma d\sigma\right|\right) \left[\int_{t-\Delta}^{t} f\left(x_\sigma(t-\Delta,x)\right)d\sigma + \int_{t-\Delta}^{t} n_\sigma d\sigma\right]^2$$

where, by Taylor's theorem, $\epsilon(|\varsigma|) \to 0$ as $|\varsigma| \to 0$, and we note that the argument of ϵ goes to zero as $\Delta \to 0$.

Hence

$$\lim_{\Delta\to 0} \frac{1}{\Delta} \left(E\left[g\left(x_t(t-\Delta,x)\right)\right] - g(x)\right)$$

$$= g'(x) \left[\lim_{\Delta\to 0} \frac{1}{\Delta} E \int_{t-\Delta}^{t} f\left(x_\sigma(t-\Delta,x)\right)d\sigma\right]$$

$$+ \frac{1}{2} g''(x) \left[\lim_{\Delta\to 0} \frac{1}{\Delta} \left(E\left(\int_{t-\Delta}^{t} f\left(x_\sigma(t-\Delta,x\right)d\sigma\right)^2 + E\left(\int_{t-\Delta}^{t} n_\sigma d\sigma\right)^2\right.\right.$$

$$\left.\left. + 2E\left(\int_{t-\Delta}^{t} f\left(x_\sigma(t-\Delta,x)\right)\right)\left(\int_{t-\Delta}^{t} n_\sigma d\sigma\right)\right]\right. + R_3,$$

where R_3 denotes the remainder term. Now the first term is

$$\lim_{\Delta\to 0} \frac{1}{\Delta} E \int_{t-\Delta}^{t} f\left(x_\sigma(t-\Delta,x)\right)d\sigma$$

$$= E \lim_{\Delta\to 0} \frac{1}{\Delta} \int_{t-\Delta}^{t} f(x_\sigma(t-\Delta,x))d\sigma$$

$$= E\left[f\left(x_t(t,x)\right)\right] = f(x)$$

where the interchange operation between limit and expectation is justified by the growth and continuity assumptions on f.

The term

$$\frac{1}{\Delta} E \left(\int_{t-\Delta}^{t} f(x_\sigma(t-\Delta,x)\right)^2 \le E \int_{t-\Delta}^{t} f\left(x_\sigma(t-\Delta,x)\right)^2 d\sigma$$

and hence

$$\lim_{\Delta\to 0} \frac{1}{\Delta} E \left(\int_{t-\Delta}^{t} f\left(x_\sigma(t-\Delta,x)\right) d\sigma\right)^2 = 0$$

Similarly,

$$\lim_{\Delta\to 0} \frac{1}{\Delta} \left(\int_{t-\Delta}^{t} n_\sigma d\sigma\right)^2 = 1$$

and

$$\frac{1}{\Delta} \left| E \left[\int_{t-\Delta}^{t} f\left(x_\sigma(t-\Delta,x)\right) d\sigma \int_{t-\Delta}^{t} n_\sigma d\sigma\right] \right|$$

$$\le \left(E \int_{t-\Delta}^{t} f\left(x_\sigma(t-\Delta,x)\right)^2 d\sigma\right)^{1/2} \left(\frac{1}{\Delta} E \left(\int_{t-\Delta}^{t} n_\sigma d\sigma\right)^2\right)^{1/2}$$

$$\to 0 \qquad \text{as } \Delta \to 0.$$

The preceding results and the fact that $\epsilon(|\varsigma|) \to 0$ as $\Delta \to 0$ imply that $R_3 \to 0$ as $\Delta \to 0$. Hence the result follows □

Theorem 4.3. $u(t,x)$ defined by eq^n. (4.3) satisfies

$$\frac{\partial}{\partial t} u(t,x) + f(x) \frac{\partial}{\partial x} u(t.x) + \frac{1}{2} \frac{\partial^2}{\partial x^2} u(t,x) = 0 \tag{4.5a}$$

$$\lim_{t\uparrow s} u(t,x) = g(x) \tag{4.5b}$$

Proof. Since $x_s(t,x)$ is the solution of the differential equation (4.1) with $x_t(t,x) = x$ (or equivalently, the solution of the integral equation (4.2)), it follows that

$$x_s(t-\Delta,x) = x_s\left(t,\ x_t(t-\Delta,x)\right), \qquad s \ge t$$

Furthermore,

$$u(t-\Delta,x) = E\ g\left(x_s(t-\Delta,x)\right)$$

so that

$$E\ g(x_s(t-\Delta,x)) = E\ g(x_s(t,x_t(t-\Delta,x)))$$

$$= E[E[g(x_s(t,x_t(t-\Delta,x))|x_t(t-\Delta,x)]]$$

$$= E\ u(t,x_t(t-\Delta,x))$$

The conditional expectation used above can be interpreted by using the fact that the solution x of eq^n. (2.3) is a physical random variable as a mapping of $L_2[0,T]$ into itself. It, therefore, induces a countably additive measure μ_x on $L_2[0,T]$. Define the process $\tilde{x}_t$ as the coordinate process on $(L_2[0,T], \mathcal{B}, \mu_x)$, where $\mathcal{B}$ is the Borel σ-algebra on $L_2[0,T]$. $(\tilde{x}_t)$ has the same finite dimensional distributions as (x_t). The conditional expectation above is, by definition, the same as the similar conditional expectation for the $(\tilde{x}_t)$ process.
Therefore,

$$\frac{u(t,x) - u(t-\Delta,x)}{\Delta} + E\,\frac{u(t,x_t(t-\Delta,x)) - u(t,x)}{\Delta} = 0 \tag{4.6}$$

It can be shown, as in the corollary the Theorem 2.3.1 in [17], that $u(t,x)$ is twice continuously differentiable in x. Using lemma 4.2 with $u(t,\cdot)$ in place of $g(\cdot)$, we get

$$\lim_{\Delta\to 0} \frac{1}{\Delta}\left(E\ u(t,x_t(x,t-\Delta)) - u(t,x)\right)$$

$$= f(x)\frac{\partial}{\partial x}u(t,x) + \frac{1}{2}\frac{\partial^2}{\partial x^2}u(t,x).$$

From (3.6) we then get

$$\lim_{\Delta\to 0}\frac{u(t,x) - u(t-\Delta,x)}{\Delta} = -f(x)\frac{\partial}{\partial x}u(t,x) - \frac{1}{2}\frac{\partial^2}{\partial x^2}u(t,x).$$

From Proposition 4.1 and assumptions on $f(\cdot)$ and $g(\cdot)$, it can be shown that $u(t,x)$ is continuous in t. It follows then that

$$\frac{\partial u}{\partial t}(t,x) + f(x)\frac{\partial u}{\partial x}(t,x) + \frac{1}{2}\frac{\partial^2}{\partial x^2}u(t,x) = 0$$

and $\lim_{t\uparrow s} u(t,x) = g(x)$. □

Remark. The result still holds when the drift term is of the form $f(t,x_t)$, provided $f(\cdot,\cdot)$ is globally Lipschitz in t and x both.

Since $x_s(t,x)$ is a physical random variable, it induces a countably additive measure μ_{x_s} on $\mathbb{R}$ and we have the following:

Definition 4.4. The function

$$P(t,x;s,B) = \mu_{x_s}(B), \qquad B \text{ Borel set in } \mathbb{R}$$

is called the *transition probability function* of the solution process (x_t) of (4.1). If we can write

$$P(t,x;s,B) = \int_B p(t,x;s,y)dy$$

the function $p(t,x;s,y)$ is called the *transition density* of the process (x_t). The following result, the Kolmogorov forward equation, can be proved by standard argument [17, p. 102]:

Theorem 4.5. Assume that the process (x_t) satisfying (4.1) has a transition density $p(t,x;s,y)$ and that $\frac{\partial f}{\partial x}, \frac{\partial p}{\partial s}, \frac{\partial p}{\partial y}$ and $\frac{\partial^2 p}{\partial y^2}$ exist. Then for $0 < t < s$, p satisfies the Kolmogorov forward equation

$$\frac{\partial p}{\partial s}(t,x;s,y) = -\frac{\partial}{\partial y}\left[f(y)p(t,x;s,y)\right] + \frac{1}{2}\frac{\partial^2}{\partial y^2}p(t,x;s,y) \tag{4.7a}$$

with

$$\lim_{s \downarrow t} p(t,x;s,y) = \delta(y-x) \tag{4.7b}$$

Remark. Conditions for the existence of $p(t,x;s,y)$ can be obtained along the same line of arguments as in [17, Chap. 3, pp. 96-99].

Proposition 4.1 may be extended to the case with a state dependent diffusion coefficient:

$$\dot{x}_t = f(x_t) + \sigma(x_t)n_t$$

where $\sigma(\cdot) \in C^\infty$ and $\sigma^2 > 0$ (see [16] for details). This is, of course, rather restrictive. More specific results have been obtained in [8] for bilinear equations, even for infinite dimensional systems. The method used there can be applied to establish the exist- ence of solution for random Schrödinger equation [13]. For appli- cation to laser propagation and other recent results in this area, see [28].

5. OPEN PROBLEMS

Inspite of the fact that the "white noise model" discussed in this paper gives results in the "robust form" enabling one to use them directly on real data, the mathematical complications of the theory are daunting. Although many important mathematical difficulties in this theory have been resolved, there still remain a large number of open problems which have to be satisfactorily solved.

A general sufficient condition guaranteeing a function on H to be a physical random variable is still not known. One has to develop in each case specific approach to prove that a certain function is, indeed, a physical random variable. This is highly unsatisfactory. It is also desirable to model the state equation in the finitely additive white noise framework. As mentioned already, this is so far possible in rather restrictive situations. This brings us to another limitation; namely, that the signal and the noise processes have been assumed to be independent. This assumption can be relaxed only in some special cases. It is important to be able to include correlation between the signal and noise in the "white noise model".

We started this paper with the parameter estimation problem. The likelihood ratio formula obtained in this paper may be used to obtain the maximum likelihood estimate of the unknown parameter vector. One important reason of using maximum likelihood estimate in statistics is that such an estimate is, under general conditions, consistent. In the "white noise model" it is still not clear how to study the consistency question. This necessitates studying the limiting behaviour of the estimate as $T \to \infty$. But this means that we need to have a model of white noise in an appropriate function space on $[0,\infty)$. It is still not clear how to do this in a physically realistic manner. One may have to work with Gauss measure on some appropriate locally convex space.

The other area that deserves serious study is the nonlinear transformation of white noise and calculating the Jacobian of this transformation. The transformation of particular interest is that determined by the differential equation (4.1). The question is whether $\dot{x}$ induces a cylindrical measure $\mu_{\dot{x}}$ on $L_2[0,T]$, whether $\mu_{\dot{x}}$ is absolutely continuous with respect to μ_G and, if so, what is the corresponding Radon-Nikodym derivative. This is being analyzed at present and some preliminary results *have already been obtained*. This is closely related to the work of Gross [18]. Here also one gets the derivative in the "robust form" and the difference with the corresponding result in the "Wiener process model" is precisely the Wong-Zakai correction term [37] as we would expect.

Anticipative Wiener integrals and stochastic differential equations with boundary conditions are active areas of research at present ([32],[33]). Since white noise integrals do not need any nonanticipativity condition, finitely additive white noise theory seems particularly suited for studying stochastic differential equations with boundary conditions. The more general theory of stochastic elliptic partial differential equations may also be developed in the "white noise model". The corresponding likelihood ratio formulas are interesting areas of further research [5].

Finally, mention should be made of the "innovations conjecture" of Kailath. In the "Wiener process model" this stipulates the equivalence of the σ-algebras generated by the observation and the "innovation" processes. Under the assumption of bounded expected energy of the signal process, Allinger and Mitter [1] proved the validity of this result. In the "white noise model" innovations conjecture boils down to proving the invertibility of the highly nonlinear Volterra integral equation expressing the innovation process in terms of the observation process. Under stronger conditions on the signal process than in [1], Mazumdar and Seo ([30],[31]) proved that this invertibility holds, implying the validity of the innovations conjecture in the "white noise model" as well. It is important to investigate whether the condition on the signal process can be relaxed further. This result has independent interest in the theory of nonlinear Volterra integral equation.

REFERENCES

[1] D. Allinger and S.K. Mitter, "New results on the innovations problem for nonlinear filtering", Stochastics, 4, pp. 339-348, 1981.

[2] A. Bagchi, "Approximation of Itô integrals arising in stochastic time-delayed systems", SIAM J. Control and Optimization, 22, pp. 878-888, 1984.

[3] A. Bagchi, "Cylindrical measures in Hilbert space and likelihood ratio for two-parameter signals in additive white noise", **Control-Theory and Advanced Technology**, 1, pp. 139-153, 1985.

[4] A. Bagchi, "Nonlinear smoothing algorithms using white noise model", **Stochastics**, **17**, pp. 289-312, 1986.

[5] A. Bagchi, "Boundary value processes: estimation and identification", **Computers Math. Applic.**, **19**, pp. 9-19, 1990.

[6] A.V. Balakrishnan, "Identification and stochastic control of a class of distributed systems with boundary noise", in **Control Theory, Numerical Methods and Computer Systems modelling.** (A. Bensoussan and J.L. Lions, eds.), Lecture Notes in Economics and Mathematical Systems, **107**, Springer-Verlag, Berlin 1975.

[7] A.V. Balakrishnan, "On the approximation of Itô integrals using band-limited processes", **SIAM J. Control**, 12, pp. 237-251, 1974; Errato: 13, p. 975, 1975.

[8] A.V. Balakrishnan, "Stochastic partial differential equations", in **Variable Structure Systems** (R. Mohler and A. Ruberti, eds.), Lecture Notes in Economics and Mathematical Systems, **111**, Springer-Verlag, Berlin, 1975.

[9] A.V. Balakrishnan, "Likelihood ratios for signals in additive white noise", **Applied Math. and Optimization**, 3, pp. 341-356, 1977.

[10] A.V. Balakrishnan, "Parameter estimation in stochastic differential systems", in **Developments in Statistics**, Vol. 1 (P.R. Krishnaiah, ed.) Academic Press, New York, 1978.

[11] A.V. Balakrishnan, "Non-linear white noise theory", in **Multivariate Analysis-V** (P.R. Krishnaiah, ed.), North-Holland Publishing Company, Amsterdam, 1980.

[12] A.V. Balakrishnan, **Applied Functional Analysis, 2nd ed.**, Springer-Verlag, 1981.

[13] A.V. Balakrishnan, "A random Schroedinger equation: white noise model", **Differential and Integral Equations**, 1, pp. 49-70, 1988.

[14] M.H.A. Davis, "On a multiplicative functional transformation arising in nonlinear filtering theory", **Z. Wahrsch. Verw. Geb.**, **54**, pp. 125–139, 1980.

[15] M.H.A. Davis and S.I. Marcus, "An introduction to nonlinear filtering", in **Stochastic Systems: The Mathematics of Filtering and Identification and Applications** (M. Hazewinkel and J.C. Willems, eds.), NATO Advanced Study Institute Series, Reidel, Dordrecht, 1980.

[16] A. Gandolfi and A. Germani, "On the definition of a topology in Hilbert spaces with applications in white noise theory", **J. Franklin Institute**, **316**, pp. 435–444, 1983.

[17] I.I. Gihman and A.V. Skorohod, **Stochastic Differential Equations**, Springer-Verlag, Berlin, 1972.

[18] L. Gross, "Integration and nonlinear transformation in Hilbert space", **Trans. American Math. Society**, **94**, pp. 404–440, 1960.

[19] L. Gross, "Abstract Wiener space", **Proc. of 5th Berkeley Symposium on Math. Stat. & Prob.**, **2**, pp. 31–42, 1965.

[20] C.W. Helstrom, **Statistical Theory of Signal Detection, 2nd Ed.**, Pergamon Press, Oxford, 1968.

[21] T. Hida, **Brownian Motion**, Springer-Verlag, New York, 1979.

[22] K.W. Iliff, **Identification and Stochastic Control with Application to Flight Control in Turbulence**, Ph. D. Thesis, University of California, Los Angeles, 1973.

[23] T. Kailath, "A general likelihood formula for random signals in Gaussian noise", **IEEE Transactions on Information Theory**, **IT-15**, pp. 350–361, 1969.

[24] G. Kallianpur and R.L. Karandikar, "A finitely additive white noise approach to nonlinear filtering", **Applied Math. and Optimization**, **10**, pp. 159–186, 1983.

[25] G. Kallianpur and R.L. Karandikar, "Some recent developments in nonlinear filtering theory", **Acta Appl. Math.**, **1**, pp. 399–434, 1983.

[26] G. Kallianpur and R.L. Karandikar, "A white noise calculus and nonlinear filtering Theory", Special Invited Paper, **Annals of Probability**, 13, pp. 1033-1107, 1985.

[27] G. Kallianpur and R.L. Karandikar, **White Noise Theory of Filtering, Prediction and Smoothing**, Gordon and Breach, London, 1988.

[28] R. Leland, **Stochastic Models for Laser Propagation in Atmospheric Turbulence**, Lecture Notes in Control and Information Sciences, 133, Springer-Verlag, Berlin, 1989.

[29] R. Mazumdar and A. Bagchi, "A finitely additive white noise approach to filtering for two-parameter diffusion processes", in **Theory and Applications of Nonlinear Control Systems** (C.I. Byrnes and A. Lindquist, eds.), North-Holland, Amsterdam, 1986.

[30] R. Mazumdar and J. Seo, "On the analytic properties of filter maps in finitely additive white noise framework", in **Recent Advances in Communication and Control Theory** (R.E. Kalman et.al. eds.), Optimization Software Publishing, New York, pp. 184-193, 1987.

[31] R. Mazumdar and J. Seo, "On the innovations problem in a finitely additive white noise approach to nonlinear filtering", **Differential and Integral Equations**, 1, pp. 231-249, 1988.

[32] D. Nualart and E. Pardoux, "Stochastic calculus with anticipating integrands", **Probability Theory and Related Fields**, 78, pp. 535-581, 1988.

[33] D. Ocone and E. Pardoux, "Linear stochastic differential equations with boundary conditions", **Probability Theory and Related Fields**, 82, pp. 489-526, 1989.

[34] E. Parzen, "Regression analysis of continuous parameter time series", **Proc. of 4th Berkeley Symposium on Math. Stat. & Prob.**, pp. 469-489, 1960.

[35] I.E. Segal, "Distributions in Hilbert space and canonical systems of operators", **Trans. American Math. Soc.**, 88, pp. 12-41, 1958.

[36] E. Wong and B. Hajek, **Stochastic Processes in Engineering Systems**, Springer-Verlag, New York, 1985.

[37] E. Wong an M. Zakai, "Riemann-Stieltjes approximations of Stochastic integrals", Z. Wahrsch. Verw. Geb., 12, pp. 87-97, 1969.

[38] E. Wong and M. Zakai, "Likelihood ratios and transformation of probability associated with two-parameter Wiener processes", Z. Wahrsch. Verw. Geb., 40, pp. 283-308, 1977.

MARKOVIAN REPRESENTATIONS OF CYCLOSTATIONARY PROCESSES

Sergio Bittanti

Dipartimento di Elettronica, Politecnico di Milano

Piazza Leonardo da Vinci 32, 20133 Milano, Italy

Giuseppe De Nicolao

Centro Teoria dei Sistemi - CNR

c/o Dipartimento di Elettronica, Politecnico di Milano

1. INTRODUCTION

In the realm of signal processing and system identification, stochastic processes with periodic structure constitute an important family of nonstationary processes. While a wide and coherent corpus of results is available in the stationary case, see e.g. [1-5], much has still to be done to develop a theory for these processes. On the application side, such an effort is motivated by the variety of seasonal time series encountered in the real world, as well as by the assortment of signal and systems with periodic structure arising in communications and control, see e.g. [6-7]. The purpose of this paper is to present a survey of recent results settling the foundations of state space theory for stochastic processes with periodic structure. In particular, we investigate the connections between the (sign-definite as well as sign-indefinite) periodic solutions of a Periodic Riccati Equation (PRE) and the important class of Markovian models given by the so-called innovation representations. Resorting to suitable definition of zeros of periodic systems, it is also possible to provide conditions for the existence of a minimum-phase (canonical) innovation representation. Only discrete-time systems are considered herein.

The paper is organized as follows. The definition of cyclostationary process together with the basics of periodic system theory is provided in Section 2. The PRE as well as the periodic innovation representation are introduced in Section 3 by resorting to Kalman prediction theory considerations. Then, the main properties of the periodic solutions of the PRE are concisely overviewed (Section 4). Finally, in Section 5, it is shown that there is a one-to-one correspondence between the periodic innovation representations and the periodic solutions of the PRE. Furthermore, the condition for the existence of the canonical innovation representation is discussed.

To achieve an expository style, priority is given to motivations and interpretations, leaving the technical details to the quoted references. For ease of reference we end this section with a list of acronyms used throughout the paper

ALE:	Algebraic Lyapunov Equation
PLE:	Periodic Lyapunov Equation
PRE:	Periodic Riccati Equation
SPPS:	Symmetric Periodic Positive Semidefinite
TIR:	Time Invariant Representation

2. PERIODIC SYSTEMS AND CYCLOSTATIONARY PROCESSES

2.1. Cyclostationary processes.

Let $y(\cdot)$: $Z \rightarrow R^p$ be a zero-mean stochastic process with autocorrelation function $\Lambda(t,\tau) = E[y(t)y(\tau)']$. The process is said to be *cyclostationary* if there exists $T > 0$ such that

$$\Lambda(t,\tau) = \Lambda(t+T,\tau+T), \qquad \forall\, t,\, \tau\, .$$

T is the period of the process. Although a fully developed theory is not available yet, cyclostationary processes have been the object of several investigations reported even in classical textbooks such as [8, Chapter 9].

Going from signals to systems, one could expect that a natural representation of cyclostationary processes is provided by periodic state space models such as

$$x(t+1) = F(t)\,x(t) + G(t)\,v(t) \tag{2.1.a}$$

$$y(t) = H(t)\,x(t) + w(t)\,, \tag{2.1.b}$$

where $F(\cdot)$: $Z \rightarrow R^{nxn}$, $G(\cdot)$: $Z \rightarrow R^{nxm}$, and $H(\cdot)$: $Z \rightarrow R^{pxn}$, are periodic matrices of period T; $v(\cdot)$ and $w(\cdot)$ are periodic white noises with zero expected value and covariance matrix:

$$Var\ [v(t)'w(t)']' = \begin{bmatrix} Q(t) & S(t) \\ S(t)' & R(t) \end{bmatrix}, \forall t.$$

Matrices $Q(\cdot)$, $S(\cdot)$, and $R(\cdot)$ are periodic of period T, and $R(t) > 0$, $\forall t$.

For the study of cyclostationary processes, it is natural to consider only stable and observable representations. Under the stability assumption, system (2.1) is indeed a Markovian representation of a cyclostationary process. The observability assumption on the periodic pair $(F(\cdot),H(\cdot))$ avoids unnecessarily oversized representations.

In order to proceed with our analysis, some basic facts on periodic systems are firstly overviewed.

2.2 *Periodic systems.*

The transition matrix $\Psi_F(t,\tau) = F(t-1)F(t-2)\ldots F(\tau)$ of system (2.1) is such that

$$\Psi_F(t+T,\tau+T) = \Psi_F(t,\tau)\,, \tag{2.2}$$

Hence, the so-called *monodromy matrix* at t, namely

$$\Phi_F(t) = \Psi_F(t+T,t)\,,$$

is periodic. However, its eigenvalues, which are called *characteristic multipliers* , turn out to be time-independent, see e.g. [9]. Since the T-sampled free motion is given by $x(kT) = \Phi_F(0)^k x(0)$, it is apparent that the characteristic multipliers determine the behaviour and stability of such a motion. In particular, system (2.1) is (asymptotically) stable iff all its multipliers belong to the open unit disk. Thus, the characteristic multipliers can be regarded as the periodic equivalents of the *poles* of a time-invariant system.

The structural properties of periodic systems, i.e. reachability, controllability, observability, reconstructibility, stabilizability and detectability, have been thoroughly investigated in last decade, see e.g. [9]. Among the various characterizations, the following one is worth mentioning.

Reachability Criterion. System (2.1) is reachable if $rank\,[\mathcal{R}(t)] = n$, $\forall t$, where

$$\mathcal{R}(t) = \left[\begin{array}{cccc} G(t-1) & \Psi_F(t,t-1)G(t-2) & \ldots & \Psi_F(t,t-nT+1)G(t-nT) \end{array}\right]. \;\square$$

By duality, one obtains the

Observability Criterion. System (2.1) is observable iff $rank\,[\mathcal{O}(t)] = n$, $\forall t$, where

$$\mathcal{O}(t) = \left[\begin{array}{cccc} H(t)' & \Psi_F(t+1,t)'H(t+1)' & \ldots & \Psi_F(t+nT-1,t)'H(t+nT-1)' \end{array}\right]. \;\square$$

In the time-invariant case, $\mathcal{R}(t)$ and $\mathcal{O}(t)$ become the celebrated Kalman reachability and observability matrices [10]:

$$\mathcal{R} = [\; G \quad FG \quad \dots \quad F^{n-1}G \;]$$
$$\mathcal{O} = [\; H' \quad F'H' \quad \dots \quad (F')^{n-1}H' \;].$$

In the periodic case, even if $\mathcal{R}(t)$ [or $\mathcal{O}(t)$] is maximum rank for some t, the rank may not be maximum for a different time point. This corresponds to the fact that the dimensions of reachability and observability subspaces of system (2.1) are, in general, time-varying. On the contrary, as shown in [11], the controllability and reconstructibility subspaces have constant dimensions. In view of this, it is not surprising that a canonical decomposition of the periodic system (2.1) can be performed, in general, only in terms of controllability and reconstructibility [12].

Another important characterization of reachability and observability of periodic systems is the modal one.

Reachability Modal Characterization. A characteristic multiplier λ of $F(\cdot)$ is said to be $(F(\cdot),G(\cdot))$-unreachable at time τ, if there exists $\eta \neq 0$, such that

$$\Phi_F(\tau)'\eta = \lambda\eta, \qquad G(j-1)'\Psi_F(\tau, j)'\eta = 0, \qquad \forall j \in [\tau - T + 1, \tau].$$

System (2.1) is reachable if no characteristic multiplier is unreachable.

Observability Modal Characterization. A characteristic multiplier λ of $F(\cdot)$ is said to be $(F(\cdot),H(\cdot))$-unobservable at time τ, if there exists $\zeta \neq 0$, such that

$$\Phi_F(\tau)\zeta = \lambda\zeta, \qquad H(j)\Psi_F(j, \tau)\zeta = 0, \qquad \forall j \in [\tau, \tau+T-1].$$

System (2.1) is observable if no characteristic multiplier is unobservable. ❑

These notions result in the so-called PBH (Popov-Belevitch-Hautus) test in the time invariant case, see e.g. [13]. As discussed in [14], the unreachability notion is time-independent for the nonzero characteristic multipliers. As for the equivalence of these modal characterization with the previous criteria, the interested reader is referred to [11].

From reachability and observability to stabilizability and detectability the step is short. With reference to the canonical decomposition issue previously mentioned, one can say that a periodic system is stabilizable [detectable] if and only if its uncontrollable [unreconstructible] part is stable [9]. This is equivalent to the following

Stabilizability Modal Characterization. System (2.1) is stabilizable if the unreachable multipliers belong to the open unit disk.

Detectability Modal Characterization. System (2.1) is detectable if the unobservable multipliers belong to the open unit disk. ❑

In the sequel, we will also say that $(F(\cdot),G(\cdot))$ is reachable to mean that system (2.1) is reachable. An analogous terminology will be adopted with respect to observability, stabilizability and detectability.

2.3. Zeros of periodic systems.

As is well known, the zeros of time-invariant systems can be characterized in terms of blocking property: associated to any zero there is an input function for which the output is identically zero for a suitable choice of the initial state. A consistent definition of zeros of discrete-time periodic systems has been first given in [15]. To be precise, consider the system

$$x(t+1) = A(t)\,x(t) + B(t)\,u(t) \qquad (2.3.a)$$

$$y(t) = C(t)\,x(t) + D(t)\,u(t)\,, \qquad (2.3.b)$$

with $A(\cdot)$, $B(\cdot)$, $C(\cdot)$, $D(\cdot)$ periodic matrices of period T. To such a periodic system one can associate a Time-Invariant Representation (TIR), which traces back to 1959, [16]. Letting

$$A_\tau = \Phi_A(\tau), \qquad A_\tau \in R^{nxn}$$

$$B_\tau = [\ \Psi_A(\tau+T,\tau+1)B(\tau) \quad \Psi_A(\tau+T,\tau+2)B(\tau+1) \quad \dots \quad B(\tau+T-1)\ ], \qquad B_\tau \in R^{nxmT}$$

$$C_\tau = [\ C(\tau)' \quad \Psi_A(\tau+1,\tau)'C(\tau+1)' \quad \dots \quad \Psi_A(\tau+T-1,\tau)'C(\tau+T-1)'\ ]', \qquad C_\tau \in R^{pTxn}$$

$$D_\tau = \{(D_\tau)_{ij}\}, \qquad i, j = 1, 2, \dots, T, \qquad D_\tau \in R^{pTxmT}$$

$$(D_\tau)_{ij} = \begin{cases} 0 & i < j \\ D(\tau+i-1) & i = j \\ C(\tau+i-1)\Psi_A(\tau+i-1,\tau+j)B(\tau+j-1) & i > j \end{cases}$$

$$u_\tau(k) = [\ u(\tau+kT)' \quad u(\tau+kT+1)' \quad \dots \quad u(\tau+(k+1)T-1)'\ ]', \qquad (2.4)$$

consider the time-invariant system

$$x_\tau(k+1) = A_\tau(k)\,x_\tau(k) + B_\tau(k)\,u_\tau(k) \qquad (2.5.a)$$

$$y_\tau(k) = C_\tau(k)\,x_\tau(k) + D_\tau(k)\,u_\tau(k)\,. \qquad (2.5.b)$$

It is easy to see that, if $u_\tau(\cdot)$ is constructed according to (2.4), and $x(\tau)$ is taken as $x_\tau(0)$,then

$$\begin{aligned} x_\tau(k) &= x(\tau+kT) \\ y_\tau(k) &= [\ y(\tau+kT)' \quad y(\tau+kT+1)' \quad \dots \quad y(\tau+(k+1)T-1)'\ ]', \ \forall k \geq 0. \end{aligned}$$

Therefore, the time-invariant system (2.5) can be seen as a state-sampled representation of system (2.3), fed by augmented noise vectors and producing an augmented output vector.

The definition of zeros of a periodic system relies on such a TIR.

Definition 2.1 *(zeros of a periodic system)* [15]. z is a zero at time τ of system (2.3), if it is a zero of the corresponding TIR (2.5), i.e. if $det\, N(z) = 0$, where

$$N(z) = \begin{bmatrix} zI-A_\tau & -B_\tau \\ C_\tau & D_\tau \end{bmatrix}. \ \square$$

As discussed in [15] and [17], the non-zero zeros together with their multiplicities are in fact independent of τ. In [15] it has also been proven that one can associate to any zero of the system a suitable initial state and input function that result in the null output (transmission-blocking property). Precisely, given a zero $\bar{z} \neq 0$, there exists an initial state $x(0)$ such that, for any Tth root m of $\bar{z}$, there exists a T-periodic function $a(t)$ such that the input $u(t) = a(t)m^t$ results into $y(t) = 0,\ \forall t$.

An important property of zeros is their invariance with respect to output feedback. By output feedback, we mean the addition to the state vector of an extra signal proportional to the output vector, as precisely stated in the following

Lemma 2.1 [18]. Consider the following periodic system generated from system (2.3) by means of a periodic output feedback

$$\begin{aligned} \tilde{x}(t+1) &= A(t)\,\tilde{x}(t) + E(t)\tilde{y}(t) + B(t)\,u(t) \\ \tilde{y}(t) &= C(t)\,\tilde{x}(t) + D(t)\,u(t)\,, \end{aligned}$$

where $E(\cdot)$ is an arbitrary periodic matrix of suitable dimension. Then, the zeros at time τ of such system coincide with the zeros at τ of (2.3). ❑

This result will be useful in the analysis of the so-called innovation representation of cyclostationary processes (Section 5).

2.4 The PLE and its periodic solution.

The covariance matrix $\Sigma(\cdot)$ of the state of system (2.1) obeys the periodic Lyapunov equation

$$\Sigma(t+1) = F(t)\Sigma(t)F(t)' + G(t)Q(t)G(t)' \, , \qquad \text{(PLE)}$$

with initial condition $\Sigma(t_0) = E[x(t_0)x(t_0)']$. Furthermore, the state covariance function $\Gamma(t,\tau) = E[x(t)x(\tau)']$ can be derived from $\Sigma(t)$ as follows

$$\Gamma(t,\tau) = \Psi_F(t,\tau)\Sigma(\tau) \, , \; t > \tau.$$

For $x(\cdot)$ (and hence $y(\cdot)$) to be a cyclostationary process one has to impose that

$$\Gamma(t+T,\tau+T) = \Gamma(t,\tau) \, .$$

Since $\Gamma(t+T,\tau+T) = \Psi_F(t+T,\tau+T)\Sigma(\tau+T)$ and $\Gamma(t,\tau) = \Psi_F(t,\tau)\Sigma(\tau)$, in view of (2.2) the previous conditions is tantamount to

$$\Sigma(\tau+T) = \Sigma(\tau) \, . \qquad (2.6)$$

In conclusion, associated to any SPPS solution of the PLE, there is a cyclostationary output of system (2.1).

Now, the solution of the PLE is given by

$$\Sigma(t) = \Psi_F(t,t_0)\Sigma(t_0)\Psi_F(t,t_0)' + W(t_0,t) \, ,$$

where

$$W(t_0,t) = \sum_{j=t_0+1}^{t} \Psi_F(t,j)G(j-1)Q(j-1)G(j-1)'\Psi_F(t,j)'$$

is the reachability Grammian matrix. The periodicity constraint $\Sigma(t_0+T) = \Sigma(t_0)$ leads to

$$\Sigma(t_0) = \Psi_F(t_0+T,t_0)\Sigma(t_0)\Psi_F(t_0+T,t_0)' + W(t_0,t_0+T) \, . \qquad (2.7)$$

Eq. (2.7) is a discrete-time Algebraic Lyapunov Equation (ALE). Since $F(\cdot)$ is stable, all the eigenvalues of $\Psi_F(t_0+T,t_0)$ lie in the open unit disk, so that (2.7) admits a unique solution $\bar{\Sigma}_{t_0}$, (which is in fact positive semidefinite). Such a solution $\bar{\Sigma}_{t_0}$ acts indeed as a periodic generator for the PLE, i.e. the solution $\bar{\Sigma}(\cdot)$ of the PLE with initial condition $\bar{\Sigma}(t_0) = \bar{\Sigma}_{t_0}$ satisfies the periodicity constraint (2.6). Moreover, it is easy to see that

$$\Sigma(t) - \bar{\Sigma}(t) = \Psi_F(t,t_0)\,[\Sigma(t_0) - \bar{\Sigma}_{t_0}]\,\Psi_F(t,t_0)' \ ,$$

where $\Sigma(\cdot)$ is any solution of the PLE. Obviously, the stability of system (2.1) implies the convergence of the difference $\Sigma(t) - \bar{\Sigma}(t)$ to zero.

We can summarize the ideas of this section as follows.

MORAL: Under the assumption that $F(\cdot)$ is stable, the periodic Lyapunov equation admits indeed a unique T-periodic solution which is in fact positive semidefinite at each time point. Furthermore, such is a solution is globally attractive, i.e. any solution of the Lyapunov equation converges to such a periodic equilibrium whatever the initial condition $\Sigma(t_0)$ be. In other words, stable periodic state-space models such as (2.1) are indeed Markovian representations of cyclostationary processes.

3. PREDICTION OF CYCLOSTATIONARY PROCESSES AND MARKOVIAN REPRESENTATIONS

The classical linear prediction problem for a stochastic process $y(\cdot)$ consists in finding a suitable linear functional of past data yielding optimal estimates of the subsequent values of the process. When optimality is given the mean-squares sense and the process is described by a Markovian representation, then the natural framework for the solution of the prediction problem is provided by Kalman filtering theory [19], [20]

Denoting by $x(t+1/t)$, the optimal one-step-ahead prediction of the state $x(t)$ based on $\{y(j), j \leq t\}$, the optimal predictor is given by

$$x(t+1/t) = F(t)\,x(t/t-1) + K(t)\,\eta(t) \quad (3.1.a)$$

$$y(t/t-1) = H(t)\,x(t/t-1) \quad (3.1.b)$$

$$\eta(t) = y(t) - y(t/t-1)\ . \quad (3.1.c)$$

The Kalman gain $K(t)$ is given by

$$K(t) = [G(t)S(t) + F(t)P(t)H(t)'][H(t)P(t)H(t)' + R(t)]^{-1} \quad (3.2)$$

where $P(t)$ is the solution of the Riccati equation with periodic coefficients

$$\begin{aligned} P(t+1) &= F(t)P(t)F(t)' + G(t)Q(t)G(t)' \\ &- [G(t)S(t) + F(t)P(t)H(t)'][H(t)P(t)H(t)' + R(t)]^{-1}[G(t)S(t) + F(t)P(t)H(t)']', \end{aligned} \quad \text{(PRE)}$$

with initial condition $P(t_0) = \Sigma(t_0) = E[x(t_0)x(t_0)']$. The PRE can also be given the equivalent expression

$$P(t+1) = A(t)P(t)A(t)' + B(t)B(t)' \\ \quad - A(t)P(t)H(t)'[H(t)P(t)H(t)' + R(t)]^{-1}H(t)P(t)A(t)',$$

where
$$A(t) = F(t) - G(t)S(t)R(t)^{-1}H(t)$$
$$B(t)B(t)' = G(t)Q(t)G(t)' - S(t)R(t)^{-1}S(t)'$$

This alternative expression of the PRE has the feature of being associated with the stochastic system

$$x(t+1) = A(t)\,x(t) + B(t)\,\tilde{v}(t) \qquad (3.3.a)$$
$$y(t) = H(t)\,x(t) + \tilde{w}(t)\,, \qquad (3.3.b)$$

where $\tilde{v}(t)$ and $\tilde{w}(t)$ are *independent* white noises with zero mean and covariance matrices given by I and $R(t)$, respectively.

In general, the solution $P(\cdot)$ will not be periodic (unless particular initial conditions are taken). The asymptotic behaviour of the solutions may include a variety of cases, such as asymptotic convergence to a periodic equilibrium of period T, asymptotic convergence to a periodic equilibrium of period kT or convergence to a quasi-periodic solution. Our attention will be focused on the first case. Note that, due to the semidefiniteness of the initial condition $P(t_0) = \Sigma(t_0)$, whenever the solution of the PRE converges to a periodic equilibrium, it actually converges to an SPPS solution. Correspondingly, the innovation $\eta(\cdot)$ defined by (3.1.c) converges to a white noise with periodic variance $H(\cdot)P(\cdot)H(\cdot)' + R(\cdot)$. Thus, as discussed in deeper detail in the next section,

$$x(t+1/t) = F(t)\,x(t/t-1) + K(t)\,\eta(t) \qquad (3.4.a)$$
$$y(t) = H(t)\,x(t/t-1) + \eta(t)\,, \qquad (3.4.b)$$
$$\eta(\cdot) \sim WN(0,H(\cdot)P(\cdot)H(\cdot)' + R(\cdot))\,, \qquad (3.4.c)$$

can be seen as an alternative Markovian representation of the cyclostationary process $y(\cdot)$ defined in (2.1).

MORAL: Given a positive semidefinite *periodic solution of the Riccati Equation, one can devise a Markovian representation of the type (3.4) of the cyclostationary process.*

The purpose of subsequent Section 5 is to show that *any* periodic solution (not necessarily positive semidefinite) of the PRE generates a Markovian representation. With this objective in mind, we will overview in Section 4 some basic facts on the periodic solutions of the PRE.

4. THE PERIODIC RICCATI EQUATION

The periodic Riccati equation has been the subject of several investigations in the last two decades, see [21] for an extended survey. Herein, we will only recall a handful of useful results. For, the characterization of some important solutions of the PRE is firstly provided.

Let $\hat{A}(t)$ denote the so-called closed-loop matrix:

$$\hat{A}(t) = F(t) - K(t)H(t) .$$

Definition 4.1. A symmetric periodic solution $\bar{P}(\cdot)$ of the PRE is said to be
maximal if $P(t) \le \bar{P}(t)$, $\forall t$, for all symmetric periodic solutions $P(\cdot)$ of the PRE;
strong if the characteristic multipliers of $\hat{A}(\cdot)$ belong to the closed unit disk;
stabilizing if the characteristic multipliers of $\hat{A}(\cdot)$ belong to the open unit disk. ❑

In the following theorem, the existence of the maximal, strong and stabilizing solution of the PRE is analyzed. The existence conditions are stated in terms of structural properties of system (3.3).

Theorem 4.1. [14], [22], [23].

(i) If $(A(\cdot),H(\cdot))$ is detectable, then the PRE admits a maximal symmetric periodic solution, which is in fact positive semidefinite and strong.

(ii) If $(A(\cdot),B(\cdot))$ is stabilizable and $(A(\cdot),H(\cdot))$ detectable, then the PRE admits a unique SPPS solution, which is in fact stabilizing.

(iii) The PRE admits a stabilizing solution iff $(A(\cdot),H(\cdot))$ is detectable and there is no unit-modulus $(A(\cdot),B(\cdot))$-unreachable characteristic multiplier. Moreover, such a stabilizing solution is unique and coincides with the maximal one. ❑

Note that observability is preserved under output feedback. Therefore, if the pair $(F(\cdot),H(\cdot))$ is observable, then $(A(\cdot),H(\cdot))$ is observable (and thereby detectable) too. On the contrary, the stability of $F(\cdot)$ by no way entails the stability of $A(\cdot)$. Therefore, in our context, thanks to the observability assumption on system (2.1), one can conclude from Thm. 4.1 that the PRE admits a maximal solution. However, there may exist other SPPS solutions and the maximal solution does not necessarily result into a stable optimal predictor. To achieve uniqueness of the SPPS solution and to guarantee closed-loop stability, additional conditions on the pair $(A(\cdot),B(\cdot))$ are to be met with as pointed out at points (ii) and (iii) of Thm. 4.1.

MORAL: System (2.1) being observable, the periodic Riccati equation admits a maximal periodic solution. Then, if the unit circle is free of $(A(\cdot),B(\cdot))$-unreachable characteristic multipliers, the predictor associated to the maximal solution is stable. Moreover, if the $(A(\cdot),B(\cdot))$-unreachable characteristic multipliers are confined within the open unit disk, the maximal solution is the only periodic positive semidefinite solution.

5. PERIODIC INNOVATION REPRESENTATION OF CYCLOSTATIONARY PROCESSES

Given a Markovian representation (2.1) of a cyclostationary process, one can obtain the equivalent representation (3.4) as described in Section 3. Representation (3.4) is associated with an SPPS solution of the PRE. We will now show that, there is a one-to-one correspondence between all symmetric periodic solutions (not necessarily positive semidefinite) of the PRE and a certain class of Markovian representations. Precisely, we consider the representations

$$\bar{x}(t+1) = F(t)\bar{x}(t) + L(t)\,\varepsilon(t) \tag{5.1.a}$$

$$y(t) = H(t)\bar{x}(t) + \varepsilon(t)\,, \tag{5.1.b}$$

where $\varepsilon(t)$ is a zero-mean White Noise with a periodic covariance matrix $\tilde{R}(t) = \tilde{R}(t+T)$. Moreover, $L(t)$ is a T-periodic matrix. If, for some $\tilde{R}(\cdot)$ and $L(\cdot)$, the cyclostationary process defined as the output of (5.1) coincides with the cyclostationary process obtained as the output of (2.1), then (5.1) will be said to be an *innovation representation* of the process. If, in addition, the zeros of (5.1) belong to the open unit disk, such a representation will be termed *canonical innovation representation.*

In the attempt to associate a representation such as (3.4) with a periodic solution which is not necessarily positive semidefinite, one may wonder whether $H(t)P(t)H(t)' + R(t)$ still represents a covariance matrix in accordance with expression (3.4.c). In Thm. 5.1 below it is shown that, under mild conditions, $H(t)P(t)H(t)' + R(t)$ is indeed positive definite for each time point t.

Theorem 5.1 [18]. Assume that the PRE admits a maximal solution. Then, for *any* symmetric periodic solution $P(\cdot)$ of the PRE, $H(t)P(t)H(t)' + R(t) > 0,\ \forall t.$ ❑

From this result and Thm. 4.1 above, one can draw the following important conclusion.

Corollary 5.1. Suppose that $(F(\cdot),H(\cdot))$ is observable. Then, for any symmetric periodic solution $P(\cdot)$ of the PRE, $H(t)P(t)H(t)' + R(t) > 0,\ \forall t.$ ❑

With this result at hand, it is possible to explore the connection between periodic solutions of the PRE and innovations representations.

Theorem 5.2. Suppose that $F(\cdot)$ is stable and $(F(\cdot),H(\cdot))$ is observable.

(i) Consider any symmetric periodic solution $P(\cdot)$ of the PRE, with the associated gain $K(\cdot)$ defined by (3.2). Then, letting $L(t) = K(t)$ and $\tilde{R}(t) = H(t)P(t)H(t)' + R(t)$, eq. (5.1) provides an innovation representation of the cyclostationary process $y(\cdot)$ defined by (2.1). Moreover, different solutions of the PRE lead to different innovation representations.

(ii) Consider an innovation representation (5.1) for the cyclostationary process generated by (2.1). Then, there exists a symmetric periodic solution $P(\cdot)$ of the PRE such that $L(t) = K(t)$ and $\tilde{R}(t) = R(t) + H(t)P(t)H(t)'$.

Proof.

(i) Consider the covariance matrix function $\Lambda(t,\tau) = E[y(t)y(\tau)']$ of the cyclostationary process defined by (2.1). This function is given by:

$$\Lambda(t,\tau) = \begin{cases} H(t)\Psi_F(t,\tau+1)[F(\tau)\Sigma(\tau)H(\tau)' + G(\tau)S(\tau)] & t > \tau \\ H(t)\Sigma(t)H(t)' + R(t) & t = \tau \end{cases} \tag{5.2}$$

where the covariance matrix of the state, $\Sigma(t) = E[x(t)x(t)']$, is the unique symmetric periodic solution of the PLE (Section 2.4).

On the other hand, the innovation representation (5.1) defines a unique cyclostationary process too. Its covariance matrix is given by

$$\tilde{\Lambda}(t,\tau) = \begin{cases} H(t)\Psi_F(t,\tau+1)[F(\tau)\tilde{\Sigma}(\tau)H(\tau)' + L(\tau)\tilde{R}(\tau)] & t > \tau \\ H(t)\tilde{\Sigma}(t)H(t)' + \tilde{R}(t) & t = \tau \end{cases} \tag{5.3}$$

where $\tilde{\Sigma}(t) = E[\tilde{x}(t)\tilde{x}(t)']$ is the unique symmetric periodic solution of:

$$\tilde{\Sigma}(t+1) = F(t)\tilde{\Sigma}(t)F(t)' + L(t)\tilde{R}(t)L(t)'. \tag{5.4}$$

It is easy to see that the difference $\Sigma(t) - \tilde{\Sigma}(t)$ is nothing but matrix $P(t)$. This observation in turn leads to the conclusion that $\tilde{\Lambda}(t,\tau) = \Lambda(t,\tau)$.

Finally, it is proven that two different solutions $P_1(\cdot)$ and $P_2(\cdot)$ of the PRE cannot give rise to the same innovation representation (5.1). Indeed, by contradiction, assume that there exist two symmetric periodic solutions $P_1(\cdot)$ and $P_2(\cdot)$ of the PRE such that $\tilde{R}_1(t) = \tilde{R}_2(t)$ and $K_1(\cdot) = K_2(\cdot)$, where $\tilde{R}_i(t) = R(t) + H(t)P_i(t)H(t)'$ and $K_i(t) = [G(t)S(t) + A(t)P_i(t)H(t)'][H(t)P_i(t)H(t)' + R(t)]^{-1}$, $i = 1, 2$. Then, let $\tilde{\Sigma}_i(\cdot)$, $i = 1,2$, be the state covariance matrices of the innovation representations associated with solutions $P_i(\cdot)$, $i = 1,2$, which satisfy the PLE's

$$\tilde{\Sigma}_i(t+1) = F(t)\tilde{\Sigma}_i(t)F(t)' + K_i(t)\tilde{R}_i(t)K_i(t)', \; i = 1, 2.$$

Since $F(\cdot)$ is stable, the PLE admits a unique periodic solution (Section 2.4). Therefore, the coincidence of the known terms of these last two equations entails that $\tilde{\Sigma}_1(t) = \tilde{\Sigma}_2(t)$. On the other hand, $P_i(t) = \Sigma(t) - \tilde{\Sigma}_i(t)$, so that $P_1(\cdot)$ and $P_2(\cdot)$ must in fact coincide too.

(ii). Let $\Sigma(t)$ and $\tilde{\Sigma}(t)$ be the covariance matrices of the state $x(t)$ of the original representation and the state $\tilde{x}(t)$ of the innovation representation, respectively. By assumption, the associated output covariance matrices $\Lambda(t,\tau)$ and $\tilde{\Lambda}(t,\tau)$ do coincide. By comparing their expressions, (5.2) and (5.3), for $t = \tau, \tau+1, \ldots, \tau+NT$, it follows that

$$\tilde{R}(t) = R(t) + H(t)[\Sigma(t) - \tilde{\Sigma}(t)]H(t)'$$
$$V(\tau+1)\,[F(\tau)\Sigma(\tau)H(\tau)' + G(\tau)S(\tau)] = V(\tau+1)\,[F(\tau)\tilde{\Sigma}(\tau)H(\tau)' + L(\tau)\tilde{R}(\tau)],$$

where matrix $V(t)$ was defined in Section 2.2. In view of the observability of $(F(\cdot),H(\cdot))$, the Observability Criterion guarantees that $V(\tau+1)$ is full-rank, $\forall\tau$. Therefore,

$$L(t) = [G(t)S(t) + F(t)\,(\Sigma(t) - \tilde{\Sigma}(t))H(t)']\tilde{R}(t)^{-1}.$$

By substituting the expressions of $\tilde{R}(t)$ and $L(t)$ in eq. (5.4), the conclusion is drawn that $\Sigma(\cdot) - \tilde{\Sigma}(\cdot)$ satisfies the PRE. Consequently, $\tilde{R}(t)$ and $L(t)$ can be seen as generated by a solution of the PRE, i.e. $\tilde{R}(t) = R(t) + H(t)P(t)H(t)'$ and $L(t) = [G(t)S(t) + F(t)P(t)H(t)']\tilde{R}(t)^{-1}$ with $P(t) = \Sigma(t) - \tilde{\Sigma}(t)$. □

This theorem tells us that, for system (5.1) to be an innovation representation, matrix $L(t)$ must coincide with a Kalman gain $K(t)$ associated to a periodic solution of the PRE. Hence, by adding and subtracting $K(t)y(t)$ to the right hand side of (5.1.a), the innovation representation (5.1) can be also written as

$$\tilde{x}(t+1) = \hat{A}(t)\,\tilde{x}(t) + K(t)\,y(t)$$
$$y(t) = H(t)\,\tilde{x}(t) + \varepsilon(t)\,,$$

where $\hat{A}(t)$ is the closed-loop matrix. Regarding now $y(\cdot)$ as input and $\varepsilon(\cdot)$ as output, the so-called *whitening filter* is obtained:

$$\tilde{x}(t+1) = \hat{A}(t)\,\tilde{x}(t) + K(t)\,y(t)$$
$$\varepsilon(t) = -H(t)\,\tilde{x}(t) + y(t)\,.$$

This whitening filter is stable if and only if the innovation representation is minimum-phase, i.e. its zeros belong to the open unit disk. Indeed, the application of the output feedback $-K(t)y(t)$ to eq. (5.1.a) (with $L(t) = K(t)$) leads to the system

$$x(t+1) = \hat{A}(t)\,x(t) \qquad (5.5.a)$$
$$y(t) = H(t)\,x(t) + \varepsilon(t)\,. \qquad (5.5.b)$$

According to the definition of zeros of a periodic system given in Section 2.3, the zeros of system (5.5) turn out to be the characteristic multipliers of $\hat{A}(\cdot)$, which are the poles of the periodic whitening filter. On the other hand, by Lemma 2.1, the zeros of (5.5) coincide with those of (5.1).

In view of the above discussion, it is apparent that an innovation representation is canonical (minimum-phase) if and only if $\hat{A}(\cdot)$ is stable. This is equivalent to saying that the canonical innovation representation exists if and only if the PRE admits a stabilizing solution. Therefore, taking also into account the Moral of Section 4, we come to the last (but not least)

MORAL: The innovation representations (5.1) of the cyclostationary process $y(\cdot)$ defined by (2.1) are in a one-to-one correspondence with the (sign definite or indefinite) periodic solutions of the periodic Riccati equation. The canonical innovation representation, the zeros of which belong to the open unit disk, exists if and only if the unit circle is free of $(A(\cdot),B(\cdot))$-unreachable characteristic multipliers. Precisely, such a representation is generated by the unique stabilizing periodic solution.

6. CONCLUDING REMARKS

In this paper, the relationship between the periodic solutions of the periodic Riccati equation and the innovation representations of cyclostationary stochastic processes has been investigated. Among other things, a consistent notion of canonical innovation representation has been given. Since stationary processes are just a subset of cyclostationary ones, it is not surprising that the well known spectral factorization theory, see e.g. [24], fits into our picture as a particular case.

Besides its intrinsic interest, the innovation representation theory of cyclostationary processes has proven useful in solving the prediction problem for seasonal time series modelled by means of ARMA models with periodic coefficients (PARMA models). This issue has been addressed in [25] and [18], where the PARMA model is first given an innovation representation. If such a representation is canonical, then the optimal predictor can be given a simple input-output expression. Otherwise, the derivation of the predictor calls for the determination of the canonical innovation representation, which therefore plays the role of a "cyclospectral canonical factor". Recently, such a cyclospectral factorization theory has been used [26] to devise a suitable identification algorithm for PARMA models through the classical "minimum prediction error" approach.

Acknowledgements. This paper has been supported by MURST -Project "Model Identification, Systems Control, Signal Processing"- and Centro di Teoria dei Sistemi of the CNR, Milano, Italy.

REFERENCES

[1] A.N. Kolmogorov, *Fundamental Concepts of Probability*. Berlin: Springer Verlag, 1933.

[2] K. Astrom, *Introduction to Stochastic Control Theory*. New York: Academic Press, 1970.

[3] M.B. Priestley, *Spectral Analysis and Time Series*. London: Academic Press, 1981.

[4] P.E. Caines, *Linear Stochastic Systems*. New York: Wiley, 1988.

[5] E.J. Hannan and M. Deistler, *The Statistical Theory of Linear Systems*. 1988, New York: Wiley, 1988.

[6] W. A. Gardner and L.E. Franks, "Characterization of cyclostationary random processes," *IEEE Trans. Inform. Theory*, vol. IT-21, pp. 4-14, 1975.

[7] G. Guardabassi, "The optimal periodic control problem", Journal A, vol. 17, pp. 75-83, 1976.

[8] A. Papoulis, *Probability, Random Variables and Stochastic Processes*. Singapore: McGraw Hill, 1984.

[9] S. Bittanti, "Deterministic and stochastic linear periodic systems", in , *Time Series and Linear Systems*, S. Bittanti, Ed. Berlin: Springer-Verlag, LNCIS Series vol. 86, 1986.

[10] R.E. Kalman, P.L. Falb and M.A. Arbib, *Topics in Mathematical System Theory*. New York: McGraw Hill, 1969.

[11] S. Bittanti and P. Bolzern, "Discrete-time linear periodic systems: Gramian and modal criteria for reachability and controllability," *Int. J. Control*, vol. 41, pp. 909-928, 1985.

[12] S. Bittanti and P. Bolzern, "On the structure theory of discrete-time linear systems," *Int. J. Systems Sc.*, vol. 17, pp.33-47, 1986.

[13] T. Kailath, *Linear Systems*. Englewood Cliffs: Prentice-Hall, 1980.

[14] S. Bittanti, P. Colaneri and G. De Nicolao, "An algebraic Riccati equation for the discrete-time periodic prediction problem," *Syst. & Control Letters*, vol. 14, pp. 71-78, 1990.

[15] P. Bolzern, P. Colaneri and R. Scattolini, "Zeros of discrete-time linear periodic systems," *IEEE Trans. Autom. Control*, vol. AC-31, pp. 1057-1058, 1986.

[16] E.J. Jury and F.J. Mullin, "The analysis of sampled-data control systems with a periodically time-varying sampling rate," *IRE Trans. Autom. Control*, vol. 5, pp. 15-21, 1959.

[17] O.M. Grasselli and S. Longhi, "Zeros and poles of linear periodic multivariable discrete-time systems," *Circuits Systems Signal Processing*, vol. 7, pp. 361-380, 1988.

[18] S. Bittanti and G. De Nicolao, "Spectral factorization of linear periodic systems with application to the optimal prediction of periodic ARMA models," Int. Rep. 90-013, Centro Teoria dei Sistemi del C.N.R., Dipartimento di Elettronica del Politecnico di Milano, Italy, 1990.

[19] R.E. Kalman, "New methods in Wiener filtering," *Proc. of First Symp. on Eng. Applications of Random Function Theory and Probability* (J. Bogdanoff and F. Kozin eds.), J. Wiley, pp.270–388, 1963.

[20] B.D.O. Anderson and J.B. Moore, *Optimal Filtering*. Englewood Cliffs: Prentice-Hall, 1979.

[21] S. Bittanti, P. Colaneri and G. De Nicolao, "The periodic Riccati equation," *The Riccati Equation* (S. Bittanti, A.J. Laub, J.C. Willems, eds.), Springer-Verlag, Berlin, Communication and Control Series, 1991.

[22] S. Bittanti, P. Colaneri and G. De Nicolao, "The difference periodic Riccati equation for the periodic prediction problem," *IEEE Trans. Autom. Control*, vol. AC-33, pp. 706-712, 1988.

[23] C.E. de Souza, "Existence conditions and properties for the maximal periodic solution of periodic Riccati difference equations," *Int. J. Control*, vol. 50, pp. 731-742, 1989.

[24] P. Lancaster, A.C.M. Ran, and L. Rodman, "Hermitian solutions of the discrete algebraic Riccati equation," Int. J. Control, vol. 44, pp. 777-802, 1986.

[25] S. Bittanti and G. De Nicolao, "Periodic ARMA models: Optimal prediction and minimum-phase condition", *Preprints 11th IFAC World Congress, August 1990, Tallinn, USSR*, vol. 6, pp. 200-205.

[26] S. Bittanti, P. Bolzern, G. De Nicolao, L. Piroddi e D. Purassanta, "A minimum prediction error algorithm for estimation of periodic ARMA models," *Proc. European Control Conference, July 1991, Grenoble, France*, to appear.

PARAMETRIZATIONS OF LINEAR STOCHASTIC SYSTEMS

Bokor J.* and L. Keviczky**

* MIT LIDS, 77 Massachusetts Ave. Cambridge, MA 02139 USA, visiting from

**Computer and Automation Institute, Hungarian Academy of Sciences, H1502 Budapest, Kende u. 13-17, Hungary

Abstract. The paper discusses the state space, matrix fraction and autoregressive moving average representations of stochastic systems parametrized by observability and constructibility invariants of state space representations. Forward and backward representations are derived and their relations are investigated.

1. INTRODUCTION

Multivariable linear systems can be described in some alternative forms, see transfer matrices, state space representations /SSR/, matrix fraction descriptions /MFD/, or autoregressive moving average /ARMA/ forms.

The use of transfer matrices, SSR-s and MFD-s is more common in the control literature, and the ARMA representations are widely applied in econometrics, statistics, time series analysis, system identification fault detection, and are also used in certain control problems involving predictions, see the references in the last paragraph.

The construction of representations for multivariable systems was originally studied in realization theory see [28,29,46,52,53] for the deterministic and [1,2,9,11,38,39,40,41] for the stochastic realization problems.

The properties of MFDs derived from factorization of transfer matrices or obtained from SSRs has been analysed e.g. in [30,31,34,51]. These references shows that the MFDs, SSRs and their relations became well understood but it is somewhat surprising that similar results concerning ARMA representations became known only in the past few years.

The derivation of ARMA models from MFDs has been studied in [3,6,22,37,49,56], and are also analysed in [14,49,50]. These results refer to the deterministic case, stochastic ARMA representations of second order stochastic processes that were derived from stochastic SSRs are available in [7,8,11,26].

Using these alternative representations it becomes possible to select a particular form representing a specific system property (e.g. observability, controllabily) that is the best suited to formulate or to solve a given filtering, prediction, identification, contol or signal processing problem. This possibility indicates a need for finding relationships among these representations in order to transfer one form into another while preserving the sytem property of interest.

In many situations there is a need for having unique representations of stochastic systems. This is the case in the stochastic approach to system identification, where the identifiability of the parameters describing a system requires that these parameters should be uniquely related to the second order statistics of the realization of a stochastic process. The problem of finding appropriate system descriptions represented by a set of parameters ensuring the above requirement is called parametrization.

The construction of unique representations can be solved by parametrizing multivariable systems through system invarians under a given equivalence relation .Then it is possible to construct canonic forms for the equivalence classes that are uniqely defined by these invariants. The formal definitions of system invariants and canonic forms can be found in [44], see also [21,23,24,28,36,42,46] for this approach.

Since these results appeared in a broad field of professional interest ranging from statistics to control and some of results on ARMA forms appeared only recently, the purpose of this paper is to collect those that give an understanding of basic system concepts represented by the above representations and show the relations among the alternative forms.

The approach applied here is based on state space concepts, where the invariants defining particular canonic forms will reflect certain system properties like observability or constructability if constructing left MFD and ARMA forms, or reachability and controllability if constructing right MFD and ARMA representations. We discusse here the left polynomial (MFD, ARMA) representations and the pure stochastic systems without control or exogenous inputs only. Dualizing the results the right polynomial representations can be derived. The generalization to consider a certain class of inputs can be performed following [11,27].

The structure of this paper is the following. In the next paragraph the state space representations will be constructed from the Wold-representation of a stochastic process, and it will be shown how a given basis selection in the predictor space leads to defining complete set of independent invariants for equivalent state space representations of a given order. The derivation of both the forward and backward state space representations of a stochastic process will be discussed.

The MFD and ARMA representations are discussed in the third paragraph. Instead of deriving system invariants under the equivalence relation on MFD and ARMA models of a given order, the MFD and ARMA representations will be constructed from state space canonic forms utilising the system properties (observability, constructibility) that these canonic forms are based on. By deriving transfer relations between a SSR and MFD, ARMA forms, it will be shown that one always can obtain canonic MFD and ARMA forms that are uniquely defined by the invariants of the SSR.

Constructing a forward and backward SSRs and the MFD,ARMA forms based on the same system property, a complete correspondence among these representations will be given.

2. MODELS OF STOCHASTIC SYSTEMS

Consider a p-component full rank wide sense stationary purely nondeterministic stochastic process z_t, $t \in Z$, $Z=\{0,+1,+2,....\}$, with zero mean $Ez_t=0$ and covariances $R_k=Ez_{t-1}z_t'$, $t \in Z$. Assume that the discrete Fourier transform of the covariance sequence $\{ R_k \}$ i.e. the spectral density exists and denote this by $\Phi(\omega)$, $\omega \in [-\pi, \pi]$.

The problem of finding a realization of the stochastic process z_t is to construct a linear system such that the output process y_t, $t \in Z$ generated by this system should have the same second order statistical properties as z_t has, i.e.

$$Ey_{t-k}y_t'=R_k, \quad t,k\in Z, \quad \Phi(\omega)=\Phi_y(\omega), \quad \omega\in[-\pi,\pi],$$

or y_t, z_t should have the same sample path, i.e. $y_t=z_t$ **P** a.s. , where **P** is an appropriately defined probability measure.

Linear systems can be described in some alternative forms like defining transfer functions, state space representations /SSR/, matrix fraction descriptions /MFD/ or autoregressive moving average /ARMA/ forms. When describing stochastic systems it is possible to define forward and backward representations, where the time proceeds forward or backward respectively. These representations have the following typical forms.

$$\textbf{FSSR:} \quad x_{t+1}=Fx_t+v_t, \qquad y_t=Hx_t+w_t \tag{2.1}$$

where x_t is the state vector, v_t, w_t are white noise processes, $\dim x_t=\dim v_t=n$, $\dim y_t=\dim w_t=p$, and

$$Ev_t=0,\; Ew_t=0,\; Ev_iv_j'=Q\delta_{ij},\; Ev_iw_j'=R\delta_{ij},\; Ex_tx_t'=P>0 \tag{2.2a}$$

$$Ex_sv_t'=0,\; Ex_sw_t'=0,\; t\geq s . \tag{2.2b}$$

$$\textbf{BSSR:} \quad x_t^B=F_B x_{t+1}^B+v_t^B, \quad y_t^B=H_B x_{t+1}^B+w_t^B, \tag{2.3}$$

where x_t^B is called the backward state variable, and the white noise processes v_t^B, w_t^B satisfy

$$Ev_t^B=0,\; Ev_t^B=0,\; Ev_i^B(v_j^B)'=Q_B\delta_{ij},\; Ev_i^B(v_j^B)'=R_B\delta_{ij} \tag{2.4a}$$

$$Ex_t^B(x_t^B)'=P_B \quad Ex_s^B(v_t^B)'=0,\; Ex_s^B(w_t^B)'=0,\; t\leq s. \tag{2.4.b}$$

The forward and backward MFD-s can be described as follows:

$$\textbf{FMFD:} \quad y_t=P^{-1}(q)Q(q)e_t+e_t, \qquad P(0)=I_p \tag{2.5a}$$

$$\textbf{BMFD:} \quad y_t=P_B^{-1}(q)Q_B(q)e_t^B+e_t^B,\; P_B(0)=I_p \tag{2.5b}$$

where q denotes the forward shift operator, $qy_t=y_{t+1}$, and we assumed that the polynomial matrices P(q), $P_B(q)$ were nonsingular. The forward and backward white noise processes e_t and e_t^B have covariance matrices $Ee_ie_j'=\delta_{ij}\Sigma$, $Ee_i^B(e_j^B)'=\delta_{ij}\Sigma_B$.

Similarly, introducing the backward shift operator $d=q^{-1}$, the forward and backward ARMA representations can be described as

$$\textbf{FARMA:} \quad y_t=A^{-1}(d)B(d)e_t+e_t, \quad A(0)=I_p \tag{2.6a}$$

$$\textbf{BARMA:} \quad y_t=A_B^{-1}(d)B_B(d)e_t^B+e_t^B, \quad A_B(0)=I_p \tag{2.6b}$$

where we assumed that the polynomial matrices A(d), $A_B(d)$ were nonsingular. The covariance matrices of the white noises e_t and e_t^B are $Ee_ie_j'=\delta_{ij}\Sigma$, $Ee_i^B(e_j^B)'=\delta_{ij}\Sigma_B$.

Consider the situation when the covariance sequence $\{ R_k \}$ is given. It is known from the theory of (weak) stochastic realizations see e.g. [18], that applying covariance factorization in the form:

$$R_k=HF_{k-1}L, \quad k \in Z \tag{2.7}$$

one can obtain FSSR $(F,H,L, R_o/2)$ and BSSR $(F_B,H_B,L_B,R_o/2)$ such that the relationship to the FSSR in (2.1-2) is given by the formulae:

$$P-FPF'=Q, \qquad L-FPH'=S, \qquad R_o-HPH'=R, \tag{2.8}$$

and similar relations are valid for the BSSR in (2.3-4). Taking the two-sided Z-transform of $\{R_k \}$ we can write

$$\Phi(z) = \sum_{k=-\infty}^{\infty} R_k z^{-k} = Z_-(z) + Z_+(z) \tag{2.9}$$

and it can be deduced that the FSSR $(F,H,L, R_o/2)$ is a realization of the spectral summand $Z_+(z)$, and the BSSR $(F_B,H_B,L_B,R_o/2)$ is a realization of the spectral summand $Z_-(z)$, i.e.

$$Z_+(z)=H(zI-F)^{-1}L+R_o/2,\ \ Z_-(z)=H_B(zI-F_B)^{-1}L_B+R_o/2. \tag{2.10}$$

State space representations are called minimal, if their state dimension is the smallest one among all the state space representations giving the same transfer function. It can be proved that if the stochastic process y_t is rational, the rank of the block Hankel matrix **H**

$$\mathbf{H}=\begin{bmatrix} R_1 & R_2 & R_3 & \dots \\ R_2 & R_3 & R_4 & \dots \\ & & & \end{bmatrix} \tag{2.11}$$

formed from the covariance matrices is finite. Assume that rank**H**=n, then the state space representations in (2.1),(2.3) are minimal iff $n=\dim x_t=\dim x_t^B$.

Another approach to obtain representations of stochastic systems is to factorize the spectral density matrix $\Phi(\omega)$, see the assumptions for doing this e.g. in [11,57],

$$\Phi(\omega)=T(z)\Sigma T'(z^{-1}),\ \ \Phi(\omega)=T_B(z)\Sigma_B T_B'(z^{-1}),\ \ z=\exp(i\omega), \tag{2.12}$$

$$\lim_{z\to\infty} T(z)=\lim_{z\to\infty} T_B(z)=I_p$$

where $T(z),T_B(z)$ are rational matrices. Forward and backward MFD-s can be obtained from coprime factorizations of the forward and backward left spectral factors $T(z)$ and $T_B(z)$ respectively,

$$T(z)=P^{-1}(z)Q(z)+I_p\ ,\qquad T_B(z)=P_B^{-1}(z)Q_B(z)+I_p\ , \tag{2.14}$$

$P(0)=P_B(0)=I_p$, $\operatorname{degdet}[P(z)]=\operatorname{degdet}[P_B(z)]=n$, where n is the McMillan degree of T,T_B respectively.

It is possible to deduce, that reproducing the second order statistics, the covariance factorization approach naturally leads to the construction of a state space representation of the spectral summands, since the spectral factorization followed by the factorization of the rational transfer matrices T,T_B leads to the construction of MFD-s. The construction of ARMA forms is not so obvious in this way and we come back to this point later.

The two approaches applied sofar are illustrated on Figure 1., where the dotted line indicates that no direct relations between SSR-s and MFD-s has been elaborated yet.

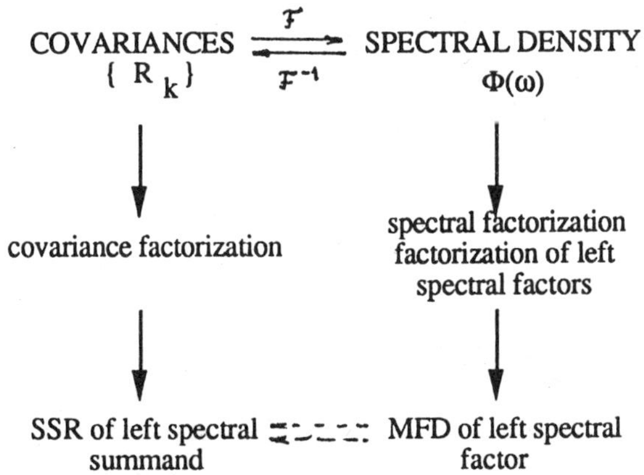

Figure 1. SSR and MFD realizations of second order stochastic processes

It follows from the covariance factorization that the resulting SSR-s are not unique, i.e. one can obtain many FSSR-s or BSSR-s of the same dimension n such that they generate the same covariances $\{ R_k \}$. It can also be seen that the SSR-s (F_i,H_i,L_i,Ro_i) generating the same $\{ R_k \}$ are related by coordinate transformation, i.e. there exisits a nonsingular scalar matrix T, such that

$$F_i=TF_jT^{-1}, \quad H_i=H_jT^{-1}, \quad L_i=TL_j, \quad R_{oi}=R_{oj}. \tag{2.15}$$

The MFD-s obtained through spectral factorization are also not unique but they are related by an unimodular matrix U(z), such that if

$$T(z)=P_i^{-1}(z)Q_i(z)+I=P_j^{-1}(z)Q_j(z)+I, \quad \mathrm{degdet}[P_i(z)]=\mathrm{degdet}[P_j(z)]=n\,, \tag{2.16}$$

then

$$P_i(z)=U(z)P_j(z)\,, \quad Q_i(z)=U(z)Q_j(z), \tag{2.17}$$

and $(P_i(z), Q_i(z),\Sigma)$ and $(P_j(z), Q_j(z),\Sigma)$ generate the same covariance sequence $\{ R_k \}$, where Σ is the covariance matrix appearing in the spectral factorization (2.12).

The construction of unique representations can be solved by utilizing that (2.15) and (2.17) define equivalence relations and a partitioning into equivalence classes on the set of all n-order SSR-s and MFD-s respectively. Then defining certain invariants under these equivalence relations, it is possible to construct canonic forms for the equivalence classes that are uniqely defined by these invariants. This approach will be discussed in the next paragraph.

3.STOCHASTIC STATE SPACE CANONIC FORMS

There are also some alternative ways to obtain canonic forms for either state space representations or for MFD-s, see e.g. [28,34,36,42,46],[19]. The approach applied here is based on state space concepts, where the invariants defining particular canonic forms will reflect certain system properties like observability or constructability if constructing left MFD and ARMA forms, or reachability and controllability if constructing right MFD and ARMA representations. Only left MFD and ARMA forms wil be discussed, the right representations can be obtained by dualizing the results.

The state space representations will be constructed from the Wold-representation of the stochastic process y_t, and it will be shown how a given basis selection in the predictor space leads to defining complete set of independent invariants for equivalent state space representations of a given order.

It is known eg.[47], that every purely nondeterministic process y_t can be written in the infinite moving-average (MA) form:

$$y_t = \sum_{k=0}^{\infty} K_k \varepsilon_{t-k}, \qquad \sum_{k=0}^{\infty} \|K_k\|^2 < \infty, \tag{3.1a}$$

$$y_t = \sum_{k=1}^{\infty} K_k^B \varepsilon_{t+k}^B, \qquad \sum_{k=1}^{\infty} \|K_k^B\| < \infty, \tag{3.1b}$$

called forward and backward Wold - representations respectively where e_t, e_t^B are white noises, $\mathbf{Y}_t = \mathbf{E}_t = \mathbf{E}_t^B$, i.e. the Hilbert spaces spanned by y_t and e_t, e_t^B are identical, and e_t, e_t^B, $t \in Z$ are the linear normalized forward and backward innovations:

$$e_t = \Sigma^{-1/2}\left(y_t - y_{t/t-1}\right), \qquad e_t^B = \Sigma_B^{-1/2}\left(y_t - y_{t/t+1}\right). \tag{3.2}$$

where $y_{t/t-1}=E\{y_t/Y_t^-\}$, denotes the conditional expectation that is identical to the orthogonal projection of y_t (Gaussian-case) to the Hilbert space $Y_t^-=\{y_t, t<0\}$ spanned by the past of y_t, and $y_{t/t+1}=E\{y_t/Y_{t+1}^+\}$ denotes the conditional expectation i.e. the orthogonal projection of y_t to the Hilbert space $Y_t^+=\{y_t, t\geq 0\}$ spanned by the present and future of y_t, respectively. The matrices Σ, Σ_B in (3.2) denote the covariances of forward and backward innovations respectively.

Define the following spaces (state-spaces of y_t):

$$X_t= Y_t^+/Y_t^-, \quad X_t^B= Y_{t+1}^-/Y_{t+1}^+, \tag{3.3}$$

where Y_t^+/Y_t^-, Y_t^-/Y_{t+1}^+ denote the orthogonal projections. It can be proved, that if y_t, $t\in Z$ is a rational process then X_t, X_t^B are finite dimensional, and $\dim X_t=\dim X_t^B=n$ for all $t\in Z$. The following result shows the existence of the SSR of y_t.

PROPOSITION 3.1 Let x_t and x_t^B be a basis in X_t and X_t^B respectively. Then there exist (F,G,H,Σ,e_t), $e_t=\Sigma^{1/2}\varepsilon_t$, and $(F_B,G_B,H_B,\Sigma_B,e_t^B)$, $e_t^B=\Sigma_B^{1/2}\varepsilon_t^B$ that realize y_t in forward SSR given as

$$x_{t+1}=Fx_t+Ge_t, \qquad y_t=He+e_t \tag{3.4}$$

$$Ee_t=0,\ Ee_ie_j'=\Sigma\delta_{ij},\ Ex_se_t'=0\ t\geq s,\ Ex_tx_t'=P>0, \tag{3.5}$$

and in backward SSR given as

$$x_t^B = F_B x_{t+1}^B + G_B e_t^B, \quad y_t = H_B x_{t+1}^B + e_t^B, \tag{3.6}$$

$$Ee_t^B = 0, \quad Ee_i^B(e_j^B)' = \Sigma_B\delta_{ij}, \quad Ex_s^B(e_t^B)' = 0, \quad t\leq s, \quad Ex_t^B(x_t^B)' = P_B \tag{3.7}$$

Proof. This is based on the orthogonal decomposition of the Hilbert spaces Y_{t+1}^-, Y_t^+. For forward realizations write $Y_{t+1}^-=Y_t^- + E_t$ where E_t is the Hilbert space spanned by the components of e_t and compute the projection Y_{t+1}^+/Y_{t+1}^-.

It can be seen that the SSRs in (3.4),(3.6) are not unique, and the forms of the resulting F,H,G matrices depend on the choice of basis in $\mathbf{X}_t$, $\mathbf{X}_t^B$ as it is known from deterministic realization theory. Since we are interested in establishing transfer relations, there is a need for deriving unique, canonical forms for the equivalent n-dimensional SSR-s. Canonic forward SSR-s for y_t can be derived from the Wold-representations (3.1) as follows.

Denote $X_t = [y_{t/t-1}, y_{t+1/t-1}, ...]'$ the vector of predictions that span $\mathbf{X}_t$, then using (3.1) one can deduce that

$$y_{t+k/t-1} = \sum_{i=k+1}^{\infty} K_i e_{t-i}\,, \quad k = 0,1,... \tag{3.8}$$

and

$$X_t = \mathbf{H} E_t \tag{3.9}$$

where $E_t = [e_{t-1}, e_{t-2},...,]$ and $\mathbf{H}$ is a Hankel matrix formed from the coefficient matrices in the forward Wold decomposition as

$$\mathbf{H} = \begin{bmatrix} K_1 & K_2 & \cdots \\ K_2 & K_3 & \\ \vdots & & \end{bmatrix}. \tag{3.10}$$

Equation (3.5) shows the relationship between the vectors in the space of forward predictions and in the space of past innovations. Similar relationship can be obtained for the backward predictions and innovations. In order to find a particular state space representation, one has to select a basis in $\mathbf{X}_t$ (or in $\mathbf{X}_t^B$) i.e. from the components of X_t.

PROPOSITION 3.2 Assume that the rational process y_t has a finite dimensional realization of minimum order n=rank **H**. Let the vectors $h_1',...,h_n'$ be a basis selected from the rows of **H**. Then $y_{t+k/t-1}^i$, the components of X_t associated with $h_1',...,h_n'$ also form a basis for $\mathbf{X}_t$.

The strategies for basis selection from the rows of the Hankel matrix (or equivalently, if **H** is factored as **H**=OC, from the rows of the observability matrix O) are called selection plans in the realization literature, see e.g. Scheme I., Scheme II selection plan of [8]. As a result of unique basis selection one obtaines a complete set of independent invariants for all the equivalence classes of the n-order SSR-s. Denote these invariants by (v_i, α_{ijk}) i=1,...,p, where v_i are called the observability (left Kronecker) indices, $v_1+...+v_p=n$.

The construction of stochastic SSR-s in canonic forms can be performed as follows. Consider the factorization of the Hankel matrix **H** in the form **H**=OC. Denote by O = $[H',(HF)',...,(HP^{n-1})']'$ the observability matrix and assume that rank O=n.Then the stochastic FSSR in canonical form can be obtained as follows.

PROPOSITION 3.3 Let the basis vectors selected by Scheme II selection plan from the rows of the observability matrix O be arranged in the following chain:

$$h_1', h_1'F, ..., h_1'F^{v_1-1}; \quad h_2',..., \quad ,...h_p'F^{v_p-1} \tag{3.11}$$

where $v_1,...,v_p$ denote the observability indices, and define the associated basis fqr X_t as

$$\bar{x}_t = \left[y^1_{t/t-1}, y^1_{t+1,t-1}, ..., y^1_{t+v_1-1/t-1}, y^2_{t/t-1}, ..., y^p_{t+v_p-1/t-1}\right]. \tag{3.12}$$

In this basis the rational process y_t has the forward Markovian SSR (F_c,H_c,G_c,I_p,e_t):

$$\bar{x}_{t+1} = F_c x_t + G_c e_t, \quad y_t = H_c \bar{x}_t + e_t, \tag{3.13}$$

where

$$F_c = \begin{bmatrix} 0 & 1 & & & 0 & & 0 \\ \vdots & & \ddots & & \vdots & & \vdots \\ 0 & & & 1 & 0 & & 0 \\ \alpha_{110}, & \alpha_{111}, ... & & ... & \alpha_{1p0}, & \alpha_{1p1}, ... & \\ \vdots & & & & & & \\ 0 & 0 & & & 0 & 1 & \\ \vdots & \vdots & & & \vdots & \ddots & \\ 0 & 0 & & & 0 & & 1 \\ \alpha_{p10}, & \alpha_{p11}, ... & & ... & \alpha_{pp0}, & \alpha_{pp1}, ... & \end{bmatrix}; \quad G_c = \begin{bmatrix} K^1_1 \\ \vdots \\ K^1_{n_1} \\ \vdots \\ K^p_1 \\ \vdots \\ K^p_{n_p} \end{bmatrix} \tag{3.14a}$$

$$H_c = \begin{bmatrix} 1 & 0 & ... & 0 & 0 & 0 \\ \vdots & \vdots & & \vdots & & \vdots \\ 0 & 0 & ... & 1 & 0 ... & 0 \end{bmatrix}, \tag{3.14b}$$

and K_j^i denotes the i-th row of K_j, $e_t=y_t-y_{t/t-1}$ and $Ee_ie_j'=\delta_{ij}\Sigma$.

Remark. Since the structure and the elements of the matrices F_c, G_c, H_c are uniquely defined by the $\{v_i\}$, $\{\alpha_{ijk}\}$ observability invariants, the SSR given in (3.9)-(3.10) is called observability state space canonical form.

Another canonic form used later is the constructibility state-space canonic form. Assume, that a SSR with F invertible is given, and define the constructibility matrix as

$$C_o = [\ (HF^{-1})',......,(HF^{-n})']' \tag{3.15}$$

Assume that the system is constructible, i.e. rank C_o =n. Applying again the Scheme II. selection plan to find the first linearly independent vectors from the rows of C_o, one obtaines a set of integers μ_i, i=1,...,p and scalars $\{\beta_{ijk}\}$ that are the (image of) constructibility invariants of (H,F^{-1}), see [6]. These invariants specify a unique $(H_{co},F_{co},G_{co},\Sigma,e_t)$ state space form called constructibility canonic form.

Proposition 3.1 states the exsistence of a bacward SSR of y_t too, but there is no indication how a specific forward SSR can be related to a backward one. This relation is provided by the following result.

PROPOSITION 3.3 Let the forward minimal SSR (F,G,H,D,e_t), $E\{e_ie_j'\}=\delta_{ij}I_p$ with state covariance $E\{x_tx_t'\}=P>0$ be given. Then the associated backward SSR is given by (F_B,G_B,H_B,D_B,e_t^B), where

$$F_B=F',\ H_B=HPF'+DG',\ G_B=[P^{-1}-F'P^{-1}F]^{1/2},$$

$$D_B=(H-H_BP^{-1})G_B^{-1} \tag{3.16}$$

the backward state vector x_t^B satisfies $x_t^B=P^{-1}x_t$, $E\{x_t^B(x_t^B)'\}=P^{-1}>0$, and e_t^B, x_t^B satisfy the backward orthogonality condition of (3.7).

4. STOCHASTIC MFD AND ARMA REPRESENTATIONS

In this paragraph we collect the results that relate the SSR-s in observability or constructability canonic forms to MFD and ARMA forms. The results are stated in the form of transfer relations that specify SSR-s and the associated MFD or ARMA forms in terms of system invariants. The proofs are omitted but the concepts to derive the transfer relations are described.

PROPOSITION 4.1 (FFSR-FMFD transfer relation). Let the n-dimensional FSSR of y_t, $t\in Z$ be given by (F,G,H,Σ,e_t), where (F,H) is observable. Then the $(P(z),Q(z),\Sigma,e_t)$ FMFD of y_t is given by

$$y_t=P^{-1}(q)Q(q)e_t+e_t \tag{4.1}$$

$$P(q)=\mathrm{diag}(q^{\nu_1},\ldots,q^{\nu_p})-F_pS_p(q),\quad Q(q)=[S_Q(q)-F_pS(q)]G \tag{4.2}$$

where $\nu_1,\ldots,\nu_p$ are the observability indices of (F,H), $\nu_1+\ldots+\nu_p=n$, the elements of the matrix F_p are the observability parameters $\{\alpha_{ijk}\}$,

$$F_p=\begin{bmatrix}\alpha_{110},\ldots,\alpha_{11,\nu_1-1},\ldots,\alpha_{1p0},\ldots & & \\ & & \\ \alpha_{p10},\ldots & ,\ldots,\alpha_{pp0},\ldots,\alpha_{pp\nu_p-1}\end{bmatrix},$$

and

$$S_p(q)=\mathrm{blockdiag}\{(1,q,\ldots,q^{\nu_1-1})',\ldots,(1,q,\ldots,q^{\nu_p-1})'\} \tag{4.3a}$$

$$S_Q(q)=\mathrm{blockdiag}\{(q^{\nu_1-1},\ldots,q,1)',\ldots,(q^{\nu_p-1},\ldots,q,1)'\} \tag{4.3b}$$

$$S(q)=S+qS^2+\ldots+q^{\nu-2}S^{\nu-1},\qquad \nu=\max\nu_i$$

$$S=\mathrm{blockdiag}\{S_{\nu_1},\ldots,S_{\nu_n}\}$$

and S_k is $\nu_k x \nu_k$ dimensional Toeplitz matrix:

$$S_k=\begin{bmatrix}0 & \ldots & & 0\\ 1 & 0 & & \vdots\\ \vdots & 1 & & \vdots\\ 0 & \vdots & 1 & 0\end{bmatrix}.$$

Proof. This is constructive and it is based on finding a bijection between the state vector of the FSSR and a basis of the "future" space $\mathbf{Y}_t^+$. Then utilize the property of state observability from most recent future observations.

Remark. The FMFD in (4.1) is uniquely defined by the observability invariants of the FSSR and by the forward innovation process and it is usually called controllability FMFD canonic form.

Using observability property of BSSR-s it is possible to show that the previous construction naturally leads to obtain transfer relation between BSSR and BARMA representetations.

PROPOSITION 4.2 (BSSR-BARMA transfer relation). Let the n-dimensional BSSR of y_t, $t \in Z$ be given by $(F_B, G_B, H_B, \Sigma_B, e_t^B)$, where (F_B, H_B) is observable. Then the $(A_B(d), C_B(d), \Sigma_B, e_t^B)$ BARMA representation of y_t is given by

$$y_t = A_B^{-1}(d)\, C_B(d)\, e_t^B + e_t^B, \tag{4.4}$$

where

$$A_B(d) = I_p - \bar{H}_B\, \bar{F}_B^{-1}\, S_A(d), \qquad C_B(d) = \bar{H}_B\, \bar{F}_B^{-1}\, S_c(d)\, \bar{G}_B, \tag{4.5}$$

and

$$S_A(d) = \text{block diag}\left\{\left[d, d^2, \ldots, d^{\mu_1}\right]', \ldots, \left[d, d^2, \ldots, d^{\mu_p}\right]'\right\} \tag{4.6a}$$

$$S_c(d) = \sum_{i=1}^{\mu} S^{i-1} d^i, \qquad \mu = \max_i \mu_i \tag{4.6b}$$

$$S = \text{block diag}\left\{S_{\mu_1}, \ldots, S_{\mu_p}\right\}$$

The indices $\mu_1, \ldots, \mu_p$ are the observability indices of (H_B, F_B) and $(\bar{H}_B, \bar{F}_B, \bar{G}_B)$ are in constructibility canonic form.

Remark. The BARMA representation in (4.3),(4.5) is uniqely specified by the backward observability invariants of the BSSR and the backward innovation process, and this BARMA representation is called controllability BARMA canonic form.

The relations between the above discussed representations are illustrated on Fig. 4.1.

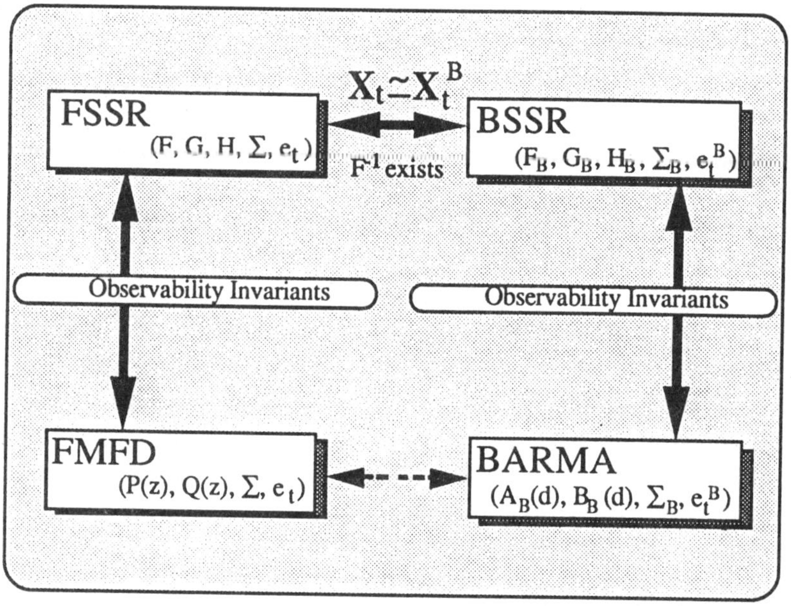

Figure 4.1 SSR, MFD and ARMA representations parametrized by observability invariants

Using constructability property of the FSSR-s and BSSR-s, it is possible to derive the associated FARMA and BMFD representations.

PROPOSITION 4.3 (FFSR-FARMA transfer relation). Let the n-dimensional FSSR (F,G,H,Σ,e_t) of y_t, $t\in Z$ be given by (F,G,H,Σ,e_t) where F is nonsingular and (H,F^{-1}) is constructible. Then the FARMA representation of y_t is given by

$$y_t = A_F^{-1}(d)\, C_F(d)\, e_t + e_t\,, \tag{4.7}$$

where

$$A_F(d)=I_p-HS_A(d), \qquad C_F(d)=-HS_C(d)G \tag{4.8}$$

and the polynomial matrices $S_A(d)$,$S_C(d)$ are given by (4.6) by replacing the indices $\mu_1,\ldots,\mu_p$ by the constructibility indices of (H,F^{-1}). The matrices (F,G,H) are in constructibility canonic form.

Proof. This is constructive and based on finding a bijection between the state vector of the FSSR and a basis of the space of past outputs Y_t^-. Then utilize the property of state observability from most recent past output observations.

Remark. The construction and the properties of constructibility state space canonic forms can be found in [6]. Since the FARMA representation in (4.7)(4.8) is completely defined by the constructibility invariants and the forward innovations, it is called constructibility canonic FARMA form.

PROPOSITION 4.3 (FFSR-BMFD transfer relation). Let the n-dimensional BSSR of y_t, $t \in Z$ be given by $(F_B, G_B, H_B, \Sigma_B, e_t^B)$, where F_B is nonsingular and (F_B^{-1}, H_B) is constructible. Then the $(P_B(q), Q_B(q), \Sigma_B, e_t^B)$ BARMA representation of y_t is given by

$$y_t = P_B^{-1}(q)\, Q_B(q)\, e_t^B + e_t^B, \tag{4.9}$$

$$P_B(q) = \mathrm{diag}(q^{\rho_1}, \ldots.., q^{\rho_p}) - F_P^B S_P^B(q), \quad Q_B(q) = -[S_Q^B(q) - F_P^B S(q) + \bar{H}_B]\bar{G}_B \tag{4.10}$$

where $\rho_1, \ldots, \rho_p$ are the constructibility indices of (F_B^{-1}, H_B), $\rho_1 + \ldots + \rho_p = n$, $\bar{G}_B, \bar{H}_B$ are the constructibility canonic forms of G_B, H_B, the elements of F_p^B are the constructibility parameters and the polynomial matrices $S_P^B(q), S_Q^B(q)$ are given in (4.3) by replacing the indices v_i by the constructibility indices ρ_i.

Remark. The BMFD in (4.9)-(4.10) is uniqely defined by the constructibility invariants of the BSSR and it is called constructibility BMFD canonic form.

The results stated in Propositions 3.3-3.4 are illustrated on Fig.4.3.

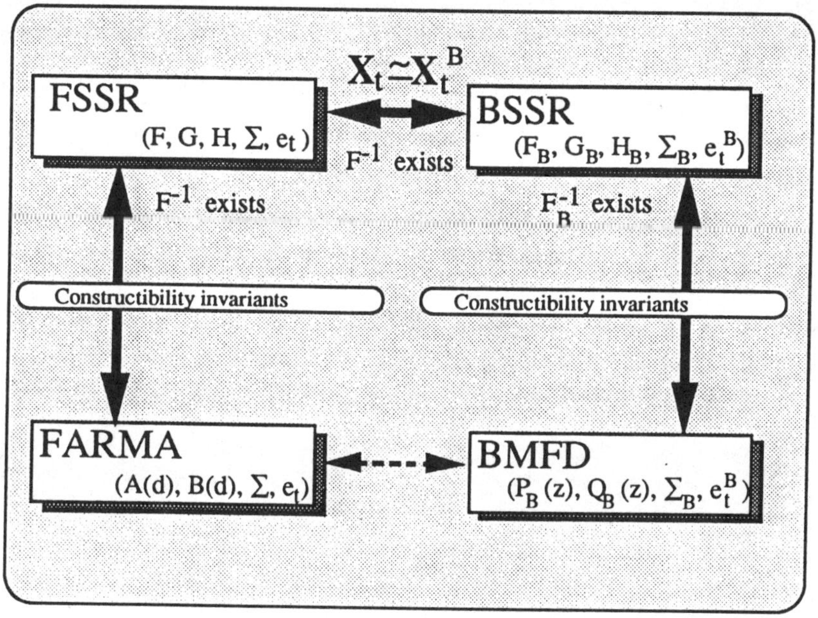

Figure 4.2 SSR, MFD and ARMA representations parametrized by constructibility invariants

CONCLUSIONS

This paper discussed the parametrization of state space /SSR/, matrix fraction description /MFD/ and ARMA representations of second order stochastic processes through system invariants. These invariants were defined under the equivalence relation on the set of SSRs with a given order. The MFD and ARMA forms were derived from SSRs and were shown to be uniqely defined by observability or constructibility invariants defining also canonic forms for SSRs. Transfer relations among the SSR,MFD and ARMA representations reflecting the same system property has been elaborated. Both forward and backward representations were discussed and their relationships was investigated.

REFERENCES

1 Akaike,H.(1974). Stochastic theory of minimal realization. IEEE Trans. Aut. Control, AC-19, 667.

2 Anderson,B.D.O. and Kailath,T.(1979). Forward, Backward and Dynamically Reversible Markovian Models of Second-Order Processes, IEEE Trans. Circuits and Systems, Vol. CAS-26, pp.956-965.

3 Bokor,J. and L.Keviczky(1982). Structural properties and structure determination of vector difference equations. Int. J. of Control, 36,3,461-475.

4 Bokor,J. and L.Keviczky(1984). Structure and parameter estimation of MIMO systems using elementary subsystem representation. Int. J. of Control, 39,5,965-986.

5 Bokor,J. and M. Ziermann(1987). Application of canonic ARMA models in the identification of multivariable systems. In Dynamic Modelling and Control of National Economies, Proc. of the 5th IFAC/IFORS Conference, Ed. M. Martos, L.F. Pau, M. Ziermann, Oxford, Pergamon Press, 363-366.

6 Bokor,J. and Keviczky,L.(1987). ARMA Canonical forms obtained from constructibility invariants. Int.J. of Control, 45, (3), 861-873.

7 Bokor,J., L.Keviczky and Cs.Bányász (1989a). Realization of stochastic processes in state space, MFD and ARMA forms. IEEE Conf. on Control and Applications, ICCON'89, Jerusalem, Wa-5-4.

8 Bokor,J., Keviczky,L., Cs.Bányász(1989b). Monic ARMA realizations of second order stochastic processes. 21st ISCIE Symp. on Stochastic Systems Theory and its Applications, Tokyo.

9 Bokor,J., Tanyi,M., Terdik, Gy.(1990). Forward and backward Markowian state space models of second order processes. Computers Math. Applic. Vol. 19, No. 1, 21-29.

10 Box,G.E.P., Jenkins,G.M.(1970). Time series analysis- -forecasting and Control. Holden-Day. San Francisco.

11 Caines,P.E.(1988). Linear Stochastic System, John Wiley & Sons.

12 Clark,J.M.C.(1976). The Consistent selection of parametrization in systems identification. Joint Automatic Control Conference, Purdue University.

13 Deistler,M., Hannan,E.J.(1981). Some properties of the parametrization of ARMA systems with unknown order. J. Multivariate Analysis, 11, 474-484.

14 Deistler,M. and Gevers M.(1989). Properties of the parametrization of monic ARMA systems. Automatica, 25,1,87-96.

15 Desai,U.B. and Pal,D.(1984). A transformation approach to stochastic model reduction. IEEE Aut.Control, Vol. AC-29, No. 12, pp.1097-1100.

16 Denham,M.J.(1974). Canonical Forms for the Identification of Multivariable Linear Systems. IEEE Trans. Aut. Control, AC-19, 646-656.

17 Dunsmuir,W., Hannan,E.J.(1976). Vector linear time series models. Adv. Applied Probability 8, 339-364.

18 Faurre,P.L.(1976). Stochastic Realization Algorithms. In System Identification: Advances and Case Studies, Ed. R.K.Mehra and D.G.Lainiotis Academic Press, New York, 1-25.

19 Forney,D.G.(1975). Minimal bases of rational vector spaces with applications to multivariable linear systems. SIAM J. Control 13, 493-520.

20 Gerencsér,L. and J.Baikovicius(1989). Detection of change in ARMA parameters. Proc. 28th CDC, Tampa, Florida,

21 Gevers, M.R. and V.Wertz(1984). Uniquely identifiable state space and ARMA parametrizations for multivariable linear systems. Automatica, 20, 333-347.

22 Gevers,M.R.(1986). ARMA models, their Kronecker indices and their McMillan degree. Int.J.Control, 43, 1745-1761.

23 Gevers,M.R. and V.Wertz(1987). Techniques for the selection of identifiable parametrizations for multivariable linear systems. In Control and Dynamic Systems, Ed. C.T. Leondes, Vol. 24. Academic Press, New York, 35-86.

24 Guidorzi,R.P.(1975). Canonical structures in the identification of multivariable, systems. Automatica 11, 361-374.

25 Guidorzi,R.P.(1981). Invariants and canonical forms for systems structural and parametric identification. Automatica, 17, 117-133.

26 Hannan,E.J.(1976). The identification and parametrization of ARMAX and state space forms. Econometrica, 44, 713-723.

27 Hannan,E.J. and M. Deistler(1988). The statistical theory of linear systems. John Wiley and Sons, New York.

28 Hazewinkel,M., Kalman,R.E.(1975). On Invariants, canonical forms, and moduli for linear constant, finite-dimensional dynamical systems. In:Lecture Notes Econ.-Math. System Theory. 131, 48-60, Springer-Verlag, Berlin.

29 Ho,B.L. and R.E.Kalman(1966). Effective construction of linear state-variable models from input/output functions. Regelungstechnik, 14, 545-548.

30 Janssen,P.H.M.(1987). MFD models and time delays; some consequences for identification. Int. J. Control, 45, 1179-1196.

31 Janssen,P.H.M. and A.A.H.Damen(1987). Comments on MFD models and time delays; some consequences for identification:. Int. J. Control, 46, 1857-1859.

32 Janssen,P.H.M.(1988a). On Model Parametrization and Model Structure Selection for Identification of MIMO-Systems. Dr.-Thesis, Dept.Electr. Eng., Eindhoven Univ. Techn., The Netherlands.

33 Janssen,P.H.M.(1988b). General results on the McMillan degree and the Kronecker indices of ARMA and MFD models. Int.J.Control, 48, 591-608.

34 Kailath,T.(1980). Linear Systems Prentice Hall Inc., Englewood Cliffs,NJ.

35 Kalman,R.E., Falb,P.L. and M.A.Arbib(1969). Topics in Mathematical System Theory. McGraw Hill, New York.

36 Kalman,R.E.(1974). Algebraic geometric description of the class of linear systems of constant dimension. 8th Annual Princeton Conf. on Information Sciences and Systems, Princeton, N.J.

37 Keviczky, L. and J. Bokor(1986). On new classes of canonic ARMA models. Statistical Methods in Experimental Researches and Quality Control Symposium, Varna.

38 Lindquist,A. and Picci,G.(1977). On the structure of minimal splitting subspaces in Stochastic realization Theory. Proc. of the 1977 Conf. on Decision and Control, New Orleans, La., 42-48.

39 Lindquist,A. and Picci,G.(1979). On the stochastic realization problem. SIAM J. Control and Optimization, 17, 361-389.

40 Lindquist,A. and Pavon, M.(1984). On the Structure of State-Space Models for Discrete Time Stochastic Vector Processes. Trans. Aut.Control, Vol. AC-29, No.5, 418-432.

41 Ljung,L. and Kailath,T.(1976). Backwards Markovian Models for Second-Order Stochastic Processes, IEEE Trans. Information Theory, Vol.IT-22, 488-491.

42 Luenberger,D.G.(1967). Canonical forms for linear multivariable systems. IEEE Trans. Autom. Control, AC-12, 290-293.

43 McFarlane,A.G.J. and N.Karcanias(1976). Poles and zeros of linear multivariable systems: a survey of the algebraic, geometric and complex-variable theory. Int.J.Control, Vol.24, No.1, 33-74.

44 McLane, S. and G. Birkhoff(1967). Algebra. MacMillan, London.

45 Mihaletzky,Gy, Tusnády,G.(1987-88). State space description of multivariable time series, Alkalmazott Matematikai Lapok, Budapest 13, 231-234.

46 Popov,V.M.(1972). Invariant description of linear time invariant controllable systems. SIAM J. Control, Vol.10, No.2, 252-264.

47 Rozanov,Yu.A.(1967). Stationary random processes. San Francisco, Holden Day.

48 Sidhu,G.S and Desai,U.B.(1976). New smoothing algorithms based on reversed-time lumped model. IEEE Trans.Aut.Control, 538-541.

49 Van den Hof, P.(1989). On residual based parametrization and identification of multivariable systems. Ph.D.Thesis. Technical University of Eindhoven.

50 Van den Hof, P.(1990).On the order and structural indices of linear systems represented in polynomial form. Proc. 11th IFAC World Congress, Tallin, USSR.

51 Verghese,G. and Kailath,T.(1979). A Further Note on Backwards Markovian Models, IEEE Trans. Information Theory, Vol. IT-25, 121-124, Correction to the above paper in ibid, Vol. IT-25,501.

52 Willems,J.C.(1986b). From tine series to linear system - Part II. Exact modelling. Automatica, 22, 675-694.

53 Willems,J.C.(1987). From time series to linear system - Part III. Approximate modelling. Automatica, 23, 87-115.

54 Wiener,N. and Masani,P.(1958). The prediction theory of multivariate stochastic processes, II. (The linear predictor), Acta Math. 99, 93-137.

55 Wolowich,W.A.(1974). Linear multivariable systems. Applied Mathematical Sciences, Vol. 11. Springer-Verlag, New York.

56 Wolovich, W.A. and H. Elliott(1983). Discrete models for linear multivariable systems. Int. J. Control, Vol.38, 337-357.

57 Youla,D.C.(1961). On the factorization of rational matrices, IEEE Trans. Inf. Theory, IT-7(3), 172-189.

Stochastic realization for approximate modeling

Andrea Gombani
LADSEB - CNR
Corso Stati Uniti 4
35020 PADOVA
ITALY

Abstract: we discuss here how to apply stochastic realization theory to the problem of constructing an approximate model. The result is a generalization of the Lindquist-Picci algorithm to derive a minimal stochastic realization from a nonminimal one. An algorithm to obtain an approximate realization is presented. Applications are in identification of stochastic systems.

§1. Introduction

This paper deals with the problem of "explaining" a given process in an approximate but simple way. There exists quite a wide literature on the problem, known as Stochastic Realization Problem. The basic reference is the work of Lindquist and Picci (see [LP] and reference therein). The question is the following: given a gaussian, stationary process $\{y(t)\}_{t\in \mathbf{Z}}$, is it possible to represent it as a function of a finite dimensional markov process $\{x(t)\}_{t\in \mathbf{Z}}$? The advantage of such a representation is clearly the possibility of using a finite dimensional statistics (the process x) to manipulate or reproduce or predict the process y. The process x represents, in other words, all is needed to know about, for example, the past of y to predict its future. The basic questions are then: how many of these processes there exists, which are the minimal ones, and how do we construct them.

There is, though, one basic difficulty in the construction: the process y is assumed to be given exactly, and the whole construction is based on this assumption. Nevertheless, what happens in practice is that the process y is only approximatly known; in general, it is computed from some long sequence of observed data constructing an autoregressive model from the estimated covariance coefficients. This procedure is known to be consistent (see [HD]). Nevertheless to obtain an exact model from a covariance sequence of lenght n, we have to use with probability 1 a model which has itself dimension n, even if the generating system has a much lower dimension. So the idea is to extend the stochastic realization algorithm to the case when the modeling is approximate. The basic idea is the following: suppose to start with the model

$$y(t) = \varepsilon u(t) + u(t-1) + \varepsilon u(t-2)$$

This process requires a two dimensional model to be represented exactly. But it is quite natural to say that y is approximated, when ε is very small, by the zero-dimensional model:

$$y(t) = u'(t)$$

where $u'(t) = u(t-1)$. A slightly more complex example is $y(t) = -\varepsilon u(t) + (1-\varepsilon^2) \sum_{n=1}^{\infty} \varepsilon^{n-1} u(t-n)$, which is is converging a.s. to $u(t-1)$ as ε goes to zero.

This presentation is elementary: we do not introduce the theory of Hardy spaces, and refer to other works for the proofs. (The basic references are ([LP], [G1], [G2], [G3]). We concentrate instead on a detailed time domain presentation of the minimal realization algorithm (in a form suitable for genaralization) and of the derived model reduction algorithm. We illustrate the procedure with an example. The computations have been made using an inplementation of the algorithm in MATLAB.

§2. A minimal realization algorithm

We consider $\{y(t)\}_{t\in \mathbb{Z}}$ to be a purely nondeterministic (p.n.d.), stationary, gaussian, centered stochastic processes taking values in $\mathbb{C}^m$, defined on a probability space $\{\Omega, \mathfrak{F}, \mathbb{P}\}$. We define the space generated by the process y, $Y := \overline{\text{span}}\ \{y_i(t);\ t \in \mathbb{Z},\ 1 \le i \le m\}$, where the inner product is $\langle \xi,\eta \rangle := E\eta^*\xi$, bar denotes closure with respect to the corresponding norm $\|\,.\,\|$, and E is the expectation. The past of y(t) is denoted by $Y^-(t) = \overline{\text{span}}\ \{y_i(s);\ s \le t,\ 1 \le i \le m\}$, and similarly, the future of y(t) by $Y^+(t) = \overline{\text{span}}\ \{y(s);\ s \ge t,\ 1 \le i \le m\}$. The space Y is naturally endowed with a shift U defined on the process as $Uy(t) := y(t+1)$, and extended by linearity and continuity to the whole space (see [Ro] for details). (We are considering, in other words, the problem of approximate "internal" realizations of y, see [LP]).

By markovian representation (or realization) of y we mean a stochastic system of the form

$$(\Sigma) \quad \begin{cases} x(t+1) = Fx(t) + Gu(t) \\ y(t) = Hx(t) \end{cases} \tag{2.1}$$

where $\{u(t)\}_{t\in \mathbb{Z}}$ is a white noise and F is stable. Clearly the process x is a markov process. We define $X := \text{span}\{x_i(0);\ 1 \le i \le n\}$ as the *state space* of the realization. A realization (Σ_1) is *contained* in (Σ) if the state space of (Σ_1) is contained in the state space of (Σ). We say that (Σ) is minimal (in a stochastic sense) if any markovian representation of y contained in (Σ) necessarily coincides with (Σ). The space X contains all the "information" from the past to predict the future (or viceversa); more formally, we say that $Y^-(0)$ and $Y^+(0)$ are conditionally independent given X, and we write $Y^-(0) \perp Y^+(0) | X$.

The dimension of the space X is called dimension of the realization. It is clearly minimal if the realization is minimal. If the realization is stochastically minimal, then it is also minimal in a deterministic sense (see [LP]). It is not true, though, that deterministic and stochastic minimality do coincide, as the following example shows: let u(t) be a white noise, and let $y(t) := u(t-1)$. Then y(t) has realization of dimension 1 given by:

$$\begin{aligned} x(t+1) &= u(t) \\ y(t) &= x(t) \end{aligned} \tag{2.2}$$

and this systems is clearly controllable and observable, i.e. minimal in a deterministic sense. Nevertheless, if we define u'(t):= u(t-1), we get the representation y(t) = u'(t), which is zero dimensional, and thus (2) is not minimal in a deterministic sense.

State spaces have a very interesting characterization in terms of invariant subspaces. Given a subspace X of Y, we define past and future of X as $X^- = \overline{\text{span}}\ \{U^tX;\ t \le 0\}$ and $X^+ = \overline{\text{span}}\ \{U^tX;\ t \ge 0\}$, respectively. If we denote by $E^B A$ the projection of A onto B, we can say that the space X is markovian (see [LP]) if one of the following equivalent conditions is satisfied:

$$\begin{aligned} &a) \quad E^{X^+}X^- = X \\ &b) \quad E^{X^-}X^+ = X \\ &c) \quad X^- \cap X^+ = X \end{aligned} \tag{2.3}$$

All the above conditions are alternative ways of saying that X makes conditionally independent its own past and future, i.e. $X^- \perp X^+ | X$. It is shown in [LP] that minimal subspaces can be characterized in terms of the two concepts of observability and constructibility.

A subspace X of Y is said to be *observable (constructible)* for y, if $E^X Y^+(0) = X$ ($E^X Y^-(0) = X$), and *unobservable (unconstructible)* otherwise. It is *completely unobservable (completely unconstructible)* if $E^X Y^+(0) = 0$ ($E^X Y^-(0) = 0$). The key result is the following:

Theorem *[LP]: X is minimal iff it is observable and constructible.*

Proof: see ([LP]).

In terms of matrices constructibility can be caracterized as follows: Denote by $c_s = Ey(t+s)y^*(t)$ the covariance coefficients of y, and let F, K, H be a realization of these coefficients, $c_s = HF^{n-1}K$.

Theorem *[Ru]: X is constructible if and only if the pair (F*,K*) is observable.*

The algorithm we present here is based on the following idea. We have introduced a shift U on Y.. Then are well defined the operator $U(X) := E^X U_{|X}$, and its adjoint $U^*(X)$. We say that X contains a *completely unconsructible* subspace if $X = X' \oplus E^X Y^+$ and X' is a non trivial subspace. Similarly, we say that X contains a *completely unobservable* subspace if $X = X'' \oplus E^X Y^-$ and X'' is nonzero. We have the following

Theorem 2.1: *X is constructible iff the projection of y(0) on every invariant subspace of U(X) is non zero; similarly, it is observable if and only if the projection of y(0) on every invariant subspace of $U^*(X)$ is non zero.*

Proof: we first show that if $X_1 \subset X$ is invariant for U(X), then X_1 is completely unconstructible if and only if $X_1 \perp y(0)$: since $U(X)\, X_1 \subset X_1$, for $\xi \in X_1$, we have for $n \ge 0$:

$$\langle U^{-n}y(0),\xi\rangle = \langle y(0),U^n\xi\rangle = \langle y(0),E^X U^n\xi\rangle = \langle y(0),E^{X_1}U^n\xi\rangle$$

the last term is zero if and only if $X_1 \perp y(0)$, as wanted. It follows that, if there exists a subspace X_1 in X invariant for U(X) and orthogonal to y(0), then this space is completely unconstructible, and thus X itself is not constructible. Conversely, X unconstructible means that $X_1 := X \ominus E^X Y^-(0)$ is nonempty. The space $E^X Y^-(0)$ is invariant for the backward shift $U^*(X)$: in fact, let $\xi \in Y^-(0)$; then,

$$E^X U^* \xi - E^X U^* E^X \xi = E^X U^* (\xi - E^X \xi) = 0$$

Hence, its orthogonal complement X_1 is invariant for U(X). Since X_1 is by definition orthogonal to y(0), the proof of the first statement is complete. A similar argument holds for observability. ■

Suppose now we start with the system (2.3). For minimality we need to make it observable and constructible. We take care of constructibility first. In view of the above theorem this is done taking the maximal invariant subspace of U(X) which is orthogonal to y(0).

We compute these subspaces by reducing F in (2.1) to triangular form with a convenient change of basis. F can be triangularized through a unitary similarity transformation (see [S]), i.e. there exists V unitary such that $\tilde{F} = VFV^*$ is lower triangular. To orthogonalize the basis, let $\tilde{G} := VG$, and solve the Lyapunov equation for the new basis

$$P = \tilde{F} P \tilde{F}^* + \tilde{G} \tilde{G}^* \tag{2.4}$$

which has a solution $P > 0$, since F is stable. We assume here that P is well conditioned to simplfy the exposition. It will be quite straightforward to take care of ill conditioned P with minor changes. Take now the Cholewski decomposition of P into LL^*, L lower triangular, and define $A := L^{-1} VFV^* L$, $B := L^{-1} VG$ and $C := HV^* L$. Equation (2) in this new basis becomes:

$$I = AA^* + BB^* \tag{2.5}$$

Then the new basis x(t) is clearly orthonormal, and since A is lower triangular, $\text{span}\{x_1, x_2, \ldots, x_j\}$ is an invariant subspace for U(X). It can be shown [G3] that every invariant subspace of U(X) can be represented in this form. Moreover, since the new basis is orthonormal, the coefficient c_i of C is given by $c_i = \langle x_i(0), y(0) \rangle$, and thus, in particular x_1 is orthogonal to y(0) iff the first coefficient c_1 of C is zero. So all we need to compute for constructibility is a matrix representation A of U(X) for which the number of leading zero coefficients of C is maximal, and then eliminate the unnecessary components of the state. Observability is even simpler: since span $\{x_n, x_{n-1}, \ldots, x_j\}$ is invariant for $U^*(X)$ for all j, and every invariant subspace can be represented in this form, we get from Theorem 2.1 that there exists a basis for which x_n is orthogonal to y(0) if and only if X is not observable. So, again all we need to compute is a basis in which the number of the ending zero coefficients of C is maximal (observe that this part coincides with the Ho-Kalman procedure to find a minimal realization in the deterministic case). We have thus the following lemma:

Lemma 2.2: *let X be a state space for y. Then X is constructible if and only if all the representations of y*

$$(\Sigma) \quad \begin{cases} x(t+1) = Ax(t) + Bu(t) \\ y(t) = Cx(t) \end{cases} \qquad (2.6)$$

with A lower triangular, and A, B satisfying (2.5), have the leading coefficient of C different from zero. Similarly, X is observable if all representations (2.6) have ending cofficient different from zero.

Corollary: *let A, B, C be a realization of y satisfying (2.5). Then the realization is constructible if none of the eigenvectors of A is orthogonal to the rows of C. It is observable if none of the eigenvectors of* A^* *is orthogonal to the rows of C.*

How many representations there exist of the above form, or equivalently, how many basis in X yield A, B satisfying (2.5)? If A has only simple eigenvalues, the answer is easy: it is the number of permutations of the eigenvalues of A, i.e. n! (for the complete proof we refer to [G3]; we exhibit below the unitary matrices which perform these changes of basis). If A has multiple eigenvalues, but all of geometric multiplicity 1, then the number of representations is still finite and we have no additional problems; if the geometric multiplicity is greater than 1 then things become more technical. For sake of simplicity we rule out this case here.

How do we use this information to find a minimal realization?

Suppose, as we said, that A has only eigenvalues of geometric multiplicity 1, and partition A as

$$A = \begin{bmatrix} A_r & & \\ x & A_k & \\ x & x & A_s \end{bmatrix} \qquad (2.7)$$

where the blocks on the diagonal are square of order r,k,s respectively.

Partition now the state process conformally as

$$\begin{bmatrix} x^r(t+1) \\ x^j(t+1) \\ x^s(t+1) \end{bmatrix} = A \begin{bmatrix} x^r(t) \\ x^k(t) \\ x^s(t) \end{bmatrix} + Bu(t)$$

where A and B are as in (2.5) and A is lower triangular. The main result is the following:

Theorem 2.3: *Suppose X contains a minimal subspace of dimension k. Let y(t) be given by*

$$(\Sigma) \quad \begin{cases} x(t+1) = Fx(t) + Gu(t) \\ y(t) = Hx(t) \end{cases}$$

Then there exists a similiarity transformations R

$$A=R^{-1}FR, \quad B=R^{-1}G, \quad C=HR$$

with A lower triangular, A, B satisfying (2.5) and $C=[0_r, C_k, 0_s]$, *with* $n-r-s = k$. *Then y(t) can be represented as*

$$(\Sigma) \qquad \begin{cases} x_k(t+1) = A_k x_k(t) + B_k u_k(t) \\ y(t) = C_k x_k(t) \end{cases}$$

where A_k is as above, B_k is a p×k matrix satisfying

$$I_k = A_k A_k^* + B_k B_k^*$$

and u_k is a white noise.

For the proof we refer to [G3].

From (2.6) we have A_k; as for B_k, it can be calculated recursively as follows: set $B^1 := B$ and denonte by b_i^1 the i-th row of B. Then, for i > j set

$$b_i^{j+1} := \frac{a_{ij}(1+\bar{a}_{jj})}{1-|a_{jj}|^2}\, b_j^j + b_i^j \qquad B^j := \begin{bmatrix} b_j^j \\ b_{j+1}^j \\ \cdot \\ \cdot \\ b_n^j \end{bmatrix} \tag{2.8}$$

We then set B_k to be given by the first k rows of B^{k+s}. Similarly, for the recursive construction, we set (observe the superscripts!)

$$A^j = \begin{bmatrix} a_{j,j} & & & \\ a_{j+1,j} & a_{j+1,j+1} & & \\ \cdots & & \cdot & \\ \cdots & & & \cdot \\ a_{nj} & & & a_{nn} \end{bmatrix} \tag{2.9}$$

so that A_k will be given by the first k×k block of A^{k+s}. So the problem is to select a pair A_k, B_k such that the largest number of leading and ending components of C vanish.

The matrix A determines a certain ordering of the eigenvalues, which is the one in which they appear on the diagonal (we recall that we assume that the poles have geometric multiplicity equal to 1); we can then consider the set S_n of the permutations of these eigenvalues. To each permutation $\sigma \in S_n$ we can attach matrices A_σ, B_σ as follows: let σ and τ differ for the transposition of the two adjacent elements in position j and j+1, i.e. $\tau = (\sigma(j+1), \sigma(j))\,\sigma$. Denote by $T_{\tau\sigma}$ the unitary matrix (Givens rotation)

$$T_{\tau\sigma} := \begin{bmatrix} I_{j-1} & & \\ & T^{\sigma}_{j,j+1} & \\ & & I_{n-j-1} \end{bmatrix} \tag{2.10}$$

where we set $D := \sqrt{|a^{\sigma}_{j+1,j+1} - a^{\sigma}_{jj}|^2 + |a^{\sigma}_{j,j+1}|^2}$, and define $T^{\sigma}_{j,j+1}$ as:

$$T^{\sigma}_{j,j+1} := D^{-1} \begin{bmatrix} a^{\sigma}_{j,j+1} & a^{\sigma}_{j+1,j+1} - a^{\sigma}_{jj} \\ -(a^{\sigma}_{j+1,j+1} - a^{\sigma}_{jj}) & a^{\sigma}_{j+1,j} \end{bmatrix} \tag{2.11}$$

If $D = 0$ we set $T^{\sigma}_{j,j+1} = I$ (it can be seen that the definition is consistent). We set $A_{\iota}:=A$, $B_{\iota}:=B$, where ι is the identityu, and $A_{\tau}:= T_{\tau\sigma}A_{\sigma}T_{\sigma\tau}$ and $B_{\tau}:= T_{\tau\sigma}B_{\sigma}$. Since any permutation is the product of adjacent transpositions, A_{τ} and B_{τ} are well defined. It is clear that all pairs A',B' satisfying (2.5) with A lower triangular are obtainable from A and B using a convenient permutation σ. Therefore, if X has a completely unconstructible subspace of dimension n-j, the particular basis of Theorem 2.3 is associated to a particular permutation σ^* and it is therefore obtainable as a product of Givens rotations (which is a Hausholder transformation).

Algorithm 1: let an initial realization of y be given

$$\begin{cases} x(t+1) = Fx(t) + Gu_X(t) \\ y(t) = Hx(t) \end{cases}$$

1. Reduce F to lower triangular form, via an LQ algorithm.

2. Solve the Lyapunov equation $P = FPF^* + GG^*$.

3. Solve $P = LL^*$, with L lower triangular. Set $A:=L^{-1}FL$, $B:=L^{-1}G$, $C:=LH$

4. Set $j = 1$
5. Using (2.10), compute if there exists a permutation σ for which the quantitiy $|c^{\sigma}_j|$ is 0.

If the permutation exists define A^{j+1}_{σ} and B^{j+1}_{σ} as in (2.8), (2.9) and repeat 5. Otherwise go to the next step.

6. If no such permutation exists, set r=n-j+1 and s=0 and compute if there exists a permutation, *among those leaving invariant the first r×r diagonal block and the last s×s diagonal block of A*, such that $|c^{\sigma}_{n-s}| = 0$. If such a permutation exists, choose it, set s:=s+1, and repeat step 6. Otherwise stop: the realization A_k, B_k, C_k is minimal.

It should be noted that the above algorithm picks a particular minimal realization. In fact, we first suppress a largest unconstructible component of X, and then a largest unobservable component in what is left. By reversing the order of these operations, a different minimal realization is obtained (at least in the generic case); or we could have chosen another strategy to eliminate unobservable and unconstructible components, and obtained an again different realization. We refer to [LP] for a description of how all these realizations are related. Nevertheless, since all the minimal realizations have the same dimension, this fact does not have much relevance in our algorithm. It will become important though in the model reduction procedure.

We consider now the following example: suppose we are give a realization of y of the form (2.1) where Γ,G,H are given below:

$$F = \begin{bmatrix} 0.3509 & -0.4375 & -0.0539 \\ 0.1244 & -0.0625 & -0.0210 \\ -0.0527 & 1.3125 & 0.4949 \end{bmatrix}$$

$$G = \begin{bmatrix} 0.3015 & 0.0178 \\ -0.0686 & 0.1439 \\ 0.0382 & -0.0533 \end{bmatrix}$$

$$H = \begin{bmatrix} 1 & 7 & 2 \\ 2 & 14 & 4 \end{bmatrix}$$

With some computation (steps 1-3 of the algorithm) it can be seen that with a change of basis throught the matrix

$$T = \begin{bmatrix} 3 & 0 & 1 \\ 1 & 7 & 2 \\ 0 & 1 & 5 \end{bmatrix}$$

we get the following realization (satisfying (2.5)) for y.

$$A = \begin{bmatrix} 0.3333 & 0 & 0 \\ 0.2887 & 0.2500 & 0 \\ -0.3464 & 0.9000 & 0.2000 \end{bmatrix}$$

$$B = \begin{bmatrix} 0.9428 & 0 \\ -0.1021 & 0.9186 \\ 0.1225 & -0.1225 \end{bmatrix}$$

$$C = \begin{bmatrix} 0 & 1 & 0 \\ 0 & 2 & 0 \end{bmatrix}$$

which yields, as can be easily checked using (2.8) and (2.9) the minimal realization:

$$A_2 = [0.2500]$$

$$B_2 = [\ 0.3062 \quad 0.9186\ ]$$

$$C_2 = \begin{bmatrix} 1.0000 \\ 2.0000 \end{bmatrix}$$

§3. The model reduction algorithm

It is quite natural, at this point, to extend the algorithm to the case when we want to operate a reduction of our model. In practice, we will always have to deal with roundoff errors which make any rapresentation minimal, and the algorithm needs a condition like "being zero within machine precision"; for instance, in the above example, this error is 1.3×10^{-16}. The above algorithm can then be given some more tolerance, so that components whch are relatively close to zero are neglected. There are two versions of the problem

Problem 1: given X markovian, with dim X = n, and k > 0, find X' ⊂ X markovian, dim X' = k, such that the error

$$\sum_{i=1}^{m} \| y_i - E^{X'} y_i \|^2 \tag{3.1}$$

is minimized.

A dual of Problem 1 is obviously the following:

Problem 2: given X markovian, with dim X = n, and $\varepsilon > 0$, find X' ⊂ X markovian of minimal dimension, such that

$$\sum_{i=1}^{m} \| y_i - E^{X'} y_i \|^2 < \varepsilon \tag{3.2}$$

It is easy to see that when $\varepsilon = 0$, this question reduces to the previous problem finding all minimal realizations of y contained in X.

We derive here the algorithm for problem 1. It is similar to the algorithm presented for the minimal realization: the basic difference lies in the choice of the subspace to be deleted, which now makes no difference among unobservable and unconstructible componenents: unlike the exact realization procedure, here there is a natural candidate for deletion, and it is the one yielding the smallest approximation error. The formulas here are modified to make the algorithm recursive.

Algorithm 2: let an initial realization of y be given

$$\begin{cases} x(t+1) = Fx(t) + Gu(t) \\ y(t) = Hx(t) \end{cases}$$

1. Reduce F to lower triangular form, via an LQ algorithm.

2. Solve the Lyapunov equation $P = FPF^* + GG^*$.

3. Solve $P = LL^*$, with L lower triangular. Set $\mathbf{A}:=L^{-1}FL$, $\mathbf{B}:=L^{-1}G$, $\mathbf{C}:=LH$

4 Set $r=n$, $\mathbf{A}_n=A$, $\mathbf{B}_n=B$, $\mathbf{C}_n=C$.

5. Using (2.10), compute the permutation σ for which the quantity $\min(|c_1^\sigma|, |c_r^\sigma|)$ is minimal. Compute $\mathbf{A}_r$, $\mathbf{B}_r$, $\mathbf{C}_r$ relative to this permutation (the dependence of $\mathbf{A}_r$, $\mathbf{B}_r$, $\mathbf{C}_r$ on the permutation is not denoted).

If the minimum is $|c_1^\sigma|$ *define*

$$\mathbf{A}_{r-1} = \{\text{lower (r-1) diagonal block of } \mathbf{A}_r\} = \begin{bmatrix} a^r_{2,2} & & & \\ a^r_{3,2} & a^r_{3,3} & . & \\ \cdots & & . & \\ \cdots & & & . \\ a^r_{r,2} & & & a^r_{r,r} \end{bmatrix}$$

Denoting by b_i^r the i-th row of $\mathbf{B}_r$, define:

$$b_{i-1}^{r-1} := \frac{a^r_{i1}(1+\bar{a}^r_{11})}{1-|a^r_{11}|^2}\, b_1^r + b_i^r$$

and

$$\mathbf{B}_{r-1} := \begin{bmatrix} b_1^{r-1} \\ b_2^{r-1} \\ . \\ . \\ b_r^{r-1} \end{bmatrix}$$

Similarly, denoting by c_i^r the i-th column of $\mathbf{C}_r$, set:

$$\mathbf{C}_{r-1} = [c_2^r, c_3^r, \ldots, c_r^r]$$

In the other case, set

$$\mathcal{A}_{r-1} = \{\text{upper (r-1) diagonal block of } \mathcal{A}_r\} = \begin{bmatrix} a^r_{1,1} & & & \\ a^r_{2,1} & a^r_{2,2} & & \\ \cdots & . & & \\ \cdots & & . & \\ a^r_{r-1,1} & & & a^r_{r-1,r-1} \end{bmatrix}$$

Denoting by b^r_i the i-th row of $\mathbf{B}_r$, define:

$$\mathbf{B}_{r-1} := \begin{bmatrix} b^r_1 \\ b^r_2 \\ . \\ . \\ b^r_{r-1} \end{bmatrix}$$

Similarly, denoting by c^r_i the i-th column of $\mathbf{C}_r$, set:

$$\mathbf{C}_{r-1} = [c^r_1 , c^r_2 , \ldots, c^r_{r-1}]$$

6. If $r > k$, set r:=r-1 and go to 5.

Theorem 3.1: *let $\mathcal{A}_k$, $\mathbf{B}_k$, $\mathbf{C}_k$ be give by the above algorithm; then there exists a white noise u_k such that the process y_k defined by*

$$\begin{cases} x_k(t+1) = \mathcal{A}_k x_k(t) + \mathbf{B}_k u_k(t) \\ y_k(t) = \mathbf{C}_k x_k(t) \end{cases}$$

has error $\sum_{i=1}^{m} \| y_i(0) - y_{i,k}(0) \|$ *given by:*

$$\left(\sum_{i \in \mathcal{I}} | c_i^T |^2 \right)^{1/2}$$

where $\mathcal{I}$ is the set of indeces of the components c_i which have been discarded at each step.

For a formal proof we refer to [G3].

We give now an example based on a small perturbation of the coefficients of the matrix C above. Let

$$F = \begin{bmatrix} 0.3509 & -0.4375 & -0.0539 \\ 0.1244 & -0.0625 & -0.0210 \\ -0.0527 & 1.3125 & 0.4949 \end{bmatrix}$$

$$G = \begin{bmatrix} 0.3015 & 0.0178 \\ -0.0686 & 0.1439 \\ 0.0382 & -0.0533 \end{bmatrix}$$

$$H = \begin{bmatrix} 1.0150 & 7.0100 & 2.0550 \\ 2.0060 & 14.0050 & 4.0270 \end{bmatrix}$$

Then, using the algorithm 2 (steps 1-3) we get

$$A = \begin{bmatrix} 0.3333 & 0 & 0 \\ 0.2887 & 0.2500 & 0 \\ -0.3464 & 0.9000 & 0.2000 \end{bmatrix}$$

$$B = \begin{bmatrix} 0.9428 & 0 \\ -0.1021 & 0.9186 \\ 0.1225 & -0.1225 \end{bmatrix}$$

$$C = \begin{bmatrix} 0.0050 & 1.0000 & 0.0100 \\ 0.0020 & 2.0000 & 0.0050 \end{bmatrix}$$

It is easy to check that the system is now minimal in a stochastic sense: , and no eigenvector of A or A^* is orthogonal to the rows of C (cfr. the corollary to Lemma 2.2). Still, applying the algorithm we get a one dimensional realization which is quite satisfactory, i.e.:

$$\mathcal{A}_2 = [0.2500]$$

$$\mathcal{B}_2 = [\ 0.3062 \quad 0.9186\ ]$$

$$\mathcal{C}_2 = \begin{bmatrix} 1.0000 \\ 2.0000 \end{bmatrix}$$

with a mean square error given by

$$\|y(0) - y_2(0)\| = 0.0112$$

We would like to remark that other approximation methods, like truncated balancing or Hankel-norm approximation yield in this case a much larger error (see [G3]).

[C] Caines, P.E. (1988), *Linear Stochastic Systems*, Wiley, New York.

[G1] Gombani A., Consistent approximations of linear stochastic models, *SIAM Journal on Control and Optimization*, **27**, 1989, 83-107.

[G2] Gombani A., On approximate stochastic realizations, to appear on *Mathematics of Control, Signals and Systems*..

[G3] Gombani A., On the multivariable approximate stochastic realization problem, *Stochstics and Stochastics Reports*, vol 34, pp.1-27, 1991

[HD] Hannan E.J., Deistler M., *The Statistical Theory of Linear Systems*, Wiley, 1988.

[GP] Gombani A, Pavon M., A general Hankel-norm approximation scheme for linear recursive filtering,*Automatica*, Vol 26, No. 1, pp 103-112,1990

[LP] Lindquist A., Picci G., Realization Theory for Multivariate Stationary Gaussian Processes, *SIAM J. Control and Optimization*, Vol 23, No. 6, 1985, 809-857.

[Ro] Rozanov Y.A., *Stationary Random Processes*, Holden-Day, 1987

[Ru] Ruckebusch G., Theorie geometrique de la representation Markovienne, *Ann. Inst. Henry Poincare* XVI, 1980, 225-297.

[S] Stewart G., *Introduction to Matrix Computation*, Academic Press, New York, 1973

REPRESENTATION OF INNER PRODUCTS AND STOCHASTIC REALIZATION

Gy. Michaletzky
Department of Probability Theory and Statistics
Loránd Eötvös University,
Muzeum krt 6–8. 1088 Budapest
Hungary

G. Tusnády
Mathematical Institute,
Hungarian Academy of Sciences,
Reáltanoda u. 13–15, 1053 Budapest
Hungary

1. Introduction

The representation of inner products of two subspaces of a Hilbert space is a pair of bounded linear operators mapping the subfields into a joint space in such a way that their inner products are preserved. This concept is used here for statistical investigation of state space models of time series. This generalizes the notion of splitting subspaces used by Ruckebusch [1], Lindquist–Picci [2] and Lindquist–Pavon [3] for stationary time series and exploited by Picci [4], Van Schuppen [5] in several other problems.

The basic mathematics used are given in [6], [7], [8] (and [9], the main ideas come from the seminal paper [10]).

2. Representation in Hilbert space

Definition 2.1 Let A and B be subspaces of a Hilbert space H and let K and L be bounded linear operators mapping A and B, respectively, into a Hilbert space R. The pair (K,L) will be called a *representation of inner products* of (A,B), if

$$(a,b)=(Ka,Lb) \tag{2.1}$$

holds for any $a \in A$, $b \in B$. In the left hand side of (2.1) (a,b) denotes the inner product in H, while in the right hand side (Ka,Lb) denotes the inner product in R.

In our definition H and R may be finite or infinite dimensional separable Hilbert spaces. Shortly we shall speak of representation if it is clear what the underlying spaces and subspaces are. Let us denote by A_0 the closure of the range of the orthogonal projection P_A of B onto A, i.e.

$$A_0 = \mathrm{cl}\{\, a : \exists\, b \in B,\ a = P_A b \,\}\ , \tag{2.2}$$

and similarly

$$B_0 = \mathrm{cl}\{\, b : \exists\, a \in A,\ b = P_B a \,\}\ . \tag{2.3}$$

(The range itself of these orthogonal projections is denoted by $P_A(B)$ and $P_B(A)$, respectively.)

First we give a special example of this representation. The crucial point of this example is the following Lemma which shows the more or less well known fact how to express a projection operator in an arbitrary base.

<u>Lemma 2.2.</u> Let $\{x(n),\ n \in N\}$ be an orthonormal base in B_0 and let $z(n)$ be defined as

$$z(n) = P_A x(n)\ , \qquad n \in N\ , \tag{2.4}$$

where N is finite or infinite set of indices. Then for any $b \in B$ we have

$$P_A b = \sum_{n \in N} (b,x(n))z(n)\ . \tag{2.5}$$

<u>Proof:</u> The sum

$$b_0 = \sum_{n \in N} (b,x(n))z(n)$$

is convergent by Bessel's inequality. The right hand side of (2.5) is $P_A b_0$, so it is enough to prove that

$$(a,b-b_0)=0$$

holds for every $a \in A$. This follows from the fact that

$$P_B a = \sum_{n \in N} (a,x(n))x(n)\ .$$

The operators

$$Ka = \{ (a,z(n)),\ n\in N \} \tag{2.6}$$

$$Lb = \{ (b,x(n)),\ n\in N \} \tag{2.7}$$

map A and B into l_2 if N is infinite and into the p–dimensional Euclidean space, if p, the number of elements of N, is finite. From (2.5) it follows that

$$(a,b) = \sum_{n\in N} (a,z(n))(x(n),b) \tag{2.8}$$

holds for any $a\in A$, $b\in B$. Thus (K,L) is an inner product representation. The projection of a non zero element of B_0 on A is non zero, thus $\{z(n),\ n\in N\}$ is a base in $P_A(B)$, so it generates A_0.

If (K,L) is arbitrary representation and R is l_2 or a finite dimensional Euclidean space, then the Riesz representation theorem implies that there are elements $z(n)\in A$, $x(n)\in B$, $n\in N$, such that (2.6) and (2.7) hold.

<u>Definition 2.3</u> The pair of systems of elements

$$\{ z(n)\in A,\ x(n)\in B,\ n\in N \} \tag{2.9}$$

is called *representative* of the subspaces A,B if the sequences (a, z(n)), (b, x(n)), $n\in N$ are finite or in l_2, and the sums

$$\sum_{n\in N} (a,z(n))x(n)\ , \tag{2.10}$$

$$\sum_{n\in N} (b,x(n))z(n) \tag{2.11}$$

are weakly convergent for any $a\in A$, $b\in B$, and also

$$(a,b) = \sum_{n\in N} (a,z(n))(x(n),b) \tag{2.12}$$

holds for any $a\in A$, $b\in B$.

Observe that a representative vector system defines a representation by means of (2.6)–(2.7).

<u>Lemma 2.4</u> If the pair (2.9) is a representative of the subspaces A, B then the element (2.10) is the orthogonal projection of $a\in A$ onto B and (2.11) is that of $b\in B$ onto A.

<u>Proof:</u> The given elements are in the appropriate subspaces and their inner products equal with that of the projections.

<u>Definition 2.5.</u> A representation (K, L) of A, B is *minimal* if for any representation (K_1, L_1) of (A,B) there are linear mappings K_2, L_2 such that $K = K_2K_1$, $L = L_2L_1$ and the pair K_2, L_2 preserves also the inner product.

Observe that K_2 and L_2 are not necessarily continuous. It is easy to see that a representation is minimal if and only if the mappings (K, L) have the largest possible nullspaces. But this pair of functions preserve also the inner products between A, B, so they can map an element of one of these subspaces into zero only if this elements is orthogonal to the other subspace. Thus the largest possible nullspaces are $A \ominus A_0$ and $B \ominus B_0$. This implies also that a minimal representation is one–to–one on A_0 and B_0. The elements of the representative vector system of a minimal representation are in A_0 and B_0, respectively.

<u>Remark.</u> We did not put any condition on the Hilbert space R in the definition of "minimal" representation. But projecting the elements in the ranges K(A), L(B) successively onto each other finally we can reduce R to a Hilbert space containing K(A) and L(B) as dense linear subspaces. In the finite dimensional case this obviously means that $K(A) = L(B) = R$. The following definition contains a further restriction in the infinite dimensional case.

<u>Definition 2.6.</u> A minimal representation is called *regular representation* if $R = l_2$ and the range of K and L contains the unit vectors e(n) of l_2, $n \in N$.

Suppose that (K,L) is a regular representation. Denote by x'(n) the element of $K^{-1}(e(n)) \cap A_0$ and by $z'(n) = L^{-1}(e(n)) \cap B_0$. Then

$$P_B x'(n) = x(n), \qquad P_A z'(n) = z(n), \qquad n \in N,$$

is the pair of representative vector systems of (K,L). Observe that they are in the range of P_B and P_A, respectively, and

$$(x'(n),z'(k)) = (x'(n),z(k)) = (x(n),z'(k)) = \delta_{nk}, \quad n, k \in N.$$

In particular there exist biorthogonal vector systems of { z(n), $n \in N$ } and { x(n), $n \in N$ }.

<u>Definition 2.7.</u> A representative vector system is called regular if the corresponding representation is regular.

Observe that the regularity is a property of the representation and not of the subspaces A and B. Introducing a new coordinate system in the space R corresponding to an arbitrary minimal representation we can always achieve that the unit vectors be in the range of K and L

At the same time we can ask which properties of the subspaces A and B assure that every minimal realization should be regular. We know that this is true if for example $A_0 = P_A(B)$ or $B_0 = P_B(A)$.

3. Representation of Gaussian processes

Let $(y_t, t = 0, \pm 1, \pm 2, \ldots\}$ be a q–dimensional stationary real Gaussian process with moments

$$Ey_t=0, \quad C_t=Ey_t y_0^T , \tag{3.1}$$

and let A be the closed subspace Y_t^+ generated by the elements of the form

$$\eta = \sum_{j=0}^{n} a_j^T y_{t+j} , \tag{3.2}$$

and let B be the closed subspace Y_t^- generated by the elements

$$\eta = \sum_{j=0}^{n} b_j^T y_{t-j} , \tag{3.3}$$

where n is arbitrary natural number, a_j, b_j are arbitrary q–dimensional vectors and ^{T} stands for transposition. The main Hilbert space H is the space Y generated by Y_t^+ and Y_t^-. In Y there is a unitary map U, the shift operator defined by

$$Uy_t = y_{t+1} . \tag{3.4}$$

To be precise, U is defined on the coordinates of y_t as follows

$$Uy_t(k) = y_{t+1}(k), \ k=1,\ldots,q, \ t\in\mathbb{Z}.$$

Denote by P_t^+ and P_t^- the orthogonal projections onto Y_t^+ and Y_t^-, respectively. Let $\{ z_0(n), x_0(n), n\in N \}$ be regular representative of Y_0^+ and Y_0^-, and for arbitrary t let $z_t(n)$, $x_t(n)$ be defined by

$$z_t(n) = U^t z_0(n), \quad x_t(n) = U^t x_0(n), \quad n\in N. \tag{3.5}$$

Then for arbitrary $s,t\geq 0$, $a\in Y_0^+$, $b\in Y_0^-$ we have that

$$(a,b) = \sum_{n\in N} (a,z_0(n))(x_0(n),b) \tag{3.6}$$

and

$$(U^t a,U^{-s}b) = \sum_{n\in N} (a,P_{A_0}U^{-t}z_0(n))(P_{B_0}U^s x_0(n),b) .$$

Denote by $T_z:A_0 \to A_0$ the operator $T_z = P_{A_0}U^{-1}$ and $S_x:B_0 \to B_0$ the operator $S_x = P_{B_0}U$. Then

$$(U^t a,U^{-s}b) = \sum_{n\in N} (a,T_z^t z_0(n))(S_x^s x_0(n),b) . \tag{3.7}$$

Let us express these operators in appropriate coordinate systems.

$$S_x x_0(n) = \sum_{k\in N} f_{nk}x_0(k) , \quad T_z z_0(n) = \sum_{k\in N} g_{nk}z_0(k) . \tag{3.8}$$

Here we have used the assumption that $x_0(n)$ and $z_0(n)$ can be written as projections onto B_0 (A_0) of appropriate vectors from A (B). In view of the commuting relations

$$P_{A_0}S_x^* = T_z P_{A_0} \qquad \text{and} \qquad S_x P_{B_0} = P_{B_0}T_z^* , \tag{3.9}$$

we have that

$$f_{nk} = g_{kn} . \tag{3.10}$$

So the matrix $F = (f_{nk})_{n,k\in N}$ belongs to S_x in the base $x_0(n)$ and F^T belongs to T_z in the base $z_0(n)$. This gives that

$$P_0^- x_t(n) = F^t x_0(n) \quad \text{and} \quad P_0^+ z_{-s}(n) = (F^T)^s z_0(n) . \tag{3.11}$$

Applying (3.6) to $y_t(k)$, $y_{-s}(j)$, $k,j=1,\dots,q$, $t,s\geq 0$, we obtain that

$$C_{t+s} = \sum_{n\in N} c_t(n)b_s^T(n) ,$$

where

$$c_t(n) = Ey_t z_0(n) ,$$

and

$$b_s(n) = Ey_{-s}x_0(n) .$$

Observe that $c_t(n)$ and $b_s(n)$ are the prediction and postdiction coefficients, i.e.

$$P_0^- y_t = E(y_t | Y_0^-) = \sum_{n\in N} c_t(n)x_0(n) , \tag{3.12}$$

$$P_0^+ y_{-s} = E(y_{-s} | Y_0^+) = \sum_{n\in N} b_s(n)z_0(n) . \tag{3.13}$$

Writing (3.11)–(3.13) for special t and s values [t=1, s=–1 in (3.11); t=0 in (3.12); s=0 in (3.13)] we get the usual system and observation equations.

It is possible to express these equations in the usual form introducing the so called prediction and postdictions errors, u_t, v_t, $t\in\mathbb{Z}$, respectively.

$$u_t = (I-P^-_{t-1})y_t, \qquad v_t = (I-P^+_{t+1})y_t$$

where I is the identity operator, since the coordinates of $x_{t+1}-Fx_t$ can be written as linear combinations of $u_{t+1}(k)$, $k=1,..,q$ and similarly $z_t-F^Tz_{t+1}$ is a linear function of v_t. So there exists a sequence of q–dimensional vectors $g(n)$, $n\in N$ such that

$$x_{t+1}(n) = \sum_{k\in N} f_{nk}x_t(k) + g^T(n)u_{t+1} ,$$

or shortly

$$x_{t+1} = Fx_t + Gu_{t+1} ,$$

where the vectors $g^T(n)$ are the row vectors of G.

Similarly

$$z_t = F^Tz_{t+1} + Kv_t .$$

Arranging the sequences $c_t(n)$ and $b_s(n)$ as column vectors of the matrices c_t, b_s we obtain that

$$y_t = c_0x_t$$

$$y_t = b_0z_t$$

and also

$$c_t = c_0F^t , \quad t\geq 0, \quad b^T_s = F^sb^T_0 , \quad s\geq 0, \tag{3.14}$$

where $F = (f_{nk})_{n,k\in N}$. In this way we have got the covariance factorization

$$C_{t+s} = c_0F^{t+s}b^T_0 , \qquad t,s\geq 0. \tag{3.15}$$

Applying these considerations to the elements of the representative vector systems we get that

$$Ez_tx^T_{-s} = (Ez_tz^T_t)F^{t+s}(Ex_{-s}x^T_{-s}) \tag{3.16}$$

where the expectations are interpreted as matrices containing $Ez_t(n)x_{-s}(k)$,$Ez_t(n)z_t(k)$ and $Ex_{-s}(n)x_{-s}(k)$, respectively.

<u>Remark.</u> The matrix F, although its rows and columns are – in view of (3.8) – in ℓ_2,

does not necessarily determine an operator from l_2 to l_2, so the expressions standing on the right hand sides of (3.14), (3.15) are understood as matrix multiplications. Applying F to the column vectors of b_0^T we obtain again l_2 sequences, so we can multiply it again by F several times, and finally by c_0.

Nevertheless in some special cases we can say a little bit more about the matrix F, or even the transformation determined by F on l_2. First of all, assuming that K(A), L(B) are dense in l_2, we can see that F is defined on a dense subspace of l_2, since L(B) is generated by the column vectors of $F^s b_0^T$, $s \geq 0$, – these are the $L(y_{-s}(k))$, $k=1,\ldots,q$ vectors. Furthermore, if for example the vectors $x_0(n)$, $n \in N$, form an orthonormal basis in B_0, then the operator which maps $x_0(n)$ into the n–th unit vector $e(n)$ of l_2 is an isometric transformation, so F determines a bounded operator on l_2, whose norm is not greater than 1. Generalizing this idea, if instead of an orthonormal basis in B_0 we choose another representative system, but the linear mapping $x_0(n) \longrightarrow e(n)$, $n \in N$, defines a bounded operator with bounded inverse, then the corresponding matrix F defines a bounded operator, moreover this operator has the same spectrum. In particular, its spectrum is inside the complex unit circle.

The condition on the linear operator $x_0(n) \longrightarrow e(n)$, $n \in N$, can be expressed by means of the "covariance" matrix $Ex_0x_0^T$, namely, it means that this matrix defines a bounded operator on l_2 with bounded inverse.

4. Kalman filter revisited

If we are given the matrices b_0, c_0, F corresponding to a regular innerproduct representation then we know that the representation is already determined, so all the other matrices in the system equations are fixed.

In this section we shortly present a version of the Kalman filter, which defines a recursive scheme for determining these unknown matrices.

Let us recall that from the regularity of the representation it follows that $x_0(n)$ can be written in the form

$$x_0(n) = P_{B_0} x'(n),$$

where $x'(n) \in A_0$. Shifting these vectors we can define $x_k'(n) = U^k x'(n)$.

Let us assume that y_t is of full rank, i.e. the covariance matrix of u_t is nonsingular. Let $\tilde{y}_t$ be defined by

$$\tilde{y}_t = E(y_t | y_1,...,y_{t-1}) \quad , t=2,3,...$$

and set

$$D_t = E(y_t - \tilde{y}_t)((y_t - \tilde{y}_t)^T \ , \quad t=2,3,...$$

and $\tilde{y}_1 = 0$, $D_1 = C_0$. Clearly, $D_t \geq E u_t u_t^T$, thus D_t is also regular.

<u>Definition 4.1.</u> The background process $\hat{x}_t$ defined by the factorization (3.15) is an infinite dimensional Gaussian process with coordinates in the space spanned by the coordinates of $y_1,...,y_t$ and with the moments

$$E\hat{x}_t = 0 \ , \ E\hat{x}_t y_{t-s}^T = F^s b_0^t \ , \ 0 \leq s < t \ . \tag{4.1}$$

Let us observe that

$$\hat{x}_t = E(x_t | y_1,...,y_t)$$

From this observation it follows immediately the

<u>Lemma 4.1.</u> For any $s \geq 0$, $t > 0$ we have

$$E(\hat{x}_{t+s} | y_1,...,y_t) = F^s \hat{x}_t \tag{4.2}$$

$$E(y_{t+s} | \ y_1,...,y_t) = c_0 F^s \hat{x}_t \tag{4.3}$$

<u>Theorem 4.3.</u> Starting with $\hat{x}_0 = 0$, $\hat{P}_0 = 0$ the following recursions hold:

$$\hat{P}_t = E(\hat{x}_t \hat{x}_t^T) \ , \tag{4.4}$$

$$\rho_t = \hat{P}_{t-1} A^T c_0^T \ , \tag{4.5}$$

$$D_t = C_0 - c_0 A \rho_t \ , \tag{4.6}$$

$$b_t^T = b^T - A \rho_t \ , \tag{4.7}$$

$$k_t = D_t^{-1/2} b_t \ , \tag{4.8}$$

$$\hat{u}_t = D_t^{-1/2} (y_t - c_0 A \hat{x}_{t-1}) \tag{4.9}$$

$$\hat{x}_{t+1} = A\hat{x}_t + k_t^T \hat{u}_t \tag{4.10}$$

Proof. Since the elements in $c_0A\hat{x}_{t-1}$ are defined by the weak convergence of the appropriate sums, we obtain that

$$E(c_0A\hat{x}_{t-1}\cdot\hat{x}^T_{t-1}) = c_0A\hat{P}_{t-1} = \rho^T_t .$$

The difference

$$\tilde{u}_t = y_t - c_0A\hat{x}_{t-1}$$

is independent of $y_1,\dots,y_{t-1}$ and of $\hat{x}_{t-1}$. Thus D_t is the covariance matrix of $\hat{u}_t$ and

$$C_0 = c_0A\rho_t + D_t \tag{4.11}$$

which is (4.6) and also

$$\begin{aligned} b^T &= E(\hat{x}_t y^T_t) = E(A\hat{x}_{t-1}y^T_t) = E((\hat{x}_t - A\hat{x}_t)\cdot y^T_t) \\ &= A\rho_t + E((\hat{x}_t - A\hat{x}_t)\cdot\tilde{u}^T_t) . \end{aligned} \tag{4.12}$$

Since the elements of $(\hat{x}_t - A\hat{x}_t)$ are linear combinations of the coordinates of $\tilde{u}_t$, (4.12) gives that with $\hat{u}_t$, k_t defined by (4.7)–(4.8) the equations (4.9)–(4.10) hold.

Shifting back $\hat{x}_t$ by U^{-t} i.e. considering the conditional expectations $U^{-t}\hat{x}_t = E(x_0 \,|\, y_0, y_{-1}, \ \ ,y_{1-t})$ we get the following

Corollary 4.1.

$$U^{-t}\hat{x}_t(n) \to x_0(n) \quad \text{strongly},$$
$$U^{-t}\tilde{u}_t(n) \to u_0(n) \quad \text{strongly},$$
$$b^T_t \to G \quad \text{elementwise},$$

if t tends to infinity.

5. Extended representation in Hilbert space

It is possible to generalize the notion of the inner product representation which led to a particular pair of realizations of a stationary time series, namely to predictive and postdictive realization. Using this so called extended representations of inner products we can obtain all the minimal realizations.

Let us fix the closed subspaces A and B of a Hilbert space H.

Definition 5.1. Suppose that $\bar{A}$ and $\bar{B}$ are closed subspaces of H, and $A \subset \bar{A}$, $B \subset \bar{B}$. The bounded linear mappings $K:\bar{A} \longrightarrow R$, $L:\bar{B} \longrightarrow R$, where R is a Hilbert space, define an *extended representation of inner products* of A,B, if

$$(a,b) = (Ka,Lb) \quad \text{for any } a\in\bar{A},\ b\in\bar{B} \ . \tag{5.1}$$

In other words (K,L) define a representation of $\bar{A}$, $\bar{B}$.

Definition 5.2. We call an extended representation minimal, if it is a minimal representation of A,B and also

$$P_{\bar{A}}(\bar{B}) = P_{\bar{A}}(B) \ , \tag{5.2}$$

$$P_{\bar{B}}(\bar{A}) = P_{\bar{B}}(A) \ . \tag{5.3}$$

Thus in some sense the subspaces $\bar{A},\bar{B}$ are minimal extensions of A,B from the point of view of inner products between the elements of A and B.

Remark. Conditions (5.2)–(5.3) can be expressed also in the following way. For every $b\in\bar{B}$ there exists a vector $b'\in\bar{B}$ such that $b-b'$ is orthogonal to $\bar{A}$. Thus denoting by $\bar{A}^{\perp}$ the orthocomplementer of $\bar{A}$ in the Hilbert space generated by $\bar{A}$ and $\bar{B}$, i.e. $\bar{A}^{\perp} = [\bar{A} \vee \bar{B}]\ominus\bar{A}$, we have that

$$\bar{B} \subset B \vee \bar{A}^{\perp} \ , \tag{5.2'}$$

and similarly

$$\bar{A} \subset A \vee \bar{B}^{\perp} \ , \tag{5.3'}$$

It is again possible to reduce also the Hilbert space R projecting the ranges of K,L successively onto each other. Thus we can assume that $K(\bar{A}),L(\bar{B})$ are dense in R. In this case from $b-b'\perp\bar{A}$ it follows that $Lb = Lb'$, i.e. $L(\bar{B}) = L(B)$, and also $K(\bar{A}) = K(A)$. This means that (K,L) is a minimal representation of $\bar{A},\bar{B}$.

In case when the minimal representation of A,B is finite dimensional, i.e. R is of finite dimension, then an extended representation is minimal if and only if it has the same dimension.

The following Lemma shows that every nonminimal extended representation can be reduced to a minimal one.

Lemma 5.1. Suppose that $(\bar{A},\bar{B},K,L)$ define an extended representation. Let

$$\tilde{A} = \bar{A} \cap (A \vee \bar{B}^{\perp}) \ , \quad \tilde{B} = \bar{B} \cap (B \vee \tilde{A}^{\perp}) \ . \tag{5.4}$$

The restrictions of K,L to $\tilde{A}$ and $\tilde{B}$, respectively, determine a minimal extended representation.

Proof: We have to prove that

$$\tilde{B} \subset B \vee [(\tilde{A} \vee \tilde{B}) \ominus \tilde{A}] \ ,$$

$$\tilde{A} \subset A \vee [(\tilde{A} \vee \tilde{B}) \ominus \tilde{B}] \ .$$

The first assertion is an obvious consequence of the definition of $\tilde{B}$, the second one follows from the observation that $\tilde{A} \cap [(\bar{A} \vee \bar{B}) \ominus \bar{B}] \subset \tilde{A} \cap [(\bar{A} \vee \bar{B}) \ominus \tilde{B}] = \tilde{A} \cap [(\tilde{A} \vee \tilde{B}) \ominus \tilde{B}]$.

As we shall see later the similarity of this procedure to the reduction of splitting subspaces (see for example Lindquist – Picci [2]) is not only a happenstance.

Denote by $\bar{A}_0$ the closure of the range of the orthogonal projection of $\bar{B}$ onto $\bar{A}$, i.e.

$$\bar{A}_0 = \mathrm{cl}\{a \in \bar{A} : \exists\, b \in \bar{B},\ a = P_A b\} \tag{5.5}$$

and similarly

$$\bar{B}_0 = \mathrm{cl}\{b \in \bar{B} : \exists\, a \in \bar{A},\ b = P_B a\}\,. \tag{5.6}$$

Observe that in case of minimal extended representations $P_{\bar{A}}(B)$ is dense in $\bar{A}_0$ and similarly $P_{\bar{B}}(A)$ is dense in $\bar{B}_0$.

First we would like to characterize the possible pairs of subspaces $\bar{A},\bar{B}$ for which there exist minimal extended representations of inner products of the fixed subspaces A,B. From the assumed minimality it follows that conditions (5.2')–(5.3') must hold. It is easy to see that these also assure the existence of a required representation. For example, the projections $P_{\bar{B}}:\bar{A} \longrightarrow \bar{B}_0$, $P_{\bar{B}_0}:\bar{B} \longrightarrow \bar{B}_0$ define an extended representation which is minimal.

There is a stronger version of this question. Namely, suppose that we have fixed a minimal representation of inner products of A,B. What are those subspaces $\bar{A},\bar{B}$ for which there exists a minimal extended representation of inner products of A,B, which is an extension of the previously fixed representation ? This is an open problem in the general case.

At the same time, if $\bar{A},\bar{B}$ determine a minimal extended representation, then so do the pairs $A,\bar{B}$ and $\bar{A},B$. So first we examine the subspaces $\bar{A},\bar{B}$ separately and after that we try to connect them. This possibility, that the subspaces $\bar{A}$ and $\bar{B}$ are not glued together but they can be examined separately, seems to be an advantage of this construction comparing with the procedures using splitting subspaces. The following Lemma and Corollary is essentially a coordinate–free formulation of the ideas used in Picci [4] for analysing different factor analysis models.

<u>Lemma</u> <u>5.2.</u> Suppose that $B \subset \bar{B} \subset A \vee B$ and let $\bar{B}_0 = \mathrm{cl}\{P_{\bar{B}}(A)\}$. Then

$$\text{(i)}\ \bar{B} = B \vee \bar{B}_0\,, \tag{5.7}$$

$$\text{(ii)}\ \bar{B}_0 = (A \cap \bar{B}_0) \vee (B \cap \bar{B}_0)\,. \tag{5.8}$$

Moreover, if $\bar{B} \subset B \vee A^{\perp}$ then $A \cap \bar{B}_0 \subset A_0$ and $B \cap \bar{B}_0 \subset B_0$.

Proof: (i) If an element $b \in \bar{B}$ is orthogonal to $\bar{B}_0$ and B then $b \perp A \vee B$, so $b=0$.

(ii) If $b \in \bar{B}$ then it can be written in the form $b=\alpha+\beta$, where $\alpha \in A$, $\beta \in B \subset \bar{B}$. So α is necessarily an element of $\bar{B}$. Thus $\bar{B} = B \vee (A \cap \bar{B})$. The second term is clearly contained in $\bar{B}_0$. Consequently $A \cap \bar{B} = A \cap \bar{B}_0$. Let us take again element $b \in \bar{B}_0$ and write as $b=\alpha+\beta$, $\alpha \in A \cap \bar{B}_0$, $\beta \in B$. We get that $\beta \in \bar{B}_0$. Thus $\bar{B}_0 = (A \cap \bar{B}_0) \vee (B \cap \bar{B}_0)$.

Projecting both sides of this last identity to A and B we obtain under the condition $\bar{B} \subset B \vee A^{\perp}$ that $A \cap \bar{B}_0 \subset A_0$ and $B \cap \bar{B}_0 \subset B_0$.

Corollary 5.1 Let be $\bar{B}$ be a subspace of $A \vee B$ satisfying the condition $\bar{B} \subset B \vee A^{\perp}$. Then $B \cap \bar{B}_0$ is generated by the vectors of the form $P_{B_0} a$, where $a \in A_0$ and

$$(a-P_{B_0}a, a'-P_{B_0}a') = 0 \qquad \text{for any } a' \in A \cap \bar{B}_0. \tag{5.9}$$

Proof: Let us observe that (5.9) is equivalent to

$$(a-P_{B_0}a, a') = 0 \qquad \text{for any } a' \in A \cap \bar{B}_0. \tag{5.10}$$

The vector $a-P_{B_0}a$ is orthogonal to $\bar{B}$. which contains B_0, so if $P_{\bar{B}_0}a \in B \cap \bar{B}_0 \subset B_0$, then $P_{B_0}a=P_{\bar{B}}a$, thus (5.10) holds. At the same time the vectors in $B \cap \bar{B}_0$ of the form $P_{\bar{B}}a$ are dense in $B \cap \bar{B}_0$.

On the other hand, if (5.10) holds then $a-P_{B_0}a$ is orthogonal to $A \cap \bar{B}_0$ and also orthogonal to B_0 containing $B \cap \bar{B}_0$, thus $(a-P_{B_0}a, b) = 0$ for any $b \in \bar{B}_0$. Using that $P_{\bar{B}}a = P_{B_0}a$, if $a \in A$, we get that $P_{B_0}a = P_{\bar{B}_0}a$, so $P_{B_0}a \in \bar{B}_0 \cap B$.

An immediate consequence of this Corollary is that

$$[\mathrm{cl}\{P_{B_0}(A \cap \bar{B}_0)\}] \cap [B \cap \bar{B}_0] = A \cap B. \tag{5.11}$$

Observe that on the other hand

$$[P_{B_0}(A \cap \bar{B}_0)] \vee [B \cap \bar{B}_0] = B_0. \tag{5.12}$$

Lemma 5.3. Suppose that $\bar{B} \subset B \vee A^{\perp}$. Let $\bar{A}=A \vee \bar{B}_0$. Then there exist minimal extended representations of $\bar{A},B$ and also of $\bar{A},\bar{B}$.

Proof: It is enough to check conditions (5.3)–(5.4). Since $\bar{A}=A\vee\bar{B}_0$ and $cl\{P_B(\bar{B}_0)\}=cl\{P_B(P_{\bar{B}}(A))\}=cl\{P_B(A)\}=B_0$ we have that $cl\{P_B(\bar{A})\}=B_0$ which gives that $\bar{A},B$ satisfy (5.2)–(5.3).

As concerns $\bar{A},\bar{B}$ we have obviously that $cl\{P_{\bar{B}}(A)\}=cl\{P_{\bar{B}}(A\vee\bar{B}_0)\}=\bar{B}_0$. At the same time if $b\in\bar{B}$ then $(b-P_{\bar{B}_0}b)\in\bar{B}$ and it is orthogonal to $A\vee\bar{B}_0$, thus

$$cl\{P_{A\vee\bar{B}_0}(B)\} = cl\{P_{A\vee\bar{B}_0}(\bar{B})\} = \bar{B}_0$$

Remark. The subspaces $\bar{A},\bar{B}$ in the previous Lemma give a special example of minimal extended representations, since in this case $P_{\bar{A}}(\bar{B}) = P_{\bar{B}}(\bar{A}) = \bar{A}\cap\bar{B} = \bar{B}_0$. Thus $\bar{B}_0$ is a so called splitting subspace of A and B (cf. Ruckebusch [1], Lindquist–Picci [2], Lindquist–Pavon [3]). So in some sense the extended representation of inner products is a generalization of the splitting subspaces.

Of course the subspace $\bar{A}$ defined in the previous Lemma is not the only one for which the pair $\bar{A},\bar{B}$ can determine a minimal extended representation of inner products of A,B. For example any subspace of $\bar{A}$ containing A is suitable for this.

Lemma 5.4. Suppose that $\bar{B}\supset B$ satisfying $\bar{B}\subset B\vee A^{\perp}$ is given. A subspace $\hat{A}$ –together with $\bar{B}$– satisfies the conditions (5.2)–(5.3) if and only if

$$\hat{A} = C \vee A_0\,, \tag{5.13}$$

where $A_0=cl\{P_A(B)\}$ and C is a closed subspace of H containing $A\ominus A_0$ and orthogonal to $\bar{B}$.

Moreover, in this case

$$\hat{A} \cap (A \vee \bar{B}) \subset \bar{A}\,, \tag{5.14}$$

where $\bar{A}$ is the subspace constructed in the previous Lemma.

Proof: First suppose that (5.3) holds. This means that $\hat{A} \subset A \vee \bar{B}^{\perp}$. Let us take an element $\alpha\in\hat{A}$. It can be written in the form $\alpha=\alpha'+\beta$, where $\alpha'\in A,\beta\in\bar{B}^{\perp}$. This gives that $\beta=\alpha-\alpha'\in\hat{A}$. Also, $\alpha'=\alpha_0+\alpha_1$, $\alpha_0\in A_0$, α_1 is orthogonal to B so also to $\bar{B}$. Let $C = \{\ \gamma\in\hat{A} : \gamma$ is orthogonal to $\bar{B}\ \}$. Then $C \supset A \ominus A_0$, and from the decomposition $\alpha=\alpha_0+(\alpha_1+\beta)$ we have that $\alpha_1+\beta\in C$ and $\alpha_0\in A_0$. Thus $\hat{A} = C \vee A_0$.

Conversely, if $\hat{A} = C \vee A_0$ then $P_{\bar{B}}(\hat{A}) = P_{\bar{B}}(A_0) = P_{\bar{B}}(A)$, and $cl\{P_{\hat{A}}(\bar{B})\}=A_0=cl\{P_{\hat{A}}(B)\}$.

Finally $\hat{A} \cap (A \vee \bar{B}) = [C \cap (A \vee \bar{B})] \vee A_0$ and $\{\gamma \in A \vee \bar{B} : \gamma \text{ is orthogonal to } \bar{B}\}$ $= \{ \alpha \in \bar{A} ; \alpha \text{ is orthogonal to } \bar{B}_0 \}$, which gives that $\hat{A} \cap (A \vee \bar{B}) \subset \bar{A}$.

Considering again the special case $\bar{B} \subset A \vee B$, $\hat{A} \subset A \vee B$, it is possible to express the assertion of Lemma 5.4 in terms of $\bar{B}_0$ and $\hat{A}_0 = \mathrm{cl}\{P_{\hat{A}_0}(\bar{B})\}$.

Lemma 5.5. Assume that $A \subset \hat{A} \subset A \vee B$, $B \subset \bar{B} \subset A \vee B$. The following conditions are equivalent:

(i) Conditions (5.2)–(5.3) hold,

(ii) $(a - P_{B_0}a, a' - P_{B_0}a') = 0$, if $P_{B_0}a \in B \cap \hat{A}_0$, $a' \in A \cap \bar{B}_0$,

(iii) $[\mathrm{cl}\{P_{B_0}(A \cap \bar{B}_0)\}] \cap [B \cap \hat{A}_0] = A \cap B$.

Proof: This is an immediate consequence of Lemma 5.4 and Corollary 5.1.

6. Representatives of extended representations

Now let us turn our attention to the representative vector systems. Consider a minimal extended representation of inner products of A,B, i.e.

$$K:\bar{A} \to R, \quad L:\bar{B} \to R$$

and assume that R is either a finite dimensional Euclidean space or l_2. Using again the Riesz representation theorem we get *representatives* $\bar{z}(n) \in \bar{A}$, $\bar{x}(n) \in \bar{B}$, such that

$$Ka = \{ (a,\bar{z}(n)),\ n \in N \}, \tag{6.1}$$

$$Lb = \{ (b,\bar{x}(n)),\ n \in N \}, \tag{6.2}$$

and

$$(a,b) = \sum_{n \in N} (a,\bar{z}(n))(\bar{x}(n),b), \quad a \in \bar{A},\ b \in \bar{B}. \tag{6.3}$$

We can see again that $\bar{z}(n)$ and $\bar{x}(n)$, $n \in N$, form bases in $\bar{A}_0$ and $\bar{B}_0$, respectively. We have also that

$$P_{\bar{B}}a = P_{\bar{B}_0}a = \sum_{n \in N} (a,\bar{z}(n))\bar{x}(n), \quad \text{for any } a \in \bar{A}, \tag{6.4}$$

where the series on the left hand side converges weekly. Also

$$P_{\bar{A}}b = P_{\bar{A}_0}b = \sum_{n\in N} (b,\bar{x}(n))\bar{z}(n), \quad \text{for any } b \in \bar{B}, \tag{6.5}$$

Restricting K and L to A and B, respectively, we get a representation of A,B, thus we can construct representative vectors $z(n)$, $x(n)$, $n\in N$ in A_0, B_0. Clearly

$$P_{A_0}\bar{z}(n) = z(n), \quad P_{B_0}\bar{x}(n) = x(n). \tag{6.6}$$

Applying (6.3) to $a=\bar{z}(k)$, $b=\bar{x}(j)$ we get that

$$(\bar{z}(k),\bar{x}(j)) = \sum_{n\in N} (\bar{z}(k),\bar{z}(n))(\bar{x}(n),\bar{x}(j)). \tag{6.7}$$

These identities can be expressed shortly using the matrices

$$\bar{W} = (\bar{w}_{kj}), \quad \bar{Q} = (\bar{q}_{kn}), \quad \bar{P} = (\bar{p}_{nj})$$

where

$$\bar{w}_{kj} = (\bar{z}(k),\bar{x}(j)), \quad \bar{q}_{kn} = (\bar{z}(k),\bar{z}(n)), \quad \bar{p}_{nj} = (\bar{x}(n),\bar{x}(j)),$$

namely

$$\bar{W} = \bar{Q}\bar{P}. \tag{6.8}$$

Following the steps in the previous "coordinate–free" construction, now let us consider a subspace $\bar{B}$ of $A \vee B$ satisfying the condition $B \subset \bar{B} \subset B \vee A^{\perp}$. In order to be able to formulate the equation (5.9) in terms of representative vectors, we have to assume that the extended representation of inner products of A,B, $K:A \longrightarrow R$, $L:\bar{B} \longrightarrow R$ as a representation of $A,\bar{B}$ is regular. This implies the existence of the vectors $z'(n)\in\bar{B}_0$, $\bar{x}'(n)\in A_0$ such that

$$\bar{x}(n) = P_{\bar{B}}\bar{x}'(n), \qquad z(n) = P_A z'(n), \quad n\in N \tag{6.9}$$

Now let us consider the decomposition (5.8) and try to express it in appropriate coordinate systems. Let the closed subspace $C \subset B \cap \bar{B}_0$ be an arbitrary complement of $A \cap B$ in $B \cap \bar{B}_0$. Any element $\bar{x}(n)$ can be written in the form

$$\bar{x}(n) = \xi(n) + \eta(\nu), \tag{6.10}$$

where $\xi(n)\in A\cap\bar{B}_0$, $\eta(\nu)\in C$. This gives that

$$\xi(n) = P_A\xi(n) = \sum_{j\in N} (\xi(n),\bar{x}(j))z(j).$$

Let $\eta'(n) = \bar{x}'(n) - \xi(n)$. Then $\eta'(n)\in A$, and

$$\eta(n) = P_{\bar{B}}\eta(n) = \sum_{j\in N} (\eta'(n),z(j))\bar{x}(j),$$

and also

$$\eta(n) = P_{\bar{B}}\eta'(n) = \sum_{j\in N} (\eta'(n),z(j))x(j) .$$

Denote by

$$\alpha'_{nj} = (\xi(n),\bar{x}(j)) , \quad \beta_{nj} = (\eta'(n),z(j))$$

and

$$\alpha' = (\alpha'_{nj}) , \quad \beta = (\beta_{nj}) .$$

Since $\eta(n)=P_{B_0}\eta'(n)$, where $\eta'(n)\in A_0$, and $\xi(k)\in A_0\cap\bar{B}_0$, we can apply Corollary 5.1 which implies that

$$(\xi(k),\eta'(n) - P_{B_0}\eta'(n)) = 0 .$$

But

$$((\xi(k),\eta'(n)) = \sum_{j\in N} (\xi(k),x(j))(z(j),\eta'(n)) ,$$

and

$$(\xi(k),\eta(n)) = (\sum_{l\in N} (\xi(k),\bar{x}(l))z(l), \sum_{j\in N} (\eta'(n),z(j))x(j)) .$$

So

$$\alpha'\beta^T = \alpha' QP\beta^T. \tag{6.11}$$

Applying the decomposition (6.10) to the series

$$\eta(n) = \sum_{j\in N} \beta_{nj}\bar{x}(j)$$

and multiplying it by $z'(l)$, $l\in N$, we obtain using the uniqueness of (6.10) that

$$\beta = \beta^2 . \tag{6.12}$$

Let us project now both sides of (6.10) onto B. We get that

$$x(n) = P_B\xi(n) + \eta(n).$$

Consequently

$$\bar{x}(n) = x(n) + \xi(n) - P_B\xi(n) .$$

Computing the inner products $(\bar{x}(n),\bar{x}(j))$ it follows that

$$\bar{P} = P + \alpha'(Q-QPQ^T)\alpha'^T .$$

Finally, projecting $\xi(n)$ onto $\bar{B}$ we get that

$$\xi(n) = P_{\bar{B}}\xi(n) = \sum_{j\in N} (\xi(n),z(j))\bar{x}(j) .$$

Denoting the coefficients in this series by $\alpha_{nj} = (\xi(n),z(j))$, and using the notation $\alpha = (\alpha_{nj})$ we obtain that

$$\alpha = \alpha'Q\,, \quad \text{and obviously} \quad \alpha + \beta = I\,.$$

Summarizing these considerations we can conclude to the following Theorem.

Theorem 6.1. There exist α' and β matrices such that the decomposition

$$\bar{x} = \alpha\bar{x} + \beta\bar{x}$$

where $\bar{x}$ denotes the $(\bar{x}(n),\ n\in N)$ sequence and α,β acts on its vector structure, $\alpha = \alpha'Q$, defines the decomposition corresponding to

$$\bar{B}_0 = (A \cap \bar{B}_0) \vee C\,.$$

These matrices satisfy the following identities

$$\beta = \beta^2\,, \quad \alpha = \alpha^2\,, \quad \alpha + \beta = I\,,$$
$$\alpha'(I-QP)\beta^T = 0$$
$$\bar{P} = P + \alpha'(Q-QPQ^T)\alpha'^T\,.$$

Corollary 6.1. Suppose that we are given two minimal extended representations of the form

$$K{:}A \longrightarrow R\,,\ \bar{L}{:}\bar{B} \longrightarrow R\,, \quad \text{and } K{:}A \longrightarrow R\,,\ \tilde{L}{:}\tilde{B} \longrightarrow R$$

where $\tilde{B} \subset A \vee B$ and $\bar{L}, \tilde{L}$ coincide on B. The following assertions are equivalent:

(i) $\bar{P} \geq \tilde{P}$

(ii) $\bar{B} \supset \tilde{B}$

(iii) $P_{\bar{B}}\tilde{x}(n) = \tilde{x}(n)$.

Proof: The inequality in (i) means that if α_k is such a sequence of coefficients for which $\Sigma\alpha_k\bar{x}(k)$ converges weakly then $\Sigma\alpha_k\tilde{x}(k)$ is also weakly convergent, and

$$(\Sigma\alpha_k\bar{x}(k),\Sigma\alpha_k\bar{x}(k)) \geq (\Sigma\alpha_k\tilde{x}(k),\Sigma\alpha_k\tilde{x}(k))\,. \tag{6.13}$$

(i) ⇒ (ii). Suppose that $\xi\in A\cap\tilde{B}_0$. Applying (6.3) we get

$$\xi = P_{\tilde{B}_0}\xi = \sum_{n\in N} (\xi,z(n))\tilde{x}(n)\,,$$

and

$$P_{\bar{B}_0}\xi = \sum_{n\in N} (\xi,z(n))\bar{x}(n)\,.$$

From (6.13) it follows that

$$(\xi,\xi) \geq (P_{\bar{B}}\xi,P_{\bar{B}}\xi)\,,$$

so $\xi\in\bar{B}$. This implies that $\tilde{B} = B \vee (A \cap \tilde{B}_0) \subset \bar{B}$.

The assertions (ii) ⇒ (iii) and (iii) ⇒ (i) are obvious.

As a next step let us try to formulate the conditions for $\bar{A}$ and $\bar{B}$ in order to define extended representations in terms of matrices. The following Corollary is an immediate consequence of Lemma 5.5 and the previous Corollary.

Corollary 6.2. Suppose that $K:A \to R$, $\bar{L}:\bar{B} \to R$ and $\bar{K}:\bar{A} \to R$, $L:B \to R$ are two minimal extended representations for which K is the restriction of $\bar{K}$ to A, and similarly $\bar{L}$ is an extension of L. Assume that $\bar{A} \subset A \vee B$ and L can be extended continuously to $B \vee \bar{A}_0$. The following conditions are equivalent:

(i) $\bar{Q}^{-1} \geq \bar{P}$

(ii) $\bar{W} = \bar{Q}\bar{P}$

(iii) $(\bar{K},\bar{L})$ define a minimal extended representation.

7. Markovian extended representations.

Let us apply the results of the previous section in the special case when – as in Section 3 – $A = Y_0^+$, $B = Y_0^-$. The shift operator U acts on $A \vee B$. In the sequel when we consider an extended representation of inner products of A,B, i.e. $K:\bar{A} \to R$, $L:\bar{B} \to R$ we implicitly assume that the operator U can be extended unitarily to $\bar{A} \vee \bar{B}$. We denote this extension also by U. For any subspace C denote by $H_-(C)$ the closed subspace generated by the subspaces $U^{-k}(C)$, $k \geq 0$, and $H_+(C)$ is generated by $U^k(C)$, $k \geq 0$.

Definition 7.1. The subspace $\bar{B}$ is called Markovian subspace of $\bar{A} \vee \bar{B}$ if

$$Ub - P_{\bar{B}_0} Ub \text{ is orthogonal to } \bar{B} \vee H_-(\bar{B}_0),$$

and

$$U^{-1}b - P_{\bar{B}_0} U^{-1}b \text{ is orthogonal to } \bar{A} \vee H_+(\bar{B}_0),$$

for every $b \in \bar{B}_0$.

If the subspaces $\bar{A}$, $\bar{B}$ are clear from the context, we call the corresponding subspace simply Markovian.

Definition 7.2. The extended representation is Markovian if the subspaces $\bar{A}$, $\bar{B}$ are Markovian subspaces.

Lemma 7.1. Let $K:\bar{A} \to R$, $L:\bar{B} \to R$ be an extended representation. It is Markovian if and only if

$$U(\bar{A}) \subset \bar{A} \quad \text{and} \quad U^{-1}(\bar{B}) \subset \bar{B}.$$

Proof: Set

$$L_0 = \{\, \xi \in \bar{A} \vee \bar{B} : \xi \text{ is orthogonal to } \bar{A} \vee \bar{B}_0 \,\},$$

$$K_0 = \{\, \eta \in \bar{A} \vee \bar{B} : \eta \text{ is orthogonal to } \bar{B} \vee \bar{A}_0 \,\}.$$

Observe that $\bar{B} = \bar{B}_0 \oplus L_0$, $\bar{A} = \bar{A}_0 \oplus K_0$. It follows that $P_{\bar{B}}Ub = P_{\bar{B}_0}Ub + P_{L_0}Uv$. So if $Ub - P_{\bar{B}_0}Ub$ is orthogonal to $\bar{B}$ for every $b \in \bar{B}_0$, then $U^{-1}(L_0) \subset L_0$. At the same time, if $b \in \bar{B}_0$ and $\eta = U^{-1}b - P_{\bar{B}_0}U^{-1}b$ is orthogonal to $\bar{A}$ then $\eta \in L_0$, consequently $U^{-1}b \in \bar{B}$. Thus the Markov property of $\bar{B}$ implies that it is invariant under U^{-1}.

Conversely, if $U(\bar{A}) \subset \bar{A}$ and $U^{-1}(\bar{B}) \subset \bar{B}$ then $U^{-1}(L_0) \subset L_0$ and $U(\bar{A} \vee \bar{B}_0) \subset \bar{A} \vee \bar{B}_0$, which implies that $\bar{A}$ and $\bar{B}$ are Markovian subspaces.

Remark. Observe that we have also got that the conditions $U(\bar{A}) \subset \bar{A}$, $U^{-1}(\bar{B}) \subset \bar{B}$ assure that $U(\bar{A} \vee \bar{B}_0) \subset \bar{A} \vee \bar{B}_0$, $U^{-1}(\bar{B} \vee \bar{A}_0) \subset \bar{B} \vee \bar{A}_0$, i.e. $\bar{A} \vee \bar{B}_0$ and $\bar{B} \vee \bar{A}_0$ are Markovian subspaces.

From the shift invariance of Markovian subspaces it follows using Lemma 7.1 that reducing a Markovian extended representation to a minimal one applying the procedure of Lemma 5.1 we obtain again a Markovian extended representation.

First we would like to characterize the Markovian extensions. The subspaces A,B are obviously Markovian. In particular, $U^{-1}(B \vee A_0) \subset B \vee A_0$. Let D be the closed subspace generated by the vectors $\xi = a - P_{B_0}a$, $a \in A_0$. Similarly, if $B \subset B \vee A_0$, then

$$\bar{D} = \langle\, \xi = a - P_{B_0}a : a \in A \cap \bar{B}_0 \,\rangle.$$

Then $B \vee A_0 = B \oplus D$, $\bar{B} = B \oplus \bar{D}$. Set

$$\Gamma = P_D U^{-1} : D \to D.$$

Lemma 7.2. Let $\bar{B}$ be a subspace of H, for which $B \subset \bar{B} \subset B \vee A_0$. Then $\bar{B}$ is U^{-1}–invariant if and only if $\bar{D}$ is Γ–invariant.

Proof: Set $U_0 = B \ominus U^{-1}(B)$. Then $\bar{B} = \bar{D} \oplus U_0 \oplus U^{-1}(B)$. Since $U^{-1}(D) \subset D \oplus U_0$, $\bar{B}$ is U^{-1}–invariant if and only if $U^{-1}(\bar{D}) \subset \bar{D} \oplus U_0$ which is equivalent to $\Gamma(\bar{D}) \subset \bar{D}$.

We can find Γ–invariant subspaces even in that case when the condition $\bar{B} \subset A \vee B$ is not fulfilled, but $U^{-1}(\bar{B}) \subset \bar{B}$. Namely, $\mathrm{cl}\{P_{A \vee B}\bar{B}\}$ and $\bar{B} \cap [A \vee B]$ are also

U^{-1}–invariant subspaces which are already in $A\vee B$, determining in this way Γ–invariant subspaces.

We can repeat the previous considerations with the necessary modifications for any subspace $\bar{A}$, $A \subset \bar{A} \subset A\vee B_0$. Let the subspace $\bar{E}$ be generated by the vectors $\eta = b - P_{A_0} b$, $b \in B \cap \bar{A}_0$, also $E = \langle \eta = b - P_{A_0} b,\ b \in B_0 \rangle$ and the operator $\Lambda = P_E U : E \to E$. The subspace $\bar{A}$ is U–invariant if and only if $\bar{E}$ is Λ–invariant. In view of Corollary 5.1 we can connect the Λ– and Γ–invariant subspaces of E and D, respectively. Namely, $(a - P_{B_0} a, a' - P_{B_0} a')$ defines a bilinear form on D. Using the above mentioned Corollary we get that if we project a Γ^*–invariant subspace of D onto B_0 then the closure of this subspace is Λ–invariant.

8. Extended representations of Gaussian processes

Consider a regular extended Markovian representation of inner products of Y_0^+ and Y_0^- and denote by $\{\bar{z}_0(n), \bar{x}_0(n), n \in N\}$ its representatives. For arbitrary t let $\bar{z}_t(n)$, $\bar{x}_t(n)$ be defined by

$$\bar{z}_t(n) = U^t \bar{z}_0(n), \quad \bar{x}_t(n) = U^t \bar{x}_0(n) .$$

In the same way as we did in Section 3 it is possible to construct a pair of realizations of the process $y(t)$, $t \in \mathbb{Z}$, using these representatives. We repeat it briefly.
Let the orthogonal projections onto $U^t(\bar{A})$ and $U^t(\bar{B})$ be denoted by $\bar{P}_t^+$ and $\bar{P}_t^-$ and let the processes $\bar{u}_t$, $\bar{v}_t$ be defined by

$$\bar{u}_t = (I - \bar{P}_{t-1}^-)\bar{x}_t, \quad \bar{v}_t = (I - \bar{P}_{t+1}^+)\bar{y}_t \tag{8.1}$$

where I is the identity operator. For arbitrary $s,t \geq 0$, $a \in Y_0^+$, $b \in Y_0^-$ we have that

$$(a,b) = \sum_{n \in N} (a, \bar{z}_0(n))(\bar{x}_0(n), b)$$

and

$$(U^t a, U^{-s} b) = \sum_{n \in N} (a, P_{\bar{A}_0} U^{-t} \bar{z}_0(n))(P_{\bar{B}_0} U^s \bar{x}_0(n), b) . \tag{8.2}$$

Denote by $T_{\bar{z}}: \bar{A}_0 \to \bar{A}_0$ the operator $T_{\bar{z}} = P_{\bar{A}_0} U^{-1}$ and $S_{\bar{x}}: \bar{B}_0 \to \bar{B}_0$ the operator $S_{\bar{x}} = P_{\bar{B}_0} U$.

Then

$$(U^t a, U^{-s} b) = \sum_{n \in N} (a, T_{\bar{z}}^t \bar{z}_0(n))(S_{\bar{x}}^s \bar{x}_0(n), b) . \tag{8.3}$$

Let us express these operators in appropriate coordinate systems.

$$S_{\bar{x}} \bar{x}_0(n) = \sum_{k \in N} f_{nk} \bar{x}_0(k) , \quad T_{\bar{z}} \bar{z}_0(n) = \sum_{k \in N} g_{nk} \bar{z}_0(k) . \tag{8.4}$$

Here we have used the assumption that $\bar{x}_0(n)$ and $\bar{z}_0(n)$ can be written as projections onto $\bar{B}_0$ $(\bar{A}_0)$ of appropriate vectors from $\bar{A}$ $(\bar{B})$ and also the identity (6.6). So the same matrix $F = (f_{nk})_{n,k \in N}$ which we have defined in Section 3 belongs to $S_{\bar{x}}$ in the base $\bar{x}_0(n)$ and F^T belongs to $T_{\bar{z}}$ in the base $\bar{z}_0(n)$. This gives that

$$\bar{P}_0^- \bar{x}_t(n) = F^t \bar{x}_0(n) \quad \text{and} \quad \bar{P}_0^+ \bar{z}_{-s}(n) = (F^T)^s \bar{z}_0(n) . \tag{8.5}$$

Again we have the commuting relations

$$P_{\bar{A}_0} S_{\bar{x}}^* = T_{\bar{z}} P_{\bar{A}_0} \qquad \text{and} \qquad S_{\bar{x}} P_{\bar{B}_0} = P_{\bar{B}_0} T_{\bar{z}}^* \tag{8.6}$$

Applying these to $y_t(k)$, $y_{-s}(j)$, $k,j=1,\dots,q$, $t,s \geq 0$, we obtain that

$$C_{t+s} = \sum_{n \in N} c_t(n) b_s^T(n)$$

where

$$c_t(n) = E y_t z_0(n) = E y_t \bar{z}_0(n),$$

and

$$b_s(n) = E y_t x_0(n) = E y_t \bar{x}_0(n).$$

Observe that $c_t(n)$ and $b_s(n)$ are the prediction and postdiction coefficients, i.e.

$$\bar{P}_0^- y_t = E(y_t | \bar{B}) = \sum_{n \in N} c_t(n) \bar{x}_0(n) , \tag{8.7}$$

$$\bar{P}_0^+ y_{-s} = E(y_{-s} | \bar{A}) = \sum_{n \in N} b_s(n) \bar{z}_0(n) . \tag{8.8}$$

(8.6)–(8.8) are the usual system and observation equations.

We have again the covariance factorization

$$C_{t+s} = c_0 F^{t+s} b_0^T , \quad t,s \geq 0. \tag{8.9}$$

Applying these considerations to the elements of the representative vector systems we get that

$$E\bar{z}_t\bar{x}_{-s}^T = (E\bar{z}_t\bar{z}_t^T)F^{t+s}(E\bar{x}_{-s}\bar{x}_{-s}^T) \tag{8.10}$$

where the expectations are interpreted as matrices containing $E\bar{z}_t(n)\bar{x}_{-s}(k)$,$E\bar{z}_t(n)\bar{z}_t(k)$ and $E\bar{x}_{-s}(n)\bar{x}_{-s}(k)$, respectively.

Remark. We would like to emphasize again that the matrices F, c_0, b_0 do not depend on the subspaces $\bar{A},\bar{B}$, they are determined by the restrictions of K,L onto A and B, respectively. But at the same time it is not necessarily true that determining these matrices from the representative vector systems $x(n)$, $z(n)$, $n \in N$, we can describe with them the operators $S_{\bar{x}}$ and $T_{\bar{z}}$, since not every representation can be extended to an extended representation. A condition for example, which assures the extendibility of every minimal representation of inner products of A and B is that the range of the projection $P_B: A \longrightarrow B$ be closed.

References

[1] G. Ruckebusch, Theorie geometrique de la representation Markovienne, Ann. Inst. H. Poincare XVI. 1980, 225–297.

[2] A. Lindquist, G. Picci, Realization theory for multivariate stationary Gaussian processes, SIAM J. Control and Optim., 23 (6), 1985, 809–857.

[3] A. Lindquist, M. Pavon, On the structure of state–space models for discrete–time stochastic vector processes, IEEE Trans. on Automat. Control, AC–29 (5), 1984, 418–431.

[4] G. Picci, On the identifiability of factor analysis models, in Modelling and Adaptive Control, Lecture Notes in Control and Inf. Sci. 105, 1988, 297–317.

[5] J. H. Van Schuppen, Stochastic realization problems motivated by econometrics modelling, in Modelling, Identification and Robust Control, North Holland, Amsterdam, 1966.

[6] L. W. Johnson, R. D. Riess, Introduction to Linear Algebra, Addison–Wesley, Reading, 1981.

[7] P. A. Fuhrmann, Linear Systems and Operators in Hilbert Space, McGraw–Hill, New York, 1981.

[8] T. Kato, Perturbation Theory for Linear Operators, Springer Verlag, Berlin, 1984.

[9] I. A. Ibragimov, Y. A. Rozanov, Gaussian Random Processes, Springer Verlag, Berlin, 1978.

[10] K. Glover, All optimal Hankel–norm approximations of linear multivariate systems and their L^{∞}–error bounds, Int. J. Control 39, 1984, 1115–1193.

[11] L. Finesso, G. Picci, Linear statistical models and stochastic realization theory, in Analysis and Optimization of Systems, Lecture Notes in Control and Inf. Sci. 62, 1984, 445–470.

[12] G. Picci, S. Pinzoni, Dynamic factor analysis models for stationary processes, IMA J. Math. Control and Inf. 3, 1986, 185–210.

[13] C. Van Putten, J. H. Van Schuppen, The weak and strong Gaussian probabilistic realization problem, J. Multivariate Anal. 13, 1983, 118–137.

ON REALIZATION AND IDENTIFICATION OF STOCHASTIC BILINEAR SYSTEMS

György Terdik
Department of Mathematics, Univ. of Kossuth, KLTE Pf.12
Debrecen, H-4010, Hungary

0. Introduction.

After solving the optimum linear filter problem Wiener's interest almost immediately turned to non-linear processing, see Gábor et al.(1960). The Wiener approach to the nonlinear stochastic systems based on two assumptions. Not to be linear in his sense means those systems with memory which can continue to exist indefinitely in time without some catastrophe which essentially changes their dynamic character and for which a small perturbation of the input produces a small perturbation of the output Wiener(1938),(1958). Because of the importance of the nonlinear models with stochastic input in many fields of engineering and sciences several authors devoted much work to describe this general nonlinear system to derive important properties, characterization and identification, see for example Cameron and Martin(1947), Itô(1951), Brillinger(1970), Schetzen(1980) and Kallianpur(1980).

There is an other approach which started to investigate nonlinear problems in control theory by the linear system side. Kálmán(1968) suggested to consider multilinear systems and starting in early sixties Mohler used bilinear system theory at research of neutron kinetics in nuclear reactor dynamics, see Mohler and Shen(1970). Bilinear system means to be linear in both the input and the output when one of them is fixed. The godfather of bilinear systems, I believe, was R. Mohler. Thought several results are available for the realization theory of the deterministic bilinear system see Bruni et al.(1974), d'Allessandro et al.(1973), Mohler and Kolodziej(1980) and Rugh(1981) for more detailed references, the stochastic realization problem is still pending. The value of the stochastic bilinear system to model a variety of physical systems was shown by Willsky and Marcus(1976) see also Mohler and Kolodziej(1980) by the stochastic bilinear control point of view. In time series analysis Granger and Andersen(1978) and T. Subba Rao(1978) introduced the bilinear model to put a parametric model for non-Gaussian time series data.

In this study we consider the stochastic bilinear systems as a particular case of the general nonlinear Wiener model, i.e., a stationary L_2 functional on a Wiener space generated by Gaussian white noise in discrete and by Wiener process in continuous time case. The Wiener-Itô spectral representation of such processes, see Dobrushin(1979), makes possible the frequency domain analysis.

1. The bilinear model and its transfer function system.

We are considering a scalar input w_t - scalar output y_t model having the following state space form

$$X_t = A\,X_{t-1} + D\,X_{t-1}\,w_{t-1} + b\,w_{t-1} \tag{1.1}$$

$$y_t = c'X_t + w_t$$

where w_t, $t \in \mathbb{Z}=\{0,\pm1,\pm2,\dots\}$ is Gaussian white noise series with $Ew_t=0$, $Ew_t^2=\sigma^2$ and A, D, **b**, **c** are appropriate matrices and vectors. The general lower triangular bilinear model used in time series analysis is defined by the following difference equation

$$y_t = \sum_{k=1}^{P} \alpha_k y_{t-k} + \sum_{k=0}^{Q} \beta_k w_{t-k} + \sum_{\substack{k=1\\ j=0}}^{R} \gamma_{j+k,k}\, y_{t-j-k} w_{t-k} \tag{1.2}$$

where $\beta_0=1$. This equation can be transformed into the state space form, see Terdik(1990),

$$X_t = A\,X_{t-1} + D\,X_{t-1}w_{t-1} + b_1 w_{t-1} + b_2(w_{t-1}^2-\sigma^2)$$

$$y_t = c'X_t + d\,w_t + \mu.$$

This state space equation contains a Hermite polynomial of degree 2 in w_{t-1}, i.e., $H_2(w_{t-1})=w_{t-1}^2-\sigma^2$. The nth order Hermite polynomial of w_t with leading coefficient 1 will be denoted by $H_n(w_t)$. The results of this paper have natural easy generalizations for the state space model which is linear only by the state process and containing products of Hermite polynomials by the past of the input process with constant matrix and vector coefficients, i.e.,

$$X_t = \left[A + \sum_{m=1}^{M} \sum_{\substack{N>n_k>0\\ S>s_{k+1}>s_k>0}} D_{n_1,n_2,\dots,n_m}^{s_1,s_2,\dots,s_m} \prod_{k=1}^{Q_m} H_{n_k}(w_{t-s_k})\right] X_{t-1}$$

$$+ \sum_{m=1}^{M} \sum_{\substack{N>n_k>0\\ S>s_{k+1}>s_k>0}} b_{n_1,n_2,\dots,n_m}^{s_1,s_2,\dots,s_m} \prod_{k=1}^{Q_m} H_{n_k}(w_{t-s_k})$$

$$y_t = c'X_t + w_t.$$

Note that the method we use works for the so called Markovian representation considered by Pham(1985) as well.

The multiple bilinear model is also of interest

$$X_t = AX_{t-1} + \sum_{k=1}^{d}\left(D_k X_{t-1}\, w_{k,t-1} + b_k\, w_{k,t-1}\right) \quad (1.3)$$

$$Y_t = CX_t + B\, w_t$$

where $w_t' = (w_{1,t}, w_{2,t}, \ldots, w_{d,t})$ is Gaussian white noise with $Ew_t = 0$ and diagonal covariance matrix $Ew_t' w_t = diag(\sigma_1^2, \sigma_2^2, \ldots, \sigma_d^2)$, $X_t \in \mathbb{R}^q$, $Y_t \in \mathbb{R}^p$ and A, D_k, B, C, b_k are appropriate matrices and vectors.

To get the transfer function system for the observation y_t by the model (1.1) let us assume now and for the rest of this section that both the state process X_t and the observation process y_t are *stationary in second order*. Moreover X_t is *strictly physically realizable* according to the input process w_t, i.e., X_t depends only the past of w_t and the shift transformation of w_t by the time is also the shift transformation for X_t. Under these assumptions the *Wiener-Itô spectral representation* is valid, see Dobrushin(1979). There exists a vector valued transfer function system $\{f_0, f_1, \ldots, f_r, \ldots\}$ for X_t where f_0 is constant and the entries of f_r are in $L_2(\mathfrak{D}^r)$, $\mathfrak{D} = [0,1]$. They are uniquely determined up to permutations such that

$$X_t = \sum_{r=0}^{\infty} \int_{\mathfrak{D}^r} z_{(r)}^{t\mathfrak{z}} f_r(z_{(r)})\, W(d\omega_{(r)}) \quad (1.4)$$

where $z_{(r)} = (z_1, z_2, \ldots, z_r)$, $z_k = e^{i2\pi\omega_k}$, $\mathfrak{z} = (1,1,\ldots,1)$, $z_{(r)}^{t\mathfrak{z}} = \prod_{k=1}^{r} z_k^t$. The stochastic integrals involved in (1.4) are multiple Wiener-Itô integrals and the convergence is meant in mean square sense. The r-multiple stochastic spectral measure $W(d\omega_{(r)})$ is connected to the w_t by the spectral representation

$$w_t = \int_0^1 e^{it2\pi\omega}\, W(d\omega).$$

Note that these assumptions imply the strict stationarity of the processes X_t and y_t. The only problem to be proved for getting explicit formulae of the transfer functions is the Wiener-Itô spectral representation of the product $X_t w_t$ which is

$$X_t w_t = \sum_{r=0}^{\infty} \int_{\mathfrak{D}^{r+1}} z_{(r+1)}^{t\mathfrak{z}}\, f_r(z_{(r)})\, W(d\omega_{(r+1)}), \quad (1.5)$$

see Terdik and Subba Rao(1989) for details. Now one can put both sides of the equation

$$X_t = A\, X_{t-1} + D\, X_{t-1}\, w_{t-1} + b\, w_{t-1}$$

into the spectral representation and gets the following recursion for the transfer functions f_r

$$f_0 = 0$$

$$f_1(z_1) = \left(z_1 I - A\right)^{-1} b \tag{1.6}$$

$$f_r(z_{(r)}) = \left(z_{(r)}^{\mathfrak{z}} I - A\right)^{-1} D\, f_{r-1}(z_{(r-1)}), \qquad r\geq 2.$$

An extra assumption, i.e., the spectral radius of the matrix A is less then 1, $\rho(A)<1$, seems to be used expressing the transfer functions but it will be clear by the next section that it follows from the second order stationarity of the model. Denote $g_1, g_2, \ldots g_r, \ldots$ the transfer functions of the observation process y_t in the model (1.1). It is easy to see that they are given by

$$g_0 = 0$$

$$g_1(z_1) = 1 + c' f_1(z_1) \tag{1.7}$$

$$g_r(z_{(r)}) = c' f_r(z_{(r)}), \qquad r>1.$$

Such that

$$y_t = \sum_{r=1}^{\infty} \int_{\mathfrak{D}^r} z_{(r)}^{t\mathfrak{z}}\, g_r(z_{(r)})\, W(d\omega_{(r)}). \tag{1.8}$$

One of the advantages of the state space representation (1.1) is that it contains models having only finite many nonzero transfer functions. Note that all the transfer functions of the scalar bilinear model (1.6) are different from zero.

2. Second order properties, stationarity.

We have seen that the second order stationary solution of the bilinear equation (1.1), if it exists, can be given by the Wiener-Itô representation (1.4) and (1.8) with transfer functions defined by (1.6), (1.7). To find the variance for the process

$$X_t = \sum_{r=0}^{\infty} \int_{\mathfrak{D}^r} z_{(r)}^{t\mathfrak{z}} f_r(z_{(r)})\, W(d\omega_{(r)})$$

first the transfer functions need to be symmetrized

$$f_r^{\sim}(z_{(r)}) = sym_r f_r(z_{(r)}) = \frac{1}{r!} \sum_{P\in \mathfrak{P}_r} f_r(Pz_{(r)})$$

where $\mathfrak{P}_r$ denotes the group of permutations for the integers $1,2,\ldots,r$. Then the second step is taking the sum of the L_2 norm of the symmetrized transfer functions, i.e.

$$EX_t^{\otimes 2} = \sum_{r=1}^{\infty} \int_{\mathfrak{D}^r} f_r^{\sim}(z_{(r)}) \otimes f_r^{\sim}(\bar{z}_{(r)})\, \sigma^{2r} d\omega_{(r)} \tag{2.1}$$

we are using tensor product denoted by $\otimes$ instead of the standard covariance matrix because of some technical reason. Now we define the transfer functions by (1.6) with the parameters given by the state space equation (1.1) and see that the necessary and sufficient condition for the stationary solution is the convergence of the series (2.1). The recursion for the transfer functions implies a recursion for the integrals as well, in fact if r≥2

$$\int_{\mathcal{D}^r} \mathbf{f}_r^{\sim}(\mathbf{z}_{(r)}) \otimes \mathbf{f}_r^{\sim}(\bar{\mathbf{z}}_{(r)})\ \sigma^{2r} d\omega_{(r)} = \sigma^2 \left(I - A^{\otimes 2}\right)^{-1} D^{\otimes 2}$$

$$\int_{\mathcal{D}^{r-1}} \mathbf{f}_{r-1}^{\sim}(\mathbf{z}_{(r-1)}) \otimes \mathbf{f}_{r-1}^{\sim}(\bar{\mathbf{z}}_{(r-1)})\ \sigma^{2(r-1)} d\omega_{(r-1)} ,$$

see Terdik(1989). It is easy to derive from this that the *necessary and sufficient condition for the second order stationarity* of the bilinear model is that the spectral radius

$$\rho\left(A^{\otimes 2} + \sigma^2 D^{\otimes 2}\right) < 1. \tag{2.2}$$

It is also follows from (2.2) that $\rho(A)<1$. The condition (2.2) appears at M. B. Rao et al.(1983) as the necessary condition of the strictly stationarity of some bilinear time series models also at Desai(1986) in connection with the existence of the unique solution for the Ljapunov equation of some multiple bilinear models. The necessary and sufficient condition for the second order stationarity of the multiple bilinear model (1.3) is

$$\rho\left(A^{\otimes 2} + \sum_{k=1}^{d} \sigma_k^2 D_k^{\otimes 2}\right) < 1,$$

see Terdik(1989). Under the assumption (2.2) one can derive the covariance structure and the spectrum for the bilinear process y_t given by the difference equation (1.1). Namely

$$EX_t^{\otimes 2} = \left(I - A^{\otimes 2} - \sigma^2 D^{\otimes 2}\right)^{-1} \mathbf{b}^{\otimes 2}$$

$$Ey_t^2 = \sigma^2 + \sigma^2 \mathbf{c}'^{\otimes 2} EX_t^{\otimes 2}$$

$$Ey_{t+s}y_t = \sigma^2 \mathbf{c}'^{\otimes 2}\left((I\otimes A)^s\, \delta_{s>0} + (A\otimes I)^{-s}\delta_{s<0}\right) + \sigma^2 \mathbf{c}' A^{|s|-1}\mathbf{b} ,\ \text{if } s\neq 0.$$

The spectrum for y_t is given by

$$\phi(z) = \sigma^2 + \sigma^2 \mathbf{c}'^{\otimes 2}\left((zI - A)^{-1} \otimes (z^{-1}I - A)^{-1}\right) EX_t^{\otimes 2}$$

$$+ 2\sigma^2 \mathbf{c}'(\cos 2\pi\omega\, I - A)(z^{-1}I - A)^{-1}(zI - A)^{-1}\mathbf{b}, \qquad z = e^{i2\pi\omega}.$$

The spectrum ϕ is clearly rational function of z so one can not make difference between a non-Gaussian linear process and a bilinear one based on the second order spectrum or statistics.

3. Bilinear realization for polynomial systems.

It is natural to say that a stochastic series y_t, $t\in\mathbb{Z}$, is *bilinear realizable* according to the Gaussian white noise series w_t if there exist matrices A, D and vectors **b**, **c** such that the following equation system fulfills

$$X_t = A\,X_{t-1} + D\,X_{t-1}\,w_{t-1} + b\,w_{t-1} \tag{3.1}$$

$$y_t = c'X_t + w_t\ .$$

The bilinear models employed in time series analysis with Gaussian residuals are bilinear realizable. Now if we suppose that y_t is strictly physically realizable and stationary in second order then we have the Wiener-Itô spectral representation

$$y_t = \sum_{r=0}^{\infty} \int_{\mathfrak{D}^r} z_{(r)}^{t\mathfrak{z}}\, g_r(z_{(r)})\; W(d\omega_{(r)}) \tag{3.2}$$

by some transfer functions g_r. So the question is reduced to find a bilinear realization for the transfer functions. The solution is analog to the deterministic problem solved by Mitzel et al.(1979), see also Rugh(1981). Let us start with considering the transfer function system (1.7) of the bilinear model (3.1). The following properties clearly hold.

Property 1. The transfer function g_r is depending on $z_{(r)}$ through the variables $u_k = z_{(k)}^{\mathfrak{z}} = \prod_{j=1}^{k} z_j = \prod_{j=1}^{k} e^{i2\pi\omega_j}$, $k=1,2,...,r$. Change the variables and refer to the functions $g_1^r(u_1)=g_1(z_1)-1$ and if $r>1$ $g_r^r(u_{(r)})=g_r(z_{(r)})$ as *regular* version of the transfer functions.

Property 2. The regular transfer functions g_r^r are *strictly proper*, i.e., they are rational functions of their variables and the degree of the numerator polynomial in u_k is less then the degree of the denominator polynomial for each k.

Property 3. The regular transfer functions g_r^r are *recognizable*, i.e., the denominators can be expressed as product of real- coefficient single variable polynomials.

If all the transfer functions g_r are zero after an index N and N is the smallest one among them then the process

$$y_t = \sum_{r=0}^{N} \int_{\mathfrak{D}^r} z_{(r)}^{t\mathfrak{z}}\, g_r(z_{(r)})\; W(d\omega_{(r)}) \tag{3.3}$$

is called *polynomial of Hermite degree* N, see Brillinger(1970). An easy construction leads to the following result, see Terdik(1989). The *necessary and sufficient condition for the bilinear realizability of the polynomial system of Hermite degree* N is that the **Properties 1-3.** be fulfilled. The **Property 1** needs some explanation because the transfer functions are determined only up to permutations and it means that a regular one can be chosen which represents the class. The necessary and sufficient condition for the second order stationarity of a bilinear realizable polynomial system of degree N is only that $\rho(A)<1$.

There is a natural approximation procedure for the bilinear processes having infinite Hermite degree, for example for the bilinear time series models. That is take the first N transfer functions of it and put zero for the others, say truncate it by N. It turns out that the truncated model is bilinear realizable again but nothing can guarantee that the dimension of that new bilinear representation would be smaller then the original one, contrary it is larger usually.

4. State space and minimal realization.

In linear stochastic system theory the predictor space, which is the space of the best linear predictors of the future by the past of the process, provides the state space and also the minimal realization, see Akaike(1974) and Faurre(1976). One of the basic property of the state space is that it is finite dimensional if and only if there exists a linear representation with finite dimension for Gaussian stationary process. The construction of the state space is not as easy for bilinear systems. Let us suppose that y_t is a bilinear realizable *homogeneous* process of Hermite degree N, i.e.,

$$y_t = \int_{\mathfrak{D}^N} z_{(N)}^{t\mathfrak{z}} \, g_N(z_{(N)}) \, W(d\omega_{(N)}).$$

We define the state space $\mathfrak{X}_t$ of the process y_t as a direct sum of the following orthogonal subspaces,

1. $\mathfrak{X}_t^1$ is spanned by $E(y_{t+k}|\mathcal{B}_t)$, k=1,2,... where $\mathcal{B}_t$ is the σ-algebra generated by the past of the input process $w_t, w_{t-1}, \ldots$.

2. $\mathfrak{X}_t^p$, p=2,3,...is spanned by

(i) the orthogonal projection of $w_{t-1}\mathfrak{X}_t^{p-1}$ onto the subspace $\mathfrak{H}_t^{N-p+1}$ generated by the products having order N-p+1 of Hermite polynomials of the variables $w_t, w_{t-1}\ldots$ with different time and

(ii) the conditional expectation with respect to $\mathcal{B}_t$ of their forward shift.

Let

$$\mathfrak{X}_t = \sum_{p=1}^{N} \mathfrak{X}_t^p .$$

It is easy to see that $\mathfrak{X}_t^p = 0$ if p>N so for linear process when N=1 it turns out that $\mathfrak{X}_t = \mathfrak{X}_t^1$, which is nothing else then the original state space for linear systems. Note that the Hilbert spaces $\mathfrak{H}_t^r$ are generated by the random variables of the form

$$\int_{\mathfrak{D}^r} z_{(r)}^{t\mathfrak{z}-s_{(r)}} W(d\omega_{(r)}) \quad , s_{(r)}=(s_1, s_2, \ldots, s_r), \; s_k \geq 0.$$

The definition of the state space for general bilinear case which is a generalization of the linear one is based on some linear operators.

If $v \in \mathfrak{H}_t^r$ then put $L_D v = \sigma^{-2} w_{t-1} v | \mathfrak{H}_t^{r-1}$ where $w_{t-1} v | \mathfrak{H}_t^{r-1}$ denotes the orthogonal projection of $w_{t-1} v$ onto $\mathfrak{H}_t^{r-1}$ and for $v_1 \in \mathfrak{H}_t^r$, $v_2 \in \mathfrak{H}_t^p$ naturally $L_D(v_1 + v_2) = L_D v_1 + L_D v_2$. The operator L_A is defined by the following

$$L_A E(y_{t+k} | \mathcal{B}_t) = L_A E(y_{t+k+1} | \mathcal{B}_t) = L_A^{k+1} y_t$$
$$L_A^k [L_D (L_A^j y_t)] = E\left(\left[L_D E(y_{t+j} | \mathcal{B}_t)\right]_{t+k} | \mathcal{B}_t\right).$$

Now let

$$\mathfrak{X}_t^1 = \mathcal{S}pan\left\{L_A^k y_t;\ k=0,1,2,\ldots\right\}$$

$$\mathfrak{X}_t^2 = \mathcal{S}pan\left\{L_A^{k_2} L_D L_A^{k_1} y_t;\ k_1, k_2 = 0,1,2,\ldots\right\}$$

.

.

.

$$\mathfrak{X}_t^r = \mathcal{S}pan\left\{L_A^{k_r} L_D L_A^{k_{r-1}} \ldots L_D L_A^{k_1} y_t;\ k_1, k_2, \ldots, k_r = 0,1,2,\ldots\right\}$$

.

.

.

The state space $\mathfrak{X}_t$ again is defined by the direct sum of the orthogonal Hilbert spaces $\mathfrak{X}_t^r$ above, i.e.

$$\mathfrak{X}_t = \sum_{r=1}^{\infty} \mathfrak{X}_t^r .$$

Now we are in the position to give another characterization of the bilinear realizability. *The polynomial system* y_t *with Hermite degree* N *is bilinear realizable if and only if its state space* $\mathfrak{X}_t$ *is finite dimensional.* Moreover if the state space is finite dimensional then the operators L_A, L_D are also with finite dimension and having matrix form. Define the linear operators L_b and L_c by $L_b a = a y_t$ for $a \in \mathbb{R}$ and $L_c x = \sigma^{-2} E x w_{t-1}$ for $x \in \mathfrak{X}_t$. If the process y_t is polynomial and bilinear realizable then its minimal realization is given by the quadruplet (L_A, L_D, L_b, L_c) having matrix and vector form respectively, see Terdik(1989) and Rugh(1981) for more details. We mention here a good try by Desai(1986) to find a minimal realization for bilinear stochastic systems based on Isidori's(1973) deterministic construction. The model considered is actually a particular multiple bilinear one by our setup. The main difficulties are to estimate the various moments and crossmoments of y_t and w_t when only y_t is observed. If both y_t and w_t are observed then the realization should be carried out by the crossmoments of y_t and Hermite polynomials of w_t. Our conjection which is based on the results of the following two section is that the spectrum and the bispectrum, i.e., the second and the third order moments of the process y_t, are supposed to be applied for the bilinear realization procedure.

5. Identification, third order properties.

The spectrum, i.e., the second order moments, totally describes the statistical properties of a Gaussian stationary process. Therefore one could say that it is identifiable in second order. We saw in chapter 2 that it is not the case for stationary bilinear realizable processes. The question is whether one can identify the process in the third order sense, i.e., by the first, second and third order statistics. That is why we are taking into the consideration the third order properties actually the marginal bispectrum. The method employed here needs the existence of the fourth order moments of the bilinear process given in state space form

$$X_t = A\, X_{t-1} + D\, X_{t-1}\, w_{t-1} + b\, w_{t-1} \tag{5.1}$$

$$y_t = c' X_t + w_t$$

If X_t is a second order stationary solution of (5.1) and the fourth moment of it exists then the quadratic process $Q_t = X_t^{\otimes 2}$ is stationary and has Wiener-Itô spectral representation

$$Q_t = \sum_{r=0}^{\infty} \int_{\mathcal{D}^r} z_{(r)}^{t\mathfrak{z}}\, q_r(z_{(r)})\; W(d\omega_{(r)}) \tag{5.2}$$

where the constant term $q_0 = EX_t^{\otimes 2}$. Taking the tensor product of both side of the state space equation we get

$$Q_t = \left[A^{\otimes 2} + \sigma^2\, D^{\otimes 2}\right] Q_{t-1} + \left[A\otimes D + D\otimes A\right] Q_{t-1} w_{t-1} + D^{\otimes 2} Q_{t-1} H_2(w_{t-1})$$

$$+ \sigma^2 D\otimes I\; X_{t-1}\otimes b + \sigma^2 I\otimes D\; b\otimes X_{t-1} + \left[A\otimes I\; X_{t-1}\otimes b + I\otimes A\; b\otimes X_{t-1}\right] w_{t-1}$$

$$+ \left[D\otimes I\; X_{t-1}\otimes b + I\otimes D\; b\otimes X_{t-1}\right] H_2(w_{t-1}) + b H_2(w_{t-1}) + \sigma^2 b\,.$$

That quadratic process Q_t is not bilinear any more because of the product by second order Hermite polynomials and X_t are also involved into the equation. Nevertheless using the fact that the transfer functions for X_t are given we are able to get the transfer functions for Q_t as well,

$$q_1(z_1) = \mathcal{A}(z_1)\left[(A\otimes D + D\otimes A) q_0 + \sigma^2 D\otimes I\; f_1(z_1)\otimes b + \sigma^2 I\otimes D\; b\otimes f_1(z_1)\right]$$

$$q_2(z_{(2)}) = \mathcal{A}(z_{(2)}^{\mathfrak{z}})\left[(A\otimes D + D\otimes A) q_1(z_1) + D^{\otimes 2} q_0 + \sigma^2 D\otimes I\; f_2(z_{(2)})\otimes b \right.$$
$$\left. + \sigma^2 I\otimes D\; b\otimes f_2(z_{(2)}) + A\otimes I\; f_1(z_1)\otimes b + I\otimes A\; b\otimes f_1(z_1) + b\right]$$

if r>2 the following recursion is valid for q_r

$$q_r(z_{(r)}) = \mathcal{A}(z_{(r)}^{\mathfrak{z}})\left[(A\otimes D + D\otimes A) q_{r-1}(z_{(r-1)}) + D^{\otimes 2} q_{r-2}(z_{(r-2)}) \right.$$
$$+ \sigma^2 D\otimes I\; f_r(z_{(r)})\otimes b + \sigma^2 I\otimes D\; b\otimes f_r(z_{(r)})$$

$$+ \quad A\otimes I \quad \mathbf{f}_{r-1}(\mathbf{z}_{(r-1)})\otimes\mathbf{b} \quad + \quad I\otimes A \quad \mathbf{b}\otimes\mathbf{f}_{r-1}(\mathbf{z}_{(r-1)})$$

$$+ \, D\otimes I \; \mathbf{f}_{r-2}(\mathbf{z}_{(r-2)})\otimes\mathbf{b} + I\otimes D \; \mathbf{b}\otimes\mathbf{f}_{r-2}(\mathbf{z}_{(r-2)})\Big]$$

where $\mathcal{A}(z^{\flat}_{(r)}) = \left[z^{\flat}_{(r)}I - A^{\otimes 2} - \sigma^2 D^{\otimes 2}\right]^{-1}$. The fourth order moment of X_t is given by

$$EQ_t^{\otimes 2} = \sum_{r=0}^{\infty} \int_{\mathcal{D}^r} q_r^{\sim}(z_{(r)}) \otimes q_r^{\sim}(\bar{z}_{(r)}) \, \sigma^{2r} d\omega_{(r)}$$

and again similarly to the existence of the second order moments the convergence of the above series is equivalent to the existence of the fourth order moments. It turns out that the *necessary and sufficient condition* for it is

$$\rho\left\{ \left[A^{\otimes 2} + \sigma^2 D^{\otimes 2}\right]^{\otimes 2} + \left[A\otimes D + D\otimes A\right]^{\otimes 2} + \sigma^4 D^{\otimes 4}\right\} < 1, \tag{5.3}$$

see Terdik(1990). Now if we suppose (5.1) then we can use the transfer functions q_r to get the third order moments like

$$E\left[X_{t+s}^{\otimes 2} - EX_{t+s}^{\otimes 2}\right]\otimes X_t = E\left[Q_{t+s} - q_0\right]\otimes X_t$$

the Fourier transform of which is the marginal bispectrum

$$\phi_x(z) = \mathcal{A}(z)\otimes\mathcal{A}_1(z^{-1})K_1 + \mathcal{A}(z)\otimes\left[D\mathcal{A}_1(z)\mathcal{A}_1(z^{-1})\right]K_2$$

$$+ \left[\mathcal{A}(z) I\otimes D\mathcal{A}_1(z)\right]\otimes\mathcal{A}_1(z^{-1})K_3,$$

where K_j, $j=1,2,3$ are some constant vectors and $\mathcal{A}_1(z) = \left[zI - A\right]^{-1}$, $z = e^{i2\pi\omega}$. The marginal bispectrum for the observation y_t is given by

$$\phi_y(z) = c'^{\otimes 3}\phi_x(z) + \sigma^2 c'^{\otimes 2} q_1(z) + 2\sigma^2 c'^{\otimes 2} I\otimes\mathcal{A}(z^{-1})K$$

where K is a constant vector. The main importance of this marginal bispectrum is that its poles do depend on the bilinear part of the process. It seems to be likely that a complete identification can be carried out considering the vector valued process $[y_t, y_t^2]'$.

6. Identification by third order for bilinear system with Hermite degree 2.

It is very likely that the simplest nonlinear model is the bilinear realizable one with Hermite degree 2. Let us suppose that we are given a process y_t by the form

$$y_t = \sum_{k=0}^{\infty} g_k \, H_1(w_{t-k}) + \sum_{j,k=0}^{\infty} g_{j,k} \, H_2(w_{t-j}, w_{t-k}). \tag{6.1}$$

Put the Hermite polynomials into the spectral representation

$$H_1(w_t) = \int_0^1 z_1^t \, W(d\omega_1), \quad H_2(w_t, w_s) = \int_0^1\int_0^1 z_1^t z_2^s \, W(d\omega_1, d\omega_2)$$

and get for the process

$$y_t = \int_0^1 z_1^t\, g_1(z_1)\, W(d\omega_1) + \int_0^1\int_0^1 z_1^t z_2^t\, g_2(z_1,z_2)\, W(d\omega_1,d\omega_2),$$

where the transfer functions g_1 and g_2 are the Fourier transforms of the weights g_k and $g_{j,k}$ respectively, i.e., y_t is a polynomial model with Hermite degree 2. We have seen in Section 3 that the necessary and sufficient condition for the bilinear realizability of y_t is that there exists polynomials α, β, α_{12}, α_{22} of one variable with real valued coefficients and γ of two variables such that

$$g_1(z_1) = \frac{\beta(z_1)}{\alpha(z_1)}, \quad g_2(z_1,z_2) = \frac{\gamma(z_1,z_1z_2)}{\alpha_{21}(z_1)\,\alpha_{22}(z_1z_2)}.$$

Put

$$\alpha(z) = \sum_{k=0}^{P} a_k z^{-k}; \qquad a_0 = 1, \qquad \beta(z) = \sum_{k=0}^{Q} b_k z^{-k}; \qquad b_0 = 1,$$

$$\alpha_{21}(z) = \sum_{k=0}^{P_1} a_k^{(1)} z^{-k}; \quad a_0^{(1)} = 1, \qquad \alpha_{22}(z) = \sum_{k=0}^{P_2} a_k^{(2)} z^{-k}; \quad a_0^{(2)} = 1,$$

$$\gamma(z,v) = \sum_{\substack{m=1\\ n=0}}^{R,S} c_{m,m+n} z^{-n} v^{-m}.$$

It is easy to see that the following bilinear model

$$\begin{aligned}
\sum_{k=0}^{P} a_k x_{t-k} &= \sum_{k=0}^{Q} b_k w_{t-k} \\
\sum_{k=0}^{P_1} a_k^{(1)} x_{t-k}^{(1)} &= w_t \\
\sum_{k=0}^{P_2} a_k^{(2)} x_{t-k}^{(2)} &= \sum_{\substack{m=1\\ n=0}}^{R,S} c_{m,m+n} x_{t-m-n}^{(1)} w_{t-m} + \text{const.}
\end{aligned} \tag{6.2}$$

$$y_t = x_t + x_t^{(2)}$$

gives the desired transfer functions $g_1 = \frac{\beta}{\alpha}$ and $g_2 = \frac{\gamma}{\alpha_{21}\alpha_{22}}$ for y_t. It is quite straightforward to find the spectrum and the bispectrum of y_t in terms of the polynomials α, β, α_{12}, α_{22} and γ, see Terdik and Meaux(1990). The spectrum is given by

$$\phi(z_1) = \sigma^2 \left|\frac{\beta(z_1)}{\alpha(z_1)}\right|^2 + \sigma^4 \left|\frac{\gamma_0(z_1)}{\alpha_{22}(z_1)}\right|^2 + \frac{\sigma^4}{|\alpha_{22}(z_1)|^2} \int_0^1 \left|\frac{\gamma(z,z_1)}{\alpha_{21}(z)}\right|^2 d\lambda,$$

where $z = e^{i2\pi\lambda}$, $z_k = e^{i2\pi\omega_k}$, and $\gamma_0(z_1) = \sum_{m=1}^{R} c_{m,m} z_1^m$. Now the bispectrum, i.e., the Fourier transform of the third order moment like $Ey_0y_ty_s$, is given by

$$\psi(z_1,z_2) = \Psi(z_1,z_2,z_1^{-1}z_2^{-1}),$$

where

$$\Psi(z_1,z_2,z_3) = 6\sigma^4 sym_3\left[\frac{\beta(z_1)}{\alpha(z_1)}\;\frac{\beta(z_2)}{\alpha(z_2)}\;\frac{\gamma(z_1^{-1},z_3)}{\alpha_{21}(z_1)\alpha_{22}(z_1z_2)}\right]$$

$$+\;\frac{6\sigma^6}{\alpha_{22}(z_1)\alpha_{22}(z_2)\alpha_{22}(z_3)}\left[\frac{\gamma_0(z_1)\;\gamma_0(z_2)\;\gamma_0(z_3)}{3}\right.$$

$$\left.+\;sym_3\left(\int_0^1 \frac{\gamma(z^{-1},z_1)\;\gamma(z^{-1}z_1,z_2)\;\gamma(z,z_3)}{|\alpha_{21}(z)|^2\;\alpha_{21}(z^{-1}z_1)}\,d\lambda\right)\right],$$

where sym_3 denotes the symmetrization by the variable $z_{(3)}$, i.e., Ψ is a symmetric function of its variables. Brillinger(1985) and (1989) employed parametric spectrum and bispectrum of type above successfully to identify the structure of real data as Mississippi River Runoff and some endocrine data.

References

Akaike, H.,(1974) Stochastic theory of minimal realization, IEEE Trans. Aut. Contr., vol. AC-19, pp667-673.

Brillinger, D.,(1970) The identification of polynomial systems by means of higher order spectra, J. Sound Vib., 12, pp301-314.

Brillinger, D.,(1985) Fourier inference: Some methods for the analysis of array and nongaussian series data, Water Resources Bull. Vol 21, No. 5, 743-756.

Brillinger, D.,(1989) Parameter estimation for nongaussian processes via second and third order spectra with an application to some endocrine data. Adv. Meth. of Physiol. System Mod, Plenum, New York.

Bruni, C., DiPillo, G., Koch, G.,(1974) Bilinear systems: An appealing class of "nearly linear" systems in theory and applications, IEEE Trans. Aut. Contr., vol. AC-19, pp334-348.

Cameron, R., Martin, W.,(1947) The orthogonal development of non-linear functionals in series of Fourier-Hermite functionals, Ann. Math., 48, pp385-392.

d'Allessandro, P., Isidori, A., Ruberti, A.,(1973) Lectures on Bilinear System Theory, New York, Springer.

Faurre, P.,(1976) Stochastic realization algorithms, Syst. Iden.: Adv. Case Stud., ed. Mehra, A., Lainiotis, D., Acad. Press, New York.

Desai, U.,(1986) Realization of bilinear stochastic systems, IEEE Trans. Aut. Contr., vol. AC-31, pp189-192.

Dobrushin, R.,(1979) Gaussian and their subordinated selfsimilar random generalized fields, Ann. Prob., 7, pp1-28.

Gábor, D., Wilby, W., Woodcock, R.,(1960) A universal non-linear filter, predictor and simulator which optimizes itself by a learning process, The Inst. of Electr. Eng., No.3270M, July.

Granger, C., Andersen, A.,(1978) An introduction to bilinear time series models, Vandenhoek and Rjprecht, Gottingen.

Isidori, A.,(1973) Direct construction of minimal bilinear realizations from nonlinear input output maps, IEEE Trans. on Aut. Contr., vol. AC-18, pp626-631.

Itô, K.,(1951) Multiple Wiener integral, J. Math. Soc. Japan, vol. 3, pp181-193.

Kallianpur, G.,(1980) Stochastic Filtering Theory, Springer-Verlag, New York.

Kálmán, R.,(1968) Pattern recognition properties of multilinear machines, IFAC Symp. on Techn. Biol. Problems of Contr., Yerevan, USSR.

Mitzel, G., Clancy, S., Rugh, W.,(1979) On transfer function representations for homogeneous nonlinear systems, IEEE Trans. on Aut. Contr., vol. AC-24, pp242-249.

Mohler, R., Kolodziej, W.,(1980) An overview of bilinear system theory

and applications, IEEE Trans. Syst. Man Cybern., vol. SMC-10, pp683-688.
Mohler, R., Kolodziej, W.,(1980) An overview of stochastic bilinear control processes, IEEE Trans. Syst. Man Cybern., vol. SMC-10, pp913-918.
Mohler, R., Shen, C.,(1970) Optimal Control of nuclear reactors, New York, Academic.
Pham, D.,(1985) Bilinear Markovian representation and bilinear models, Stoch. Proc. Appl., vol 12, pp295-306.
Rao, M.B., Subba Rao, T., Walker, M.,(1983) On the existence of some bilinear time series models, J. Time Ser. Anal., 4, pp95-110.
Rugh, W.,(1981) Nonlinear system theory, J. Hopkins Univ. Press, Baltimore.
Schetzen, M.,(1980) The Volterra and Wiener theories of nonlinear systems, New York, Wiley.
Subba Rao, T., (1981) On the theory of bilinear time series models, J. Roy. Statist. Soc., Ser. B, 43, pp244-255.
Terdik, Gy.,(1985) Transfer functions and conditions for stationarity of bilinear models with Gaussian residuals, Proc. Roy. Statist. Soc., Ser. A, 400, pp315-330.
Terdik, Gy.,(1989) Bilinear state space realization for polynomial systems, Techn. Rep., Dept. Stat., UC, Berkeley, CA 94720. To appear in Computers and Maths. with Appl..
Terdik, Gy.,(1990) Second order properties for multiple-bilinear models, J. Multivar. Anal., vol. 35, No. 2, pp295-307.
Terdik, Gy.,(1989) Stationarity in fourth order and the marginal bispectrum for bilinear models with Gaussian residuals, Techn. Rep., Dept. Maths, UofA, Fayetteville, AR 72701.
Terdik, Gy., Meaux, L.,(1990) The exact bispectra for bilinear realizable processes with Hermite degree 2, Techn. Rep., Dept. Maths, UofA, Fayetteville, AR 72701. To appear in Adv. of Appl. Prob..
Terdik, Gy., Subba Rao, T.,(1989) On Wiener-Itô representation and the best linear predictors for bilinear time series, J. Appl. Prob.,vol. 26,pp274-286.
Wiener, N., (1938) The homogeneous chaos, Amer. Journ. Math., Vol. LV, No. 4, pp
Wiener, N.,(1958) Nonlinear problems in random theory, Cambridge, Mass.
Willsky, A., Marcus, S.,(1976) Analysis of bilinear noise models in circuits and devices, J. Franklin Inst., vol. 301, pp103-122.

ON STOCHASTIC PARTIAL DIFFERENTIAL EQUATIONS. RESULTS ON APPROXIMATIONS

ISTVÁN GYÖNGY

Department of Probability Theory and Statistics
Eötvös Loránd University Budapest,
Múzeum krt. 6–8
H–1088 Hungary

We consider second order stochastic linear partial differential equations of parabolic type driven by continuous semimartingales. Approximating the driving process by continuous semimartingales in the topology of uniform convergences we get, under some conditions, the convergence of the resulting solutions to the solution of the original equation understood in Stratonovich's sense. Hence we establish a support theorem for SPDEs which is a generalization of the well known support theorem of Stroock and Varadhan. We present our results for SPDEs where the differential operators are given in terms of derivatives along certain vector fields in place of derivatives with respect to fixed coordinate vectors. By virtue of this generality we can treat, in particular, the case of SPDEs with unbounded coefficients which is important from the point of view of applications. We apply the results presented here to the problems of approximating both the nonlinear filter of partially observed diffusion processes and the equation of the hydromagnetic dynamo. This work is based on the papers [10]–[13].

The paper is organised as follows:

1. Second order SPDEs with unbounded coefficients
2. Approximations of SPDEs
3. Theorems on supports
4. Applications to nonlinear filtering
5. An application to a stochastic model of the hydromagnetic dynamo

1. Second order SPDEs

We consider here the stochastic partial differential equation

$$\begin{aligned} du(t,x) =& \Big(\sum_{j=1}^{d_2} \mathcal{M}_j^2 u(t,x) + \mathcal{M}_0 u(t,x) + f(t,x) \Big) dt \\ &+ (\mathcal{N}_i u(t,x) + g_i(t,x)) \circ dM^i(t) \qquad {}^{*)} \\ u(0,x) =& u_0(x) \end{aligned}$$

where $((M^i(t)), \mathcal{F}_t)_{t\geq 0}$ is a d_1-dimensional continuous semimartingale given on a stochastic basis $(\Omega, \mathcal{F}, P, (\mathcal{F}_t)_{t\geq 0})$, and $\mathcal{M}_j$, $\mathcal{N}_i$ are first order differential operators in the space variable $x \in \mathbf{R}^d$. The stochastic differential on the right-side is understood in the sense of

*The summational convention is used throughout the paper with respect to repeated integer-valued indices.

Stratonovich (see the precise meaning later). In the theory of these equations, developed in the framework of the variational approach, the operators $\mathcal{M}_j$, $\mathcal{N}_i$ are given in the basis $D_{e^1} := \frac{\partial}{\partial x^1}, \ldots, D_{e^d} := \frac{\partial}{\partial x^d}$, i.e.,

$$\mathcal{M}_j := b_j^k \frac{\partial}{\partial x^k} + b_j^0 D_0$$
$$\mathcal{N}_i := c_i^k \frac{\partial}{\partial x^k} + c_i^0 D_0,$$

where D_0 denotes the identity and b_j^p, c_i^p are some functions of $t \in \mathbf{R}_+$, $\omega \in \Omega$, $x \in \mathbf{R}^d$ with the necessary measurability and smoothness conditions. (See [20]–[22], [25]–[29] and the references there). It is usually assumed that these coefficients are bounded functions of x, an assumption which does not allow the treatment of some important equations (namely, those having unbounded coefficients) within the framework of the theory. We think for example of the equations of nonlinear filtering and the equation of the magnetic field evolving in a fluid conducting with growing velocity (see [1]–[3], [9]–[12], [19], [28], [30]–[33], [35]–[37] which deal with equations having unbounded coefficients).

In order to cover those equations having unbounded coefficients b_j^k, c_i^k, we consider the operators $\mathcal{M}_j, \mathcal{N}_i$ written in the form

$$\begin{aligned} \mathcal{M}_j &:= \hat{b}_j^k D_{\Gamma^k} + b_j^0 D_0 \\ \mathcal{N}_i &:= \hat{c}_i^k D_{\Gamma^k} + c_i^0 D_0, \end{aligned} \tag{1.1}$$

where the coefficients $\hat{b}_j^k, \hat{c}_i^k$ are still bounded but $D_{\Gamma^1}, \ldots, D_{\Gamma^d}$ are derivations along some vector fields $\Gamma^1 = (\Gamma^{1l}(x)), \ldots, (\Gamma^{dl}(x))$. That means

$$D_{\Gamma^k} := \Gamma^{kl} \frac{\partial}{\partial x^l} \quad k = 1, 2, \ldots, d.$$

Obviously, if the matrix $\Gamma = (\Gamma^{kl})$ is invertible for every x, then

$$b_j^k = \hat{b}_j^l (\Gamma^{-1})^{lk}, \quad c_i^k = \hat{c}_i^l (\Gamma^{-1})^{lk}$$

give the connection between the coefficients. Thus, generalizing the theory of SPDEs having bounded coefficients in terms of derivatives with respect to fixed coordinate vectors to SPDEs with bounded coefficients given in terms of the spatial derivations $D_{\Gamma^1}, \ldots, D_{\Gamma^d}$, we can treat a large class of SPDEs with unbounded coefficients by choosing suitable vector fields $\Gamma^1, \ldots, \Gamma^d$. This idea motivates a natural generalization of the notion of Sobolev spaces below. First we need to introduce some notation.

Let r and Γ be sufficiently smooth functions defined on $\mathbf{R}^d$, such that $r(x)$ is a positive real number, and $\Gamma(x)$ is a $d \times d$ matrix with rows $\Gamma^1(x), \ldots, \Gamma^d(x)$ for every $x \in \mathbf{R}^d$. We use the notation

$$D_\Gamma^\gamma := D_{\Gamma^{i_k}} \ldots D_{\Gamma^{i_1}}$$

for a multi-number $\gamma = (i_1, \dots, i_k) \in \{1, 2, \dots, d\}^k$ of length $|\gamma| = k$. For notational convenience we accept the "empty sequence" γ as a multi-number of length $|\gamma| = 0$, and D_Γ^γ as the identity for $|\gamma| = 0$. When Γ is the $d \times d$ unit matrix, we simply write D^γ instead of D_Γ^γ.

DEFINITION 1.1 (Well-weighted Sobolev spaces.) ([10], cf. [30]–[33], [35].) Let $m \geq 0$ be an integer. Then the weighted Sobolev space $W_2^m(r, \Gamma)$ is the closure of $C_0^\infty(\mathbf{R}^d)$ in the norm

$$|\varphi|_m := \left(\sum_{0 \leq |\gamma| \leq m} \int r^2(x) |D_\Gamma^\gamma \varphi(x)|^2 dx \right)^{1/2}.$$

Obviously $W_2^m(r, \Gamma)$ is a separable Hilbert space for every integer $m \geq 0$, and $W_2^0(r, \Gamma) =: L_2(r)$ is the Hilbert space of functions $\varphi : \mathbf{R}^d \to \mathbf{R}$ which are square integrable with respect to the measure $r^2 dx$.

Let $\{H^\alpha\}_{\alpha \in \mathbf{R}}$ be the scale of Hilbert spaces generated from $H^0 := L_2(r)$ by the operator

$$\Lambda := \left(I - r^{-2} \frac{\partial}{\partial x^i} \left(r^2 \Gamma^{ik} \Gamma^{jk} \frac{\partial}{\partial x^j} \right) \right)^{1/2},$$

where I denotes the identity. Clearly

$$|\Lambda \varphi|_0^2 = (\Lambda^2 \varphi, \varphi)_{H_0} = |\varphi|_0^2 + \sum_{i=1}^d |D_{\Gamma^i} \varphi|_0^2 = |\varphi|_1^2$$

for every $\varphi \in C_0^\infty(\mathbf{R}^d)$. Consequently $H^1 = W_2^1(r, \Gamma)$ for every pair of weight functions r, Γ. In order to have $H^n = W_2^n(r, \Gamma)$ as topological vector spaces for every integer $n \geq 2$ we assume

(W_1) For every multi-number γ there exists a constant K_γ such that

$$|D_\Gamma^\gamma (r^{-2} \operatorname{div}(r^2 \Gamma^i))| \leq K_\gamma \quad \text{for } i := 1, 2, \dots, d.$$

(W_2) For every i, j there exist sufficiently smooth functions $\alpha_1^{ij}, \dots, \alpha_d^{ij}$ such that

$$[\Gamma^i, \Gamma^j] = \alpha_k^{ij} D_{\Gamma^k}$$

and

$$|D_\Gamma^\gamma \alpha_k^{ij}| \leq K_\gamma$$

for every multi-index γ and i, j, k, where K_γ is a constant and $[\Gamma^i, \Gamma^j]$ denotes the Lie bracket of the vector fields Γ^i, Γ^j.

PROPOSITION 1.2. ([10], [13], cf. [30].) *Assume* (W_1)–(W_2). *Then for every integer* $m \geq 0$ *the spaces* H^m *and* $W_2^m(r, \Gamma)$ *are the same as vector spaces and the norms in* H^m *and in* $W_2^m(r, \Gamma)$ *are equivalent.*

For the proof we refer to [10]. (In [10] we assume that Γ is invertible but this is not essential in the proof.)

Let us consider the special case

$$\Gamma(x) := \operatorname{diag}(\rho_1(x), \dots, \rho_d(x)) =: \rho(x)$$

with smooth real functions $\rho_i > 0$. Then (W_1)–(W_2) follow from

($\overline{W}$) For every multi-number γ the functions

$$\rho_i^{-1}\rho^\gamma D^\gamma \rho_i, \quad r^{-1}\rho^\gamma D^\gamma r \quad (i := 1, 2, \dots, d)$$

are bounded, where $\rho^\gamma := \rho^{i_1}, \dots, \rho^{i_k}$ and $D^\gamma := \frac{\partial}{\partial x^{i_1}} \cdots \frac{\partial}{\partial x^{i_k}}$ for a multi-number $\gamma = (i_1, \dots, i_k)$, $\rho^\gamma := 1$ and D^γ is the identity for $|\gamma| = 0$ (see [10]).

PROPOSITION 1.3. *If* ($\overline{W}$) *holds then the norm* $|\ |_m$ *and the norm defined by*

$$\|\varphi\|_m := \left(\sum_{0 \le |\gamma| \le m} \int r^2 |\rho^\gamma D^\gamma \varphi|^2 dx \right)^{1/2} \tag{1.2}$$

are equivalent for each integer $m \ge 0$.

For the proof we refer again to [10].

We remark that the norm $\|\ \|_m$ and the corresponding weighted Sobolev space was introduced (together with assumption ($\overline{W}$)) and applied to SPDEs with unbounded coefficients by Purtukhia [30]–[33] in the special case $\rho_1 = \dots = \rho_d$ (see also [35]). The following examples taken from [30] satisfy condition ($\overline{W}$).

EXAMPLES ([30]) (I denotes the unit matrix.)

1. $r(x) = (1 + x^2)^\alpha, \quad \alpha \in \mathbf{R}$
   $\Gamma(x) = (1 + x^2)^\beta I, \quad \beta \le 1/2$
2. $r(x) = \exp(\pm(1 + x^2))^\alpha, \quad 0 \le \alpha \le 1/2$
   $\Gamma(x) := (1 + x^2)^\beta I, \quad \beta \le 1/2 - \alpha$
3. $r(x) := (1 + x^2)^\alpha, \quad \alpha \in \mathbf{R}$
   $\Gamma(x) := \ln^\beta(2 + x^2) I, \quad \beta \in \mathbf{R}$
4. $r(x) := (1 + x^2)^\alpha \ln^\gamma(2 + x^2), \quad \alpha \ge 0, \quad \gamma \ge 0$
   $\Gamma(x) := (1 + x^2)^\beta I, \quad \beta \le 1/2$
   or $\Gamma(x) := \ln^\beta(2 + x^2) I, \quad \beta \ge 0$
5. $\Gamma(x) := \exp\left(-(1 + x^2)^\beta\right) I, \quad \beta \ge 0$
   $r(x) := \exp\left(\pm(1 + x^2)\right), \quad \alpha \in \mathbf{R},$
   or each weight function r from Examples 1–4 can be taken.

We can construct product weights by the next statement.

PROPOSITION 1.4. ([10]) *Let* r', Γ' *and* r'', Γ'' *be pairs of weight functions on* $\mathbf{R}^k$ *and on* $\mathbf{R}^l$, *respectively. Define*

$$r(x) := r'(y) r''(z)$$
$$\Gamma(x) := \Gamma'(y) \oplus \Gamma''(z) \quad \text{for } x = (y, z) \in \mathbf{R}^{k+l},$$

where $\Gamma' \oplus \Gamma''$ denotes the direct sum of the matrices Γ' and Γ''. Assume that r', Γ' and r'', Γ'' satisfy (W_1)–(W_2) (or $(\overline{W})$). Then r, Γ also satisfy (W_1)–(W_2) ($(\overline{W})$ resp.).

Let us now consider the problem

$$\begin{aligned} du(t,x) =& \Big(\sum_{j=1}^{d_2} \mathcal{M}_j^2 u(t,x) + \mathcal{M}_0 u(t,x) + f(t,x)\Big) dt \\ &+ \frac{1}{2}\mathcal{N}_i\mathcal{N}_j u(t,x) d\langle M^i, M^j\rangle(t) \\ &+ (\mathcal{N}_i u(t,x) + g_i(t,x)) dM^i(t) \qquad (1.3) \\ u(0,x) =& u_0(x), \qquad (1.4) \end{aligned}$$

where f, g_i are well-measurable random fields and $\mathcal{M}_j, \mathcal{N}_i$ are of the form (1.1) with bounded coefficients $\hat{b}_j^p, \hat{c}_i^p$ which are well-measurable in (t,ω) and are differentiable in x. Here the stochastic differential with respect to the continuous semimartingale $(M(t), \mathcal{F}_t)$ is understood in Ito's sense. This gives a reason of the appearance of the term

$$\frac{1}{2}\mathcal{N}_i\mathcal{N}_j u(t,x) d\langle M^i, M^j\rangle(t),$$

where $\langle M^i, M^j\rangle(t)$ is the quadratic variational process of $M = (M^i)$.

The solution to (1.3)–(1.4) can be defined as follows.

DEFINITION 1.5. A $W_2^1(r,\Gamma)$-valued weakly continuous adapted process $(u(t))_{t\in[0,T]}$ is called a generalized solution of the problem (1.3)–(1.4) if for every $\varphi \in C_0^\infty(\mathbf{R}^d)$

$$\begin{aligned} (u(t), \varphi) =& (u_0, \varphi) + \int_0^t \{(\mathcal{M}_j u(s), \mathcal{M}_j^* \varphi) + (\mathcal{M}_0 u(s), \varphi) + (f(s), \varphi)\} \, ds \\ &+ \frac{1}{2}\int_0^t (\mathcal{N}_j u(s), \mathcal{N}_i^* \varphi) d\langle M^i, M^j\rangle(s) \\ &+ \int_0^t \{(\mathcal{N}_i u(s), \varphi) + (g_i(s), \varphi)\} \, dM^i(s) \qquad \text{(a.s.)} \qquad (1.5) \end{aligned}$$

for every $t \in [0,T]$. Here $\mathcal{M}_j^*$ and $\mathcal{N}_i^*$ denote the adjoints of $\mathcal{M}_j$ and $\mathcal{N}_i$ with respect to the scalar product $(\ ,\ )$ in L_2.

REMARK 1.6. Let us consider the problem

$$\begin{aligned} du(t,x) =& \left(a^{kl} D_{\Gamma^k} D_{\Gamma^l} u(t,x) + a^k D_{\Gamma^k} u(t,x) + a^0 u(t,x) + f(t,x)\right) dt \\ &+ \left(b_i^k D_{\Gamma^k} u(t,x) + b_i^0 u(t,x) + g_i(t,x)\right) dM^i(t) \qquad (1.6) \\ u(0,x) =& u_0(x), \qquad (1.7) \end{aligned}$$

where $a^{kl} = a^{kl}(t,x)$, $a^k = a^k(t,x)$, $a^0 = a^0(t,x)$, $b_i^k = b_i^k(t,x)$, $b_i^0 = b_i^0(t,x)$ are well-measurable random fields.

One defines the generalized solution of this problem requiring

$$(u(t),\varphi) = (u_0,\varphi) + \int_0^t \Big\{ -(a^{kl} D_{\Gamma^l} u(s), D_{\Gamma^k}\varphi) + ((a^l - D_{\Gamma^k} a^{kl}) D_{\Gamma^l} u(s),\varphi)$$
$$+ (a^0 u(s) + f(s),\varphi) \Big\} ds + \int_0^t (b_i^k D_{\Gamma^k} u(s) + b_i^0 u(s) + g_i(s),\varphi) dM^i(s)$$

instead of (1.5) in Def. 1.5.

One can show that if a^{kl}, b_i^k are sufficiently smooth in x and

$$a^{kl}(t,x) z^k z^l dt \geq \frac{1}{2} b_i^k b_j^l (t,x) z^k z^l d\langle M^i, M^j\rangle(t)$$

for all $t \in \mathbf{R}$, $z = (z', \dots, z^d) \in \mathbf{R}^d$, then the problem can be cast into (1.3)–(1.4) with appropriate operators $\mathcal{M}_j$, $\mathcal{N}_i$ of the form (1.1). Thus the result formulated for the problem (1.3)–(1.4) can be translated into results concerning the problem (1.6)–(1.7).

Before presenting an existence and uniqueness theorem we introduce some notions and notations.

DEFINITION 1.7. Let $c = c(t,x)$ be a random field. We say that c has bounded derivatives with respect to D_Γ up to the order n if for every $t \in \mathbf{R}_+$, $x \in \mathbf{R}^d$

$$|D_\Gamma^\gamma c(t,x)| \leq \xi \quad \text{(a.s.) for all multi-numbers } 0 \leq |\gamma| \leq n \tag{1.8}$$

where ξ is a (finite) random variable.

If $(M(t), \mathcal{F}_t)$ is a continuous semimartingale, then $\tilde{M}$ denotes its martingale part, starting from 0, and $\overline{M} := M - \tilde{M}$, denotes its bounded variational part. The total variation of a stochastic process V over the interval $[0,t]$ is denoted by $||V||(t)$. If V is a continuous increasing stochastic process and H is a separable Banach space then $\mathcal{L}_p(dV)H$ denotes the space of H-valued progressively measurable processes v, satisfying

$$\int_0^T |v(t)|_H^p dV(t) < \infty \quad \text{a.s.},$$

where $|v|_H$ is the norm of v in H.

THEOREM 1.8. *Let $m \geq 0$ be an integer. Assume that the coefficients $\hat{b}_j^k, \hat{b}_j^0, \hat{b}_0^k, \hat{c}_i^k$ have bounded derivatives w.r. to D_Γ up to the order $m+2$ and b_0^0, c_i^0 have bounded derivatives w.r. to D_Γ up to the order $m+1$ for $k = 1,2,\dots,d$; $i = 1,2,\dots,d_1$; $j := 1,2,\dots,d_2$. Let u_0 be an $\mathcal{F}_0$-measurable random element in $W_2^{m+1}(r,\Gamma)$ and suppose that $f \in L_2(dt) W_2^{m+1}(r,\Gamma)$, $g_i \in L_2(dV) W_2^{m+2}(r,\Gamma)$, where $V := \langle \tilde{M} \rangle + ||\overline{M}||$. Then the problem (1.3)–(1.4) has a unique generalized solution u. Moreover u is a $W_2^{m+1}(r,\Gamma)$-valued weakly continuous process and it is strongly continuous in $W_2^m(r,\Gamma)$.*

PROOF: This theorem is proved in [22] when $D_{\Gamma^i} := \frac{\partial}{\partial x^i}$ $(i := 1,2\dots d)$ $r \equiv 1$ and M is a d_1-dimensional Wiener process. The new difficulties, caused by the fact that D_{Γ^i} and and D_{Γ^j} do not commute for $i \neq j$, can be overcome by the following lemma: (see [13]).

If $\mathcal{N}_i, \mathcal{N}_j$ are first order differential operators of the form

$$\mathcal{N}_i := \hat{c}_i^k D_{\Gamma^k} + c_i^0 D_{\Gamma^0}, \qquad \mathcal{N}_j := \hat{c}_j^k D_{\Gamma^k} + c_i^0 D_{\Gamma^0} \tag{1.9}$$

with coefficients $\hat{c}_i^k, c_i^0, \hat{c}_j^k, c_j^0$ satisfying the condition of the theorem, then there is a random variable ξ such that for every integer $n \in [0, m]$

$$(\mathcal{N}_i \mathcal{N}_j u, u)_n + (\mathcal{N}_i u, \mathcal{N}_j u)_n \leq \xi |u|_n^2$$

for all $u \in W_2^n(r, \Gamma)$, where $(\ , \ )_n$ denotes the scalar product in $W_2^n(r, \Gamma)$.

Using this lemma we can get the necessary estimates for proving our theorem by known methods from [22] (see also [34]). For the details we refer to [13].

REMARK 1.9. If the operators $\mathcal{M}_j, \mathcal{N}_i$ from (1.3) are given in the usual form

$$\mathcal{M}_j = b_j^l \frac{\partial}{\partial x^l} + b_j^0, \qquad \mathcal{N}_i = c_i^l \frac{\partial}{\partial x^l} + c_i^0. \tag{1.10}$$

and $\Gamma(x)$ is invertible for every $x \in \mathbf{R}^d$, then

$$\mathcal{M}_j = \hat{b}_j^k D_{\Gamma^k} + b_j^0, \qquad \mathcal{N}_i = \hat{c}_i^k D_{\Gamma^k} + c_i^0$$

with $\hat{b}_j^k = b_j^l (\Gamma^{-1})^{lk}$, $\hat{c}_i^k = c_i^l (\Gamma^{-1})^{lk}$. Thus Theorem 1.8 presents existence and uniqueness results for SPDEs with unbounded coefficients, if we choose suitable weights r and Γ. Let us consider, for example the case $\Gamma := \mathrm{diag}(\rho_1, \ldots, \rho_d)$ with sufficiently smooth real functions $\rho_i > 0$, and assume $(\overline{\mathrm{W}})$.

Let $b^k = b^k(t, x)$ be a random field and set

$$\hat{b}^k := b^l (\Gamma^{-1})^{lk} = b^k / \rho_k.$$

Assume that

$$|D^\gamma b^k| \leq \xi \rho_k / \rho^\gamma \quad \text{for all multi-numbers } 0 \leq |\gamma| \leq n, \tag{1.11}$$

where ξ is a random variable. Then, using $(\overline{\mathrm{W}})$ one can show that

$$|D_\Gamma^\gamma \hat{b}^k| \leq K\xi \quad \text{for all multi-numbers } 0 \leq |\gamma| \leq n, \tag{1.12}$$

where K is a constant (see [10]). Consequently, an assumption like (1.12) can be replaced by (1.11). Hence we have

THEOREM 1.10. *Consider the problem (1.3)–(1.4) with operators $\mathcal{M}_j, \mathcal{N}_i$ of the form (1.1). Let $\Gamma = \mathrm{diag}(\rho_1, \ldots, \rho_d)$ with smooth real functions $\rho_1 > 0, \ldots, \rho_d > 0$, and assume $(\overline{\mathrm{W}})$. Suppose that the coefficients satisfy the necessary conditions on measurability, and that*

(i) $|D^\gamma b_j^p(t, x)| \leq \xi \rho_p / \rho^\gamma \quad j = 1, 2, \ldots, d_2, \ p = 0, 1, 2, \ldots, d$

$|D^\gamma b_0^k(t, x)| \leq \xi \rho_k / \rho^\gamma \quad k = 1, 2, \ldots, d, \ i = 1, 2, \ldots, d_1$

$|D^\gamma c_i^k(t, x)| \leq \xi \rho_k / \rho^\gamma$

$|D^\beta b_0^0(t,x)| \le \xi$

$|D^\beta c_i^0(t,x)| \le \xi \rho_p/\rho^\gamma$

for all multinumbers $0 \le |\beta| \le m+1;\ \ 0 \le |\gamma| \le m+2$.

(ii) $f \in L_2(dt)W_2^{m+1}(r,\Gamma)$, $g_i \in L_2(dV)W_2^{m+2}(r,\Gamma)$ $(i = 1,2,\dots,d_1)$ *and* u_0 *is an* $\mathcal{F}_0$*-measurable random element in* $W_2^{m+1}(r,\Gamma)$.

Then there is a unique generalized solution u. *Moreover* u *is a* $W_2^{m+1}(r,\Gamma)$*-valued weakly continuous process and it is strongly continuous in* $W_2^m(r,\Gamma)$.

REMARK 1.11. This theorem is proved in [30] for the special case $\rho_1 = \rho_2 = \cdots = \rho_d$. Taking $\rho_1 = \cdots = \rho_d = \sqrt{1+|x|^2}$, for example, one gets an existence and uniqueness theorem for the case of linearly growing coefficients b_j^k, c_i^k. We make use of Theorem 1.10 with differential weights, $\rho_1,\dots,\rho_d$ in solving systems of SPDEs below.

EXAMPLE 1.12. Solving systems of SPDEs. The following system of SPDEs arose by studying the evolution of a magnetic field $H = (H^1,H^2,H^3)$ in random medium:

$$\begin{aligned} dH^i =&(\Delta H^i + \frac{1}{2} b_l^j b_l^k H^i_{x^j x^k} - b^i_{kx^j} b^l_k H^j_{x^l} \\ &+ \frac{1}{2} b^j_k b^l_{kx^j} H^i_{x^l} - \frac{1}{2} b^j_k b^i_{kx^l x^j} H^l + \frac{1}{2} b^i_{kx^j} b^j_{kx^l} H^l)dt \\ &+ (b^i_{kx^j} H^j - b^j_k H^i_{x^j})dW^k(t), \quad i = 1,2,3 \qquad (1.13)\\ H^i(0,x) =&H_0^i(x), \qquad (1.14) \end{aligned}$$

Here $(W^k(t))_{t\ge 0}$ is an m-dimensional Wiener-process, Δ is the Laplacian in the space variable $x = (x^1,x^2,x^3) \in \mathbf{R}^3$, and $b = (b_k^j(t,x))$ is a function of $(t,x) \in \mathbf{R}_+ \times \mathbf{R}^3$. The case when b is linearly growing is of special interest (see [16]). An existence and uniqueness theorem for (1.14)–(1.14) is established in [16]. Below we show that one can reduce the system (1.13) to a single equation having unbounded coefficients growing differently at the different derivations.

Set $u(t,x,z) := H^i(t,x)z^i$, where $z = (z^1,z^2,z^3) \in \mathbf{R}^3$ is a new variable. Then by multiplying the equation for H^i by z^i and summing up, one can see that u satisfies

$$\begin{aligned} du(t) =&(\Delta_x u(t) + \frac{1}{2} b^j_k b^l_k u_{x^j x^l} - b^i_{kx^j} b^l_k z^i u_{z^j x^l} \\ &+ \frac{1}{2} b^j_k b^l_{kx^j} u_{x^l} - \frac{1}{2} b^j_k b^i_{kx^l x^j} z^i u_{z^l})dt \\ &+ (b^i_{kx^j} z^i u_{z^j} - b^j_k u_{x^j})dW^k(t) \\ u(0,x) =&z^i H^i(0,x). \qquad (1.15) \end{aligned}$$

Introducing the operators

$$\begin{aligned} \mathcal{M} :=& \frac{\partial}{\partial x^1} + \frac{\partial}{\partial x^2} + \frac{\partial}{\partial x^3} \\ \mathcal{N}_k :=& b^i_{kx^j} z^i \frac{\partial}{\partial z^j} - b^j_k \frac{\partial}{\partial x^j} \qquad k := 1,2,\dots,m, \end{aligned}$$

we can write the above equation as follows

$$du(t) = (\mathcal{M}^2u(t) + \frac{1}{2}\sum_{k=1}^{m}\mathcal{N}_k^2u(t))dt + \mathcal{N}_ku(t)dW^k(t). \tag{1.16}$$

Let r and $\Gamma := \rho I$ be weights satisfying $(\overline{\mathrm{W}})$, where $I \in \mathbf{R}^{3\times 3}$ is the unit matrix, and $\rho(x) > 0$ for every $x \in \mathbf{R}^3$. Under a generalized solution of (1.13)–(1.14) we mean a stochastic process $H = (H^i(t))$ such that H^i is a $W_2^1(r,\Gamma)$-valued weakly continuous adapted process, satisfying (1.13)–(1.14) in the sense of Def. 1.5. Define

$$r'(x,z) := r(x)(1+|z|^2)^{-3/2}, \quad \Gamma'(x,z) = (\rho(x)I) \oplus (\rho'(z)I), \tag{1.17}$$

where $\rho'(z) = \sqrt{1+|z|^2}$. Then these new weights also satisfy $(\overline{\mathrm{W}})$ by virtue of Proposition 1.4. Moreover, H is a generalized solution of (1.13)–(1.14) iff u is a generalized solution of (1.16)–(1.15), with respect to the weights r', Γ'. Moreover one can show that $M^i(t)$ is strongly (weakly) continuous in $W_2^m(r,\Gamma)$ for $i = 1,2,3$, iff $u(t)$ is strongly (weakly) continuous in $W_2^m(r',\Gamma')$. Thus from Theorem 1.10 we get the following

THEOREM. (Cf. [35].) *Let $m \geq 0$ be an integer. Assume that $H_0^i \in W_2^{m+1}(r,\Gamma)$ for every $i := 1,2,3$. Suppose that*

$$|D^\gamma b_k^j(t,x)| \leq \xi\rho^{1-|\gamma|} \quad \text{for} \quad 0 \leq |\gamma| \leq m+3$$

where ξ is a random variable. Then the problem (1.13)–(1.14) admits a unique generalized solution $(H^i(t))$. Moreover H^i has a weakly continuous $W_2^{m+1}(r,\Gamma)$-valued modification which is strongly continuous in $W_2^m(r,\Gamma)$.

PROOF: ([13].) Apply Th. 1.10 to (1.16)–(1.15).

We note that this theorem can be generalized to a certain class of systems of SPDEs (see [10]).

2. Approximations of SPDEs

A family of (d_1-dimensional) continuous semimartingales $\{(M_\delta(t), \mathcal{F}_t) \colon \delta > 0\}$ is called an approximation for the continuous semimartingale $(M_t, \mathcal{F}_t)$ with accompanying process $S = (S^{ij}(t))$ if

(i) for $\delta \to 0$

$$\sup_{t\leq T}|M(t) - M_\delta(t)| \to 0, \qquad \sup_{t\leq T}|S_\delta(t) - S(t)| \to 0$$

in probability for every $T > 0$, where $S = (S^{ij}(t))$ is a matrix-valued continuous process of bounded variation and

$$S_\delta^{ij}(t) := \int_0^t (M^i - M_\delta^i)d\overline{M}_\delta^j - \langle M^i - M_\delta^j, M_\delta^j\rangle(t) + \frac{1}{2}(\langle M_\delta^i, M_\delta^j\rangle(t) - \langle M^i, M^j\rangle(t))$$

$(i,j := 1,\ldots,d_1)$;

(ii) For every $T > 0$ the families of random variables

$$\left\{\int_0^T |M^i - M^i_\delta| d\|\overline{M}^j_\delta\| : \delta > 0\right\}, \quad \{\langle M_\delta\rangle(T) : \delta > 0\}$$

are bounded in probability.
(We recall that $\overline{M}_\delta$ is the bounded variational part of M_δ and $\|\overline{M}_\delta\|(t)$ denotes the total variation of $\overline{M}_\delta$ over the interval $[0, t]$. A family of random variables $\{\xi_\delta : \delta > 0\}$ is said to be bounded in probability if $\lim_{K\to\infty} \sup_\delta P(|\xi_\delta| \geq K) = 0$.)

If M_δ is a process of bounded variation for every $\delta > 0$, then we speak about good approximations. In this case

$$S^{ij}_\delta(t) = \int_0^t (M^i - M^i_\delta) dM^j_\delta(s) - \frac{1}{2}\langle M^i, M^j\rangle(t).$$

When $S \equiv 0$ then the approximation $\{(M_\delta(t), \mathcal{F}_t) : \delta > 0\}$ is called symmetric.

EXAMPLE 2.1. Smoothing. Let $\rho : \mathbf{R} \to \mathbf{R}$ be a nonnegative function such that $\rho(x) = 0$ for $x \notin [0, 1]$ and $\int \rho(x)dx = 1$. Define M_δ by

$$M_\delta(t) = \frac{1}{\delta}\int_0^t M(t-s)\rho(s/\delta)$$

where M is a continuous semimartingale ($M(t) := M(0)$ for $t < 0$). Then one can show that M_δ is a good symmetric approximation for M (see [23]–[24]).

EXAMPLE 2.2. The approximation M_δ defined by

$$M^1_\delta(t) := \delta \sin(t/\delta^2), \quad M^2_\delta(t) := \delta \cos(t/\delta^2)$$

is a simple example for good but non-symmetric approximation for the trivial (two-dimensional) semimartingale $M \equiv 0$. (One can easily see that $S^{21}(t) = -S^{12} = \frac{1}{2}t$.)

Let $\{(M_\delta(t), \mathcal{F}_t) : \delta > 0\}$ be an approximation for the continuous semimartingale $(M(t), \mathcal{F}_t)$, and let us consider the problems

$$\begin{aligned} du_\delta(t,x) =& \left(\sum_{j=1}^{d_2} \mathcal{M}^2_j u_\delta(t,x) + \mathcal{M}_0 u(t,x) + f\right)dt \\ &+ (\mathcal{N}_i u_\delta(t,x) + g_i) \circ dM^i_\delta(t) \qquad (2.1_\delta) \\ u_\delta(0,x) =& u_0(x) \qquad (2.2_\delta) \\ du(t,x) =& \left(\sum_{j=1}^{\delta_2} \mathcal{M}_j u(t,x) + \mathcal{M}_0 u(t,x) + f\right)dt \\ &+ (\mathcal{N}_i u(t,x) + g_i) \circ dM^i(t) + \frac{1}{2}([\mathcal{N}_i, \mathcal{N}_j]u(t,x) + \mathcal{N}_i g_j - \mathcal{N}_j g_i)dS^{ij}(t) \qquad (2.1) \\ u(0,x) =& u_0(x), \qquad (2.2) \end{aligned}$$

where the first order differential operators $\mathcal{M}_j, \mathcal{N}_i$ are given by (1.1) and $[\mathcal{N}_i, \mathcal{N}_j] := \mathcal{N}_i\mathcal{N}_j - \mathcal{N}_j\mathcal{N}_i$. For the sake of simplicity of the presentation assume that $\mathcal{N}_i, g_i$ do not depend on (t, ω).

Then the stochastic differentials in the right sides of equations (2.1_δ) and (2.2_δ) are defined as follows

$$\begin{aligned}(\mathcal{N}_i u_\delta + g_i) \circ dM_\delta^i(t) :=& (\mathcal{N}_i u_\delta + g_i) dM_\delta^i(t) \\ &+ \frac{1}{2}(\mathcal{N}_i\mathcal{N}_j u_\delta + \mathcal{N}_i g_j) d\langle M_\delta^i, M_\delta^j\rangle(t) \\ (\mathcal{N}_i u + g_i) \circ dM^i(t) :=& (\mathcal{N}_i u + g_i) dM^i(t) + \frac{1}{2}(\mathcal{N}_i\mathcal{N}_j u + \mathcal{N}_i g_j) d\langle M^i, M^j\rangle(t),\end{aligned} \tag{2.3}$$

where the stochastic differential on the right-sides is understood in Ito's sense.

Now we are in the position to formulate a theorem on the approximations of SPDEs.

THEOREM 2.3. ([12]–[13], cf. [3], [14]–[15]). *Assume the conditions of Th. 1.8 with $m+2$ in place of the non-negative integer m. Then for the generalized solutions u_δ and u of (2.1_δ)–(2.2_δ) and (2.1)–(2.2) we have*

$$\sup_{t \le T} |u_\delta(t) - u(t)|_m \to 0 \quad \textit{in probability}$$

for $\delta \to 0$.

PROOF: One can prove this theorem by using the methods of [17]–[18] combining with the following generalization of the lemma from the proof of Theorem 1.8:

LEMMA 2.4. ([13].) Let $\mathcal{N}_i, \mathcal{N}_j, \mathcal{N}_k$ be first order differential operators of the form (1.9) such that their coefficients $\hat{c}_i^p, \hat{c}_j^p, \hat{c}_k^p$ have bounded derivatives with respect to D_Γ up to the order $m+3$. Suppose that the weights r, Γ satisfy (W_1)–(W_2). Then for all $\varphi \in W_2^{m+3}(r, \Gamma)$

$$\begin{aligned}(\varphi, \mathcal{N}_i\mathcal{N}_j\mathcal{N}_k\varphi)_m + (\mathcal{N}_i\varphi, \mathcal{N}_j\mathcal{N}_k\varphi)_m + (\mathcal{N}_i\mathcal{N}_j\varphi, \mathcal{N}_k\varphi) + (\mathcal{N}_j\varphi, \mathcal{N}_i\mathcal{N}_k\varphi)_m \\ \le K|\varphi|_m^2\end{aligned}$$

where $(\ ,\ )_m$ and $|\ |_m$ denote the scalar product and the norm in $W_2^m(r, \Gamma)$, respectively.

The details of the proof will appear elsewhere.

COROLLARY 2.4. *Let $\Gamma := \mathrm{diag}(\rho_1, \dots, \rho_d)$ with smooth functions $\rho_i > 0$, such that r and ρ_i satisfy $(\overline{W})$. Assume the conditions of Th. 1.10 with $m+2$ in place of m. Then*

$$\sup_{t \le T} \|u(t) - u_\delta(t)\|_m \to 0 \quad \textit{in probability}$$

for $\delta \to 0$, where the norm $\|\ \|_m$ is defined by (1.2).

PROOF: The norms $\|\ \|_m$ and $|\ |_m$ are equivalent by Proposition 1.3.

COROLLARY 2.5. *Let n be a non-negative integer such that $m > d/2 + n$. Assume the conditions of Corollary 2.4. Then for $\delta \to 0$*

$$\sup_{t\le T}\sup_{|x|\le R}\sum_{0\le|\gamma|\le n}|D^\gamma u_\delta(t,x) - D^\gamma u(t,x)| \to 0$$

in probability for every R.

PROOF: Since for balls $B_R := \{x \in \mathbb{R}^d : |x| \le R\}$,

$$\begin{aligned}\|\varphi\|_m^2 = &\sum_{|\gamma|\le m}\int r^2(x)|\rho(x)|^2|D^\gamma\varphi(x)|^2\,dx\\ \ge &K_R\sum_{|\gamma|\le m}\int_{B_R}|D^\gamma\varphi(x)|^2\,dx\end{aligned}$$

for all $\varphi \in W_2^m(r,\Gamma)$, we have

$$\sup_{|x|\le R}\sum_{|\gamma|\le n}|D^\gamma\varphi(x)| \le K\|\varphi\|_m,$$

by Sobolev's embedding theorem, where K is a constant depending on R, d, n, r, Γ.

REMARKS 2.6. In the case when $\mathcal{N}_i$ and g_i depend on (t,ω) we assume that for every $\varphi, \psi \in C_0^\infty$ the stochastic processes $(\mathcal{N}_i(t)\varphi,\psi)$, (g_i,ψ) have stochastic differentials

$$\begin{aligned}d(\mathcal{N}_i(t)\varphi,\psi) =&(\mathcal{N}_i^k(t)\varphi,\psi)dM^k(t)\\ d(g_i(t)\psi) =&(g_i^k(t),\psi)dM^k(t)\end{aligned}$$

where $\mathcal{N}_i^k$ is a first order differential operator, $g_i^k = g_i^k(t,x)$ is a random field for every $i,k := 1,2,\ldots,d_1$, and $(\ ,\ )$ denotes the scalar product in L_2. We suppose that $\mathcal{N}_i^k, g_i^k$ satisfy the same conditions as $\mathcal{N}_i$ and g_i in the above theorems. Then these theorems hold if $(\mathcal{N}_i u_\delta + g_i)\circ M_\delta^i(t)$, $(\mathcal{N}_i u + g_i)\circ dM^i(t)$ on the right sides of equations (2.1_δ) and (2.1) denote

$$(\mathcal{N}_i u_\delta + g_i)dM_\delta^i + \frac{1}{2}(\mathcal{N}_i\mathcal{N}_j u_\delta + \mathcal{N}_i g_j)d\langle M_\delta^i, M_\delta^j\rangle + (\mathcal{N}_i^k u_\delta + g_i^k)d\langle M_\delta^i, M^k\rangle\quad,$$

$$(\mathcal{N}_i u + g_i)dM^i + \frac{1}{2}(\mathcal{N}_i\mathcal{N}_j u + \mathcal{N}_i g)d\langle M^i, M^j\rangle + (\mathcal{N}_i^k u + g_i^k)d\langle M^i, M^k\rangle$$

respectively, where the stochastic differentials are understood in Ito's sense.

For generalizations when not only the semimartingale M but the operators $\mathcal{M}_j, \mathcal{N}_i$, the free data f, g_i and the initial value u_0 are also simultaneously approximated we refer to [17]–[18].

3. Theorems on supports

Let us consider the problem

$$
\begin{aligned}
du(t,x) = & \left(\sum_{j=1}^{d_2} \mathcal{M}_j^2 u(t,x) + \mathcal{M}_0 u(t,x) + f\right) dt \\
& + (\mathcal{N}_i u(t,x) + g_i) \circ dW^i(t) && (3.1) \\
u(0,x) = & u_0(x), && (3.2)
\end{aligned}
$$

where $W = (W^i(t))$ is a d_1-dimensional Wiener process. Assume in this section that $\mathcal{M}_j$, $\mathcal{M}_0$, f, $\mathcal{N}_i$, g_i, u_0 are not random, and for the sake of simple presentation, that $\mathcal{N}_i, g_i$ do not depend on t. Then the stochastic differential on the right side is defined by

$$
\begin{aligned}
(\mathcal{N}_i u(t,x) + g_i) \circ dW^i(t) := & (\mathcal{N}_i u(t,x) + g_i) dW^i(t) \\
& + \frac{1}{2}\left(\sum_{i=1}^{d_1} \mathcal{N}_i^2 u(t,x) + \mathcal{N}_i g_i\right) dt, && (3.3)
\end{aligned}
$$

where the stochastic differential on the right side is understood in Ito's sense.

Replacing W in (3.1) by an absolutely continuous deterministic function

$$w: [0,\infty) \longrightarrow \mathbf{R}^{d_1}$$

we get the deterministic Cauchy problem

$$
\begin{aligned}
\frac{\partial}{\partial t} u^w = & \left(\sum_{j=1}^{d_2} \mathcal{M}_j^2 + \mathcal{M}_0\right) u^w + f + \dot{w}^i \mathcal{N}_i u^w + \dot{w}^i g_i && (3.1_w) \\
u^w(0) = & u_0, && (3.2_w)
\end{aligned}
$$

where w^i is the i-th coordinate function of w and $\dot{w}^i(t) = \frac{d}{dt} w^i(t)$. Let r, Γ be weights satisfying (W_1)–(W_2).

Assume that $\mathcal{M}_j, \mathcal{M}_0$, and $\mathcal{N}_i$ are first order differential operators given by (1.1) and that the condition of Th. 1.8 with $m+2$ in place of m are satisfied. Then (3.1_w)–(3.2_w) has a unique generalized solution u^w and (3.1)–(3.2) has a unique generalized solution u. Moreover, by Th. 1.8 u^w belongs to $\mathbf{X} := C([0,T]; W_2^m(r,\Gamma))$, and u is a random element in $\mathbf{X}$. Let $\mathcal{H} = \{w\}$ be a subset of absolutely continuous applications

$$w: [0,\infty) \longrightarrow \mathbf{R}^{d_1}$$

such that $C^\infty(\mathbf{R}_+; \mathbf{R}^{d_1}) \subset \dot{\mathcal{H}} := \{\dot{w}\}$. Then we have

THEOREM 3.1. (Cf. [16], [18].) *Assume the conditions of Th. 1.8 with $m+2$ in place of m. Then*

$$\operatorname{supp} \mu_u^{\mathbf{X}} = \overline{\mathcal{U}}^{\mathbf{X}},$$

where supp μ_u^X *is the topological support of the distribution of* u *in*

$$X := C([0,T]; W_2^m(r,\Gamma))$$

and $\overline{\mathcal{U}}^{\mathbf{X}}$ *is the closure of* $\mathcal{U} := \{u^w : w \in \mathcal{H}\}$ *in* $\mathbf{X}$.

PROOF: By Th. 2.3 one can easily show that the closure of $\mathcal{U}$ and the closure of $\{u^w : w \in C^\infty([0,T]; \mathbf{R}^{d_1})\}$ in $\mathbf{X}$ are the same. Therefore we may (and we do) assume that $\mathcal{H} = C^\infty([0,T]; \mathbf{R}^{d_1})$. Let W_δ be obtained from W by smoothing, using the kernel $\rho_\delta(x) := \frac{1}{\delta}\rho(x/\delta)$, where ρ is a C_0^∞ function (see Example 2.1). Then by Theorem 2.3

$$\sup_{t\le T} |u(t) - u_\delta(t)|_m \to 0 \qquad \text{in probability}$$

for $\delta \to 0$, where u_δ is the generalized solution of the problem (3.1_δ)–(3.2_δ) obtained replacing W with W_δ in (3.1)–(3.2). Hence $u_\delta \to u$ weakly in $\mathbf{X} = C([0,T] : W_2^m(r,\Gamma))$ and consequently

$$P(u \in \overline{\mathcal{U}}^{\mathbf{X}}) \ge \limsup_{\delta\to 0} P(u_\delta \in \overline{\mathcal{U}}^{\mathbf{X}}) = 1.$$

This means supp $\mu_u^{\mathbf{X}} \subseteq \overline{\mathcal{U}}^{\mathbf{X}}$. To prove the inclusion in the opposite direction take an application w from $\mathcal{H}$ and approximate it by stochastic processes w_δ defined as follows

$$w_\delta(t) := W(t) + (w - W)_\delta(t)$$

where

$$(w - W)_\delta(t) := \frac{1}{\delta}\int_0^t (w - W)(t-s)\rho(s/\delta)ds =: h_\delta(t)$$

and ρ is a smooth kernel as before. Then $\{w_\delta : \delta > 0\}$ is a symmetric approximation of the "semimartingale" w. Thus for $\delta \to 0$ we have

$$\sup_{t\le T} |u_\delta(t) - u^w(t)|_m \to 0 \quad \text{in probability,}$$

by Th. 2.3, where u_δ is the generalized solution of (3.1)–(3.2) with w_δ in place of W. In particular, for every $\varepsilon > 0$

$$P(\sup_{t\le T} |u_\delta(t) - u^w(t)|_m < \varepsilon) > 0 \tag{3.4}$$

if $\delta > 0$ is sufficiently small. It is not difficult to show that there exist some constants $\lambda > 0$, K such that

$$Ee^{\lambda h_\delta^2(t)} \le K \quad \text{for all} \quad t \in [0,T],$$

which implies

$$E \exp\left(\pm \int_0^T h_\delta(t)dW(t) - \frac{1}{2}\int_0^T h_\delta^2(t)dt\right) = 1.$$

Hence, applying Girsanov's theorem one can show that the distributions of u_δ and u on $C([0,T];W_2^m(r,\Gamma))$ are absolutely continuous with respect to each other. Thus from (3.4) we get

$$(\sup_{t\leq T}|u(t)-u^w(t)|_m<\varepsilon)>0$$

for every $\varepsilon>0$. The proof is completed.

REMARK 3.2. Theorem 3.1 is a generalization of Stroock–Varadhan's support theorem (see [38], [24]) to parabolic SPDEs. The device of approximating a function w from $\mathcal{H}$ by semimartingales w_δ is used in [23]–[24] for proving Stroock–Varadhan's support theorem for stochastic differential equations.

If $\mathcal{N}_i$ and g_i depend on t then Th. 3.1 holds if in addition to the conditions we assume that for every $i=1,2,\dots,d_1$ the derivatives of $\hat{c}_i^k, g_i$ in t exist and satisfy the same conditions as c_i^k, g_i $(k:=0,1,\dots,d)$ respectively.

REMARK 3.3. Let $(M(t),\mathcal{F}_t)$ be a d_1-dimensional continuous semimartingale such that $\overline{M}^i, \langle\tilde{M}^i\rangle$ are absolutely continuous for $i:=1,2,\dots,d_1$ and for $dt\times P$-almost all $(t,\omega)\in \mathbf{R}_+\times\Omega$

$$\sum\left|\frac{d}{dt}\overline{M}^i(t)\right|\leq K,$$

$$\frac{1}{\lambda}\sum_i z_i^2\geq Q^{ij}(t)z_iz_j\geq\lambda\sum_i z_i^2$$

for all $z=(z_1,\dots,zd_1)\in\mathbf{R}^{d_1}$, where $\lambda>0$, K are constants and $Q^{ij}(t):=\frac{d}{dt}\langle\tilde{M}^i,\tilde{M}^j\rangle(t)$ for $i,j:=1,2,\dots,d_1$. ($\tilde{M}$ is the martingale part and $\overline{M}$ is the bounded variational part of M.) Then Th. 3.1 holds when W is replaced by M. For generalizations when $\mathcal{M}_j$, $\mathcal{M}_0$, $\mathcal{N}_i$, f, g_i depend on t and M we refer to [17]–[18].

Let r and $\Gamma:=\operatorname{diag}(\rho_1,\dots,\rho_d)$ be weights satisfying ($\overline{\mathrm{W}}$). Then we have

THEOREM 3.4. *Assume the conditions of Th. 1.10 with $m+2$ in place of m. Then*

$$\operatorname{supp}\mu_u^{\mathsf{X}}=\overline{\mathcal{U}}^{\mathsf{X}}$$

where $\overline{\mathcal{U}}^{\mathsf{X}}$ is the closure of $\{u^w : w\in\mathcal{H}\}$ in $\mathsf{X}=C([0,T];W_2^m(r,\Gamma))$.

PROOF: Apply Remark 1.9 to Th. 3.1.

Let $C^{0,n}$ be the space of functions $\psi\colon[0,\infty)\times\mathbf{R}^d\to\mathbf{R}$ having continuous derivatives in $x\in\mathbf{R}^d$ up to the order n, with the topology of uniform convergence of the functions and their derivatives up to the order n on compact subsets of $[0,\infty)\times\mathbf{R}^d$.

THEOREM 3.5. *Let n be a non-negative integer such that $m>d/2+n$. Assume the conditions of Th. 1.10 with $m+2$ in place of m. Then u is a random element in $C^{0,n}$, u^w belongs to $C^{0,n}$ for every $w\in\mathcal{H}$ and Theorem 3.4 on support holds also with $\mathsf{X}:=C^{0,n}$.*

PROOF: See Corollary 2.5.

4. Applications to nonlinear filtering

Let us consider the following model for a signal process $x(t)$ and an observation process $y(t)$

$$dx(t) = b(x(t))dt + \sigma(x(t))dW(t), \qquad x(0) = \xi \tag{4.1}$$

$$dy(t) = B(x(t))dt + \Sigma dW(t), \qquad y(0) = \eta, \tag{4.2}$$

where $b = (b^i) : \mathbb{R}^d \to \mathbb{R}^d$, $\sigma = (\sigma^{ij}) : \mathbb{R}^d \to \mathbb{R}^{d \times d_2}$, $B = (B^i) : \mathbb{R}^d \to \mathbb{R}^{d_1}$, are Lipschitz functions, $\Sigma \in \mathbb{R}^{d_1 \times d_2}$, ξ and η are random variables which are independent of the Wiener process W.

The basic problem of filtering is to compute the conditional distribution of $x(t)$ given the observation $y(s)$ for $0 \le s \le t$. It is well known that under general conditions this conditional distribution has a density $p(t, \cdot)$ with respect to the Lebesgue measure on $\mathbb{R}^d$, and it can be obtained by normalizing the solution of the Duncan–Mortensen–Zakaï equation. That means

$$p(t,x) = \varphi(t,x)\Big(\int \varphi(t,x)dx\Big)^{-1},$$

where φ is the solution of the problem

$$d\varphi(t,x) = \mathcal{L}^*\varphi(t,x)dt + \mathcal{N}_i^*\varphi(t,x)dy^i(t) \tag{4.3}$$

$$\varphi(0,x) = p_0(x), \tag{4.4}$$

where p_0 is the conditional density of ξ given η and

$$\mathcal{L}^*\varphi := \frac{\partial^2}{\partial x^k \partial x^l}(a^{kl}\varphi) - \frac{\partial}{\partial x^k}(b^k\varphi)$$

$$\mathcal{N}_i^*\varphi := \hat{B}_i\varphi - \frac{\partial}{\partial x^l}(\hat{\sigma}_i\varphi)$$

with $a^{kl} := \frac{1}{2}(\sigma\sigma^*)^{kl}$, $\hat{B} := B^*(\Sigma\Sigma^*)^{-1}$, $\hat{\sigma} := \sigma\Sigma^*(\Sigma\Sigma^*)^{-1}$. ($\sigma^*$ denotes the transpose of a matrix σ.)

Note that the problem (4.3)–(4.4) can be rewritten as

$$d\varphi(t,x) = (\mathcal{L}^* - \frac{1}{2}Q^{ij}\mathcal{N}_i^*\mathcal{N}_j^*)\varphi(t,x)dt + \mathcal{N}_i^*\varphi(t,x) \circ dy^i(t) \tag{4.5}$$

$$\varphi(0,x) = p_0(x), \tag{4.6}$$

where

$$Q^{ij}dt = (\Sigma\Sigma^*)^{ij}dt = d\langle y^i, y^j\rangle(t)$$

$$\mathcal{N}_i^*\varphi(t,x) \circ dy^i(t) = \mathcal{N}_i^*\varphi(t,x)dy^i(t) + \frac{1}{2}Q^{ij}\mathcal{N}_i^*\mathcal{N}_j^*\varphi(t,x)dt.$$

In the model described by (4.1)–(4.2) the observation process $y = (y(t))$ has unbounded variation. In practice, however, the observations we receive have bounded variations. These "real observations" are the results of some measurements. Thus we can regard them as symmetric good approximations $\{y_\delta : \delta > 0\}$ for the process y, since the measuring instruments have smoothing effects. Having a "real observation" y_δ we solve the problem

$$d\varphi_\delta(t,x) = \left(\mathcal{L}^* - \frac{1}{2}Q^{ij}\mathcal{N}_i^*\mathcal{N}_j^*\right)\varphi_\delta(t,x)dt + \mathcal{N}_i^*\varphi_\delta(t,x)dy_\delta(t) \tag{4.7}$$

$$\varphi_\delta(0,x) = p_0(x) \tag{4.8}$$

instead of (4.3)–(4.4), and compute

$$p_\delta(t,x) = \varphi_\delta(t,x)\left(\int \varphi_\delta(t,x)dx\right)^{-1}. \tag{4.9}$$

The theorem below states that this procedure is robust in the sense that

$$\varphi_\delta \to \varphi \quad \text{and} \quad p_\delta \to p$$

if the results y_δ of the measurements for y converge to y.

THEOREM 4.1. *Let $r, \Gamma := \rho I$ be a pair of weights satisfying $(\overline{W})$, where r and ρ are positive smooth functions and I is the $d \times d$ identity matrix. Let m be a non-negative integer and assume that*

(i) $\Sigma\Sigma^*$ is a regular matrix

(ii) $|D^\gamma\sigma(x)| \le K|\rho(x)|^{1-|\gamma|}$, $|D^\beta B(x)| \le K|\rho(x)|^{-|\beta|}$, $|D^\beta b(x)| \le K|\rho(x)|^{1-|\beta|}$

for all multi-numbers $0 \le |\gamma| \le m+5$, $0 \le |\beta| \le m+4$, where K is a constant.

Then for a symmetric good approximation $\{y_\delta : \delta > 0\}$ of y

$$\sup|\varphi_\delta(t) - \varphi(t)|_m \to 0 \quad \text{in probability when} \quad \delta \to 0, \tag{4.10}$$

where φ and φ_δ are the generalized solutions of (4.5)–(4.6) and (4.7)–(4.8) respectively. Moreover, if $\frac{1}{r} \in L_2$ then for $\delta \to 0$ we also have

$$\sup_{t \le T}|p_\delta(t) - p(t)|_m \to 0 \quad \text{in probability} \tag{4.11}$$

where $p(t,\cdot)$ is the conditional density of $x(t)$ given $\{y(s) : s \le t\}$ and p_δ is defined by (4.9).

PROOF: We get (4.10) by applying Th. 2.3 to (4.5)–(4.6) and (4.7)–(4.8). To obtain (4.11) it suffices to show that for $\delta \to 0$

$$\sup_{t\le T}\left|\int_{\mathbf{R}^d}(\varphi_\delta(s,x) - \varphi(s,x))dx\right| \to 0 \quad \text{in probability.}$$

This follows from

$$\sup_{t\le T}\left|\int_{\mathbf{R}^d}(\varphi_\delta(s,z) - \varphi(s,x))dx\right| \le K\sup_{t\le T}\left(\int r^2(x)|\varphi_\delta(s,x) - \varphi(s,x)|^2dx\right)^{1/2}$$

using (4.10) with $m = 0$, where

$$K := \left(\int_{\mathbf{R}^d} 1/r^2(x)dx\right)^{1/2}.$$

COROLLARY 4.2. *Let n be a non-negative integer such that $m > d/2 + n$. Assume the conditions of Th. 4.1. Then for $\delta \to 0$*

$$\sup_{t\leq T}\sup_{|x|\leq R} |D^\gamma \varphi_\delta(t,x) - D^\gamma \varphi(t,x)| \to 0$$

$$\sup_{t\leq T}\sup_{|x|\leq R} |D^\gamma p_\delta(t,x) - D^\gamma p(t,x)| \to 0 \qquad (\text{if } 1/r \in L_2)$$

in probability for every $R > 0$.

PROOF: Apply Corollary 2.5.

Let us now consider the deterministic Cauchy problem

$$\frac{\partial}{\partial t}\varphi^w(t,x) = \left(\mathcal{L}^* - \frac{1}{2}\mathcal{N}_i^*\mathcal{N}_j^* Q^{ij}\right)\varphi^w(t,x) + \dot{w}^i(t)\mathcal{N}_i^*\varphi^w(t,x)$$

$$\varphi^w(0,x) = p_0(x)$$

for $w \in \mathcal{H}$, where $\mathcal{H} = \{w\}$ is a subset of the absolutely continuous applications $w\colon \mathsf{R}_+ \to \mathsf{R}^{d_1}$ such that $\dot{\mathcal{H}} = \{\dot{w}\} \supset C^\infty(\mathsf{R}_+ : \mathsf{R}^{d_1})$.

THEOREM 4.3. *Assume the conditions of Theorem 4.1 and suppose that ξ is independent of η. Then the unnormalized and the normalized conditional densities φ and p are random elements in $\mathsf{X} := C([0,T]; W_2^m(r,\Gamma)$ and*

$$\operatorname{supp} \mu_\varphi^{\mathsf{X}} = \overline{\mathcal{U}}_\varphi^{\mathsf{X}}$$

$$\operatorname{supp} \mu_p^{\mathsf{X}} = \overline{\mathcal{U}}_p^{\mathsf{X}} \qquad (\text{if } 1/r \in L_2)$$

where $\operatorname{supp} \mu_\varphi^{\mathsf{X}}$, $\operatorname{supp} \mu_p^{\mathsf{X}}$ *are the topological supports of the distributions of φ, p on X; and $\overline{\mathcal{U}}_\varphi^{\mathsf{X}}$, $\overline{\mathcal{U}}_p^{\mathsf{X}}$ are the closures of*

$$\mathcal{U}_\varphi := \{\varphi^w : w \in \mathcal{H}\}$$

$$\mathcal{U}_p := \{\varphi^w \left(\int \varphi^w(\cdot,x)dx\right)^{-1} : w \in \mathcal{H}\}$$

in X. (φ is the generalized solution of (4.5)–(4.6).)

PROOF: Apply Th. 3.4.

THEOREM 4.4. (Cf. [4]–[6].) *If $m \geq d/2 + n$, where n is a non-negative integer then Th. 4.3 holds with $\mathsf{X} := C^{0,n}$.*

PROOF: See Th. 3.5.

REMARK 4.5. The results of this section can be generalized to the case when b, σ, B, Σ depend also on t and on the observation y at time t (see [16]–[18]).

Theorems on supports for the nonlinear filter are obtained in [4]–[6] by using a different approach.

5. An application to a stochastic model of hydromagnetic dynamo

The evolution of the initial magnetic field H_0 in a flow of conducting fluid is described by the system of PDEs

$$\frac{\partial H^i}{\partial t} = \Delta H^i + v^i_{x^j} H^j - v^j H^i_{x^j} \qquad i = 1,2,3 \tag{5.1}$$

$$H(0,x) = H_0(x), \tag{5.2}$$

where $H^i(t,x)$ is the i-th coordinate of the magnetic field H at time t in $x = (x^1, x^2, x^3) \in \mathbf{R}^3$, and $v = (v^1, v^2, v^3)$ is the velocity of the fluid.

There are very interesting works devoted to investigations of phenomena, occuring in turbulent flows, where the velocity $v = v(t,x)$ is replaced in (5.1) by a "δ-correlated random field $v_\delta(t,x)$", which is considered for small δ as a model for turbulent flow, and the limit of the solutions H_δ is investigated for $\delta \to 0$. Here we replace v with random fields $v_\delta = (v^i_\delta(t,x))$ defined by

$$v^i_\delta(t,x) = \int_0^t b^{ik}(s,x)\,dW^k_\delta(s),$$

where $b = (b^{ik}(s,x))$ is a function of $(s,x) \in \mathbf{R}_+ \times \mathbf{R}^3$ and $\{(W^k_\delta) : \delta > 0\}$ is a good symmetric approximation for a d_2-dimensional Wiener process W.

THEOREM 5.1. ([15]–[16], [13], cf. [36].) *Let $r \colon \mathbf{R}^3 \to \mathbf{R}$ and $\rho \colon \mathbf{R}^3 \to \mathbf{R}$ be positive smooth functions such that $r, \Gamma := \rho I$ satisfy $(\overline{W})$. Assume that*

$$|D^\gamma b(t,x)| \le K|\rho(x)|^{1-|\gamma|}, \qquad |D^\gamma b'(t,x)| \le K|\rho(x)|^{1-|\gamma|}$$

for every multi-number $0 \le |\gamma| \le m+5$, where $b'(t,x) := \frac{\partial}{\partial t} b(t,x)$ and K is a constant. Suppose that $H^i_0 \in W^{m+3}_2(r,\Gamma)$ for $i = 1,2,3$. Then

$$\sup_{t \le T} |H^i_\delta(t) - H(t)|_m \to 0 \quad \textit{in probability,}$$

where H_δ is the generalized solution of (5.1)–(5.2) with v_δ in place of v, and H is the generalized solution of (5.1)–(5.2).

PROOF: Set $u_\delta(t,x,z) := z^i H^i_\delta(t,x)$ and $u(t,x,z) = z^i H^i(t,x)$ where $z = (z^1, z^2, z^3) \in \mathbf{R}^3$ is a new variable. Then u_δ is a generalized solution of

$$\frac{\partial}{\partial t} u_\delta(t,x,z) = \Delta_x u_\delta(t,x,z) + \dot W^k_\delta(t)(b^{ik}_{x^j}(t,x) - b^{jk}(t,x) u_{\delta x^j}(t,x)$$

$$u_0(0,x,z) = H^i_0(x) z^i,$$

and u is a generalized solution of (1.15)–(1.16) with respect to the product weights r' and Γ' defined by (1.17). Hence Theorem 5.1 follows from Corollary 2.4.

For further results concerning the topic of the last section we refer to the interesting paper by Rozovskii [36].

References

[1] J. S. Baras, G. L. Blankenship and W. E. Hopkins, Existence, uniqueness and asymptotic behaviour of solutions to a class of Zakai equations with unbounded coefficients, IEEE Trans. A. C. 28 (1983), 203–214.

[2] A. Bensoussan and J. L. Lions, Applications of variational inequalities in stochastic control, North-Holland Publ. Co., Amsterdam–New York–Oxford, 1982.

[3] Z. Breźniak, M. Capiński and F. Flandori, A convergence result for stochastic partial differential equations, Vol. 24, N. 4 (1988), 423–445.

[4] M. Chaleyat-Maurel and D. Michel, The support of the law of a filter in C^∞ topology, in Stochastic differential systems, stochastic control theory and applications. The IMA volumes in mathematics and its applications, Vol. 10, Springer Verlag, 1987.

[5] M. Chaleyat-Maurel and D. Michel, A Stroock–Varadhan support theorem in non-linear filtering theory, Probability Theory and Related Fields, 84 (1990), No 1, 119–139

[6] M. Chaleyat-Maurel and D. Michel, The support of the density of a filter in the uncorrelated case, in Stochastic Partial Differential Equations and Applications II, Proceedings Trento 1988, Lecture Notes in Mathematics, 1390 (1989).

[7] P. Cannarsa and V. Vespri, Existence and uniqueness results for a nonlinear stochastic partial differential equation, in: Stochastic Partial Differential Equations and Applications Proceedings, G. Da Prato and L. Tubaro (eds.), Lecture Notes in Math. 1236, pp. 1–24, Springer Verlag, 1987.

[8] W. H. Fleming and S. K. Mitter, Optimal control and nonlinear filtering for non-degenerate diffusion processes, Stochastics 8 (1982), 63–77.

[9] P. Florchinger, Zakai equation of nonlinear filtering with unbounded coefficients, to appear in the Proceedings of the Workshop on Stochastic Partial Differential Equations and Applications III, Trento (Italy) 1990.

[10] I. Gyöngy and N. V. Krylov, Stochastic partial differential equations with unbounded coefficients and applications I, Stochastics and Stochastics Reports, Vol. 32, (1990), 53–91.

[11] I. Gyöngy and N. V. Krylov, Stochastic partial differential equations with unbounded coefficients and applications II, Stochastics and Stochastics Reports, Vol. 32, (1990), 165–180.

[12] I. Gyöngy and N. V. Krylov, Stochastic partial differential equations with unbounded coefficients and applications III, to appear.

[13] I. Gyöngy and N. V. Krylov, On stochastic partial differential equations with unbounded coefficients, to appear in the Proceedings of the Workshop on "Stochastic Partial Differential Equations and Applications III" held in Trento, 1990.

[14] I. Gyöngy, On the approximations of stochastic partial differential equations I, Stochastics, Vol. 25 (1988), 129–164.

[15] I. Gyöngy, On the approximations of stochastic partial differential equations II, Stochastics, Vol. 26 (1989), 129–164.

[16] I. Gyöngy, The stability of stochastic partial differential equations and applications. Theorems on supports, in in Stochastic Partial Differential Equations and Applications II, Proceedings, Trento 1988, Lecture Notes in Mathematics 1390 (1989).

[17] I. Gyöngy, The stability of stochastic partial differential equations I, Stochastics and Stochastics Reports, Vol. 27 (1989) 129–150.

[18] I. Gyöngy, The stability of stochastic partial differential equations II, Stochastics and Stochastics Reports, Vol. 27 (1989) 189–233.

[19] G. Kallianpur and R. L. Karandikar, The nonlinear filtering problem for the unbounded case, Stochastic Processes and their Applications 18 (1981), 57–66.

[20] N. V. Krylov and B. L. Rozovskii, Ito equations in Banach spaces, Itogi nauki, Teor. verojatn. 14, Moscow (1979), 72–147 (in Russian).

[21] N. V. Krylov and B. L. Rozovskii, On Cauchy problem for linear stochastic partial differential equations, Math. USSR Izvestija 11 (1977), 1267–1284.

[22] N. V. Krylov and B. L. Rozovskii, On characteristic of degenerate parabolic Ito equation of second-order, in "Trudy Seminara i. I. G. Petrovskogo" V. 8 (1982), 153–168 (in Russian).

[23] V. Mackevičius, S^p stability of symmetric stochastic differential equations, Lietuvos Matematikos Rinkinys 25 (1985), 72–84 (in Russian).

[24] V. Mackevičius, On the support of the solution of stochastic differential equations, Lietuvos Matematikos Rinkinys 25 (1986), 91–98 (in Russian).

[25] E. Pardoux, Équations aux dérivées partielles stochastiques non linéaries monotones. Etude des solutions forte de type Ito. Thèse Université de Paris Sud, Orsay (1975).

[26] E. Pardoux, Stochastic partial differential equations and filtering of diffusion processes, Stochastics 3 (1979), 127–167.

[27] E. Pardoux, Sur les équations aux dérivées partielles stochastiques de type parabolique, in Colloque en l'honneur de L. Schwartz, Vol. 2, Astérisque 132, 71–87, SMF 1985.

[28] E. Pardoux, Equations du filtrage non linéaire de la prédiction et du lissage, Stochastics 6 (1982), 193–231.

[29] E. Pardoux, Filtrage non-linéaire et équations aux dérivées partielles stochastiques associées, École d'été de Probabilités de Saint-Flour 1989.

[30] O. G. Purtukhia, On the equations of filtering of multi-dimensional diffusion processes (unbounded coefficients), Thesis, Moscow, Lomonosov University 1984 (in Russian).

[31] O. G. Purtukhia, Innovation problem for degenerate diffusion processes (unbounded coefficients), Uspekhi Math. Nauk 39, No 4 (1984) (in Russian).

[32] O. G. Purtukhia, On the representation of the solution of the Cauchy problem, Abstracts of the XVIIIth School-Seminar in Probability Theory and Statistics, Bakhuriani, 1984, p. 40 (in Russian).

[33] O. G. Purtukhia, On the Cauchy problem for linear stochastic partial differential equations (unbounded coefficients), in: Stochastic Analysis and Asymptotic Problems in Probability Theory and Statistics, Metziereba, Tbilisi 1984, pp. 57–71 (in Russian).

[34] B. L. Rozovskii, Stochastic Evolution Systems, The theory of linear equations with applications to the statistics of stochastic processes. Nauka, Moscow, 1983 (in Russian).

[35] B. L. Rozovskii, To the mathematical Theory of Hydromagnetic Dynamo in Random Flow, Doklady Ak. Nauk USSR, No. 6 (1987) 1311–1314 (in Russian).

[36] B. L. Rozovskii, Some results on a diffusion approximation to the induction equation, to appear in the Proceedings of the Workshop on Stochastic Partial Differential Equations and Applications III, Trento (Italy) 1990.

[37] S. J. Sheu, Solutions of certain parabolic equations with unbounded coefficients and its applications to nonlinear filtering, Stochastics 10 (1983), 31–46.

[38] D. W. Stroock and S. R. S. Varadhan, On the support of diffusion processes with applications to strong maximum principle, Proc. Sixth Berkeley Symp. Math. Statist. Prob. III, (1972), 333–359, University of California Press.

DEVELOPMENTS IN PARAMETER BOUNDING

John P. Norton and Sándor M. Veres

School of Electronic and Electrical Engineering,
University of Birmingham, Edgbaston, Birmingham B15 2TT, U.K.

".... we demand rigidly defined areas of doubt and uncertainty".

Vroomfondel, the younger philosopher
The Hitch Hiker's Guide to the Galaxy, D. Adams

1. INTRODUCTION

Since the appearance of an earlier review of parameter-bounding for identification (Norton, 1987a), there has been rapid progress in bounding algorithms, experiment design, model-validation techniques and convergence analysis. The purpose of this paper is to note some of these developments.

The use of bounds to describe uncertainty in variables and model parameters is an alternative to statistical description. Bounds may often be employed when a statistical description would be unsatisfactory. Instances are when (1) records are too short for the noise or input statistics to be estimated adequately (Norton [31]); (2) model-output error is largely structural e.g. when non-linear or high-order, linear dynamics are approximated by a low-order, linear model; (3) disturbances are inhomogeneous, e.g. mainly sporadic changes; or (4) prior information on variables or parameters already exists in the form of bounds, e.g. on instrument errors, steady-state gain, pole positions, bandwidth. Bounds also allow definite statements to be made about single realisations, a valuable feature when the performance of a model or control system has to meet a rigid specification.

The basic idea of parameter bounding is simply to translate bounds on model-output error into bounds on model parameters, via the model structure (Walter and Piet-Lahanier [61]; Norton, [32], [33]).

For example, bounds $\pm r_t$ on the error e_t between an observation y_t and the output $\underline{\phi}_t^T\underline{\theta}$ of a model

$$y_t=\underline{\phi}_t^T\underline{\theta}+e_t \tag{1}$$

give a pair of hyperplanes

$$y_t-r_t\leq\underline{\phi}_t^T\underline{\theta}\leq y_t+r_t \tag{2}$$

which bound the parameter vector $\underline{\theta}$. Here $\underline{\phi}_t$ is a vector of explanatory variables known at time t. In this linear case, the parameter bounds due to a sequence of observations form a convex polytope. In general, the parameter bounds define a region in parameter space containing all values which do not conflict with the model structure, the specified error bounds, the observations or any prior bounds. These values make up the *feasible parameter set* $\mathcal{D}$. All model structures yielding non-empty $\mathcal{D}$'s are equally valid, as are all parameter values within $\mathcal{D}$. The use of bounds neither demands nor precludes the selection of an "optimal" model or parameter value; if an optimality criterion is employed, it is additional to, and separate from, the bounding process.

There are several widely used bounded-noise models:

(I) *nonlinear regression equation-error models*

$$y_t=f(\underline{\theta},\ y_{t-1},\ y_{t-2},\ \dots,\ u_{t-1},u_{t-2},\dots)+e_t,\qquad |e_t|\leq\delta$$

where f is a known nonlinear function, all samples of the input u and output y are assumed known exactly, and the equation-error term e_t has to account for both measurement and modelling errors.

(II) *nonlinear regression errors-in-variables models*

$$y_t=f(\underline{\theta},\ y_{t-1},\ y_{t-2},\ \dots,\ u_{t-1},u_{t-2},\dots)$$

where again f is a known nonlinear function but now the inputs u and outputs y are only known to within bounds

$$|y_k-y_k^o|\leq\delta_y,\qquad |u_k-u_k^o|\leq\delta_u,\qquad k=t,t\text{-}1,t\text{-}2,\dots$$

where y_k^o and u_k^o are the measured values.

(III) *nonlinear state-space equation-error models*

$$\underline{x}_{t+1}=\underline{f}(\underline{x}_t,\ \underline{u}_t,\underline{\theta})+\underline{e}_t$$

$$\underline{y}_t=\underline{h}(\underline{x}_t,\underline{\theta})+\underline{v}_t$$

where $\underline{f}$ and $\underline{h}$ are known nonlinear vector functions with parameters $\underline{\theta}$ and bounds are specified for the components of the noises:

$$|e_t^i|\leq\delta_e^i,\quad i=1,2,\dots,n;\qquad |v_t^i|\leq\delta_v^i,\quad i=1,2,\dots,m$$

In this state-space model the error bounds cover modelling error as well as unknown forcing and measurement errors.

(IV) *nonlinear state-space errors-in-variables models*

$$\underline{x}_{t+1}=\underline{f}(\underline{x}_t,\ \underline{u}_t,\underline{\theta})$$

$$\underline{y}_t=\underline{h}(\underline{x}_t,\underline{\theta})$$

where the inputs $\underline{u}$ and outputsy $\underline{y}$ are only known within error bounds

$$|y_k^i-y_k^{oi}|\leq\delta_y^i,\quad i=1,2,\dots,m;\qquad |u_k^i-u_k^{oi}|\leq\delta_u^i,\ i=1,2,\dots,r$$

So far the great majority of work on parameter bounding has considered models of type I. However, errors-in-variables models are now receiving some attention. As with their statistical or least-squares counterparts, parameter-bounding dynamical errors-in-variables models present substantial computational problems. For example, a linear-in-parameters model in which the same sample of an uncertain variable appears in the model at two or more sampling instants generally has nonlinear parameter bounds, even though the bounds due to measurements made at a single instant are linear. Another complication is that the feasible parameter set for such a model may consist of disjoint subsets, analogous to the multiple minima typical of least-squares errors-in-variables estimation problems.

Because the parameter-bounding approach is essentially deterministic, experiment design for parameter bounding is different in character from design for statistical parameter estimation. Specifically, the natural way to measure the goodness of a parameter-bounding experiment is on a worst-case basis, considering a single experiment, rather than by average behaviour over an ensemble of experiments. Nevertheless, thus far most work on experiment design for parameter bounding has in fact taken the traditional statistical attitude.

An important aspect of parameter-bounding identifaction is its potential influence on control synthesis. A worst-case predictor of the system output may be computed from the appropriate extrema of the feasible-parameter set, and a worst-case-optimal controller may be based on such a predictor. An attractive feature of parameter-bounding identification is thus that uncertainty in the model can be taken explicitly into account in prediction or control synthesis, yielding robustness in a very straightforward way. This point is discussed further in Section 8.

Historically, bounding methods first appeared in the control-enginering literature in connection with state bounding rather than identification [46,7]. There is close correspondence between parameter- and state-bounding algorithms, just as in least-squares or statistical estimation. The practically important problem of combined state and parameter bounding (which has obvious links with the errors-in-variables problem) has not yet received much attention.

In the following sections, analogies with statistical parameter estimation will often be evident. Section 2 briefly recapitulates the best-known bounding algorithms, then describes recent extensions and alternatives. Section 3 is a brief comparison of the ideas behind the statistical and parameter-bounding identification. Methods for time-varying parameters are discussed in Section 4. Section 5 covers experiment design for parameter bounding. Section 6 discusses bounding in the context of optimal algorithms theory, and Section 7 notes recent results on structure selection for bounded-parameter models. Section 8 discusses the place of parameter and state bounding in robust control design, including some application examples.

2. PARAMETER-BOUNDING ALGORITHMS

2.1 Linear models

The parallel hyperplanes (2) due to observation y_t confine the parameters of the linear model (1) to a strip

$$\mathcal{S}_t=\{\underline{\theta} \mid y_t-r_t \leq \underline{\phi}_t^T\underline{\theta} \leq y_t+r_t\} \tag{3}$$

so observations $\{y_1, y_2, \dots, y_N\}$ yield the feasible parameter set

$$\mathcal{D}_N = \mathcal{D}_0 \cap \left(\bigcap_{t=1}^{N} \mathcal{S}_t\right) \tag{4}$$

where the initial set $\mathcal{D}_0$ embodies any prior information. The number of vertices of $\mathcal{D}_N$ could soon become too large for updating to be practicable. However, the increase with N is often slow after an initial rapid rise (Mo and Norton [28]), and several recursive algorithms for updating the exact

polytope bound $\mathcal{D}_t$ have been devised (Broman and Shensa [8], [9]; Piet-Lahanier [40]; Walter and Piet-Lahanier [62]; Mo and Norton [28]; Mo [27]). They have much in common, but differ in how they characterize $\mathcal{D}_t$, and in the details of updating.
Many algorithms are now available to approximate $\mathcal{D}$ if it does get too complicated. Fogel and Huang [15] developed a recursive algorithm, based on a state-bounding algorithm of Schweppe [45], to update an ellipsoid $\mathcal{E}_t$ which contains $\mathcal{D}_t$. On receiving scalar observation y_t, the algorithm updates

$$\mathcal{D}_{t-1}=\left\{ \underline{\theta} \mid (\underline{\theta}-\underline{\theta}_{t-1})^{T}\, P_{t-1}^{-1}(\underline{\theta}-\underline{\theta}_{t-1}) \leq 1\right\} \tag{5}$$

by finding the minimum-volume ellipsoid $\mathcal{E}_t$ containing $\mathcal{E}_{t-1}\cap\mathcal{F}_t$. The algorithm, which strongly resembles recursive least squares, is economical but generally inefficient, in the sense that $\mathcal{E}_t$ is much larger than the actual feasible parameter set $\mathcal{D}_t$. A non-recursive alternative is to construct a box (Belforte and Milanese [6]) around $\mathcal{D}_N$ by linear programming (LP), maximizing and minimizing each element of $\underline{\theta}$ in turn subject to all the bounds (2). The box gives exact bounds on individual parameters, but no other indication of the shape of $\mathcal{D}_N$.

Various improvements of the basic algorithms have been devised. When one of the hyperplanes defining $\mathcal{F}_t$ cuts $\mathcal{E}_{t-1}$ but the other does not, $\mathcal{E}_t$ may be reduced in volume by replacing the nonintersecting hyperplane by the supporting (tangent) hyperplane parallel to it. The ellipsoid may be reduced further by reprocessing the observations (Belforte and Bona [1]). Deller has derived a recursive ellipsoid-updating algorithm along different lines. In the recursion he minimized $\mathrm{Vol}(\mathcal{E}_t)/\mathrm{Vol}(\mathcal{E}_{t-1})$ under the constraint of bounding a wheighted form of the errors, instead of looking for a minimum of $\mathrm{Vol}(\mathcal{E}_t)$ among the ellipsoids fitting to the intersection $\mathcal{E}_{t-1}\cap\mathcal{F}_t$. Thus he has obtained an algorithm different from that of Fogel & Huang [15] or Belforte & Bona [1] or Norton [32].

Another useful tactic is to construct a box bound $\mathcal{B}_N$ with faces orthogonal to the axes of the final ellipsoid $\mathcal{E}_N$ (the eigenvectors of P_N) (Mo and Norton [28]). This box is generally smaller than one aligned with the coordinate axes. Moreover, its faces include bounds on those linear combinations defined best and worst by the data (in the directions in which the ellipsoid is narrowest and widest).

Approximate polytope bounds may be constructed, at low computational cost, from the relatively few hyperplanes found to cut the ellipsoids during updating (Belforte and Bona [1]; Mo and Norton [28]). Limited-complexity polytopes such as simplexes and parallelotopes, i.e. boxes with parallel but generally non-orthogonal faces, have also been investigated (Mo and Norton, [28]). Another alternative is to construct an approximate polytope-bound from the hyperplanes found to be active in determining the extrema of $\mathcal{B}_N$. Preprocessing of the data with ellipsoidal bounding and than applying box-bounding for the active hyperplanes is dicussed in Belforte, Bona and Cerone [2].

A major aspect of comparison is : how tight are the computed bounds around the exact feasibiliy parameter region. From the recursive character of ellipsoidal and simplex bounding it follows that the broadness of the bounds is influenced by the recursive calculation and the final ellipsoid depends on the order of the samples. The computation of the smallest outer boxes does not suffer from this problem, but in cases of long feasibility regions the outer box can be rather rude approximation of the true feasibilty region. This points out the need for fast procedures to compute the exact polytopes in the linear in the parameters case, which we briefly discuss in the next section.

Most linear-in-the-parameters models considered in parameter bounding have had equation error independent of the parameter error. An exception is the ARMAX model in the form of equation (1) obtained from the rational transfer function model

$$Y(z^{-1})=\frac{B(z^{-1})}{A(z^{-1})}\,U(z^{-1})+\frac{C(z^{-1})}{A(z^{-1})}V(z^{-1}) \tag{6}$$

by way of

$$A(z^{-1})Y(z^{-1}) = B(z^{-1})\,U(z^{-1}) + C(z^{-1})V(z^{-1}) \tag{7}$$

giving

$$\underline{\phi}_t=[-y_{t-1}\ \ -y_{t-2}\ \dots\ \ u_{t-1}\ \ u_{t-2}\ \dots]^T,\ \ \underline{\theta}=[a_1\ \ a_2\ \dots\ \ b_1\ \ b_2\ \dots\ c_1\ \ c_2]^T \tag{8}$$

and

$$e_t = v_t + c_1 v_{t-1} + \dots + c_n v_{t-n} \tag{9}$$

The information on the error is likely to be

$$|\,v_{t-i}\,| \le r_{t-i}\,, \qquad i = 0,\,1,\,\dots,\,n \tag{10}$$

so the bounds on {e} are uncertain because a_1 to a_n are uncertain. The parameter bounds yielded by the observations turn out to be still piecewise linear but no longer convex (Norton [32]). They are, however, convex within any one orthant, and in practice the signs of a_1 to a_n are usually known a priori, at least for low-order models, from stability and sampling-rate considerations. The other peculiarity of this model is that the two bounding hyperplanes due to an observation are not parallel. Clement and Gentil [10]) point out that the ellipsoidal bounding algorithm can still be applied, inserting a supporting hyperplane parallel to each and performing two updates.

Optimal sampling design under bounded noise assumptions was investigated in Belforte et al. [4], [5] for the case of orthotope parameter bounding.

2.2 Non-linear models

Computation of $\mathcal{D}$ amounts to identifying which bounds are active and where they intersect. For models non-linear in their parameters, the intersections may be impossible to describe explicitly, the bounds may be non-convex and the feasible sets may be disjoint (Walter and Piet-Lahanier, [58]). Non-linear parameter bounds are usually found in one of two ways: by a Monte Carlo computation of points on the bounds, or by deterministic search along the bounds.

The simplest approach is to check, at a large number of trial parameter values, whether the model output error ever exceeds its specified bounds, i.e. whether the corresponding point is inside or outside $\mathcal{D}$ (Fedra, van Straten and Beck [14]). Several iterative adjustment techniques to improve resolution of the boundary of $\mathcal{D}$ have been suggested. Keesman [19] scales, translates and axis-rotates the region within which trial values lie. (Axis rotation can be viewed as a counterpart of principal-component analysis, an idea which will recur in model-structure selection later).

Belforte and Milanese [6] use LP to find a point $\underline{\theta}^{(i)'}$ on the bounds of a model linearized about the current $\underline{\theta}^{(i)}$. If $\underline{\theta}^{(i)'}$ proves feasible, it is taken as $\underline{\theta}^{(i+1)}$; if not, $\underline{\theta}^{(i+1)}$ is found by dichotomous search from $\underline{\theta}^{(i)}$ and $\underline{\theta}^{(i)'}$. Lahanier, Walter and Gomeni [22] present a more computationally economical search technique to find point on the boundary of $\mathfrak{D}$. First a starting point $\underline{\theta}^{(0)}$ is found by maximizing the number of observations at which the model-output error is within its bounds (stopping if all of them are). Searches in random directions from $\underline{\theta}^{(0)}$ then find pairs of boundary points in opposite directions about $\underline{\theta}^{(0)}$. Similar searches from each of the furthest-apart pair give more boundary points. The start of the next iteration is moved to the mean of that pair, if it is feasible. The algorithm is effective on highly non-linear boundaries, and provides for outliers as described later. Extra subsidiary searches from the extreme boundary points in each coordinate direction improve the efficiency of the algorithm (Walter and Piet-Lahanier [57,58]). Disjoint sections of $\mathfrak{D}$ may be found by repeatedly restarting the search, excluding all points within the range of feasible parameter values determined in earlier stages (Walter and Piet-Lahanier [58]).

A point on the boundary of $\mathfrak{D}$ may be computable exactly, without a search, by fixing all parameters but one, solving the bounding inequalities as equations for the remaining parameter and picking the dominant result. Repeating over a grid of fixed parameter values gives the whole boundary (Norton [31]). This technique is too expensive for large records or models and depends on explicit solution of the bound equations. It can handle disjointness of $\mathfrak{D}$ provided non-unique solutions are not missed.

A deterministic bound-exploration algorithm has recently been developed (Veres and Norton [51]) for nonlinear state-space equation-error models. The procedure finds and follows the boundary of 2-dimensional cross sections of the feasible parameter region. The search step size is automatically adjusted according to the boundary curvature. Nonconvex and disconnected feasible parameter regions can be handled. An extension of the method for higher dimensions builds up a picture of the bounds at points on a topological equivalent of a spherical grid in the dimensions beyond two.

2.3 Exact computations of feasibility regions

Sometimes simplification of $\mathfrak{D}$ is not essential to keep the amount of computing acceptable. It has been observed empirically that records $\{y_t, \underline{\phi}_t\}$ with noise $\{e_t\}$ whose probability density function (pdf) is flat or unimodal and symmetrical between the bounds give rise only infrequently to new active parameter bounds after an initial period of frequent new bounds, with the result that the exact bounding polytope $\mathfrak{D}$ can, in practice, be computed for some hundreds of observations when the number of parameters is modest. Three exact-polytope updating algorithms have been described (Broman and Shensa, [8], [9]; Piet-Lahanier [40]; Piet-Lahanier and Walter [41]; Mo and Norton [28]), differing in their compromises between economy in characterizing the polytope and ease in updating it. The first keeps a record of the polytope's faces of every dimension; it was developed for a target-tracking application (i.e. as a state estimator) in which the number of quantities to be bounded was modest. The last two methods keep record only of the vertices, edges and *p-1*-dimensional hyperplanes bordering the polytope. The method of Piet-Lahanier and Walter treats the set of inequalities (2) for a range of t as defining a cone $\mathcal{C}$ in the (p+1)-dimensional space of $[\underline{\theta}\ 1]^T$. Updating then requires a check whether the new bounds (still hyperplanes) cut $\mathcal{C}$, then interpolation of new edges of $\mathcal{C}$ (vertices of $\mathcal{D}$) between existing adjacent edges. Edges made redundant by the new bounds are then deleted. Adjacency and redundancy testing is susceptible to topological error caused by round-off in the arithmetic, although a partial cure has been found (Walter and Piet-Lahanier [62]). No such difficulty arises in the method of Mo and Norton, which stores $\mathcal{D}$ as a list of vertices, each vertex accompanied by a list of its adjacent vertices and a list of the faces containing it. Arithmetical adjacency tests are thus avoided, but the method needs more storage and more complex programming than the cone method. Broman and Shensa go a step further, by storing $\mathcal{D}$ as lists of vertices (0-faces), edges (1-faces), 2-faces, ... , (p-1)-faces. The results are very concise program code and numerically reliable updating, but heavy computing and large storage. Simulation shows that all the procedures are numerically stable. The minor differences between the last two are that the method of Piet-Lahanier is easier to implement while that of Mo and Norton needs less checking of the adjacency of bordering hyperplanes and vertices.

2.4 Outliers

One might ask that a parameter-bounding model be tested against all the data. Any clash between observations, prior constraints and the specified bounds on model-output error, reducing the feasible parameter set to zero, is then taken to show that the model or error bounds are invalid. Alternatively, any observation conflicting with the model or error bounds may be regarded as invalid. In either case the problem is to identify every critical observation, and is not straightforward (Norton [32]). The most effective tactic yet suggested (Lahanier, Walter and Gomeni [22]) is to identify the set of parameter values compatible with a specified proportion of the observations. It is arguably more satisfactory to define outliers as observations not fitted by such a parameter set than as observations inconsistent with the "true" model and noise properties. Such partially valid models may be helpful in modelling time-varying systems, with the outlier set varying as the model passes along the records.

3. COMPARISON WITH THE STATISTICAL APPROACH

Parameter bounds convey less information than would the likelihood function or posterior probability density function (pdf) so exclusive reliance on bounding does sacrifice performance in cases where extra knowledge could be used to compute the likelihood function or posterior pdf. However, it is common to assume Gaussian densities precisely because such extra knowledge is lacking; the information converged by a Gaussian pdf, i.e. mean and covariance, is closely comparable with or even less than that converged by bounds, depending on how complex the bounds are. Use of bounds alone loses the benefit of any extra statistical information (beyond mean and covariance) which would have to be replaced by extra bounds, e.g. on properties of noise values outside a specified range, if similar model performance were to be achived. On the other hand, known bounds on system behaviour, such as gain or damping are readily mapped into parameter bounds but are not conveniently added to statistical descriptions of parameter estimates.

Statistics		Parameter Bounding
	Assumptions	
prior noise distributions prior parameter distributions		prior noise bounds
	Methods of Inference	
calculations of statistical averages		calculation of feasible parameter region
	Result of Inference	
parameter estimates or posterior density		feasible parameter set

In bounding the conclusion is drawn only for the feasibility of the sample with a model structure. Observe that the logical structure of parameter bounding is simpler. In statistical inference more assumptions are required but also more information can be gained if the assumptions are correct. In technical terms, besides its simplicity, parameter bounding can be considered as a special case of statistical inference when applying uniformly distributed independent noise variables.

Parameter bounding consists of deterministic mapping from bounds on noise to bounds on parameters. It differs from statistical parameter estimation in the following respects:

(a) there is no attempt to derive a unique estimate, so there is no need to choose an estimate-ranking procedure (or performance index and optimization method). Furthermore, parameter bounds can be computed exactly in many realistic cases, so we are not forced to choose one of a number of competing algorithms right from the start and investigate its properties, as in statistical estimation.

In principle in statistics one could first compute the exact joint probability distribution of the parameters conditioned on the data, a mapping procedure, then as a separate exercise choose an algorithm to derive from it a unique parameter estimate. However, this is very seldom practicable. Alternatively, the conditional mean and covariance could be computed exactly. If the conditional mean and covariance are regarded as the final descriptors of the data, and not merely as estimates of the true parameter value and its reliability, one does not ask about the consistency and efficiency of an algorithm which computes them exactly. If so, the situation is similar to that in parameter bounding. *The properties of exact parameter bounds are only properties of the observations (more specifically, explanatory variables and noise), not joint properties of the observations and estimation algorithms as in statistical parameter estimation.* We have the option of computing simpler approximate parameter bounds, in which case the properties of the bounding algorithm must be examined, but in many practical cases we are not forced to do so. Whether or not we use such an algorithm, the behaviour of the exact bounds is a benchmark against which to measure approximate algorithms' performance, and must be acceptable if approximate algorithms are to have any chance of being acceptable. Efficiency of an approximate parameter-bounding algorithm can be measured as the ratio of the hypervolume of the feasible set it produces to that of the exact set.

(b) We are not obliged to consider a "true" parameter value and ask how close our results come to it. *We can compare models (on the basis of the extent and shape of the parameter bounds) even when we know that all of them differ from the process which generated the observations.* The noise includes any modelling error, and the noise bounds may reflect the explanatory power we require of the model, rather than any prior knowledge of the noise behaviour. If so, we need not ask what are the effects of mis-specifying the noise bounds. On the other hand, if we do wish to regard the observations as having been generated by a process of some specific type, we can. For instance, we can find conditions for parameter-bounding to be consistent, i.e. to exclude asymptotically all values but the true ones.

(c) Uncertainty is described by bounds without reference to any probability distribution, so *model validation is completely deterministic, and does not involve hypothesis testing.* Validation consists only of checking whether the feasible-parameter set is null, that is whether the model can fit the observations with the desired precision. *Testing the quality of the model is quite separate,* and consists of checking whether (i) the parameters are well enough determined (bounds tight enough), (ii) the model is parsimonious (explanatory power sufficiently superior to that of any simpler model), and (iii) the model-output errors have any unacceptable features (unforeseen, since foreseeable unacceptable behaviour can be prevented by including bounds on that behaviour in the computation of the parameter bounds). The whiteness test for model structure is a validation test, but the volume criterion is a quality test. Quality might be tested statistically, but doesn't have to be; it can be based entirely on bounds;

(d) *any conclusion drawn from the parameter bounds is specific to the records from which they derive.* It simply does not refer to what might have happened in other circumstances. To say how far the conclusion indicates what would happen in other circumstances, we should have to make assumptions, which might be statistical, worst-case (bounding) or both;

(e) in view of (d), we should as far as possible use performance criteria, for instance in experiment design, which relate to the current experiment only, not to an ensemble. On-line experiment design, based on what has happened so far in the present experiment, is therefore particularly attractive. The next-best thing (because it involves no extra assumptions) is to employ the assumed noise bounds in a worst-case design;

(f) *we are released from the necessity of considering asymptotic behaviour in order to make probabilistic analysis tractable.* If we are to carry out a general analysis of parameter-bound behaviour, not just valid for a single case, we must define a set of possible records. The set can be described by bounds (which may cover time structure as well as amplitude structure). These bounds are on explanatory variables and noise (not necessarily mutually independent). If we say nothing about the relative frequency of occurrence of explanatory-variable and noise values within the bounds, we can only obtain weak or trivial results, of little help in choosing input sequence, record length or parameterization. We can say more about the variables and noise, without resorting to statistics, by imposing bounds on the number of occurrences of values within specified ranges. Results relating bounds on the size and location of the parameter bounds to these ***histogram bounds*** should then be obtainable.

Note that histogram bounds are quite distinct from probability densities, and do not become a probability density as the number of histogram abscissae tends to infinity. First, the record length must tend to infinity if the number of abscissae does, so the limiting case does not arise in practice. Second, histogram bounds require that the record actually has the specified relative frequencies; they do not say that the record is a sample from a population with that density.

4. BOUNDING TIME-VARYING PARAMETERS

Recursive least-squares estimation algorithms for time-varying parameters suggest how to bound time-varying parameters. Three possibilities (Norton and Mo [37]) are fixed-window processing, use of a forgetting factor and use of an explicit model for the variation of the parameters. A fixed-window algorithm simply drops the influence of the observation L samples ago at the same time as picking up the current observation. In least squares this is easy, but in either ellipsoidal or polytope bounding it is not. In ellipsoidal bounding, the influence of observation t-L cannot easily be separated from those of later observations. In polytope bounding, removal of an active bound due to observation t-L may make formerly redundant bounds become active; finding the polytope resulting from removal of the old active bound, then reimposing the formerly redundant bounds, typically involves almost as much computing as constructing from scratch the bounding polytope due to the last L observations.

The ellipsoidal-bounding equivalent of using a forgetting factor is to expand the ellipsoid by a fixed factor before each update: P is multiplied by a scalar. It is less straightforward to apply a forgetting factor to a polytope. The distance between the bounding hyperplanes from observation t-i is proportional to the corresponding noise range, so the noise bounds are relaxed exponentially in time by increasing their range by a fixed factor at each observation instant. However, the relaxation alters the shape of the polytope, and perhaps the number of vertices. Moreover, bounds due to later observations which were not active before the expansion may become active. Some heuristic alternative to exponential expansion is essential if the computing demand is to be acceptable (Mo [27]).

If an explicit parameter-variation model is applied, such as a random walk (Norton [29]) or integrated random walk (Norton [30]), the parameter bounds have to be time-updated between observations. For a simple random walk model, the time update consists of forming the vector (Minkowski) sum of the prior feasible parameter set $\mathfrak{D}_{t-1}$ and a set of all possible increments, say $\mathcal{G}$, to yield $\mathfrak{D}_t$. If $\mathfrak{D}_{t-1}$ and $\mathcal{G}$ are polytopes, complexity rises rapidly, since many combinations of a

k-face of one and a $(p\text{-}1\text{-}k)$-face of the other give rise to a new $(p\text{-}1)$-face of $\mathcal{D}_t$. If $\mathcal{D}_{t-1}$ and $\mathcal{G}$ are ellipsoidal, their vector sum is not, and a possibly loose ellipsoidal outer bound has to be calculated. Chernous'ko [11] has provided an algorithm to do so. The time update is

$$P_t=(1+\alpha)\ P_{t-1}+(1+1/\alpha)\ Q_t \tag{11}$$

where Q_t is the describing matrix of $\mathcal{G}_t$ and α is the unique positive root of

$$\sum_{i=1}^{k}\frac{1}{\alpha+d_i}=\frac{k}{\alpha(1+\alpha)} \tag{12}$$

Here

$$D=\mathrm{diag}(d_1,\ d_2,\ldots,d_k) \tag{13}$$

is the diagonal matrix

$$D=A\ Q_t\ A^T \tag{14}$$

obtained by the same transformation that gives

$$I=A\ P_{t-1}\ A^T \tag{15}$$

It is easy to think of cases where this algorithm will give a loose outer bound on $\mathcal{D}_t$. For example, when $\mathcal{G}_t$ is a line, long compared with the width of $\mathcal{D}_{t-1}$ in the same direction (indicating that one linear combination of parameters varies rapidly), the vector sum of $\mathcal{D}_{t-1}$ and $\mathcal{G}_t$ is two half-ellipsoids connected by a long ellipsoidal prism.

5. EXPERIMENT DESIGN

Statistical D-optimal experiment design minimizes $|P|$, where P is the asymptotic covariance matrix of the parameter estimates. The volume $\sqrt{|P|}$ of an asymptotic confidence ellipsoid is thus minimized. The feasible parameter set plays a similar role to a confidence region, but parameter bounding does not rely on asymptotic considerations or on ensemble averaging. In V-optimal design (Pronzato and Walter [43]), the volume V of the feasible parameter set is minimized. When an ellipsoidal outer bound $\mathcal{E}_t$ is used, the volume of $\mathcal{E}_t$ is proportional to $\sqrt{|P_t|}$.

Volume V is a function of the observation errors, so $\hat{V}$, an upper bound on the actual V, is minimized. $\hat{V}$ is found in the linear case by substituting $\underline{\phi}_t^T\hat{\underline{\theta}}$ for y_t at each t, with $\hat{\underline{\theta}}$ a nominal value (of which $\hat{V}$ is independent). The resulting worst-case design may, of course, be far from optimal because $\hat{V}$ may be very different from V. The method can be generalized for non-linear models, by minimizing with respect to the experimental conditions the largest V over a prior feasible set $\mathcal{D}_0$, or maximizing the smallest value of the determinant of the Fisher information matrix over $\mathcal{D}_0$ (Pronzato and Walter [44]).

Design of sampling schedules for parameter bounding has been tackled in a similar way by Belforte, Bona and Ferdiani [4], [5] or Belforte, Bona and Cerone [3]. They compute the individual parameter uncertainty intervals by LP using $\underline{\phi}_t^T\hat{\underline{\theta}}$ for y_t at each t. The uncertainty intervals are found for a schedule which samples as often and as long as is allowed. The samples giving active parameter bounds are accepted as the optimal sampling schedule. The authors point out that the schedule obtained in this fashion may consist of too many samples, and depends on the parameterization.

6. PARAMETER BOUNDING AND OPTIMAL ALGORITHMS THEORY

One of the distinctive features of parameter bounding is the lack of any need to consider optimality; any feasible parameter value is as good as any other. Nevertheless, it is sometimes helpful to think about the optimality of estimators, such as predictors, based on feasible parameter sets, or even occasionally optimality of a point parameter value derived from the feasible set. (See Fig. 1 and Fig. 3 for the relations of estimation theory.) A general framework for doing so is optimal algorithms theory [Milanese and Tempo [26]; Tempo and Vicino [48]. The sort of optimality considered is minimax, and the algorithms are estimators of some function of the parameters. *Y-local optimality* minimizes, over the choice of estimators (algorithms), the worst-case error over the feasible parameter set for a specific realisation of the observations, and achieves this for every possible realisation. *F-local optimality* minimizes the worst-case error over the feasible output set corresponding to a specific value of the parameters, and does so for all possible parameter values. Global optimality is defined as minimization of the worst-case error over both all feasible outputs and all feasible parameters. Because this worst case is unique, whereas the worst cases in the definitions of local optimality are not, global optimality is a weaker property than local optimality. Optimality and near-optimality results are available for the Chebychev centre of the set of estimates given by the feasible parameter set, and for estimates based on the parameter values giving the best fit to the observed output in whatever norm is employed. For prediction, a Y-locally optimal algorithm consists of taking the mean of the extremes of the predictions over the feasible parameter set. Vicino, Tempo, Genesio and Milanese [55] have investigated its performance on some well known time series and found, in particular, that it is robust with regard to performance deterioration at larger prediction leads, compared with a number of statistical predictors.

Next we aim to give some more insight into the relation of parameter bounding to estimation within the framework of Traub et al. [50] created for optimal algorithms. An earlier version of the theory by Traub et al. [49] has been applied recently by Milanese and Vicino [25] to estimation theory under the assumption of bounded noise. First we briefly describe the theory of Traub et al. [50] and then give an example of how parameter bounding can fit into this framework.

Three basic abstract sets are considered by Traub et al. [50]:

- *the problem element set* F
- *the solution set* G
- *the data set* H

The operators acting between these sets are as follows. The *solution operator* is a mapping

$$S : F \times \mathbb{R}_+ \rightarrow 2^G,$$

where $\mathbb{R}_+ = [0,\infty)$ and 2^G denotes the family of subsets of G. It is assumed that the operator S has two properties:

$$S(f,0) \neq \emptyset, \qquad \forall f \in F$$
$$\delta_1 \leq \delta_2 \textit{ implies } S(f,\delta_1) \subset S(f,\delta_1)$$

for all $\delta_1, \delta_2 \in \mathbb{R}_+$ and all $f \in F$, *where* $\delta>$ is used to measure the modelling accuracy, which incorporates both modelling error and observation noise. Thus the only requirement is that an increase in $\delta>0$ should increase the solution set $S(f,\delta)$, *starting from a nonempty set* $S(f,0)$ for each problem element $f \in F$. The next operator is the *information operator, defined over* F and taking values in H, $N: F \rightarrow H$. The information operator N is called *partial* if N is many-to-one in F and *complete* if it is one-to-one in F. The concept of algorithm then appears as a map ϕ : $H \rightarrow 2^G$.

Let us now see how the theory applies to parameter bounding for the model

$$y_t = g(\underline{\theta}, y_{t-1}, \ldots, u_{t-1}, \ldots, t) + e_t \tag{16}$$

with noise bounds $|e_t| \leq r_t$, where g is assumed to be a deterministic function. G can be represented by the set of parameter vectors $\underline{\theta} \in \mathbb{R}^k$. The set of problem elements consist of the vectors composed from the coefficients of $\underline{\theta}$ and from the error terms: $f = (\theta^{\mathrm{T}}, e_{t_0}, \ldots, e_T)$. Finally, the data set is composed of $((u_{t_0}, y_{t_0}), \ldots, (u_T, y_T)) \in H$. The information operator N is the sampling of the system, resulting in $((u_{t_0}, y_{t_0}), \ldots, (u_T, y_T))$. Using these definitions of problem elements, sampling gives partial information for dynamical systems in the sense defined above. The solution operator can be defined as the set of feasible parameters

$$S(g,\delta) = \{\underline{\theta} : | y_t - g(\underline{\theta}, y_{t-1}, \ldots, u_{t-1}, \ldots, t) | \leq \delta,\ t \leq T\} = \mathfrak{D}_T .$$

Thus the theory of Traub et. al. [50] applies to this example with the modification that the results of algorithms are not parameter points but sets of parameter points and δ is a vector of positive real bounds.

7. STRUCTURE SELECTION FOR PARAMETER BOUNDING MODELS

Bounding simplifies model-structure selection conceptually. In least-squares estimation or probabilistic parameter estimation, selection is a compromise between model complexity and explanatory power. Statistical information criteria such as AIC, BIC, ϕ or F-testing weigh a measure of model quality against a threshold dependent on the number of parameters. In bounding this dual choice is split, by specifying in advance the explanatory power (output error bounds) then selecting the simplest model which meets the specification.

Four methods of structure selection for parameter-bounding models have been suggested. One (Veres and Norton [52]) picks the simplest model satisfying both the output-error bounds and bounds on the sample output-error autocorrelations for a range of lags. It is motivated by the wish to exclude too-structured errors, and could be adapted to other constraints on error behaviour. The remaining methods are based on the shape of the feasible parameter region or its approximations. The first selects the model giving the smallest reduction in volume from $\mathfrak{D}_0$ to $\mathfrak{D}_N$ (Veres and Norton [53]). Overparameterization ultimately leads to $\mathfrak{D}$ occupying little more than a subspace, and underparameterization to $\mathfrak{D}$ becoming empty; in both cases the volume soon becomes relatively small. The last two methods (Norton [35]) are suggested by principal-component analysis and singular-value decomposition.

8. CONTROL SYNTHESIS APPLYING PARAMETER BOUNDING

Conventionally, optimal estimation (of parameter or state) and control have been two distinct processes; an optimal model and state estimate has been computed, then usual as if it were exact in the optimal-control synthesis. In restricted circumstances (state estimation and control with linear dynamic, quadratic control cost, Gaussian noise) the overall result is optimal: *the separation principle*. In other cases (as minimum variance self tuning control), parameter estimates are used as if exact because of the lack of any effective way to take account of this uncertainty: *certainty equivalence*. The evaluability of parameter or state bounds offers the prospect of conservative control design taking full account of parameter or state uncertainty with acceptable computing demands. It also permits inclusion of known constraints on state and/or control straitforwardly as extra bounds.
In stochastic dual control the parameter uncertainties and control performance are balanced. There is a perspective to extend the ideas of dual control to the bounding circumstances by setting up an appropriate criterion function.

In parameter-bounding optimal control the separation principle applies differently. Since feasibile- parameter sets are used there is no use of estimated parameter values in place of true values in the controller. Instead of using single model-parameter values to calculate the optimal control input, optimization is carried out for the worst case over the feasible-parameter set. A performance guarantee (subject to correctness of the specified error bounds) can then be given, on a deterministic and not statistical basis.

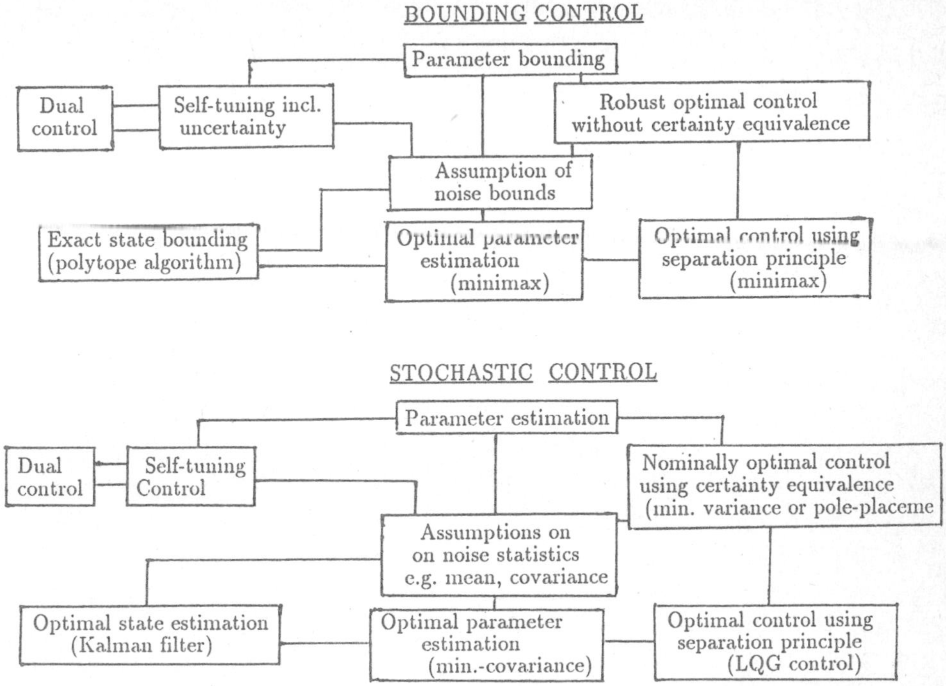

Fig 1. Parallels between bounding and stochastic identification and control

9. STATE BOUNDING AND ROBUST CONTROL

The earliest work in the control-engineering literature substituting simple noise bounds for statistical noise specifications (Schweppe [45], [46]; Bertsekas and Rhodes [7]) concerned state estimation. Schweppe [45] considered the bounded-output-error model

$$\dot{x}(t) = Fx(t) + Gu(t)$$

$$y(t)=Hx(t)+ v(t)$$

where F, G, H are known and $v(t)$ satisfies

$$v(t)^{\mathrm{T}}R^{-1}v(t) \leq 1$$

Bounds are specified for $u(t)$:

$$u(t_k)Q^{-1}u(t_k) \leq 1$$

or

$$\int_{t_k}^{t_{k+1}} u(t)^{\mathrm{T}}Q^{-1}(t)u(t)dt \leq C(t_{k+1}-t_k)$$

Initial states are in an ellipsoid:

$$S(t_o) = \left\{x : [x-\hat{x}(t_o)]^T\Sigma^{-1}(t_o)[x-\hat{x}(t_o)]\right\}$$

and the problem is to compute the smallest possible ellipsoid which includes the reachable state under the given conditions, i.e. an ellipsoidal outer bound for the feasible-state set. In Schweppe [45] an application to tracking of a manoeuvring target was mentioned. The target accelerations are the inputs to the dynamic system, and it is assumed that bounds on the accelerations are known. There are also known bounds on the observation errors. From these, bounds are computed on the state at each new sample instant by virtually the same ellipsoid-updating algorithm described earlier; the only substantial extra feature is time updating, which necessitates vector summing of two ellipsoids. In this and many other applications it is more realistic and safer to assume deterministic bounds on unknown inputs and error variables than to try to specify their statistical distributions. Stochastic estimation methods are vulnerable to mistaken or oversimplified statistivcal assumptions, while deterministic bounding techniques are vulnerable to mistaken values for the specified noise and error bounds. An advantage of deterministic bounding is that its assumptions and the conclusions drawn fron them are so simple that they are relatively easy to assess and if necessary correct or augment.

The same comments apply to parametric bounding in robust control, which is beginning to receive seriuos attention. For example, a control application of bounding methods is given by Parlos et al. [38]: multivariable control of a nuclear power plant. A linear, time-invariant model is assumed, in continuous-time, state-space form:

$$\dot{x}(t) = Ax(t) + Bu(t) + G\omega(t)$$
$$y(t) = Hx(t) + Du(t) + E\omega(t)$$
$$z(t) = Cx(t)$$

where $x(t)$, $u(t)$, $y(t)$ are n-, r- and m-dimensional state, control and output vectors, respectively, and $\omega(t)$ is a scalar disturbance satisfying

$$| \omega(t) | \leq B$$

The *output constraints* specified for the plant are of the form

$$y \in \Omega_{yb} = \{y: | y_i - y_{oi} | \leq S_i \ , \ i = 1,\ldots, m\},$$

and the *control is required to be in the set*

$$u \in \Omega_{ub} = \{u: | u_i - u_{oi} | \leq T_i \ , \ i=1,\ldots, r\},$$

wioth additionally *control-rate constraints*

$$\dot{u} \in \Omega_{\dot{u}b} = \{\dot{u}: | \dot{u}_i - \dot{u}_{oi} | \leq \dot{T}_i \ , \ i=1,\ldots, r\},$$

Under these conditions the set of reachable states can be proved to be included in a set

$$\Omega_x(t) = \{ x : (x-x_o)^T\Gamma^{-1}(t)(x-x_o) \leq 1 \}$$

where the positive-definite $\Gamma(t)$ satisfies a Riccati differential equation

$$\frac{d\Gamma(t)}{dt} = (A+BK)\Gamma(t) + \Gamma(t)(A+BK)^T + \beta(t)\Gamma(t) + GB^2G^T$$
$$\Gamma(0)=\Psi, \ \beta(t) > 0,$$

where K is an $r \times l$ feedback gain matrix: $u = Kx$, Ψ is a positive-definite matrix defining the ellipsoidal set of possible initial states and $\beta(t)$ is a free parameter. Since the output constraints can be converted into state constraints, the problem is to design a control system using at all times an admissible control $u \in \Omega_{ub}$ and control rate $\dot{u} \in \Omega_{\dot{u}b}$, while keeping the system states within specified constraints. The next problem is to find a control gain matrix K^* that maximizes the least feasible B. Parlos et. al. [38] show that this control problem can be reduced to an unconstrained nonlinear optimization, where first nonlinear models are written for the neutronics and the thermal-hydraulics part of the core as well as the rest of the primary loop. Then an actuator-sensor-augmented linearized plant model is created, to which the unconstrained nonlinear optimization (to optimize B) was applied. For further details see Parlos et al. [38].

10. FURTHER APPLICATIONS

The main reasons for the development of techniques for state and parameter bounding were the demands of some applications. This section records some applications for which bounding methods have proved to be more suitable than statistical methods with dubious prior probability distributions.

In many situations, such as the analysis of chemical reactions [57], outliers may occur in the observations. An approach computing feasible-parameter sets and feasible-output sets (model bars) has been developed by Walter & Piet-Lahanier [56] to allow assessment of how the model fits such data. The method is robust and gives the set of models fitting a specified proportion of the observations to within the specified accuracy. The same method has been applied in image processing to automated comparison of two images and also to image restoration by Piet-Lahanier [40]. The exact polytope algorithm presented in Piet-Lahanier [40] found successful application in film compensator design. Ellipsoidal bounding has been used successfully by Deller [12] in signal-processing applications. He emphasized the bounding algorithms' selectivity, incorporating only the informative samples.

Parameter bounding has been applied to biomedical models where the number of samples is very small by Norton [34]. In this work methionine tolerance-test responses were studied by nonlinear-in-the-parameters models. For small samples the parameter bounds of nonlinear models can be computed point by point and the results are more informative than a statistical approach necessarily based on a small sample.

This is far from an exhaustive list, and parameter bounding will find other applications where records are short, error statistics are unknown but the errors can be bounded, and where bounds can be imposed on modelling errors for the model to be usable. In many cases, however, where the system analysed is expected to behave in a stationary way, possibly after some transformations, statistical methods of systems modelling will still offer the best solution.

REFERENCES

[1] Belforte, G. and Bona, B. An improved parameter identification algorithm for signals with unknown-but-bounded errors. *Proc.7th IFAC/IFORS Symp. on Identification & System Parameter Estimation, 1985, York,1507-1512.*

[2] Belforte, G. , Bona, B. and Cerone, V. A bounded error approach to the tuning of a digital voltmeter. *Proc. 12th IMACS World Congr., 1988, Paris.*

[3] Belforte, G., Bona, B. and Cerone Optimal sampling design for parameter estimation: the bounded error approach. *Biomed Meas Infor Contr,* **2** (1988), 191-199.

[4] Belforte, G. , Bona, B. and Ferdiani, Optimal sampling schedule with replications for linear models with unknown but bounded errors. In: *Proceedings of the International Conference on System Research, Informatics and Cybernetics, 1984, Baden Baden.*

[5] Belforte, G. , Bona, B. and Ferdiani, S. Optimal sampling Schedule for Parameter Estimatio of Linear Models with Unknown But Bounded Measurement Errors. *IEEE Trans. on Autom. Control,* **AC-32** (1987), 179-182.

[6] Belforte, G. and Milanese, M. Uncertainty intervals evaluation in presence of unknown but bounded errors: nonlinear families of models. *Proc. 1st IASTED Symp. Modelling, Identification and Control, 1981, 75-79.*

[7] Bertsekas, D.P. and Rhodes, I. B. Recursive state estimation for a set membership description of uncertainty. *IEEE Trans. on Autom. Control,* **AC-16** (1971), 117-128.

[8] Broman, V. and Shensa, M.J. Polytopes, a novel approach to tracking. *Proc. of 25th Conference on Decision and Control, Athens, Greece, December 1986, 1749-1752.*

[9] Broman, V. and Shensa, M.J. A compact algorithm for the intersection and approximation of n-dimensional polytopes. *Proc. 12th IMACS World Congr., Paris, July 1988, 473-476.*

[10] Clement, T. and Gentil, S. Recursive membership set estimation for ARMAX models: an output-error approach. *Proc. 12th IMACS World Congr., Paris, July 1988, 484-486.*

[11] Chernous'ko, F. L. Optimal garanteed estimates of indeterminacies with the aid of ellipsoids. I-II, *Engin. Cybernetics,* **18** (1981)

[12] Deller, J. R. Set Membership Identification in Digital Signal Processing. *IEEE ASSP Magazine,* **6** (1989), no. 4., 4-20.

[13] Deller, J. R. A "systolic array" formulation of the optimal bounding ellipsoid algorithm, *IEEE Trans. Acoust., Speech, and Signal Process.,* **37** (1989), 1432-1436.

[14] Fedra, K., Van Straten, G. and Beck, M. B. Uncertainty and arbitrariness in ecosystems modelling: a lake modelling example, *Ecol. Model.,* **13** (1981), 87-110.

[15] Fogel, E. and Huang, Y. F. On the value of information in system identification–bounded noise case. *Automatica, 18 (1982), 229-238.*

[16] Happel, J., Walter, E. and Lahanier, H. Application to Heterogenous catalysis. In: *Identification of Biometric Models, ed. E. Walter, 1987, Pergamon Press.*

[17] Kacewicz, B.Z., Milanese, M. Tempo, R. and Vicino,A. Optimality of central and projection algorithms for bounded uncertainty. *Systems & Control Letters,* **8** (1986), 161-171.

[18] Keesman, K. J. A set-membership approach to the identification and prediction of ill-defined systems: application to a water quality system. *Doctoral Dissertation, University of Twente,*

1989, The Netherlands.

[19] Keesman, K.J. and Van Straten, G. Embedding of random scanning and principal component analysis in set theoretic approach to parameter estimation. *Proc. 12th IMACS World Congr., Paris, July 1988.*

[20] Kouvaritakis, B. and M. S. Trimboli A frequency response approach to the robust design of multivariable bounded-parameter models. *Proc. 12th IMACS World Congr., Paris, July 1988.*

[21] Kurzhanski, A. B. Identification – A Theory of Guaranteed Estimates. *IIASA Working Paper, July 1988.*

[22] Lahanier, H., Walter, E. and Gomeni, R. OMNE: A ne robust membership-set estimator for the parameters of nonlinear models. *J. of Pharmacocinetics and Biopharmaceutics,* **15** (1987), 203-219.

[23] Lahanier, H., Walter, E. and J. Happel A new approach for parameter estimation, applicatio to heterogenous catalysis. *Procedings of DICORD - 86, IFAC Symposium on Dynamic Control of Chemical Reactors and Destillatio Columns, 8-10 December, Bournemouth, UK.*

[24] Lozano-Leal, R. and Ortega, R. Reformulation of the Parameter Identification Problem for Systems with Bounded Disturbances. *Automatica,* **23** (1987), 247-251.

[25] Milanese, M. and Vicino, A. Estimation theory for nonlinear models and set membership uncertainty. *Automatica,* **27** (1991).

[26] Milanese, M. and Tempo, R. Optimal algorithms theory for robust estimation and prediction *IEEE Trans. Autom. Control,* **AC-30** (1985), no. 8.

[27] Mo, S. H. Parameter bounding identification algorithms for bounded-noise records. *IEE Proceedings,* **135** (1989), 127-132.

[28] Mo, S. H. and Norton, J.P. Recursive parameter-bounding algorithms which compute polytope bounds. *Proc. 12th IMACS World Congr., Paris, July 1988, 477-480.*

[29] Norton, J.P. Optimal smoothing in the identification of linear time-varying systems. *Proc. IEE,* **122** (1975), 663-668.

[30] Norton, J.P. Identification by optimal smoothing using integrated random walks. *Proc. IEE,* **123** (1976), 451-452.

[31] Norton, J. P. Problems of identifying the dynamics of biological systems from very short records. *Proc. 25th CDC Conf, 1986, Athens, Greece, 286-290.*

[32] Norton, J.P. Identification and application of bounded-parameter models. *Automatica, 23 (1987), 497-507.*

[33] Norton, J. P. Identification of parameter bounds for *ARMAX* models from records with bounded noise. *Int. J. Control,* **45** (1987), 375-390.

[34] Norton, J.P. Parameter bounding for biomedical models based on small sets of observations. *Biomedical Meas. Infor. Contr.,* **2** (1987), 101-107.

[35] Norton, J. P. Bounding techniques for model-structure selection. In: *Robustness in Identification and control, 1989, Plenum Press.*

[36] Norton, J. P. Recursive computation of inner bounds for the parameters of linear models. *Internat. Journal of Control,* **51** (1989).

[37] Norton, J. P. and Mo, S. H. Parameter bounding for time varying systems. *Proc. 12th IMACS World Congr., Paris, July 1988.*

[38] Parlos, A. G., Henry, A. F., Schweppe, F.C. , Gould, L. A. amd D.D. Lanning Nonlinear Multivariable Control of Nuclear Power Plants Based on the Unknown-but-Bounded Disturbance Model. *IEEE Trans. Automat. Contr.*, **AC-33** (1988), 130-137.

[39] Pearson, R. K. Approximate moel identification via set-theoretic estimation. *Proc. 4th American Control Conf, 1985, Boston.*

[40] Piet-Lahanier, H. Estimation de parametres pour des modeles a erreur bornee. *Doctoral Thesis, Universitè de Paris-Sud, Centre D'Orsay, 1987 .*

[41] Piet-Lahanier, H. and E. Walter Practical implementation of an exact and recursive algorithm for characterizing likelihood sets. *Proc. 12th IMACS World Congr., Paris, July 1988, 481-483.*

[42] Piet-Lahanier, H., Veres, S.M. and E. Walter On solutions of a set of linear inequalities for parameter bounding, 1990, *Manuscript available from authors.*

[43] Pronzato, L. and Walter, E. Robust experiment design via maximin optimization. *Math. Biosci.,* **89** (1988), 161-176.

[44] Pronzato, L. and Walter, E. Experiment design in a bounded-error context: comparison with D-optimality. *Automatica,* **25** (1989), 383-391.

[45] Schweppe, F. C. Recursive state estimation: Unknown-but-bounded errors and system inputs. *IEEE Trans. Automat. Contr.*, **AC-13** (1968)

[46] Schweppe, F. C. *Uncertain Dynamic Systems, 1973,* Englewood Cliffs: Prentice Hall.

[47] Smit, M. K. A novel approach to the solution of indirect measurement oriblems with minima error propagation. *Measurement,* **1** (1983) , 181-190.

[48] Tempo, R. and Vicino, A. Optimal algorithms for system identification: a review of some recent results. *Proc. 12th IMACS World Congr., Paris, July 1988.*

[49] Traub, J. F. and Woźniakowski, H. *A general Theory of Optimal Algorithms, 1980, Academic Press, New York.*

[50] Traub, J. F. , Wasilkowski and Woźniakowski, H. *Information, Uncertainty, Complexity, 1983,* Addison-Wesley, Reading, Massachusetts.

[51] Veres, S. M. and Norton, J. P. Parameter bounding identification by deterministic search. *Proc. of 9th IASTED, Innsbruck, February, 1990.*

[52] Veres, S. M. and Norton, J. P. Structure identification of parameter-bounding models by us of noise-structure bounds. *Internationa Journal of Control,* **50** (1989), 639-649.

[53] Veres, S. M. and Norton, J. P. Consistency and Volume Criterion for Structure Selection of Models. *IEEE Autom. Control,* **AC-36** (1989).

[54] Verhoof, J. W. and Smit, M. K. Error analysis for ellipsometry. *Proc. 12th IMACS World Congr., Paris, July 1988.*

[55] Vicino, A., Tempo, R., Genesio, R. and M. Milanese Optimal error and GMDH predictors: a comparison with some statistical techniques, *The Int. J. of Forcasting,* **2** (1987), 313-328.

[56] Walter, E. and Piet-Lahanier, H. Robust nonlinear parameter estimation in the bounded nois case. *Proc. 25th CDC Conf, 1986, Athens, Greece, 1037-1042.*

[57] Walter, E. and Piet-Lahanier, H. and J. Happel A new approach for parameter estimation, application to heterogeneous catalysis. *Procedings of DICORD - 86, IFAC Symposium on Dynamic Control of Chemical Reactors and Destillatio Columns, 8-10 December,*

Bournemouth, UK.

[58] Walter, E. and Piet-Lahanier, H. OMNE versus least squares for estimating the parameters o a biological model from short records. *Proc. 12th IMACS World Congr., Paris, July 1988.*

[59] Walter, E. and Piet-Lahanier, H. Estimation of parameter bounds from bounded-error data: survey. *Proc. 12th IMACS World Congr., Paris, July 1988 ,467-472.*

[60] Walter, E. and Piet-Lahanier, H. Estimation of the parameter uncertainty resulting from bounded-error data. *Math. of Biosci., 1989.*

[61] Walter, E. and Piet-Lahanier, H. Bounded-error alternative to least squares for estimating the parameters of a nonlinear model from short record. *In IMACS Transcations on Scientific Computations. Eds J Eisenfeld and D. Levine, J.C. Baltzer, J.C. Baltzer AG, Basel, 1989.*

[62] Walter, E. and Piet-Lahanier, H. Exact recursive plyhedral description of the feasible parameter set for bounded-error models. *IEEE Trans. Autom. Control,* **AC-34** (1989), 911-914.

APPENDIX

Time update of ellipsoidal parameter bound using explicit model of time variation.

For notational simplicity consider $\mathfrak{D}_{t-1}$ and $\mathfrak{G}_t$ to be centred at the origin, so that

$$\mathfrak{D}_{t-1} = \{\underline{\rho} \mid \underline{\rho}^T P_{t-1}\underline{\rho} \le 1\}, \quad \mathfrak{G}_t = \{\underline{\rho} \mid \underline{\rho}^T Q_t \underline{\rho} \le 1\}$$

Find the non-singular but not generally orthogonal matrix

$$A = N\,L\,M$$

where N is the modal matrix of $P_{t-1}^{-1/2} Q_t P_{t-1}^{-T/2}$, M is the modal matrix of P_{t-1},

$$L = \mathrm{diag}(\lambda_1^{-1/2}, \lambda_2^{-1/2}, \ldots, \lambda_k^{-1/2})$$

and λ_i is eigenvalue i of P_{t-1}. Matrix A simultaneously transforms $\mathfrak{D}_{t-1}$ to the unit ball centred at the origin and $\mathfrak{G}_t$ to an ellipsoid with axes aligned with the coordinate axes, i.e.

$$A\,P_{t-1}\,A^T = I, \quad A\,Q_t\,A^T = D$$

with D= $\mathrm{diag}\{d_1, d_2, \ldots, d_k\}$. The transformed $\mathfrak{D}_t$ is centred at the origin and diagonal, by symmetry, since it is the vector sum of two sets both symmetrical about every axis of the transformed space. Thus its describing matrix K is diagonal, and as $\mathfrak{D}_t$ is an ellipsoid K is positive-definite. Any vector on the boundary of $\mathfrak{D}_t$ must be at least as far from the origin as the sum of the distances of the boundary points of $\mathfrak{D}_{t-1}$ and $\mathfrak{G}_t$ from the origin in the same direction, so

$$\sqrt{(\underline{\psi}^T K \underline{\psi})} \ge \sqrt{(\underline{\psi}^T \underline{\psi})} + \sqrt{(\underline{\psi}^T D \underline{\psi})} \quad \forall\, \underline{\psi}$$

We can make $\mathfrak{D}_t$ touch the vector sum of $\mathfrak{D}_{t-1}$ and $\mathfrak{G}$ by setting the minimum of

$$x = \sqrt{(\underline{\psi}^T K \underline{\psi})} - \sqrt{(\underline{\psi}^T \underline{\psi})} + \sqrt{(\underline{\psi}^T D \underline{\psi})}, \quad \underline{\psi}^T\underline{\psi} = 1$$

to zero. Lagrangian minimisation shows that the minimum occurs at $\underline{\psi}_0$, where

$$\sqrt{(\underline{\psi}_0^T K \underline{\psi}_0)} = 1 + \sqrt{(\underline{\psi}_0^T D \underline{\psi}_0)}$$

and

$$\frac{K\underline{\psi}_0}{\sqrt{(\underline{\psi}_0^T K \underline{\psi}_0)}} - \frac{D\underline{\psi}_0}{\sqrt{(\underline{\psi}_0^T D \underline{\psi}_0)}} + \underline{\psi}_0 \sqrt{(\underline{\psi}_0^T D \underline{\psi}_0)} - \underline{\psi}_0 \sqrt{(\underline{\psi}_0^T K \underline{\psi}_0)} = \underline{0}$$

Denoting $\sqrt{(\underline{\psi}_0^T D \underline{\psi}_0)}$ by α,

$$K\underline{\psi}_0 = ((1+\tfrac{1}{\alpha})\ D + (1+\alpha)I)\underline{\psi}_0$$

Hence, with all the matrices diagonal and positive-definite, and each element of $\underline{\psi}_0$ generally non-zero,

$$K = (1+\tfrac{1}{\alpha})\ D + (1+\alpha)\ I$$

It remains only to minimize the volume of $\mathfrak{D}_t$ by choice of α (i.e. by choice of where it touches the vector sum of $\mathfrak{D}_{t-1}$ and $\mathfrak{G}$). Since

$$|\ P_t\ | = |\ A^{-1}K\ A^{-T}| = |\ A\ |^{-2}\ |\ K\ |$$

the volume of $\mathfrak{D}_t$ is minimized by minimizing $|\ K\ |$. Now

$$|\ K\ | = \prod_{i=1}^{k}(1+\tfrac{d_i}{\alpha})(1+\alpha) = (1+\alpha)^k \prod_{i=1}^{k}(1+\tfrac{d_i}{\alpha})$$

so, setting the derivative with respect to α to zero,

$$\sum_{i=1}^{k}\frac{d_i}{\alpha+d_i} = \frac{k\alpha}{1+\alpha}$$

As the left-hand side falls monotonically from p to 0 as α rises from 0 to ∞, and the right-hand side meanwhile rises monotonically from 0 to p , there is a unique positive root for α. Minor rearrangement of this equation gives

$$\sum_{i=1}^{p}\frac{1}{\alpha+d_i} = \frac{p}{\alpha(1+\alpha)}\ . \square$$

RECENT PROGRESS IN PARALLEL STOCHASTIC APPROXIMATIONS

G. YIN†

Department of Mathematics
Wayne State University, Detroit, MI 48202

ABSTRACT. Parallel recursive algorithms of stochastic approximation type are discussed. The basic idea is to utilize a collection of parallel processors in lieu of using a single one alone as in the classical algorithms. The processors compute and communicate with each other asynchronously. By asynchronous implementations, we mean that each processor updates on its own pace, and uses the most recently received information from other processors. Several procedures are dealt with. These include distributed communicating schemes via convexifications, parallel algorithms with interacting processors and renewal type of computation times, and real time implementable procedures with pipeline arrangements. It appears that from the closed connection between real world applications and mathematical theories, the study of parallel stochastic approximation methods gains not only much of its special appeal, but also much of its intrinsic tension. The main emphasis and interest are the asymptotic properties of the algorithms. The parallel interacting algorithm and its variation are used as representatives for illustrating various limiting results. Convergence and rate of convergence issues are addressed, and robustness analysis is carried out.

KEYWORDS. stochastic approximation, parallel processing, decentralized stochastic optimization, martingale averaging, convergence, rate of convergence, robustness.

1. Introduction

In this work, we survey some of the recent developments in parallel stochastic approximation methods. Several classes of recursive algorithms will be reviewed. Problems such as convergence, rate of convergence, robustness etc. will be addressed.

Stochastic approximation (S.A.) methods have a long and illustrious history. The methods were initiated in [RM] and [KW] for locating unknown zeros or maximum values of a nonlinear function with noisy observations. Substantial progress has been made in the past forty years. To date the developments, we mention the work of [Sa], [Lj], [KC], [LR], [MP], [Kus]. In fact, the list includes merely a few among others. Extensive surveys can be found in [AG], [Wa], [NK], [BMP] etc. The S.A. methods have been employed widely in sequential analysis, estimation, detection, system identification, Monte Carlo optimization, adaptive control, signal processing, communication theory and many other related fields (cf. [Fa], [Ve], [Rup], [We],

---

† This research has been supported in part by the National Science Foundation under grant DMS-8814624, and in part by the Institute for Mathematics and its Applications with fund provided by the National Science Foundation.

[Tsy], [Ge], [GRC], [LS], [Ch], [Ru], [Pf], [Po], [Pol], [KV], [Ca], [Er] and the references therein). Comparing with the rapid developments in S.A. methods, the study of parallel stochastic approximations is still in its infancy. In 1984, a class of recursive algorithms of stochastic approximation type, with distributed processors and asynchronous communications, was first proposed and analyzed in [Ts] (see also [TBA]). Some asymptotic results were obtained and various potential applications in stochastic control and system identifications were discussed. Such decentralized algorithms have been attracting growing interest.

The aforementioned model was studied further in [KY1]. Utilizing the weak convergence and martingale averaging techniques, convergence properties as well as rate of convergence were established under weaker conditions. Moreover, state dependent noise was treated, communication through noisy channels was dealt with and projection procedures were considered. Later, another class of parallel S.A. algorithms was suggested in [KY2]. Such algorithms utilize parallel processing and distributed computations in a natural way. Instead of using a single processor as in the classical setting, a collection of processors is used. Each processor acts on only part of the system vector. These processors compute and communicate with each other interactively and at generalized renewal times. Some interesting asymptotic theorems were obtained. Based on this model, the problem of finding zeros of a nonlinear function with noisy observations via parallel processing methods was treated in [YZ1], and the almost sure convergence was proved. Some robustness issues were considered in [YZ2] and [GZY]. Along another line, seeking for real time implementable procedures via pipelining of communication and computation, algorithms with "delayed" observations were considered in [ZY].

We would like to take a pause and answer the following questions: Why should one be interested in the study of parallel S.A. problems? Why is it useful and important? To answer these questions, we note that:

(1) Distributed algorithms exploit the opportunities provided by parallel processing for Monte Carlo methods of system optimization or evaluation.

(2) In various applications, the input data of the computation may be physically distributed (for example, the underlying computation consists of physically separated processors, estimators, trackers, controllers etc.).

(3) By means of parallel processing, one may be able to take advantage of state space decomposition. As a result, a large dimensional system may be split into small pieces such that each of the sub-system can be handled by one of the parallel processors.

(4) By employing and utilizing the parallel structure, stochastic approximation type of recursive procedures can be executed through pipelining of computation and communication.

Parallel computations seem to have advantageous in many situations. The list, in fact, is not exclusive. For instance, by means of parallel computing, the desired final result might be obtained much faster, the algorithms may be more efficient than the classical one and more robust with respect to large burst of errors. It seems that from the closed connection between real world applications and mathematical theories, the study of parallel stochastic approximation methods gains not only much of its special appeal, but also much of its intrinsic tension.

The rest of the paper will be arranged as follows: In the next section, several parallel algorithms of stochastic approximation type are reviewed; section 3 contains some convergence and rate of convergence argument for a distributed computation model under the framework of weak convergence; almost sure convergence and robustness problems are dealt with in section 4. Finally, the paper is closed with some concluding remarks in the last section.

To begin with, a word about the notations is in order. Throughout the paper, z' will denote the transpose of z, and $z_x(x)$ denotes the gradient of $z(x)$; similar notations will be used for the second derivatives. $|\cdot|$ stands for either an absolute value of a scalar, or a norm of a vector, which will be clear from the context.

2. Several distributed algorithms

We shall describe several parallel stochastic approximation algorithms, and make effort to indicate the salient feature of each algorithm. Although not all details are spelled out, appropriate references will be provided.

2.1. A convexification algorithm

The algorithm described in this section is due to Tsitsiklis Bertsekas and Athans [TBA] although it is a variation of the original version.

A good survey of decentralized and synchronous algorithms for traditional numerical analysis problem can be found in [Mi]. In many applications, *asynchronous* and decentralized algorithms may be more preferable (cf. [Ts] and the references therein).

Motivated by the work of [Be] and [CM], an iterative scheme was proposed in [Ts] and [TBA]. Let there be q parallel processors, each with a state variable of dimension r. Let X_n^i denote the state of processor i at time n and define $X_n = (X_n^1, \ldots, X_n^q)'$. We use X to denote a qr vector with $X = (X^1, \ldots, X^q)'$, where each $X^i \in \mathbb{R}^r$. At time n, processor i, $i \leq q$ decides whether or not to communicate the current value of its state to any other processors $j \neq i$, and takes an noisy observation $b_i(X_n^i, \xi_n^i)$. If processor i does not receive any information from j, for all $j \neq i$, then it updates according to

$$X_{n+1}^i = X_n^i + \varepsilon b_i(X_n^i, \xi_n^i)$$

as in the classical procedures. If there are any communications to processor i, then the communicated information must be weighted non-trivially. Let $\{a_{ji}(n)\}$ be a sequence of diagonal $r \times r$ matrices with nonnegative entries and $\sum_j a_{ji}(n) = I$, the identity matrix in $\mathbb{R}^r$. Note that the notation for $a_{ji}(n)$ is not standard. The subscripts of the matrix is of the form (source, destination)=(column, row). The algorithm is given by

$$X_{n+1}^i = \sum_j a_{ji}(n) X_n^j + \varepsilon b_i(X_n^i, \xi_n^i). \tag{2.1}$$

Notice that $a_{ji}(n)$ can be thought of as non-homogeneous Markov transition matrices. Nonetheless, the entries of the matrices are also random. Equation (2.1) simply says that the computations are executed by giving suitable weight to each processor through convexifications. Thus, if there is no communications to processor i, then $a_{ji}(n) = \delta_{ji}I$; otherwise $a_{ji}(n) \geq \alpha_0 I$ for some $\alpha_0 > 0$.

Remark: A model with bounded communication delays can be considered. The analysis will be essentially the same. In [TBA], the dimensions of X^i are not all the same. Such a case can be treated by using the same mathematical framework as presented here, however, the notations are more involved. We are using a constant gain in this work, a sequence of decreasing step-size can be incorporated in our formulation (cf. [KY1] section 8).

In the above model, processors do not need to communicate to each other at every time instance. In addition, they do not need to wait for receiving information being sent to them from other processors before doing the actual computations. Due to the asynchronous communication, some processors may compute faster than others while some may remain idle some of the time.

Write

$$A_n = \begin{pmatrix} a_{11}(n) & \dots & a_{q1}(n) \\ \vdots & \ddots & \vdots \\ a_{1q}(n) & \dots & a_{qq}(n) \end{pmatrix}.$$

Let

$$\Phi(n|k) = \begin{cases} A_n \dots A_k, & \text{for } k \leq n; \\ I_{qr}, & \text{for } k = n+1. \end{cases}$$

Under suitable conditions, it can be shown that $\lim_n \Phi(n|k) = \Phi_k$ exists w.p.1 uniformly in k, and for each $i \leq r$, all rows i, $i+r$, ..., $i+qr-r$ of Φ_k are equal. Assume that Φ_k satisfy a law of large numbers type of condition

$$\frac{1}{n} \sum_{k=m}^{n+m-1} E_m \Phi_k \xrightarrow{n,m} \bar{\Phi} \text{ in probability,}$$

where

$$\bar{\Phi} = \begin{pmatrix} \bar{\phi}_1 & \dots & \bar{\phi}_q \\ \vdots & \ddots & \vdots \\ \bar{\phi}_1 & \dots & \bar{\phi}_q \end{pmatrix} = \begin{pmatrix} \hat{\Phi} \\ \vdots \\ \hat{\Phi} \end{pmatrix},$$

$\bar{\phi}_i$ are the diagonal $r \times r$ matrices with diagonal elements $(\bar{\phi}_{i1}, \dots, \bar{\phi}_{ir})$, and $\sum_j \bar{\phi}_{ji} = 1$. In particular, if $r = 1$, then $\bar{\Phi}$ is a matrix with identical rows. This is expected due to the structure of $a_{ji}(n)$ (recall the corresponding results in Markov chains).

To analyze the asymptotic properties, we take a piecewise constant (continuous time) interpolation defined by

$$X(t) = X_n \text{ for } t \in [n\varepsilon, n\varepsilon + \varepsilon).$$

It was shown in [KY1] that under appropriate conditions, $X^\varepsilon(\cdot) \Rightarrow X(\cdot)$ (the symbol $\Rightarrow$ stands for weak convergence), with $X(\cdot) = (x(\cdot), \dots, x(\cdot))'$, such that $x(\cdot)$ satisfies an ordinary differential equation:

$$\dot{x}_i = \sum_{j=1}^{q} \bar{\phi}_{ji} \bar{b}_j^i(x), \; i \leq r, \tag{2.2}$$

where $\bar{b}_i(\cdot)$ denotes the average of $b_i(\cdot,\xi)$ and $\bar{b}_i^j(x)$ denotes its jth component.

In addition to getting the ODE limit, we showed that

$$X^\varepsilon(t_\varepsilon+\cdot)\Rightarrow\Theta=(\theta,\ldots,\theta)',\ \text{as } \varepsilon\to 0,\ t_\varepsilon\to\infty,$$

where θ is a unique asymptotic stable point of the limiting ODE (2.2).

Next, denote $\bar{B}(x)=(\bar{b}_1(x),\ldots,\bar{b}_q(x))'$. Consider the normalized error $U_n=X_n-\Theta/\sqrt{\varepsilon}$. Let $U^\varepsilon(\cdot)$ be a suitably interpolated sequence of U_n. We can show that $U^\varepsilon(\cdot)\Rightarrow U(\cdot)=(u(\cdot),\ldots,u(\cdot))'$ a Gauss-Markov process with $u(\cdot)$ satisfying the following stochastic differential equation

$$du=\hat{\Phi}\bar{B}_x(\Theta)udt+dw, \tag{2.3}$$

where $w(\cdot)$ is an r-dimensional Brownian motion. Interested readers are referred to [KY1] for details. In view of the above discussion, these results conform that all processors asymptotically come close together, and reach an agreement (cf. [WT]).

The essence of this algorithm is that a collection of parallel processors iterates on the same (or essentially the same) system, with perhaps different dynamics and noise processes. The processors share information and occasionally communicate with each other through convexifications. Some applications of such algorithm for on-line identifications were discussed in [Ts].

2.2. An interacting algorithm

The algorithm discussed in this section was suggested in [KY2]. It is quite natural and suitable for parallel processing. Instead of using a single processor alone as in the traditional approach, a collection of processors is used interactively.

Let X be an rp-vector. Let there be r-processors, such that the ith one controls X^i, with $X^i\in\mathbb{R}^p$. Without loss of generality, we may assume that $p=1$, i.e., each processor controls one component of the state vector. In our model, the computations are done asynchronously and at random times.

To be more precise, for each $i\le r$, let processor i takes Y_n^i units of time to complete its nth iteration. For notational simplicity, we assume Y_n^i to be integer-valued random variables. Define $\tau_n^i=\sum_{j=1}^n Y_j^i$ and $\tau_0^i=0$. These are "generalized renewal" type of random computation times. It is "generalized", since we do not require Y_n^i to be i.i.d. sequences. Let $\tilde{\xi}_n^i$ be the noise incurred in the nth iteration for processor i, and $X^i_{\tau_n^i}$ denote the value of the ith component of the state at the end of nth iteration of processor i, and $X_{\tau_n^i}=\left(X^1_{\tau_n^i},\ldots,X^r_{\tau_n^i}\right)'$. The algorithm is given by:

$$X^i_{\tau^i_{n+1}}=X^i_{\tau^i_n}+\varepsilon b_i(X_{\tau_n^i},\tilde{\xi}_n^i),\ i\le r. \tag{2.4}$$

To proceed, we need the following working definitions. Define the process $\{\xi_n^i\}$ by

$$\xi_n^i=\tilde{\xi}_j^i \text{ for } n\in[\tau_j^i,\tau_{j+1}^i),$$

and define

$$N_i(n) = \sup\{j; \tau^i_j \le n\}, \; \Delta^i_n = n - \tau^i_{N_i(n)}.$$

$\{\xi^i_n\}$ is a piecewise constant interpolation of $\{\tilde{\xi}^i_n\}$ in the "renewal interval" $[\tau^i_j, \tau^i_{j+1})$. As is the case of the usual renewal process, for each i, $\{N_i(n)\}$ is a counting process. It counts the number of computations being completed up to time n. Similarly, Δ^i_n are those "current life" processes. Define augmented vectors $\tilde{X}_n$ by

$$\tilde{X}_n = \left(\tilde{X}^1_n, \ldots, \tilde{X}^r_n\right)',$$

where $\tilde{X}^i_n = (X^1_{\tau^i_{N_i(n)}}, \ldots, X^r_{\tau^i_{N_i(n)}})'$. The notation appears to be pretty heavy, but it seems to be unavoidable. We have to keep track of all the state values used in the r-processors at each time n, these in general are different for different processors. Although each component $b_i(\cdot)$ still depends on a r-vector, such vectors are in general different from processor to processor due to asynchronous computation times.

Now, put

$$X_n = \left(X^1_{\tau^1_{N_1(n)}}, \ldots, X^r_{\tau^r_{N_r(n)}}\right), \text{ and } I^i_n = I_{\{\Delta^i_n = 0\}}.$$

In view of the above definitions, equation (2.4) can also be written as

$$X^i_{n+1} = X^i_n + \varepsilon b_i(\tilde{X}^i_n, \xi^i_n) I^i_{n+1}. \tag{2.5}$$

With slight abuse of notations, we write

$$b(\tilde{X}_n, \xi_n) = (b_1(\tilde{X}^1_n, \xi^1_n), \ldots, b_r(\tilde{X}^r_n, \xi^r_n))',$$

and

$$I_n = \begin{pmatrix} I^1_n & \ldots & 0 \\ \vdots & \ddots & \vdots \\ 0 & \ldots & I^r_n \end{pmatrix}.$$

Thus, (2.5) can in turn be written as

$$X_{n+1} = X_n + \varepsilon I_{n+1} b(\tilde{X}_n, \xi_n). \tag{2.6}$$

In the above algorithm, each processor takes a random time to complete its iteration. For different processors, the "renewal" times are different. This reflects the salient nature of the current algorithm, but it also brings about much of the difficulty. Due to the asynchronization, the iteration number can no longer be used as a common index of time indicator for all the processors. We thus have to use the elapsed processing time as the time indicator.

For more detailed motivations and discussions, we refer to the paper of [KY2]. This algorithm and its specialized version, which is described in section 2.3, will be used as archetype examples to illustrate various asymptotic approaches in the following sections. The analysis of asymptotic properties of the algorithm formulated in this section will be discussed in section

3 by using weak convergence methods; the algorithm described in section 2.3 will be studied in section 4.

2.3. A parallel RM type of algorithm

We concern ourselves with a specialized version of the algorithm given in section 2.2. The objective here is to locate the zeros of a function with random noise corrupted observations by RM type of algorithms with many processors.

The use of parallel processors comes from the following consideration. Suppose that the dimension of the underlying problem is large. Then, enormous memory and extensive computation would be needed. Instead of using the classical RM algorithms, we use a number of processors to solve the problem in a cooperative way. By means of state space decomposition, a large dimensional system can be partitioned into sub-systems with relatively small dimensions. Each processor iterates on only a small portion of the state vector. As a result, micro-computers may be used to carry out the computations.

As in the previous section, let $X \in \mathbb{R}^r$, and let there be r processors such that each processor controls one component of the vector X. Let $b(\cdot) : \mathbb{R}^r \mapsto \mathbb{R}^r$ be a function with $b(\cdot) = (b_1(\cdot), \ldots, b_r(\cdot))'$, such that $b_i(\cdot) : \mathbb{R}^r \mapsto \mathbb{R}$. The basic algorithm is given by

$$X^i_{\tau^i_{n+1}} = X^i_{\tau^i_n} + a_{\tau^i_n}(b_i(X_{\tau^i_n}) + \tilde{\xi}^i_n), \; i \leq r. \tag{2.7}$$

Similarly, we can derive other equations corresponding to (2.5) and (2.6), respectively. Starting from an arbitrary initial value, X^i is computed recursively based on the most recently determined values of X^j, $j \leq r$. The newly computed values are passed to all other processors. For simplicity, we assume that the passage of information does not involve any time delay. This does not seem to be restrictive. In fact, time delays can be incorporated into our model as long as they are bounded.

In order to get the almost sure convergent result, we shall consider a modified version of the algorithm. In proving w.p.1 convergence theorems of stochastic approximation type, normally, one assumes that either some compact set in the domain of attraction of an averaged ODE (ordinary differential equation) is entered infinitely often, or the function $b(\cdot)$ has certain growth rate. To ensure the boundedness condition to be satisfied, one often uses a projection or truncation scheme, which in turn requires some *a priori* information on the bounded region. To relax these conditions, a procedure with randomly varying truncations was suggested in [CZ]. Their result was further extended to the non-additive noise case in [YZ3].

Now, we describe the randomly truncated version of (2.7) as follows. Let $\{M(n)\}$ be a sequence of monotone increasing positive real numbers, such that $M(n) \xrightarrow{n} \infty$. Define σ_n as:

$$\sigma_0 = 0, \; \sigma_{n+1} = \sigma_n + I_{\{|X_n + a_n I_{n+1}(b(\tilde{X}_n) + \xi_n)| > M(\sigma_n)\}},$$

and write

$$J_n = I_{\{|X_n + a_n I_{n+1}(b(\tilde{X}_n) + \xi_n)| \leq M(\sigma_n)\}}$$
$$J^c_n = I_{\{|X_n + a_n I_{n+1}(b(\tilde{X}_n) + \xi_n)| > M(\sigma_n)\}},$$

where

$$b(\tilde{X}_n) = (b_1(\tilde{X}^1), \ldots, b_r(\tilde{X}^r_n))',$$

and I_n, ξ_n are similar to the definitions in section 2.2. The algorithm with random truncation bounds is:

$$X_{n+1} = \left(X_n + a_n I_{n+1}(b(\tilde{X}_n) + \xi_n)\right) J_n + \hat{x} J^c_n, \tag{2.8}$$

where $\hat{x} \in \mathbb{R}^r$ is a fixed but otherwise arbitrary point, and $|\hat{x}| < M$ for some $M > 0$.

Remark: It can be easily seen that the truncations are so designed that they force the iterates to return to a bounded region whenever they exit. The main advantage of such truncations is that the usual restriction on the growth rate of the function $b(\cdot)$ can be discarded completely. In contrast with the traditional projection or truncation methods, no *a priori* information on the projection region is needed. To some extend, the random truncation bounds provide a way to construct an infinite return environment. In addition, the implementation of the procedure can be easily done.

2.4. Pipelining of computation and communication

A real time implementable algorithm is discussed in this section. We start with the classical RM algorithm

$$X_{n+1} = X_n + a_n O_n, \; O_n = b(X_n) + \xi_n, \tag{2.9}$$

where $X, \xi \in \mathbb{R}^r$ and $b(\cdot) : \mathbb{R}^r \mapsto \mathbb{R}^r$. The above procedure can be thought of as a two phase computations. The first phase is taking observations O_n, and the second phase is forming new approximations by adding X_n to $a_n O_n$.

In a wide range of applications, most of the computation time is spent on the phase 1 operation. The second phase is normally less time consuming, especially, for the case when we have very large dimensional systems, or the places where the data collected are far away from the data processing center. This feature cannot be seen from (2.9) owing to the fact that n is the iteration number rather than the real time elapsed in one iteration.

Taking the aforementioned fact into account, we shall design another algorithm in this section. A number of processors are lined up as on a production line, and update interactively. After one iteration is completed, the available data are passed to the next processor on the line.

To give some motivations, we consider the following example first. Suppose that the time required for taking a measurement is 3 units of real time, and the time needed for the addition of X_n and $a_n O_n$ is 1 unit. Instead of using a single processor alone to implement the computation, four parallel processors will be employed in a pipeline fashion. All processors perform the same kind of computation, but with different starting times. At a given instance n, suppose that processor 1 has completed its measurement taken at the time of three units ago, i.e., O_{n-3} has been obtained, and suppose that X_n is available which was communicated by processor 3. Then, step 2 is executed by processor 1, i.e.,

$$X_{n+1} = X_n + a_{n-3} O_{n-3}.$$

At the next time instance $n+1$, processor 1 sends its content X_{n+1} to processor 2. For simplicity, we assume that the passage of information is instantaneous, i.e., there is no time delay (again, bounded delays can be considered in the model). By now, processor 2 has completed its measurement started at $n-2$. Then,

$$X_{n+2} = X_{n+1} + a_{n-2}O_{n-2}$$

is obtained, and the result X_{n+2} is passed to processor 3, etc.

Now, we are in a position to design the algorithm in a more general setting. In view of the above illustration, the overall system consists of a number of parallel processors connected through asynchronous communications. A sequence of estimates is generated by using a number of processors cyclically and periodically. The length of a cycle is equal to the number of processors participating the computation, which is also equal to the time required for a single processor to complete its phase one operation.

To be more specific, let d be a positive integer, which is the unit of time for a single processor to complete phase one operation. $d+1$ identical processors are selected. All processors work on the same system vector X interactively, and communicate with each other through pipeline structure. In spite of different time indices, the schemes are the same for all the processors taking part in the computation. Thus, it suffices to use a single formula to describe the procedure, i.e., all d processors iterate according to the same format, but with different starting time. The initial conditions X_0, X_2,..., X_{d-1} are given to all processors. In place of (2.9), we consider

$$X_{n+1} = X_n + a_{n-d}O_{n-d} \tag{2.10}$$

with $O_n = b(X_n) + \xi_n$. Equivalently, we can write

$$X_{n+1} = X_n + a_{n-r}b(X_{n-d}) + a_{n-d}\xi_{n-d}.$$

Some of the crucial questions are: Whether or not the algorithm actually works, and do the convergence properties and asymptotic normality still hold? These questions were answered in [ZY]. In that paper, we showed that X_n converges almost surely to the set $E = \{x \in \mathbb{R}^r; b(x) = 0\}$. Assuming further that $E = \{\theta\}$ a singleton set, $a_n = \frac{1}{n}$, and $b(x) = H(x-\theta) + o(|x-\theta|)$, we then established the asymptotic normality, i.e., $\sqrt{n}(X_{n+1} - \theta)$ converges in distribution to a normal random variable. This is done, by virtue of an functional invariant theorem in [EK] and examining the interpolation

$$M_n(t) = \frac{[nt]}{\sqrt{n}} \sum_{k=1}^{[nt]} \frac{1}{k} A_{[nt],k}\xi_{k-d}, \text{ for } t \in [0,1],$$

where $[z]$ denotes the integral part of z, and

$$A_{jk} = \begin{cases} \prod_{l=k+1}^{j} \left(I + \frac{1}{l}H\right), & \text{if } j > k; \\ I, & \text{otherwise.} \end{cases}$$

To some extent, the algorithm discussed in this sub-section can be viewed as an aggregation method. It uses extensive concurrency by means of pipelining of computation and communication for each individual processor. The computing times (for both phase 1 and phase 2) are uniformly across processors because we have to keep load balancing in our mind (cf. [BT]). Moreover, the fact that when one processor transfers its partial results to another processor, all other processors, including the one being communicated, are still in operation, makes the communication penalty relatively insignificant. As a result, the communication penalty will not prevent the concurrent utilization of a large number of processors in parallel when the underlying computing task is large. If the classical procedure is used with a single processor, then a certain amount of time must be spent on the waiting for the required data to become available. In the newly developed algorithm, at any given time, there is always one processor doing the phase 2 computation (addition), and all the others are in phase 1–the data collection mode. Proceeding in this way, the amount of real time required for the completion of one iteration is reduced from $d+1$ units to one unit only. It seems that the regular pipeline and the "delayed" observation structure provides a way to speed up the underlying computation.

3. The study of asymptotic properties via the weak convergence methods

In this section, we review some asymptotic results by means of the weak convergence methods. We shall concentrate on the algorithm described in 2.2. This algorithm is served as a representative for illustrating the weak convergence approach. In fact, we shall try to make the main idea clear without giving much of the technical details. For verbatim proofs, we refer to the papers [KY1], [KY2].

3.1. ODE limit

To study the convergence properties of algorithm (2.2), we define the continuous time interpolation $X^\varepsilon(\cdot)$ by $X^\varepsilon(t) = X_n$ on $[n\varepsilon, n\varepsilon+\varepsilon)$. In this section, we treat the weak convergence of $\{X^\varepsilon(\cdot)\}$ as $\varepsilon \to 0$. The weak convergence methods have been found very useful for studying many recursive stochastic algorithms (cf. [KH], [KS], [Kus] and the references therein). We shall use various notion of weak convergence freely. Consult the work of [Bi], [Kus] and [EK] among others, for references of definitions, notations, basic facts and theorems.

The weak convergence arguments usually require that first tightness of the underlying sequence be established, and then the limit process be identified or characterized. Roughly speaking, tightness is equivalent to sequential compactness in any complete separable metric space. This is known as the Prohorov's Theorem (cf. [Pr]).

For simplicity, we shall consider the case that the noise processes $\{\tilde{\xi}_n^i\}$, $i \le r$ are "exogenous" in this paper. For more general state dependent noise cases, see the treatment of [KY2] (cf. also [KS]).

We assume that:

(A.1.1) For each $i \le r$, $b_i(\cdot,\xi)$ is continuous in x, uniformly on each bounded (x,ξ) set. For each $N<\infty$, $\sup_{|x|\le N}\{|b_i(x,\tilde{\xi}_n^i)|; i,n\}$ is uniformly integrable.

(A.1.2) $\{Y_n^i\}$, $i \le r$ are bounded, and there are $\tilde{\mu}_i(x,\xi)$, continuous in x uniformly in each bounded (x,ξ) set, such that $E^\varepsilon_{\tau_n^i} Y_n^i = \tilde{\mu}_i(X_{\tau_n^i}, \tilde{\xi}_n^i)$, where $E^\varepsilon_{\tau_n^i}$ denotes the conditioning on the σ-algebra generated by $\mathcal{F}^\varepsilon_{\tau_n^i}$.

(A.1.3) For each $i \le r$, there exist continuous functions $\bar{b}_i(\cdot)$ and $\mu_i(\cdot)$, such that for each x,

$$\frac{1}{n}\sum_{j=m}^{m+n} \tilde{E}^i_m b_i(x,\tilde{\xi}^i_j) \xrightarrow{n,m} \bar{b}_i(x) \text{ in probability,} \tag{3.1}$$

and

$$\frac{1}{n}\sum_{j=m}^{m+n} \tilde{E}^i_m \tilde{\mu}_i(x,\tilde{\xi}^i_j) \xrightarrow{n,m} \mu_i(x) \text{ in probability.} \tag{3.2}$$

Remark: The uniform integrability in (A.1.1) is a standard condition in weak convergence methods. The continuity of $b_i(\cdot,\xi)$ can be replaced by a "weak" continuity (cf. [KS], [Kus] and [KY2]) condition. (A.1.2) allows the renewal interval to depend on both the state variable and the noise. (A.1.3) is the main averaging assumption. Eqn. (3.1) concerns about the average on the noise and Eqn. (3.2) takes care of the average on the random computation time.

From the definition of the interpolated process $\{X^\varepsilon(\cdot)\}$, $X^\varepsilon(\cdot) \in D^r[0,\infty)$ the space of functions that are right continuous and have left-hand limits endowed with the Skorohod topology (cf. [Bi], [Kus], [EK] and the references therein).

Theorem 3.1. *Suppose that (A.1.1) to (A.1.3) are satisfied and suppose the ODE (3.3) has a unique solution on $[0,\infty)$ for each initial condition $x = (x_1,\ldots,x_r)'$,*

$$\dot{x}_i = \frac{\bar{b}_i(x)}{\mu_i(x)},\ i \le r. \tag{3.3}$$

If $X^\varepsilon(0) \Rightarrow x(0)$, then $\{X^\varepsilon(\cdot)\}$ is tight, and the limit of any convergent subsequence satisfies Eqn. (3.3).

Idea of proof: Without loss of generality, we may assume that X_n^i are bounded, otherwise, we can always use a truncation device as in [Kus]. It can be shown that $\{X^\varepsilon(\cdot)\}$ is tight and all limits are uniformly Lipschitz continuous. We thus need only characterize the limiting process. This can be divided into the following steps:

(1) We show that a limit theorem holds for the random computation time. For each $i \le r$, define Z^ε_{ni}, $Z^\varepsilon_i(\cdot)$, B^ε_{ni} and $B^\varepsilon_i(\cdot)$ by

$$Z^\varepsilon_{ni} = \varepsilon\sum_{j=1}^{n} Y^i_j,\ Z^\varepsilon_i(t) = Z^\varepsilon_{ni} \text{ on } [n\varepsilon, n\varepsilon+\varepsilon)$$

$$B^\varepsilon_{ni} = \varepsilon\sum_{j=1}^{n} b_i(X_{\tau^i_j}, \tilde{\xi}^i_j),\ B^\varepsilon_i(t) = B^\varepsilon_{ni} \text{ on } [n\varepsilon, n\varepsilon+\varepsilon).$$

The set $\{X_i^\varepsilon(\cdot), Z_i^\varepsilon(\cdot), B_i^\varepsilon(\cdot),\ i \le r\}$ is tight in the space $D^{3r}[0,\infty)$ and all limits are uniformly Lipschitz continuous. Extract a convergent subsequence, also indexed by ε and with limit $(x_i(\cdot), Z_i(\cdot), B_i(\cdot)), i \le r$.

(2) We claim that for each $i \le r$,

$$Z_i(t) = \int_0^t \mu_i(x(Z_i(u)))du. \tag{3.4}$$

To verify this, we need only show that

$$M_i(t) = Z_i(t) - \int_0^t \mu_i(x(Z_i(u)))du$$

is a martingale. Since $M_i(\cdot)$ is Lipschitz continuous, the martingale property will imply that $M_i(t) =$ constant. Then $M_i(0) = 0$ yields that $M_i(t) \equiv 0$ and (3.4) holds.

In order to verify the martingale property, we show that for each bounded and continuous function $h(\cdot)$, arbitrary integer ν and $t_j < t < t+s$,

$$Eh(x(Z_i(t_j)), Z_i(t_j), j \le \nu)\left(Z_i(t+s) - Z_i(t) - \int_t^{t+s} \mu_i(x(Z_i(u)))du\right) = 0.$$

To begin with, we examine the process $Z_i^\varepsilon(\cdot)$ and show that the limit satisfies the above equation.

Choose a sequence $n_\varepsilon \to \infty$ as $\varepsilon \to 0$, such that $\varepsilon n_\varepsilon = \delta_\varepsilon \to 0$. It can be shown that

$$\begin{aligned} 0 &= \lim_\varepsilon Eh(X^\varepsilon(Z_i^\varepsilon(t_j), j \le \nu) \\ &\times \left(Z_i^\varepsilon(t+s) - Z_i^\varepsilon(t) - \delta_\varepsilon \sum_{l=t/\delta_\varepsilon}^{(t+s)/\delta_\varepsilon} \frac{1}{n_\varepsilon} E^\varepsilon_{\tau^i_{ln_\varepsilon}} \sum_{j=ln_\varepsilon}^{ln_\varepsilon+n_\varepsilon-1} \tilde{\mu}_i(X^\varepsilon(Z_i^\varepsilon(\varepsilon j)), \tilde{\xi}_j^i)\right). \end{aligned}$$

Letting $l\delta_\varepsilon \to u$, and by virtue of the weak convergence of $Z_i^\varepsilon(\cdot)$ and the Skorohod imbedding (cf. [Sk]), we need only show

$$\frac{1}{n_\varepsilon} E^\varepsilon_{\tau^i_{ln_\varepsilon}} \sum_{j=ln_\varepsilon}^{ln_\varepsilon+n_\varepsilon-1} \tilde{\mu}_i(X^\varepsilon(Z_i^\varepsilon(\varepsilon j)), \tilde{\xi}_j^i) \xrightarrow{\varepsilon} \mu_i(x(Z_i(u))) \text{ in probability.}$$

The details can be found in [KY2].

(3) By the same token, we can show

$$B_i(t) = \int_0^t \bar{b}_i(x(Z_i(u)))du.$$

(4) Owing to the definition of the interpolations $X_i^\varepsilon(\cdot)$,

$$X_i^\varepsilon(t) - X_i^\varepsilon(0) = \varepsilon \sum_{j=0}^{N_i(t/\varepsilon)-1} b_i(X_{\tau^i_j}, \tilde{\xi}_j^i) = \varepsilon \sum_{\varepsilon j=0}^{Z_i^{\varepsilon,-1}(t)} b_i(X^\varepsilon(Z_i^\varepsilon(\varepsilon j)), \tilde{\xi}_j^i),$$

where $Z^{\varepsilon,-1}(\cdot)$ and later on $Z_i^{-1}(\cdot)$ denote the inverse of $Z^\varepsilon(\cdot)$ and $Z_i(\cdot)$, respectively (the existence of these inverses is guaranteed by the conditions of Y_n^i and the definition of $Z_i^\varepsilon(\cdot)$). Combining the above results (1)-(3), we have the following

$$x_i(t) = x_i(0) + \int_0^{Z_i^{-1}(t)} \bar{b}_i(x(Z_i(u)))du.$$

This yields the desired result. A few details are omitted.

3.2. Some order of magnitude estimates

To proceed, we shall assume that there is a point θ which is globally attracting for the ordinary differential equation (3.3) and is stable in the sense of Liapunov.

Owing to the Skorohod topology of $D^r[0,\infty)$, the weak convergence provides information on arbitrarily large but bounded intervals. For many applications, it is very useful to have a convergence result which is uniform in t. To proceed further, the following asymptotic estimates are of particular importance.

Suppose that there is a twice continuously differentiable Liapunov function $V(\cdot)$ for (3.3), such that $V(\cdot) > 0$ for $x \neq \theta$, $V(\theta) = 0$; $V(x) \to \infty$ as $|x| \to \infty$; $V_{xx}(\cdot)$ is bounded;

$$\sum V_{x_i}(x)\bar{b}_i(x)/\mu_i(x) < -\lambda V(x), \text{ for some } \lambda > 0.$$

Suppose that some "appropriate" conditions are satisfied for the functions $b_i(\cdot)$, the noise processes etc. Then, we have

(a) $\lim_n EV(X_n) = O(\varepsilon)$;

(b) The set $\{X_n; n < \infty, \varepsilon > 0\}$ is tight;

(c) The set $\{(X_n - \theta)/\sqrt{\varepsilon}; n \geq N_\varepsilon, \varepsilon > 0\}$ is tight for $0 < N_\varepsilon \xrightarrow{\varepsilon} \infty$.

To avoid excessive notations, we are being pretty vague about the "appropriate" conditions. The precise statements can be found in [KY2]. The proof of the argument uses a technique known as perturbed Liapunov functions (cf. [Kus], [KH], [BP], [PSV] among others).

3.3. Further limit results

Theorem 3.2. *Let the conditions of Theorem 3.1 be satisfied, M_m^i be a sequence of stopping times and Eqns. (3.1) (3.2) hold with m replaced by M_m^i. If (b) in the above subsection holds, then*

$$X^\varepsilon(t_\varepsilon + \cdot) \Rightarrow \theta \text{ for any sequence } t_\varepsilon \xrightarrow{\varepsilon} \infty.$$

We give indications on how this theorem can be proved. Define

$$\hat{Z}_{ni}^\varepsilon = \varepsilon \sum_{j=N_i(t_\varepsilon/\varepsilon)}^{N_i(t_\varepsilon/\varepsilon)+n-1} Y_j^i,$$

$$\hat{X}_{ni}^\varepsilon = \varepsilon \sum_{j=N_i(t_\varepsilon/\varepsilon)}^{N_i(t_\varepsilon/\varepsilon)+n-1} b_i(X_{\tau_j^i}, \tilde{\xi}_j^i) + X_{t_\varepsilon/\varepsilon}^i,$$

$$\hat{X}_i^\varepsilon(t) = \hat{X}_{ni}^\varepsilon, \quad \hat{Z}_i^\varepsilon(t) = \hat{Z}_{ni}^\varepsilon \text{ on } [n\varepsilon, n\varepsilon + \varepsilon),$$

$$\hat{X}_n^\varepsilon = (\hat{X}_{n1}^\varepsilon, \ldots, \hat{X}_{nr}^\varepsilon), \quad \hat{Z}_n^\varepsilon = (\hat{Z}_{n1}^\varepsilon, \ldots, \hat{Z}_{nr}^\varepsilon).$$

In the above, $t_\varepsilon/\varepsilon$ is interpreted as a positive integer. This is no loss in generality since we can always take the integral part of it.

It can be easily seen that $\hat{X}^\varepsilon(\cdot) = X^\varepsilon(t_\varepsilon + \cdot)$ and $\{\hat{X}^\varepsilon(\cdot)\}$ is tight. Extract a weakly convergent subsequence, and denote it (with no loss of generality) by $\{\hat{X}^\varepsilon(\cdot)\}$ with limit $\hat{x}(\cdot)$. The proof is quite similar to that of Theorem 3.1. In fact, we only need to show $\hat{x}(0) = \theta$. To this end, for any $T > 0$,

$$\{\hat{X}^\varepsilon(\cdot), \hat{Z}_i^\varepsilon(\cdot), \hat{X}^\varepsilon(\cdot - T), \hat{Z}^\varepsilon(\cdot - T)\} = \{X^\varepsilon(t_\varepsilon + \cdot), Z^\varepsilon(t_\varepsilon + \cdot), X^\varepsilon(t_\varepsilon + \cdot - T), Z^\varepsilon(t_\varepsilon + \cdot - T)\}$$

with limit denoted by $(\hat{x}(\cdot), \hat{Z}(\cdot), \hat{x}_T(\cdot), \hat{Z}_T(\cdot))$, we have $\hat{x}(0) = \hat{x}_T(T)$, and all possible values of $\hat{x}_T(0)$ belong to a set which is tight. For any $\eta > 0$, there is a T_η large enough, such that for all $T \geq T_\eta$,

$$P\{|\hat{x}_T(T) - \theta| > \eta\} < \eta.$$

The theorem thus follows.

Remark: To elaborate further on the above theorem, we consider the special case $b_i(x, \xi^i) = b_i(x) + \xi^i$. In this case, suppose that θ is the unique zero of the function $b(x)$. The problem reduces to the Robbins-Monro setting with parallel processors.

Next, we discuss the rate of convergence issue. Define

$$U_n = \frac{X_n - \theta}{\sqrt{\varepsilon}}.$$

We show that the interpolated process of U_n converges to a Gauss-Markov process. The covariance of this process provides a measure of the asymptotic dependence of $X_n - \theta$ on ε. Thus, the study on the normalized error provides a way to analyze rates of convergence.

By virtue of a truncated Taylor expansion,

$$\begin{aligned} U_{n+1}^i = U_n^i &+ \varepsilon b_{ix}'(\theta, \xi_n^i) U_n I_{n+1}^i \\ &+ \sqrt{\varepsilon} b_i(\theta, \xi_n^i) I_{n+1}^i + o(\varepsilon) O(|U_n|^2). \end{aligned} \tag{3.5}$$

Define the continuous time interpolation of U_n by

$$U^\varepsilon(t) = U_n \text{ on } [\varepsilon n - \varepsilon N_\varepsilon, \varepsilon n - \varepsilon N_\varepsilon + \varepsilon).$$

For any n, let

$$z_{ni}^\varepsilon = \varepsilon \sum_{j=N_\varepsilon}^{N_\varepsilon+n-1} Y_j^i, \quad z_i^\varepsilon(t) = z_{ni}^\varepsilon \text{ on } [n\varepsilon, n\varepsilon + \varepsilon),$$

$$W_{ni}^\varepsilon = \sqrt{\varepsilon} \sum_{j=N_\varepsilon}^{N_\varepsilon+n-1} b_i(\theta, \xi_j^i) I_{j+1}^i, \quad W_i^\varepsilon(t) = W_{ni}^\varepsilon \text{ on } [n\varepsilon, n\varepsilon + \varepsilon),$$

$$\tilde{W}_{ni}^\varepsilon = \sqrt{\varepsilon} \sum_{j=N_\varepsilon}^{N_\varepsilon+n-1} b_i(\theta, \tilde{\xi}_j) \text{ and } \tilde{W}_i^\varepsilon(t) = \tilde{W}_{ni}^\varepsilon \text{ on } [\varepsilon n, \varepsilon n + \varepsilon).$$

We assume:

(A.3.1) $b_i(\cdot, \cdot)$ are bounded and continuously differentiable in x (with bounded derivatives) uniformly in ξ, and $Eb_i(\theta, \xi_n^i) = 0$.

(A.3.2) The following holds.

(1) For any sequence $N_\varepsilon \to \infty$,

$$\frac{1}{n} \sum_{j=N_i(n)}^{N_i(N_\varepsilon)+n-1} b_{ix}(\theta, \tilde{\xi}_j^i) \xrightarrow{n} \bar{b}_{ix}(\theta) \text{ in probability.} \tag{3.6}$$

(2) (3.1) (3.2) hold with m replaced by $N_i(N_\varepsilon)$.

(A.3.3) The sets $\{\xi_n^i; n < \infty\}$, $i \le r$ are mutually independent. For each $i \le r$, $\tilde{W}_i^\varepsilon(\cdot) \Rightarrow \tilde{W}_i(\cdot)$ a Brownian motion with covariance $\sigma_i^2 t$, for any sequence of stopping times $\{N_\varepsilon\}$ with respect to the σ-algebra $\mathcal{F}_n^\varepsilon$, and $\{\xi_n^i\}$ is uniformly mixing with rate $\phi_k \xrightarrow{k} 0$.

Theorem 3.3. *Under (A.3.1) to (A.3.3), the Brownian motions $\tilde{W}_i(\cdot)$, $i \le r$ are mutually independent. Suppose the tightness condition (c) of section 3.2 is satisfied, $U^\varepsilon(0) \Rightarrow U(0)$, and the equation (3.7) has a unique solution (in the sense of distribution) for each initial condition $U(0)$. Then, $\{U_i^\varepsilon(\cdot)\}$ $i \le r$ are tight and the limits satisfy*

$$dU_i = \frac{\bar{b}_{ix}'(\theta)}{\mu_i(\theta)} U dt + \frac{\sigma_i}{\sqrt{\mu_i(\theta)}} d\hat{W}_i, \; i \le r, \tag{3.7}$$

where $\hat{W}_i(\cdot)$, $i \le r$ are standard mutually independent Brownian motions.

We omit the proof here. The computed asymptotic covariance can be used for comparing different algorithms. For some numerical comparisons of the asymptotic performance of this algorithm and the classical procedures, see [KY2].

4. With probability one convergence and robustness

In this section, we review the results obtained in [YZ1] and [GZY]. The computation model was given in section 2.3. The approach we are taking is the random truncation methods (cf. [CZ]). In the first subsection, we state the global convergence theorem, and in the second section we deal with the problem of joint robustness with respect to the Liapunov function and noise processes.

4.1. Convergence property

We let the set of zeros of $b(\cdot)$ be denoted by $E = \{x \in \mathbb{R}^r; b(x) = 0\}$. For simplicity, we take the step-size to be of the simplest form $a_n = \frac{1}{n}$. All the subsequent developments remain to be valid for a_n satisfying conditions of the form: $\limsup_n na_n < \infty$, and

$$1 - \alpha_1/(n+1) \leq \frac{a_{n+1}}{a_n} \leq 1 - \alpha_2/(n+1),$$

for some $0 < \alpha_2 < \alpha_1 < 2$, or $a_n = \frac{1}{n^\beta}$ for some $0 < \beta \leq 1$. Of cause, some modifications are necessary.

The following assumptions are needed in the sequel.

(B.1.1) For each $i \leq r$, $\{Y_n^i\}$ is bounded. There exist a sequence of random variables μ_n^i and a $\mu_i > 0$, such that

$$E(Y_n^i - \mu_n^i | Y_l^i - \mu_l^i) = 0, \quad \text{for } l < n, \ \mu_n^i \xrightarrow{n} \mu_i \text{ w.p.1.}$$

Moreover, for some $0 < \alpha < 1$,

$$\sum_{n=1}^{\infty} n^{\frac{1}{2}(1-\alpha)}(\mu_n^i - \mu_{n-1}^i) \text{ converges w.p.1.}$$

(B.1.2) For each $i \leq r$,

$$\tilde{\xi}_n^i = \tilde{\phi}_n^i + \tilde{\psi}_n^i,$$

with $\tilde{\psi}_n^i \xrightarrow{n} 0$ w.p.1 and $\sum a_{\tau_j^i} \tilde{\phi}_j^i$ converges w.p.1.

Remark: Similar to the definition of ξ_n^i, we can define ϕ_n^i and ψ_n^i in an obvious manner.

(B.1.3) There are a twice continuously differentiable Liapunov function $V(\cdot)$, a real number $M > 0$ (as in the definition of the truncations), such that

$$V(\hat{x}) < \inf\{V(x); |x| = M\} = d, \ [V(\hat{x}), d] \cap V(E) \neq [V(\hat{x}), d]$$

$$V(x) \neq V(y),\ V_x'(x)\mu^{-1}b(x) < 0,\ \forall x \notin E, y \in E,$$

where

$$\mu^{-1} = \begin{pmatrix} \frac{1}{\mu^1} & \cdots & 0 \\ \vdots & \ddots & \vdots \\ 0 & \cdots & \frac{1}{\mu^r} \end{pmatrix}.$$

Theorem 4.1. *Under (B.1.1) to (B.1.3), for any initial condition X_1, algorithm (2.5) is strongly convergent, i.e., $\lim_{n\to\infty} d(X_n, E) = 0$ w.p.1, where $d(\cdot,\cdot)$ denotes the distant function.*

The proof of this theorem consists of four steps.

Step 1. We prove a strong law of large numbers type of lemma for the random computation times. This is stated as:

Lemma 4.1. *Under (B.1.1), $\frac{1}{n}\tau_n^i - \mu_i = o\left(\frac{1}{n^{\frac{1}{2}(1-\alpha)}}\right)$.*

In view of the above lemma,

$$\frac{\tau^i_{N_i(n)}}{N_i(n)} \leq \frac{n}{N_i(n)} \leq \frac{\tau^i_{N_i(n)+1}}{N_i(n)+1}\frac{N_i(n)+1}{N_i(n)}.$$

As a consequence, w.p.1,

$$\frac{N_i(n)}{n} \xrightarrow{n} \frac{1}{\mu_i},\ \frac{N_i(n)}{n} - \frac{1}{\mu_i} = o\left(\frac{1}{n^{\frac{1}{2}(1-\alpha)}}\right). \tag{4.1}$$

In fact, this lemma should be compared with the well-known renewal theorem. It has certain resemblance to the conventional renewal theorem. Owing to (4.1), not only convergence, but also rate of convergence are given.

Step 2. To proceed, we observe that since σ_n is an increasing sequence, either $\sigma_n \to \sigma < \infty$, or $\sigma = \infty$. The finiteness of σ actually imply the boundedness of the algorithm. If $\sigma < \infty$ w.p.1, then for sufficiently large n,

$$|X_{n+1}| = \left|X_n + \frac{1}{n}I_{n+1}(b(\tilde{X}_n) + \xi_n)\right| \leq M(\sigma).$$

Hence,

$$X_{n+1} = X_n + \frac{1}{n}I_{n+1}(b(\tilde{X}_n) + \xi_n),$$

i.e., truncations are employed for only finitely many times. Thus, we need only prove the following lemma.

Lemma 4.2. *Under (B.1.1) to (B.1.3), $\sigma < \infty$ w.p.1.*

Denote $m(n,\eta) = \max\{m; \sum_{j=n}^{m} \frac{1}{j} \leq \eta\}$. Let $\{X_{n_k}\}$ be a convergent subsequence of $\{X_n\}$. Suppose that $\{\tilde{X}_n\}$ is bounded whenever $\{X_n\}$ is. We show that there is a $\delta > 0$, for any $0 < \eta < \delta$, and some $K_i > 0$ $i = 1,2$, $K_\eta > 0$, and any $k \geq K_\eta$,

$$\begin{aligned}\left|\sum_{j=n_k}^{m(n_k,\eta)} \frac{1}{j} I_{j+1}(b(\tilde{X}_j) + \xi_j)\right| &\leq K_1, \forall m \in [n_k, m(n_k,\eta)+1], \\ |X_m - X_{n_k}| &\leq K_2\eta, \ \forall m \in [n_k, m(n_k,\eta)+1].\end{aligned} \tag{4.2}$$

Then, we use these estimate to prove Lemma 4.2. The verification is mainly a proof by contradiction. The approach of the truncation methods is closely related to that of [CZ]. Nevertheless, in our problem, the averaging procedure is in fact two folds. One is the average of the noise processes and another one is the average of the computation times. As in the approach of weak convergence discussed in section 3, the two averaging procedures are intertwined, and hence the proofs are more involved and complicated.

Let $\{X_k\}$ denote a convergent subsequence with limit $\bar{x}$ (this can be shown to exist). A typical argument in the averaging involves dealing with a term of the form

$$\sum_{j=N_i(k)}^{N_i(m(k,\eta))} \frac{1}{\tau_j^i} b_i(X_{\tau_j^i}).$$

We refer to the paper [YZ1] for detailed discussions.

Step 3. Put

$$t_n = \sum_{j=1}^{n} a_j = \sum_{j=1}^{n} \frac{1}{j}, \ m(t) = \sup\{n; t_n \leq t\}.$$

Let $x^0(\cdot)$ be the piecewise linear interpolation of X_n, such that

$$x^0(t_n) = X_n, \ \ x^0(t) = \frac{t_{n+1} - t}{a_n} X_n + \frac{t - t_n}{a_n} X_{n+1}, \ t \in [t_n, t_{n+1}),$$

let $\bar{x}(\cdot)$ be the piecewise constant interpolation of X_n with $\bar{x}(t) = X_n$, for $t \in [t_n, t_{n+1})$, and define $x^n(\cdot) = x^0(\cdot + t_n)$. We first show that $\{x^n(\cdot)\}$ is uniformly bounded and equicontinuous. Then, by virtue of the Ascoli-Arzela's Lemma, we can extract a convergent subsequence $\{x^k(\cdot)\}$ with limit $x(\cdot)$. Note that for each $i \leq r$,

$$x^{k,i}(t) - x^{k,i}(0) = \sum_{j=N_i(k)}^{N_i(m(t+t_k))-1} \frac{1}{\tau_j^i} b_i(X_{\tau_j^i}) + \sum_{j=N_i(k)}^{N_i(m(t+t_k))-1} \frac{1}{\tau_j^i} \tilde{\xi}_j^i + O\left(\frac{1}{k}\right).$$

Using this representation and similar kind of argument as in step 2, we can show that $x^i(\cdot)$, $i \leq r$ satisfy the ODE $\dot{x}^i = \frac{1}{\mu_i} b_i(x)$, where $x = (x^1, \ldots, x^r)'$.

Step 4. By virtue of the well-known stability results of ordinary differential equations and condition (B.1.3),

$$\lim_{t \to \infty} d(x(t), E) = 0,$$

where $x(\cdot)$ is the limit of $x^k(\cdot)$. Now, by using an argument as in the proof of Lemma 4.2, we can derive a contradiction. This in turn completes the proof of the theorem.

Remark: We point out that the convergence property can also be established by the martingale averaging techniques described in [Ku]. There are certain advantages in each approach. Using the approach in [Ku], functions with discontinuity can be treated, but some conditions on the growth rate of the function needs to be imposed. The treatment discussed in this section can deal with functions having arbitrary nonlinearity, however, the continuity seems to be crucial.

4.2. Robustness

Recently, the robustness issue has brought much attention. In [CGG], it was proved that a strong law of large numbers type of condition for measurement errors with 0 limit is a necessary and sufficient condition for with probability one convergence of the Robbins-Monro method. Moreover, the robustness issue for stochastic approximations was brought up.

In a broad informal sense, robustness analysis for S.A. algorithms gives an account on the allowable tolerance and relate deviations from idealized assumptions. In other words, it intends to answer the question: "If the idealized model is only approximately true, can one still get reasonably good estimates for the true parameter within certain bounds?" It is a common practice to prove convergence of stochastic approximation algorithms by using the Liapunov function approach. As a result, the following questions are of particular interest: What kind of measurement errors can be tolerated? What kind of deviations can be allowed for the corresponding Liapunov function.

It seems that the analysis of robustness plays an important role in organizing information about the algorithms to a manageable form, and appears to be quite informative and helpful for various applications.

The approach which we are taking is somewhat different from that of [YZ1]. In lieu of the ODE (ordinary differential equation) method (cf. [Lj], [KC]), elementary inequalities will be employed to establish the desired results. With the assumption that a strong law of large numbers with non-zero limit holds for the noise processes, and the Liapunov function evaluated at θ (the true parameter which we are seeking), is different from zero, we obtain asymptotic upper bounds for the absolute deviations. It turns out that the upper bound depends jointly on the local behavior of the Liapunov function (near the true parameter), and the large time properties of the measurement noise. In this sense, the result obtained is interpreted as joint robustness. In addition, the marginal or individual robustness with respect to either the noise or the Liapunov function is also derived as a direct consequence. Finally, the almost sure convergence as well as the necessary and sufficient condition for convergence are obtained by virtue of the robustness property.

The following assumptions are needed in the subsequent development.

(B.2.1) For each $i \le r$, $|Y_n^i| \le \nu$ a.s. for some $\nu > 0$, and there is a $\mu_i > 0$, such that

$$a_n \sum_{j=1}^{n} Y_j^i \xrightarrow{n} \mu_i \text{ a.s.}$$

Denote $\mu = \max_{i \le r} \mu_i$.

(B.2.2) For each $i \le r$, there exist a twice continuously differentiable function $V_i(\cdot) : \mathbb{R} \mapsto \mathbb{R}$ and a positive real number ε_1, such that

$$|V_i(\theta^i)| = \bar{\varepsilon}^i < \varepsilon_1, \ |V_{i,x^i}(\theta^i)| = \hat{\varepsilon}^i < \varepsilon_1$$

$$\sup_{|x-\theta| > \varepsilon_1} V_{i,x^i(x^i)} b_i(x) < 0$$

where $V_{i,x^i}(\cdot)$ denotes the partial derivative of $V_i(\cdot)$ with respect to x^i. Let $V(\cdot) : \mathbb{R}^r \mapsto \mathbb{R}$, be given by

$$V(x) = \sum_{i=1}^{r} V_i(x^i).$$

Suppose

$$V(x) > 0, \text{ for all } |x - \theta| > \varepsilon_1 \text{ and } \lim_{|x| \to \infty} V(x) = \infty.$$

(B.2.3) For each $i \le r$, there is a $\varepsilon_2^i \ge 0$, such that

$$\limsup_{n \to \infty} a_{\tau_n^i} |\sum_{j=1}^{n} \tilde{\xi}_j^i| = \varepsilon_2^i \text{ a.s.}$$

Denote $\varepsilon_2 = \max_{i \le r} \varepsilon_2^i$, and

$$d_0 = V(x^*) < \inf\{V(x); |x| > M\} = d$$

$$\frac{d - d_0}{2} > \sup\{V(x); |x - \theta| < \varepsilon_1\}$$

where $M > 0$ is as given in section 2.3. In fact, the existence of M is guaranteed by the above condition (B.2.2).

(B.2.4) The sequence $\{a_n\}$ satisfies: $0 < a_n \xrightarrow{n} 0$, $\sum_{n=1}^{\infty} a_n = \infty$, $a_n \ge a_{n+1}$, and assume (without loss of generality) that there is a $0 < b \le 1$, such that

$$a_n \le a_{n+1}(1 + b a_n).$$

Remark: The above assumptions are the conditions on the random computation times, Liapunov functions and noise processes, respectively. (B.2.1) is essentially a law of large numbers type of condition. It is satisfied for example, if $\sum_{j=1}^{n} Y_j^i$ are renewal processes and $a_n = \frac{1}{n}$. It also holds for certain stationary processes, and many other processes.

(B.2.3) is a law of large numbers type of condition on the measurement noise. Such assumptions are needed in the robustness analysis. For more discussion on this point and various examples, we refer to [CGG].

(B.2.4) gives conditions on the gain or step size sequence. The regularity requirement on a_n includes a wide variety of sequences. In particular, (B.2.4) is satisfied by the usual choice $a_n = \frac{A}{n}$ for some $A > 0$. It is also satisfied when $a_n = \frac{A}{\sqrt{n}}$, $a_n = \frac{A}{\log n}$ etc.

In most of the previous convergence analyses, the noise processes were assumed to satisfy a law of large numbers type (or the like) condition with limit 0, and $V(\theta) = 0$. The robustness analysis deals with the situation where such conditions are violated or disturbed. In our assumption, the Liapunov function takes positive values when it is away from the true parameter. Whenever it is close to the true parameter, a certain amount of variation is allowed. As for the noise processes, the limit of the law of large numbers type can be different from zero. Loosely speaking, this takes care of the case that the noise is not completely wiped out (not averaged out to 0).

The robustness theorems are given as follows.

Theorem 4.2.

(1) Under assumption (B.2.1)-(B.2.4), there exist $\varepsilon_1^* > 0$, $0 < \varepsilon_2^* \le 1$, and a function $g(\cdot,\cdot) : [0,\infty) \times [0,\infty) \mapsto [0,\infty)$, which is monotone non-decreasing and left continuous with respect to each variable t_1, t_2, and

$$\lim_{t_2\to 0} g(t_1,t_2) = g(t_1,0)$$
$$\lim_{t_1\to 0} g(t_1,t_2) = g(0,t_2)$$
$$\lim_{\substack{t_1\to 0\\ t_2\to 0}} g(t_1,t_2) = 0;$$

for all $\varepsilon_1 \in (0,\varepsilon_1^*]$, $\varepsilon_2 \in (0,\varepsilon_2^*]$, and X_n given by (2.8),

$$\limsup_{n\to\infty} |X_n - \theta| \le g(\varepsilon_1,\varepsilon_2). \tag{4.3}$$

(2) If in addition, for each $i \le r$, $\bar{\varepsilon}^i$, $\hat{\varepsilon}^i$ given in (B.2.2) satisfy $\bar{\varepsilon}^i = \hat{\varepsilon}^i = 0$, then for all $\varepsilon_2 \in (0,\varepsilon_2^*]$,

$$\limsup_{n\to\infty} |X_n - \theta| \le g_2(\varepsilon_2) = g(0,\varepsilon_2). \tag{4.4}$$

(3) If the conditions are fulfilled with $\varepsilon_2 = 0$, then for all $\varepsilon_1 \in (0,\varepsilon_1^*]$,

$$\limsup_{n\to\infty} |X_n - \theta| \le g_1(\varepsilon_1) = g(\varepsilon_1,0). \tag{4.5}$$

(1) gives an account on the joint robustness; (2) and (3) are the corresponding marginal robustness. $g(\cdot,\cdot)$ is a measure on the joint robustness property. Similarly, $g_1(\cdot)$ and $g_2(\cdot)$ are the marginal measures.

Further more, we can prove the following theorem.

Theorem 4.3. *Assume that (B.2.1)-(B.2.4) are satisfied with the following modifications:*

(1) *(B.2.2) holds for any $\varepsilon_1 > 0$, i.e., $\bar{\varepsilon}^i = \hat{\varepsilon}^i = 0$, $i \le r$; (B.2.3) holds with $\varepsilon_2 = 0$. Then, algorithm is strongly consistent, i.e., $X_n \xrightarrow{n} \theta$ a.s.*

(2) *Suppose assumptions (B.2.1)-(B.2.4) are satisfied with an additional condition on the step-size:*

$$\limsup_{n\to\infty} na_n < \infty$$

$$1 - \frac{\beta_1}{n+1} \le \frac{a_{n+1}}{a_n} \le 1 - \frac{\alpha_1}{n+1}$$

where $0 < \alpha_1 < \beta_1 < 2$. If $X_n \xrightarrow{n} \theta$ a.s., then for each $i \le r$,

$$a_{\tau_n^i} \sum_{j=1}^{n} \tilde{\xi}_j^i \xrightarrow{n} 0 \text{ a.s.} \tag{4.6}$$

Idea of proofs: As in the previous section, to prove Theorem 4.3, we have to show the number of truncations is finite first. This is done by establishing the following facts in succession.

(i) Suppose (B.2.3) holds for some $\varepsilon_2 \le \varepsilon_2^* \le 1$, and suppose that for each $i \le r$ and K with $\tau^i_{N_i(K)} > K$, there exists a $n \ge K$, $|\tilde{X}_n^i| \le M$. Then, there is a T such that for any $\hat{T} \in [0, T]$, and any m satisfying

$$n \le m \le m(n, \hat{T}) \equiv \max\left\{m;\ \sum_{j=n}^{m} a_j \le \hat{T}\right\}$$

$$\left|\sum_{j=n}^{m+1} a_j I_{j+1}(b(\tilde{x}_j) + \xi_j)\right| \le 6\sqrt{r\varepsilon_2^*} + 2.$$

(ii) Under the conditions of (i), for each $\hat{T} \in [0, T]$, and all $m \in [n, m(n, T) + 1]$,

$$|X_m - X_n| \le 3\sqrt{r\varepsilon_2^*} + \hat{T}(b_1 + 2\sqrt{r\varepsilon_2^*}).$$

(iii) Under (B.2.1)-(B.2.4) with $\varepsilon_1 \in (0, \varepsilon_1^*]$ and $\varepsilon_2 \in (0, \varepsilon_2^*]$, there is an integer-valued random variable σ, such that

$$\sup_n \sigma_n \le \sigma < \infty \text{ a.s.}$$

After (iii) is proved, we then carry out analysis as in [CGG], nevertheless, the averaging is more involved due to the renewal times. The difficulties also stem from the fact that the iteration number is no longer a common index of all the processors.

As for the proof of Theorem 4.3, we note that (1) in Theorem 4.3 can be proved by letting $\varepsilon \to 0$ in Theorem 4.2, whereas the proof of (2) depends on an observation that for sufficiently large n, the truncations stop, i.e., there is a $\tilde{N}$, such that for all $n \ge \tilde{N}$,

$$X_{n+1} = X_n + a_n I_{n+1}(b(\tilde{X}_n) + \xi_n), \forall n \ge \tilde{N}.$$

Consequently,

$$(X_{n+1}-\theta)+a_n\sum_{j=\tilde N-1}^{n-1}\left(\frac{1}{a_j}-\frac{1}{a_{j+1}}\right)(X_{j+1}-\theta)$$
$$-\frac{a_n}{a_{\tilde N-1}}(X_{\tilde N}-\theta)-a_n\sum_{j=\tilde N}^{n}I_{j+1}b(\tilde X_j)=a_n\sum_{j=\tilde N}^{n}I_{j+1}\xi_j.$$

The desired result follows from the above expression.

5. Concluding remarks

A survey on some of the recent developments in parallel stochastic approximations is made in this work. The purpose is to facilitate the understanding of the potential of using parallel and distributed computation for stochastic iterative schemes, which arise naturally in Monte Carlo optimization and many other related fields. We have focused on those algorithms that are suited for large scale parallelization and represent the hope for solving problems which are much larger than what can be solved at present.

Several algorithms have been discussed here. The asymptotic properties of the algorithms are of our main concerns. We believe that a thorough understanding of an algorithm (by that we mean not only the understanding of its structure and procedure, but also the asymptotic behavior) is essential for applications to various problems. Based on this point of view, we presented several limit results for a class of parallel S.A. algorithms and their variations.

As was mentioned at the beginning, the research in parallel S.A. is quite new. There are still many interesting and challenging problems need to be solved. To be a little bit more specific, the following items are worthwhile considering.

(1) In implementing any recursive algorithm, one needs a good stopping criterion. There is no exception for the parallel S.A. methods. One of the possible approaches is to use the construction of sequentially determined confidence ellipsoids (cf. [Yi] and the references therein). We can design a stopping rule for each individual processor first by constructing appropriate ellipsoidal regions, and then take the maximum value of those individual stopping times as the stopping rule for the overall system. Among other things, the Bonferroni's inequality is needed in establishing the limit theorem.

(2) Recently, progress has been made for analyzing S.A. type of recursive algorithms via large deviation approach (cf. [DK] and the references therein). It seems that similar approach can be taken to treat the parallel S.A. problems.

(3) An interesting problem is the actual implementation of parallel algorithms, especially the VLSI type of implementations. They will help us to get deeper insight into the algorithms and lead to further improvement.

(4) We have made some numerical comparisons of asymptotic performance for different algorithms in [KY1] and [KY2]. Further investigation in optimal design can be taken. It is clear

that the design of efficient algorithms is not only theoretically necessary but also practically important.

Parallel and decentralized S.A. algorithms offer promise because they make sufficient utilization of computing power on many problems possible. Our study is actually just started. It is conceivable that parallel stochastic approximations will play more and more important role in the future, especially in those areas such as systems theory and optimization.

References

[AG] A.E. Albert and L.A. Gardner, *Stochastic Approximation and Nonlinear Regression*, MIT Press, Cambridge, MA, 1967.

[BMP] A. Benveniste, M. Metivier and P. Priouret, *Algorithmes Adaptatifs et Approximations Stochastiques*, Masson, Paris, 1987.

[Be] D.P. Bertsekas, Distributed computation of fixed points, *Mathematical Programing* **27** (1983), 107-120.

[BT] D.P. Bertsekas and J.N. Tsitsiklis, *Parallel and Distributed Computing*, Prentice-Hall, New Jersey, 1989.

[BTs] D.P. Bertsekas and J.N. Tsitsiklis, Parallel and distributed iterative algorithms: A selective survey, Technical Report, Laboratory for Information and Decision Systems, LIDS-P-1835, MIT, 1988.

[Bi] P. Billingsley, *Convergence of Probability Measures*, J. Wiley & Sons, New York, 1968.

[BP] G.B. Blankenship and G.C. Papanicolaou, Stability and control of stochastic systems with wide band noise, *SIAM J. Appl. Math.* **34** (1978), 437-476.

[Ca] P. E. Caines, *Linear Stochastic Systems*, J. Wiley, New York, 1988.

[CM] D. Chazan and W. Miranker, Chaotic relaxation, *Linear Algebra and Appl.* **2** (1969), 199-222.

[Ch] H.F. Chen, *Recursive Estimation and Control for Stochastic Systems*, J. Wiley, New York, 1985.

[CGG] H.F. Chen, L. Guo and A.J. Gao, Convergence and robustness of the Robbins-Monro algorithm truncated at randomly varying bounds, *Stochastic Process. Appl.* **27** (1988), 217-231.

[CZ] H.F. Chen and Y.M. Zhu, Stochastic approximation procedures with randomly varying truncations, *Scientia Sinica (series A)* **XXIX** (1986), 914-926.

[DK] P. Dupuis and H.J. Kushner, Stochastic approximation and large deviations: upper bounds and w.p.1 convergence, *SIAM J. Control Optim.* **27** (1989), 1108-1135.

[Er] Yu. Ermoliev, Stochastic quasigradient Methods and their applications to system optimization, *Stochastics* **9** (1983), 1-36.

[EK] S. Ethier and T.G. Kurtz, *Markov Processes: Characterization and Convergence*, J. Wiley & Sons, New York, 1986.

[Fa] V. Fabian, On asymptotic normality in stochastic approximation, *Ann. Math. Statist.* **39** (1968), 1327-1332.

[Ge] L. Gerencsér, Rigorous results in the theory of recursive identification, in this volume.

[Ger] L. Gerencsér, On a class of mixing processes, *Stochastics* **26** (1989), 165-191.

[GRC] G. Goodwin, P. Ramadge and P. Caines, Discrete time stochastic adaptive control, *SIAM J. Control Optim.* **19** (1981), 829-853.

[GZY] A.J. Gao, Y.M. Zhu and G. Yin, Joint robustness on noise and Liapunov function for parallel stochastic approximation algorithms, Preprint, 1989.

[KW] J. Kiefer and J. Wolfowitz, Stochastic estimation of the maximum of a regression function, *Ann. Math. Statist.* **23** (1952), 462-466.

[KV] P.R. Kumar and P.P. Varaiya, *Stochastic Systems: Estimation, Identification and Adaptive Control*, Prentice-Hall, Englewood Cliffts, NJ, 1986.

[Ku] H.J. Kushner, Stochastic approximation with discontinuous dynamics and state dependent noise: w.p.1 and weak convergence, *J. Math. Anal. Appl.* **82** (1981), 527-542.

[Kus] H.J. Kushner, *Approximation and Weak Convergence Methods for Random Processes with Applications to Stochastic Systems Theory*, MIT Press, Cambridge, MA, 1984.

[KC] H.J. Kushner and D.S. Clark, *Stochastic Approximation for Constrained and Unconstrained Systems*, Springer-Verlag, Berlin, 1978.

[KH] H.J. Kushner and H. Huang, Asymptotic properties of stochastic approximations with constant coefficients, *SIAM J. Control Optim.* **19** (1981), 86-105.

[KS] H.J. Kushner and A. Shwartz, An invariant measure approach to the convergence of stochastic approximations with state-dependent noise, *SIAM J. Control Optim.* **22** (1984), 13-27.

[KY1] H.J. Kushner and G. Yin, Asymptotic properties of distributed and communicating stochastic approximation algorithms, *SIAM J. Control Optim.* **25** (1987), 1266-1290.

[KY2] H.J. Kushner and G. Yin, Stochastic approximation algorithms for parallel and distributed processing, *Stochastics* **22** (1987), 219-250.

[LR] T.L. Lai and H. Robbins, Consistency and asymptotic efficiency of slope estimates in stochastic approximation schemes, *Z. Wahr.* **56** (1981), 329-360.

[LB] S. Li and T. Basar, Asymptotic agreement and convergence of stochastic algorithm, *IEEE Tran. Automat. Control* **AC-32** (1987), 612-618.

[Lj] L. Ljung, Analysis of recursive stochastic algorithms, *IEEE Trans. Automat. Control* **AC-22** (1977), 551-575.

[LS] L. Ljung and T. Söderström, *Recursive identification*, MIT Press, Cambridge, MA, 1983.

[MP] M. Metivier and P. Priouret, Applications of a Kushner and Clark Lemma to general classes of stochastic algorithms, *IEEE Trans. Inform. Theory* **IT-30** (1984), 140-151.

[Mi] W.L. Miranker, A survey of parallelism in numerical analysis, *SIAM Review* **13** (1971), 524-547.

[NK] M.B. Nevel'son and R.Z. Khas'minskiĭ, *Stochastic Approximation and Recursive Estimation*, Translation of Math. Monographs, v47, AMS, Providence, 1973.

[Pf] G. Ch. Pflug, Stochastic minimization with constant step-size: asymptotic laws, *SIAM J. Control Optim.* **24** (1986), 655-666.

[Po] B.T. Polyak, Convergence and convergence rate of iterative stochastic algorithms, I, II, *Automatic Remote Control* **38** (1978), 537-542.

[Pol] B.T. Polyak, New stochastic approximation type procedures, to appear in *Automatic Remote Control.*

[Pr] Yu. V. Prohorov, Convergence of random processes and limit theorems in probability theory, *Theory Probab. Appl.* **1** (1956), 157-214.

[PSV] G.C. Papanicolaou, D. Stroock and S.R.S. Varadhan, Martingale approach to some limit theorems, *Proc. of the 1976 Duke Univ. Conference on Turbulence*, M. Reed ed., Duke University Mathematics Series, vol. 3, Durham, NC, 1977.

[Qu] M.J. Quinn, *Designing Efficient Algorithms for Parallel Computers*, McGraw-Hill, New York, 1987.

[RM] H. Robbins and S. Monro, A stochastic approximation method, *Ann. Math. Statist.* **22** (1951), 400-407.

[Ru] R.Y. Rubinstein, *Monte Carlo Optimization and Sensitivity of Queueing Networks*, J. Wiley & Sons, New York, 1986.

[Rup] D. Ruppert, A Newton-Raphson version of the multivariate Robbins-Monro Procedure, *Ann. of Statist.* **13** (1985), 236-245.

[Sa] J. Sacks, Asymptotic distribution of stochastic approximation procedures, *Ann. Math. Statist.* **29** (1958), 159-167.

[Sk] A.V. Skorohod, Limit theorems for stochastic processes, *Theory Probab. Appl.* **1** (1956), 261-290.

[Ts] J. Tsitsiklis, Problem in decentralized decision making and computation, Ph.D thesis, Elect. Eng. Dept., MIT, Cambridge, MA, 1984.

[TBA] J.N. Tsitsiklis, D.P. Bertsekas and M. Athans, Distributed asynchronous deterministic and stochastic gradient optimization algorithms, *IEEE Trans. Automat. Control* **AC-31** (1986), 803-812.

[Tsy] Ya. Z. Tsypkin, *Adaptation and Learning in Automatic Systems*, Academic Press, New York, 1971.

[Ve] J.H. Venter, An extension of the Robbins-Monro procedure, *Ann. Math. Statist.* **38** (1967), 181-190.

[Wa] M.T. Wasan, *Stochastic Approximation*, Cambridge Press, London, 1969.

[WT] R.B. Washburn and D. Teneketzis, Asymptotic agreement among communicating decision makers, *Stochastics* **13** (1984), 103-129.

[We] C.Z. Wei, Multivariate adaptive stochastic approximation, *Ann. Statist.* **15** (1987), 1115-1130.

[Yi] G. Yin, A stopping rule for the Robbins-Monro method, *J. Optim. Theory. Appl.* **67** (1990), 151-173.

[YZ1] G. Yin and Y.M. Zhu, On parallel processing approach to adaptive stochastic approximation, *Proc. 27th Conference on Decision and Control.*

[YZ1] G. Yin and Y.M. Zhu, On w.p.1 convergence of a parallel stochastic approximation algorithm, *Probab. Eng. Inform. Sci.* **3** (1989), 55-75.

[YZ2] G. Yin and Y.M. Zhu, On robustness of the Robbins-Monro method for parallel processing, *System Control Lett.* **12** (1989), 77-86.

[YZ3] G. Yin and Y.M. Zhu, On H-valued Robbins-Monro processes, *Journal of Multivariate Analysis*, Vol. **34** (1990), 116-140.

[ZY] Y.M. Zhu and G. Yin, Stochastic approximation in real time: pipelining computation and communication, Preprint, 1989.

ON THE ADAPTIVE STABILIZATION AND ERGODIC BEHAVIOUR OF STOCHASTIC SYSTEMS WITH JUMP-MARKOV PARAMETERS VIA NONLINEAR FILTERING

Peter E. Caines[†‡] **Karim Nassiri-Toussi**[†]

[†] McGill Research Center for Intelligent Machines–McRCIM,
McGill University, McConnel Engr. Bldg.,
3480 University St., Montreal, CANADA H3A 2A7

[‡] Canadian Institute for Advanced Research

1. Introduction

For the past two decades, the problem of control of jump-Markov parameter, or the so-called hybrid [6] , systems has been of considerable theoretical interest and at the same time, has found major applications, in particular to power systems and electrical networks (see for example [23] or the references in [6]).

For a finite-state jump-Markov parameter system, the set of parameters of the system is equal to one of N different sets depending on the value of $\theta_t \in \{1, 2, \ldots, N\}$, where θ_t is a finite-state jump-Markov process. In other words, θ_t indexes the current dynamics of the system.

An extremely general class of jump-Markov processes can be described by the following stochastic integral equation (see [15] , Lemma 9-2):

$$\Phi_t = \Phi_0 + \int_0^t \Pi_s \Phi_s ds + \zeta_t, \qquad (1-1)$$

where

$$\Phi_t = \left[\mathbb{1}_{\{\theta_t=1\}}, \mathbb{1}_{\{\theta_t=2\}}, \dots, \mathbb{1}_{\{\theta_t=N\}}\right]^T$$

and ζ_t is a zero-mean L^2 martingale, measurable with respect to the continuously increasing family of the σ-fields, $\{\mathcal{F}_t;\ t \geq 0\}$. Φ_0 is also an $\mathcal{F}_0$-measurable random variable where $E[\Phi_0] = p_0$. (Throughout this paper, $\mathbb{1}_B$ shall denote the indicator function of the set B.)

By taking the expectation of eq. (1-1), we find a more familiar equation for the probability distribution of the jump-Markov process, which gives $\dot{p}_t = \Pi_t p_t$.

Although in general, the probability transition matrix, Π_t, can be time-varying, it is often a realistic assumption to say that the process is homogeneous. In any case, it is only for autonomous systems that the ergodic property can generally be established. Henceforth in our discussion, we shall take $\Pi_t = \Pi$, for all $t \geq 0$. In this manner, since the process is finite-state, we know that if $\mathrm{Tr}(-\Pi) < \infty$, it will necessarily be ergodic and p_t will converge to an invariant distribution, p_∞, which may depend on the initial distribution, p_0. The quantities Π and p_0, and therefore p_∞, are assumed to be known and deterministic. Also a reasonable assumption in most cases is that the jump-Markov process θ_t is independent of the noise inputs. This simplifies the analysis significantly; otherwise, further difficult-to-verify assumptions or conditions will be needed.

To provide a brief history of the work done so far on the jump-Markov parameter systems, one must mention [14, 25, 22] as early works related to these systems, where an optimal state feedback solution based on complete-observation of the jump-Markov process has been obtained. On the other hand, [16, 17, 9] are more recent works where an optimal <u>constant</u> output feedback solution for the linear hybrid systems has been proposed.

However, none of the previously-mentioned works can be regarded as <u>adaptive</u> control solutions for hybrid systems. In fact, so far, the only general adaptive control solution for stochastic jump-Markov parameter systems has been the one provided by Caines and Chen in [2, 4] .

In this paper, we continue the investigation begun in the above-mentioned papers concerning the adaptive control of hybrid systems with unobserved parameters, using the same techniques for adaptation. Although in [2] , the jump-Markov system includes a certain class of nonlinear systems, we restrict our analysis to the more special but still important case of autonomous linear hybrid systems.

Therefore, our system equation will be of the form:

$$dx_t = (A(\theta_t)x_t + B(\theta_t)u_t)\,dt + F\,dw_t\,, \qquad y_t = x_t, \tag{1-2}$$

where w_t is an m-dimensional Wiener process measurable with respect to the continuously increasing σ-fields $\mathcal{F}_t$. θ_t is the unobserved finite-state Markov process whose indicator function, Φ_t, satisfies eq. (1-1) with $\Pi_t = \Pi$ for all $t \geq 0$. When $\theta = i$, $A(\theta) = A_i$ and $B(\theta) = B_i$, where A_i-s and B_i-s are respectively $\mathcal{R}^{n\times n}$ and $\mathcal{R}^{n\times p}$ matrices such that $\|A_i\| < \overline{a} < \infty$, $\|B_i\| < \overline{b} < \infty$ and either $A_i \neq A_j$ or $B_i \neq B_j$ for every non-equal $i, j = 1, 2, \ldots, N$. We will also assume that FF^T is a full rank matrix. It should be made clear that, in fact, the complete set of system equations includes both eq.'s (1-1) and (1-2).

(It should be mentioned that eq. (1-2) and all other stochastic differential equations discussed in this report, unless specified otherwise, are in Ito's sense.)

One way to attempt to control the system of eq.'s (1-2) and (1-1) by an adaptive controller is to estimate θ_t, the unobserved parameter process, using the appropriate nonlinear filter, and then, following the "certainty equivalence principle", to substitute this value for that of the system parameters in a controller proven to stabilize the system at the correct parameter values. Regarding the causality of the system, any such controller will necessarily be a function of $y^t \triangleq \{y_s;\ 0 \geq s \geq t\}$. In particular, if $u_t = -Ky_t$, where $K = K(y^t)$ is any bounded functional on y^t, then the conditions mentioned in the hypothesis of theorem 9-3 in [15] will be satisfied and we are able to use the optimal filtering equations to estimate Φ_t.

Although in principle, the optimal nonlinear filter may also be applied to the partially-observed system, so far, no explicit filter equations have been found for this case. Therefore, at present, our results are restricted to the case of complete observation of the states of the system, x_t. (Here, we distinguish between the state process x_t and the parameter process, θ_t or Φ_t, whereas actually, the state process for the system is the joint process (x_t, θ_t).)

Let $\hat{p}_t$ denote the conditional expectation, $E[\Phi_t \mid \mathcal{F}_t^y]$,

$$\overline{H}_{\hat{p}} = \overline{A}_{\hat{p}} - \overline{B}_{\hat{p}}K \qquad \text{and} \qquad H_j = A_j - B_jK. \tag{1-3}$$

Then if $E\|y_0\|^2 < \infty$ and the stated condition on u_t is satisfied, then, as shown in [15] , the following system of stochastic differential equations will represent the linear hybrid system and the corresponding optimal nonlinear filter:

(Notice that as proven in [12] theorem 8-1.2, the first equation is almost surely equivalent to eq.

(1-2).)

$$\begin{cases} dy_t = \overline{H}_{\hat{p}} y_t \, dt + F \, d\nu_t, \\ d\hat{p}_t^j = \Pi^j \hat{p}_t \, dt + \hat{p}_t^j y_t^T (\overline{H}_{\hat{p}} - H_j)^T F^{-1} \, d\nu_t, \\ \qquad \text{for } j = 1, 2, \ldots, N-1 \\ \text{and} \\ \hat{p}_t^N = 1 - \sum_{j=1}^{N-1} \hat{p}_t^j, \end{cases} \tag{1-4}$$

where $\overline{A}_{\hat{p}} = \sum_{j=1}^{N} \hat{p}_t^j A_j$, $\overline{B}_{\hat{p}} = \sum_{j=1}^{N} \hat{p}_t^j B_j$ and $\hat{p}_0 = p_0 = E[\Phi_0]$.

(For the Wiener process w_t, every $F_t w_t$ with non-square F_t is equivalent to $\overline{F}_t \overline{w}_t$ with square symmetric $\overline{F}_t$ and $\overline{w}_t$ another Wiener process. Moreover, if $F_t {F_t}^T$ is invertible, $\overline{F}_t$ will also be invertible. Therefore, with no loss of generality, henceforth, we shall consider F to be an invertible square symmetric matrix.)

It is easy to show that with the above initial condition for $\hat{p}_t$, the filter output will always remain inside S, the standard $(N-1)$–vertex simplex and hence, will be uniformly bounded.

In this paper, our attention is restricted to the case of adaptive linear state feedbacks, that is to say we shall henceforth assume that the feedback law has the form; $u_t = -K_{\hat{p}} y_t$, where $K_{\hat{p}} = K(\hat{p}_t)$. In this case, with appropriate state feedback laws and initial conditions for $\hat{p}$, eq. (1-2) will obviously satisfy the conditions for the global existence and uniqueness theorem, and hence, the above filtering equations will still be applicable.

In the next section of this paper, we establish the stochastic stability of this system for a certain class of state feedbacks, denoted by $\mathcal{S}(K_{\hat{p}})$, which is the set of those state feedback that asymptotically (*i.e.* strictly) stabilize the pair $(\overline{A}_{\hat{p}}, \overline{B}_{\hat{p}})$ for all $\hat{p} \in S$. For simplicity, and also to satisfy later requirements, in addition, we shall restrict our analysis to the case where $K_{\hat{p}}$ is a C^∞ function of $\hat{p}$. A specific case is where $K_{\hat{p}}$ is the optimal state feedback obtained for $(\overline{A}_{\hat{p}}, \overline{B}_{\hat{p}})$ with respect to some quadratic cost criterion.

For such a class of controllers to exist, we evidently have to assume that $(\overline{A}_{\hat{p}}, \overline{B}_{\hat{p}})$ is a stabilizable pair (*i.e.* eigenvectors corresponding to eigenvalues of $\overline{A}_{\hat{p}}$ with non-negative real parts are controllable) for every $\hat{p} \in S$, which can be a restrictive assumption. To see that such a system can always exist, one may consider the case that $(A_j, B_j) = (k_j A_0, k_j B_0)$, where (A_0, B_0) is a stabilizable pair and $k_j > 0$ for all $j = 1, 2, \ldots, N$, or the case where (A_j, B_j) is inside some sufficiently small neighborhood of (A_0, B_0), an stabilizable pair. In fact, we know that the class of stabilizing adaptive state feedbacks is by no means restricted to $\mathcal{S}(K_{\hat{p}})$.†

† To verify this, we may look at the simplest case where the jump-Markov process is completely observed, *i.e.* $\hat{p}_t =$

2. Stochastic Stability of the Controlled Linear Hybrid system

Before proving our result on the stochastic stability of the controlled linear hybrid system of eq.'s (1-4), with certain adaptive state feedback algorithms, we shall define our notion of stochastic stability and present a stochastic Lyapunov function criterion to investigate this property. (This definition follows the definition of stochastic stability introduced in [18] for discrete systems.)

Definition. Assume that X_t is the solution of an SDE evolving on M, a C^∞ submanifold of $\mathcal{R}^n$.

We call the corresponding stochastic system, or the Markov process X_t, *stochastically stable*, if X_t is *bounded in probability* for every initial state, that is to say for every $\epsilon > 0$, $x \in M$ and $s \in \mathcal{R}^+$, there exists a compact set, $K \subset M$, such that:

$$\limsup_{t \to \infty} P(s, x, t, K^c) \leq \epsilon.$$

It will be shown in the next sections that to establish the existence of invariant probability measures and ergodicity, no stronger version of stochastic stability is required.

The following theorem, which is applicable to our special case, gives sufficient conditions for this form of stochastic stability and is similar to the Lyapunov function arguments used in the corresponding deterministic case:

Let the Markov process X_t, evolving on M, satisfy the following Ito SDE:

$$dX_t = b_t(X_t)\,dt + G_t(X_t)\,dw_t, \tag{2-1}$$

and let $\mathcal{A}$ denote the infinitesimal generator of the above equation, that is to say

$$\mathcal{A} = {b_t}^T(x)\frac{\partial}{\partial x} + \frac{1}{2}\mathrm{Tr}\left(G_t(x){G_t}^T(x)\frac{\partial^2}{\partial x^2}\right), \tag{2-2}$$

where $\frac{\partial}{\partial x} = \left[\frac{\partial}{\partial x_1}, \frac{\partial}{\partial x_2}, \ldots, \frac{\partial}{\partial x_n}\right]^T$ and $\frac{\partial^2}{\partial x^2} = \frac{\partial}{\partial x}\frac{\partial}{\partial x}^T$.

Theorem 1. *Suppose that eq. (2-1) satisfies the conditions for global uniqueness and existence of a solution. Furthermore, assume that there exist a $C^1(\mathcal{R}^+) \times C^2(\mathcal{R}^n)$ function*

Φ_t. In this case, it has been shown that the optimal solution, which is a function of Θ_t, does not necessarily stabilize the (A_j, B_j) pair when Θ_t is taken equal to j, for all j-s.

$V_{\cdot}(\cdot)$, $C(\mathcal{R}^+)$ functions $W(\cdot)$ and $U(\cdot)$ and positive real numbers $\alpha > 0$ and κ, such that

$$U(\|x\|) \leq V_t(x) \leq W(\|x\|), \quad \forall x \in M, \quad \forall t \geq 0,$$
$$\min_x U(\|x\|) \geq 0, U(\|x\|) \to \infty \quad \text{as} \quad \|x\| \to \infty$$

and

$$\frac{\partial V_t(x)}{\partial t} + \mathcal{A} V_t(x) \leq -\alpha W(\|x\|) + \kappa, \qquad \forall x \in M, \quad \forall t \geq 0.$$

Then if $E[V_0(X_0)] < \infty$,

$$\limsup_{t\to\infty} E[V_t(X_t)] \leq \frac{\kappa}{\alpha}$$

and X_t is bounded in probability.

The proof is simply an application of Ito's formula. By this formula and the above inequality we are able to show that

$$E[V_t(X_t)] \leq E[V_0(X_0)] \cdot e^{-\alpha t} + \frac{\kappa}{\alpha}\left(1 - e^{-\alpha t}\right), \quad \forall t \geq 0$$

and the assertion concerning $\limsup_{t\to\infty} E[V_t(X_t)]$ follows immediately.

Since $V_t(x)$ is uniformly nonnegative and radially unbounded, uniform boundedness of its mean can happen only if $\lim_{k\uparrow\infty} \sup_{t\geq 0} \Pr\{\|X_t\| \geq k\} = 0$, which is equivalent to

$$\forall \epsilon > 0 \quad \exists K \subset M \text{ compact} \quad \ni \quad \sup_{t\geq 0} \Pr\{X_t \in K^c\} \leq \epsilon, \tag{2-3}$$

for any initial probability distribution such that $E[V_0(X_0)] < \infty$.

This proves that the process X_t is bounded in probability and that the system is stochastically stable.

Now, by using this theorem and under the stabilizability assumption made in the last section, we are able to prove stochastic stability of the linear hybrid system of eq.s (1-1) and (1-2), by some adaptive state feedback gain $K_{\hat{p}} \in \mathcal{S}(K_{\hat{p}})$;

Proposition 1. *Consider the system of autonomous stochastic differential equations (1-4) where $K_{\hat{p}}$ is any adaptive state feedback gain in the set $\mathcal{S}(K_{\hat{p}})$.*

For this system, if $\max_{i,j} \|A_i - A_j\|$ and $\max_{i,j} \|B_i - B_j\|$ are sufficiently small, $E\|y_0\|^2 < \infty$ and $\hat{p}_0 = p_0$ is a valid probability distribution, then there exists a non-empty subset of the set $\mathcal{S}(K_{\hat{p}})$ (as defined in the last section) such that for all state feedback laws in that set, the system is stochastically stable.

In particular, $K_{\hat{p}}$ *can be chosen as* $K^*_{\hat{p}}$*, the solution of Riccati equation for the following "frozen" system:*

$$dy_t(\hat{p}) = \left(\overline{A}_{\hat{p}}y_t(\hat{p}) + \overline{B}_{\hat{p}}u_t(\hat{p})\right)\, dt + F\, d\nu_t,$$

which minimizes the infinite-horizon cost function,

$$J = \lim_{t\to\infty} \frac{1}{t}\int_0^t y_s(\hat{p})^T Q y_s(\hat{p}) + u_s(\hat{p})^T R u_s(\hat{p})\, ds, \qquad (2-4)$$

for all finite initial conditions.

(One may rephrase this result by the statement that if the variations of the hybrid system parameters over the simplex are sufficiently small, then the controlled system is stochastically stable.)

We will not present the complete proof of this proposition here, but it can be said that the proof is simply an application of theorem 1 with the following function as our choice of the stochastic Lyapunov function:
(Throughout this paper, the joint state process of eq.'s (1-4) will be denoted by x_t, *i.e.* $x = (y, \hat{p})$.)

$$V(x) = y^T P_{\hat{p}} y + \frac{\epsilon}{2}\hat{p}^N(1-\hat{p}^N)y^T y \qquad \text{for some } \epsilon > 0, \qquad (2-5)$$

where $P_{\hat{p}}$ is the solution of the following Lyapunov equation:

$$\overline{H}_{\hat{p}}^T P_{\hat{p}} + P_{\hat{p}}\overline{H}_{\hat{p}} = -qI, \qquad \text{for some } q > 0. \qquad (2-6)$$

To have the above function satisfying the conditions of theorem 1, that is to say to find constants α and κ satisfying the required inequality for $\mathcal{A}V(x)$, we have been able to show that it is sufficient to have the following inequalities being true for the value of q chosen in (2-6) and some $\epsilon > 0$:

$$\|\frac{\partial P_{\hat{p}}}{\partial \hat{p}^j}\| \le \epsilon \qquad \text{and} \qquad \|\frac{\partial^2 P_{\hat{p}}}{\partial \hat{p}^i \partial \hat{p}^j}\| \le \frac{q - \overline{h}\frac{\epsilon}{4}}{(N-1)(1+2\sigma)} - \frac{\epsilon}{2}, \qquad (2-7)$$

for all $i, j = 1, 2, \ldots, N-1$ and $\hat{p} \in S$.
Let $\max_j \|A_j\| = \overline{a}$, $\max_j \|B_j\| = \overline{b}$, $\max_{i,j} \|A_i - A_j\| = \sigma_1$, $\max_{i,j} \|B_i - B_j\| = \sigma_2$ and $\max_{\hat{p}\in S} K_{\hat{p}} = \overline{k}$, then σ and $\overline{h}$ in the above inequalities will be equal to $\sigma_1 + \sigma_2\overline{k}$ and $\overline{a} + \overline{bk}$, respectively. (The above maximums can be achieved because $K_{\hat{p}}$ is assumed to be a C^∞ function of $\hat{p}$.)

On the other hand, we are able to show that the upper bounds on $\|\frac{\partial P_{\hat{p}}}{\partial \hat{p}^j}\|$ and $\|\frac{\partial^2 P_{\hat{p}}}{\partial \hat{p}^i \partial \hat{p}^j}\|$, which depend upon the system and controller parameters, are directly proportional to σ, $\gamma_1 =$

$\max_{\substack{\hat{p}\in S\\ j}} \|\frac{\partial K_{\hat{p}}}{\partial \hat{p}^j}\|$ and $\gamma_2 = \max_{\substack{\hat{p}\in S\\ i,j}} \|\frac{\partial^2 K_{\hat{p}}}{\partial \hat{p}^i \partial \hat{p}^j}\|$. Thus, to satisfy inequalities (2-7), it is sufficient to have σ and the upper bounds on the first and second derivatives of $K_{\hat{p}}$ with respect to $\hat{p}$, small enough.

Furthermore, by using the Riccati equation, we are able to prove that in the case of $K^*_{\hat{p}}$, the optimal state feedback for the above-mentioned frozen system (as well as for some other state feedback algorithms), γ_1 and γ_2 are directly proportional to σ_1 and σ_2. Notice that because of the stabilizability condition imposed on $(\overline{A}_{\hat{p}}, \overline{B}_{\hat{p}})$ for all $\hat{p} \in S$, $K^*_{\hat{p}}$ is a C^∞ function of $\hat{p}$ and hence, a member of $\mathcal{S}(K_{\hat{p}})$.

Therefore, the conclusion is that for some $K_{\hat{p}} \in \mathcal{S}(K_{\hat{p}})$, including the frozen-system optimal state feedback $K^*_{\hat{p}}$, with sufficiently small σ_1 and σ_2, which are among the system specifications, for any chosen $q > 0$, one may choose $\epsilon > 0$ such that the inequalities (2-7) are satisfied. As was mentioned earlier, inequalities (2-7) are sufficient (although not necessary) to prove the stochastic stability of the system.

Regarding the specific case of $K^*_{\hat{p}}$, in spite of the fact that in this case, $\hat{p}$ is actually varying with time and the system is not frozen, if, however, the variations of the system parameters over the simplex S are reasonably small, or in other words, if the dynamics of the system are not much different from the corresponding frozen system for every $\hat{p}$, one can regard $K^*_{\hat{p}}$ as a good approximation to the optimal solution. Thus, for every linear hybrid system which satisfies the conditions of the proposition, $K^*_{\hat{p}}$ will be called the *near-optimal adaptive state feedback*. Of course, when the conditions of proposition 1 are satisfied, application of this adaptive controller stochastically stabilizes the linear hybrid system of eq. (1-2).

In the next sections, we shall assume that the hybrid system has been stabilized using a certain adaptive state feedback and will proceed with our discussion of invariant probability measures, weak controllability and ergodicity of the system. As will be shown, after we have assumed that the system is stochastically stable, the type of (stochastically) stabilizing adaptive controller employed will not have any effect in our subsequent conclusions.

3. Existence of Invariant Probability Measures for the Controlled Linear Hybrid System

Of the important consequences of stochastic stability of a homogeneous Markov process is the existence of an invariant probability measure which under specific conditions implies the ergodicity of the process in some nontrivial set. One of the earliest results about this relationship was given in [1] . A theorem to follow gives a statement of these results completed by those obtained in [20] .

To present this theorem, we will require the following definitions and notations, where we will assume X_t to be a homogeneous Markov process, solution of an autonomous SDE with coefficients satisfying certain standard growth and Lipschitz continuity conditions, evolving on M, a C^∞ submanifold of $\mathcal{R}^n$. This process is by definition a diffusion process and for such a process, the Feller property holds.

Definitions.

- ⋆ With $P(t,x,B)$ the probability distribution function of the process X_t, define the operator U_t for all $t \geq 0$ as:

$$U_t\nu(B) = \int_M P(t,x,B)\nu(dx) \qquad \forall \nu \in \mathcal{M}^+(M),$$

where $\mathcal{M}(M)$ denotes the space of all normalized positive bounded (*i.e.* probability) measures on M.

Notice that with μ_t^X denoting the probability distribution of X_t at time t, one may rewrite the Chapman-Kolmogoroff equation as $\mu_t^X = U_t\mu_0^X$.

- ⋆ A family of probability measures, $\{\mu_i;\ i \in I\}$ is said to be (*uniformly*) *tight* if

$$\forall \epsilon > 0 \quad \exists K \subset M \text{ compact} \qquad \ni \qquad \inf_{i \in I} \mu_i(K) \geq 1 - \epsilon.$$

- ⋆ For a homogeneous Markov process, a probability measure, π, is called *an invariant probability measure* if $U_t\pi = \pi$ for all $t \geq 0$.
- ⋆ An invariant probability measure π is called *extremal* if it cannot be decomposed into a nontrivial sum (or integral) of distinct probability measures, that is to say, a sum like $\pi = \sum_{i \in I} \alpha_i \pi_i$ with each α_i a strictly positive real number (or the corresponding integral).

Theorem A (Beneš-Saperstone).[1, 20] *For a homogeneous Feller Markov process, X_t, the following conditions are equivalent:*

(i) X_t *has a nontrivial invariant probability measure,* π.

(ii) There exists a moment function f *and a probability distribution* $\nu \in \mathcal{M}^+(M)$ *such that if* $\nu = \mu_0^X$, *then*

$$\sup_{t\geq 0} E[f(X_t)] = \sup_{t\geq 0} \int_M f(x) U_t \nu(\,dx) < \infty \tag{3-1}$$

(iii) There exists $\nu \in \mathcal{M}^+(M)$ *such that the family of probability measures* $\{U_t\nu;\ \forall t \geq 0\}$ *is tight.*

Moreover, each one of the above conditions is sufficient for the following weak convergence in $\mathcal{M}^+(M)$ *to be true, where* ϑ *is any probability measure such that* $\operatorname{Supp}\vartheta \subset \overline{\operatorname{Supp}\nu}$ †, ν *is the above-mentioned initial probability measure and* π_ϑ *is an invariant probability measure unique with respect to* ϑ:

$$\lim_{t\to\infty} \frac{1}{t} \int_0^t U_s\vartheta\, ds = \pi_\vartheta, \tag{3-2}$$

or equivalently, for every $f \in C_0(M)$ *(the Banach space of all continuous functions vanishing at infinity), if* $\vartheta = \mu_0^X$, *then* ‡,

$$\lim_{t\to\infty} \frac{1}{t} \int_0^t E[f(X_t)]\, ds = E_{\pi_\vartheta}[f(x)].$$

□

Later, by using the ergodicity theorem, we are able to complete this result by showing that in fact for every initial probability measure ϑ such that $\operatorname{Supp}\vartheta \subseteq \overline{\operatorname{Supp}\pi}$, where π is an invariant probability measure,

$$\lim_{t\to\infty} \frac{1}{t} \int_0^t U_s\vartheta\, ds = \pi. \tag{3-2'}$$

A process for which the above-mentioned condition *(ii)* is satisfied for every initial condition $x \in M$, by the definition of a moment function¶, is stochastically stable. On the other hand, if X_t is the solution of a stochastically stable SDE, then by the almost sure continuity of the solutions

---

† We define $\operatorname{Supp}\mu$ or *support* of the measure μ as the set $\{x \mid \exists \epsilon_x > 0\ \forall N_x$ a neighborhood of $x \ni \frac{\mu(N_x)}{\lambda(N_x)} \geq \epsilon_x\}$, where λ is the Lebesgue measure for M.
This can also be stated more informally as $\operatorname{Supp}\mu = \{x \mid \mu(\,dx) > 0\}$. For probability distributions with a probability density function, support is the set on which the pdf is strictly positive.

‡ $E_\nu[f(x)]$ is the expectation of $f(x)$ with respect to the probability measure ν.

¶ A continuous function f is called a *moment* if $\liminf_{n\to\infty} \inf_{x\in K_n^c} g(x) = \infty$, where (K_n^c) denotes a sequence of compact sets such that $K_n \uparrow M$ as n goes to infinity.

of the SDE, it satisfies relation (2-3) for all initial conditions $x \in M$. Notice that relation (2-3) is exactly equal to the tightness of $\{P(t, x, \cdot);\ t \geq 0\}$ as a family of probability distributions, which shows that stochastic stability is a sufficient condition for the weak convergence of eq. (3-1) to hold for any initial condition. This result is stated in the following corollary:

Corollary 1. *Suppose that the Markov process X_t is the solution of a stochastically stable Ito SDE. For such a process, corresponding to every finite probability distribution ν*[§]*, there exists an invariant probability measure π_ν as the limit of the following weak convergence:*

$$\lim_{t \to \infty} \frac{1}{t} \int_0^t U_s \nu \, ds = \pi_\nu$$

or equivalently for every $f \in C_0(M)$, if $\nu = \mu_0^X$, then

$$\lim_{t \to \infty} \frac{1}{t} \int_0^t E[f(X_t)] \, ds = E_{\pi_\nu}[f(x)].$$

□

The above corollary and proposition 1 clearly result in the following statement:

○ *Consider the controlled linear hybrid system of eq.'s (1-4), with any stabilizing adaptive state feedback (including $K_{\hat{p}} \in S(K_{\hat{p}})$, provided that the conditions of proposition 1 are satisfied). Then for every finite initial probability distribution ϑ for the state process, i.e. $\vartheta = \mu_0^x$, an invariant probability measure π_ϑ exists such that*

$$\frac{1}{t} \int_0^t \int_M P(s, x, \cdot) \vartheta(dx) \, ds \to \pi_\vartheta \qquad \text{as} \quad t \to \infty. \tag{3-3}$$

With the existence of an invariant probability measure having been proved, our next step is to prove that for every initial condition, the state process of the system is ergodic. Of course, if the number of distinct extremal invariant probability measures is not finite or the ergodicity does not hold for every initial condition, then there is practically no point in using the ergodicity property for estimation of the system performance indices. The next section gives some general results about the sufficient (and necessary) conditions for the above-mentioned properties to be satisfied, which we will apply to the stochastically stabilized system of eq.'s (1-4).

---

§ Suppose weak convergence holds for every $x \in M$, then by the uniform boundedness of $\frac{1}{t} \int_0^t E[f(X_s(x))] \, ds$ on $M \times \mathcal{R}^+$ for any $C_0(M)$ function f, it also holds for every finite probability distribution on M.

4. Invariant Probability Measures, Ergodicity and Weak Controllability of Diffusion Processes

In this section, we present a few results about the relationship between invariant probability measures, ergodicity and weak controllability, for the diffusion process generated by the following standard autonomous stochastically stable Ito SDE, evolving on M, a C^∞ submanifold of $\mathcal{R}^n$:

$$dX_t = b(X_t)\,dt + G(X_t)\,dw_t, \qquad (4-1)$$

where X_t denotes the $\mathcal{R}^n$-valued diffusion process and w_t is an m-dimensional Wiener process.

By the Stroock-Varadhan support theorem (see [21]), we know that for the above stochastic differential equation, the following equality is true:
(Since eq. (4-1) is stochastically stable, its coefficients are not required to be bounded as it is the case in [21] . Also notice that support of a measure, as defined in [21] , is equivalent to the closure of that set using our definition (see the footnotes in section 3). Hence, eq. (4-2) is slightly different from the corresponding equality given by [21] .)

$$\overline{\text{Supp}\, P(t,x,\cdot)} = \overline{\Omega^+(t,x)} \qquad \forall t \geq 0, \qquad (4-2)$$

where $P(t,x,\cdot)$ denotes the transition probability function of the process X_t and $\Omega^+(x)$ is the reachable set of the trajectories of the following deterministic differential equation, originating at point x at time t, using all piecewise continuous functions ψ as inputs:

$$d\xi_t^\psi = b'(\xi_t^\psi)\,dt + G(\xi_t^\psi)\psi_t\,dt, \qquad (4-3)$$

where

$$b'(x) = b(x) - \frac{1}{2}\sum_{j=1}^{m}\left(\frac{\partial}{\partial x}G^{j^T}(x)\right)^T G^j(x),$$

whenever the derivatives exist. Notice that $dx_t = b'(x_t)\,dt + G(x_t)\delta w_t$ is the Stratonovich differential equation almost surely equivalent to eq. (4-1).

In the following definitions, we follow [13] in most cases, which gives these definitions in term of the trajectories of eq. (4-3). Of course, by using the mentioned equivalence between these deterministic trajectories and the sample paths of solutions of eq. (4-1), the definitions concerning accessibility and invariance properties, may equivalently be interpreted in term of supports of probability distributions.

Definitions. Let $\Omega^+(x) = \bigcup_{t\geq 0} \Omega^+(x,t)$. With respect to these trajectories:

* ⋆ Two points $x, y \in M$ are said to be *accessible* from each other if there exists a sequence of points $x_0, x_1, x_2, \ldots, x_l$ where $x_0 = x$, $x_l = y$ such that for each $i = 1, 2, \ldots, l$, for some finite time $t \geq 0$, $x_i \in \Omega^+(t, x_{i-1})$ or $x_{i-1} \in \Omega^+(t, x_i)$.

  We say that x and y are *accessible* from each other *with respect to the set* N, if in the above definition, we have $x_i \in \Omega^+_N(t, x_{i-1})$ or $x_{i-1} \in \Omega^+_N(t, x_i)$, instead, where $\Omega^+_N(t, x_i)$ is the reachable set considering only the trajectories strictly inside N.
* ⋆ Suppose that for every neighborhood of x, N_x, there exists another neighborhood of x, N'_x, such that for all $y \in N'_x$, x and y are accessible from each other with respect to N_x. Then, the system is said to be *locally accessible at point* x.
* ⋆ For systems with $C^\infty(M)$ coefficients, the (local) accessibility property has equivalently been called *(local) weak controllability* (see [8]), which is the term we shall use henceforth in this paper.[‡] for any $t > 0$.
* ⋆ If the system is locally weakly controllable at all points in M, it is said to be *weakly controllable* in M.
* ⋆ A set $A \subset M$ is called *invariant* with respect to eq. (4-1), if for all $x \in A$, $\Omega^+(x) \subset \overline{A}$.
* ⋆ A set $C \subset M$ is called an *invariant control set*, if for every $x \in C$, $\overline{\Omega^+(x)} = \overline{C}$.
* ⋆ The process X_t is called *Harris recurrent* or *recurrent in the sense of Harris* on set $A \subset M$ *(with respect to the probability measure* ν*)* if for every $E \subset A$ such that $\nu(E) > 0$,

$$\Pr\{X_t(x) \in E \quad i.o.\} = 1 \qquad \forall x \in A, \tag{4-4}$$

  where $X_t(x)$ denotes the process X_t when $X_0 = x$.

Considering the above definitions and assumptions on X_t, we have been able to prove the following lemma and theorem:

Lemma 1. *Any invariant control set for the process X_t satisfies the following properties:*

(i) An invariant control set C is a path-connected set which is either closed, i.e. $\overline{C} = C$, or its boundary, $\partial C = \overline{C} - C$, contains a non-empty invariant set.

(ii) If C_1 is an invariant subset of C, then $\overline{C_1} = \overline{C}$.

(iii) Every two distinct invariant control sets are disjoint. □

[‡] The reason for this terminology of weak controllability is that our definition of local accessibility is shown to be similar to the notion of local controllability defined for nonlinear deterministic systems by; $\mathrm{Int}\,\Omega^+(x, t) \neq \emptyset$, where $Int\, C$ denotes the interior of the set C with respect to the topological set M, unless specified otherwise.

Theorem 2. *For the diffusion process X_t, the following statements regarding the set C are equivalent:*

(I) C is the support of a unique extremal invariant probability measure π, recurrent with respect to the process X_t.

(II) $\overline{C}$ is an invariant control set where the process X_t is Harris recurrent on C with respect to any probability measure equivalent to π. □

If, in addition, the system is stochastically stable, the following corollary to the above theorem can easily be proved:

Corollary 2. *If the system is stochastically stable and C is a closed invariant control set, then X_t is Harris recurrent on C and there exists a unique extremal invariant probability measure π such that $C = \overline{\text{Supp}\,\pi}$.* □

The mentioned results about invariant probability measures provide the necessary means for proving a global ergodicity result for certain stochastically stable systems. This is stated in the following theorem:

Theorem 3. *Consider an extremal invariant probability measure π for the process X_t. For every probability measure ν, such that $\text{Supp}\,\nu \subset \overline{\text{Supp}\,\pi}$, and every $L^1(\pi)$ function f, if $\mu_0^X = \nu$, the following ergodic relation holds:*

$$\lim_{t\to\infty} \frac{1}{t} \int_0^t f(X_s)\, ds = E_\pi[f(x)] \qquad \text{a.s.} \tag{4-5}$$

Furthermore, suppose that X_t is the solution of a stochastically stable SDE with $C^\infty(M)$ coefficients and that every invariant probability measure for this process has an open-set support.

Then the above convergence will be true for every initial condition $x \in M$, some invariant probability measure π_x and all $L^1(\lambda)$ functions f, where λ denotes the Lebesgue measure on M.

□

To prove the above theorem, we use Doob's classic ergodicity theorem for strictly stationary processes ([5] theorems $X-2-1$ and $XI-2-1$).

If the initial probability distribution of X_t is equal to an invariant probability measure, by definition, X_t will be stationary in the strict sense. Thus, using Doob's theorem and the Chapman-Kolmogoroff equation, eq. (4-5) is proved for almost every initial condition in Supp π, where π is any invariant probability measure. Next, lemma 1 and theorem 2 prove that eq. (4-5) is true for all $x \in \text{Supp}\,\pi$.

Finally, by using the Beneš-Saperstone theorem and openness of the supports of invariant probability measures, we can extend this result to a global ergodicity result. We prove that for every $x \in M$, $X_t(x)$ enters Supp π_x, for some invariant probability measure π_x, at some finite time. Then, invariance of Supp π_x will ensure that eq. (4-5) holds.

It should be mentioned that [13] or [18] also provide results similar to theorems 2 and 3 though their approach is different from ours.

As it is observed, to establish the global ergodicity property for a stochastically stable system, one has to prove that the supports of all invariant probability measures for this system are open sets.

In our next step, assuming that $b(\cdot)$ and $G(\cdot)$, the coefficients of the eq. (4-1), are C^∞ functions, we show that openness of the supports is equivalent to the weak controllability of a stochastically stable system. If $K_{\hat{p}}$ is a C^∞ function of $\hat{p}$, clearly, the mentioned smoothness assumption will be applicable to our special case of eq.'s (1-4).

Before presenting these results, we should recall the fact that every invariant probability measure for eq. (4-1) must satisfy the Kolmogoroff forward equation; $\mathcal{A}_y^* \pi(dy) = 0$ in the distributional sense. Here, $\mathcal{A}^*$ denotes the adjoint of $\mathcal{A}$, the infinitesimal generator of the stochastic differential equation (4-1). (For a proof of this statement, see the proof of theorem 3 in [11] .)

We will prove the following theorem by using this partial differential equation:

Theorem 4. *Suppose that there exists some point x_0 such that the system is locally weakly controllable at x_0. Then any invariant probability measure for this system is either zero or has an infinitely differentiable ($C^\infty(M)$) density, at x_0. Furthermore, if the system is globally weakly controllable in M, then every invariant probability measure will have a $C^\infty(M)$ density function everywhere in M.* □

For a proof, see appendix A.

Regarding the results obtained so far in this section, we are now prepared to establish the relationship between the weak controllability of a stochastically stable system and openness of the supports of invariant probability measures. Of course, by the support theorem, the stochastic system of eq. (4-1) is weakly controllable if and only if the deterministic system of eq. (4-3) is weakly controllable.

Theorem 5. *If the system of eq. (4-1), generating the process X_t, is stochastically stable, then the following statements are equivalent:*

(i) *No extremal invariant probability measure of the process X_t can lie on a submanifold of dimension lower than* dim M.

(ii) *All invariant control sets of the system are closed and have non-empty interiors.*

(iii) *The system is weakly controllable in M.*

(iv) *For the process X_t, a non-zero invariant probability measure, π, exists such that $\pi(dx) = \phi(x)\,dx$, where $\phi(x)$ is a $C^\infty(M)$ function. In this case,* Supp π *will naturally be an open set.* □

For a proof, see appendix B.

At this stage, we ought to also mention a well-known sufficient condition for weak controllability which is the satisfaction of the controllability rank condition (see [8]). The controllability rank condition is satisfied whenever a Lie algebra of vector fields generated by the coefficients of eq. (4-3) ($b'x$) and columns of $G(x)$) is of full rank. The discussion in [10] and [11] shows that this is also a sufficient condition for hypoellipticity § of $\mathcal{A}$ and its adjoint operator, $\mathcal{A}^*$. In particular, the satisfaction of this condition will imply that for any distribution μ on M, if $\mathcal{A}\mu$ or $\mathcal{A}^*\mu$ has a $C^\infty(M)$ density function, then so does μ. Obviously, this is equivalent to saying that all the corresponding invariant probability measures are infinitely differentiable.

5. Weak Controllability and Ergodicity of the Controlled Linear Hybrid System

With the results obtained in the last section (theorems 3 and 5), it is now quite easy to prove that in our special case, the controlled linear hybrid system of eq.'s (1-4) is weakly controllable and ergodic.

As it was mentioned in the last section, one way of proving the weak controllability property is to prove that the corresponding Lie algebra is of full rank. However, regarding the generality of the problem in our case, our efforts to examine the rank of this Lie algebra have not been successful. For this reason, we have used the equivalence stated in theorem 5 instead to prove our objective. In the following discussion, we show that the first statement of theorem 5 holds for the controlled hybrid system, that is to say, every invariant probability measure for the system

§ A partial differential operator $\mathcal{P}$ is hypoelliptic in M if for every distribution μ on M, Sing Supp $\mu \subseteq$ Sing Supp $\mathcal{P}\mu$, where Sing $\equiv$ Singularity of.

of SDE's (1-4) has an $(n+N-1)$–dimensional support. (n is the dimension of the y-space and N the number of distinct states of θ_t, the Jump-Markov process.)

To prove this claim, we need the following auxiliary result which states that for any invariant probability measure for the system of eq.'s (1-4), the projection of its support on the y-space is $\mathcal{R}^n$, itself.

Let M, the submanifold of $\mathcal{R}^{n+N-1}$ subject to our analysis, be any C^∞ submanifold containing $S \times \mathcal{R}^n$.

Proposition 2. *Consider the controlled linear hybrid system of eq.'s* $(1-4)$ *where* $K_{\hat{p}}$ *is any stabilizing state feedback gain.*

Then for all $t > 0$ *and* $x \in M$, *for every* $y \in \mathcal{R}^n$, *there exists* $\hat{p} \in S$ *such that* $(y, \hat{p}) \in$ Supp $P(t, x, \cdot)$. *In other words, for any neighborhood* N *in* $\mathcal{R}^n$,

$$P(t, x, S \times N) > 0 \qquad \forall t > 0, \quad \forall x \in M.$$

□

In view of the support theorem, the proof of this proposition is virtually immediate. By this theorem, with respect to Supp $P(t,x,\cdot)$, eq.'s (1-4) can be replaced with a system of ODE's, where evidently, $\hat{p}_t$ can be shown to be a C^∞ functional on $y^t = \{y_s;\ 0 \le s \le t\}$.

Therefore, we have $\dot{y}_t = Z(y^t) + F\psi_t$, where ψ_t can be any piecewise continuous function. Obviously, an implication of F being full rank in this equation is that the reachable set in the y-space is equal to $\mathcal{R}^n$. This is equivalent to saying that $\overline{\text{Supp}\, P(t,x,S\times\cdot)} = \mathcal{R}^n$, where Supp $P(t, x, S \times \cdot)$ denotes the set $\{y \mid \exists \hat{p} \in S \ \ni (y, \hat{p}) \in \text{Supp}\, P(t, x, \cdot)\}$.

Since $\pi(\cdot) = \int_M P(t,x,\cdot)\pi(dx)$, as an immediate corollary to the above proposition, we can extend its result to any invariant probability measure for the system of eq.'s (1-4);

$\pi(S \times N) > 0$, *where* π *is any invariant probability measure and* N *is any neighborhood in* $\mathcal{R}^n$.

Now, by using the above proposition and the fact that $\hat{p}_t$ is a filter output and a conditional expectation, we are able to prove what we need: suppose that there exists some invariant probability measure which lies in a submanifold with dimension less than $n + N - 1$ and that the Jump-Markov process θ_t is irreducible. Then, y_t and $\hat{p}_t$ must be dependent at every time $t \ge 0$, which, with y_t sweeping all $\mathcal{R}^n$ and $\hat{p}_t$ being a conditional expectation on y^t, is proved to be a contradiction.

Proposition 3. *For the system of eq.'s (1-4), suppose that $K_{\hat{p}}$ is a stabilizing adaptive state feedback and a C^∞ function of $\hat{p}$, and that Π represents the probability transition matrix of an irreducible finite Jump-Markov process.*

Then every closed invariant subset of M has a non-empty interior. In other words, all the invariant probability measures for this system are $(n+N-1)$–dimensional. □

For a proof, see appendix C.

Now that the first statement of theorem 5 is proved to be true, the other statements will follow, as well:

Corollary 3. *The system of eq.'s (1-4), under the conditions of the proposition, is weakly controllable in a C^∞ submanifold containing M. Furthermore, every extremal invariant probability measure for that system has a $C^\infty(M)$ density function and an open-set support.* □

Now, by using theorem 3 and regarding the smoothness of invariant probability measures, global ergodicity of the system is clear. The only remaining part is to prove that the number of extremal invariant probability measures is finite. These properties form the final result of this paper:

Proposition 4. *Consider the system of eq.'s (1-4). Under the conditions given in the hypothesis of proposition 3, for the state process of this system, a finite number of extremal invariant probability measures, with disjoint supports in M, exist.*

Moreover, for every initial condition in M, the ergodicity property holds. That is to say, for every $f \in L^1(M)$ and every initial condition $x_0 \in M$ for the state process x_t (i.e. $\mu_0^x = \delta(x_0)$), we have

$$\lim_{t\to\infty} \frac{1}{t}\int_0^t f(x_s)\,ds = E_{\pi x_0}[f(x)] \qquad \text{a.s.}$$

□

For a proof, see appendix D.

As one specific application of this proposition, the cost function of eq. (2-4) satisfies the following equality:

$$J = \lim_{t\to\infty} \frac{1}{t}\int_0^t ({y_s}^T Q y_s + {u_s}^T R u_s)\,ds = E_{\pi x_0}[y^T(Q + {K_{\hat{p}}}^T R K_{\hat{p}})y], \qquad (5-1)$$

where $x_0 = (y_0, \hat{p}_0 = p_0)$ the initial condition, is in general a finite random variable. This equality may be used to analyze the performance of the system.

6. Conclusion

Under certain hypotheses, the stochastic stability of a proposed class of adaptive state feedbacks for the linear hybrid system of eq. (1-2) has been proved.

Furthermore, it has been established that for any (stochastically) stabilizing adaptive state feedback, the controlled system is weakly controllable and that the global ergodicity property holds.

In addition, some computer simulation results not presented here, show that the near-optimal state feedback, introduced in section 2 with respect to the cost function of eq. (2-4), not only (stochastically) stabilizes the system, but also provides a better performance than the optimal constant state feedback introduced in [9] .

For a final remark, as a conjecture, we expect the proposed adaptive controller to stabilize some linear hybrid systems for which no stabilizing constant state feedback exists.

We believe that the combination of methods used in this chapter of the volume are most promising for the analysis of the adaptive control of time varying systems. This is true for several reasons:

First, the use of the invariant measure techniques combined with types of stochastic Lyapunov functions introduced by Meyn & Caines [19] are among the very few approaches that have been used to date to successfully analyse the behaviour of control systems with time varying parameters, which could exhibit a large range of behaviour. The earlier martingale methods (see for example [3]) were restricted to parameters that had bounded sample paths, and, furthermore, the results that one could generate were less informative than those given by the invariant measures technique.

Second, the use of non-linear filtering opens the possibility of an analysis of the optimal behaviour of adaptive control systems, but , of course , at the cost of analysing the behaviour of the non-linear filters. However, it is here again that the invariant measure analysis may give us substantial information without necessarily requiring a completely detailed analysis of the filter's transient behaviour.

Third, and, finally, it seems that the techniques of this paper may be extended by stochastic controllability analysis, using Lie algebraic methods. Such methods have not been used in this paper, but we have attempted to exploit them in related works. An analysis that involves such techniques seems to be related to some recent methods involving the Malliavin calculus.

Appendix A. Proof of Theorem 4

Let the system be locally weakly controllable at x_0, then it can be shown that $\operatorname{Int}\Omega^+(x_0) \neq \emptyset$ or to be more specific, for any neighborhood of x_0, N_{x_0}, $(\Omega^+(x_0) \cap N_{x_0})$ cannot lie in a submanifold of $\mathcal{R}^n$ of lower dimension than dim M.

On the other hand, if any invariant probability measure has some singularity at x_0, then obviously, x_0 is inside an invariant set which is singular (of lower dimension) with respect to M in a neighborhood of x_0 (see the following lemma).

The above contradiction implies that no invariant probability measure can be singular at the point x_0. Now, recalling the partial differential equation for invariant probability measures; $\mathcal{A}_y^*\pi(dy) = 0$, it is a routine argument for homogeneous linear differential equations with C^∞ coefficients to show that none of the derivatives of $\phi(y) = \pi(dy)/dy$ can be singular at x_0. This proves that $\phi(y)$ is $C^\infty(M)$ (infinitely differentiable) at point x_0.

Lemma 2. *Consider X_t the diffusion process generated by the system of eq. 4-1, where $b(x) \in C^1(M)$ and $G(x) \in C^2(M)$, and let π be an invariant probability measure for this process. Then if π is singular on the closed set $A \subset \operatorname{Supp}\pi$, A is an invariant set with respect to the process X_t.*

Moreover, with $A^c = \operatorname{Supp}\pi - A$, if $A^c \neq \emptyset$, then $\pi_0 = \frac{\mathbb{1}_A}{\pi(A)}\pi$ and $\pi_1 = \frac{\mathbb{1}_{A^c}}{\pi(A^c)}\pi$ will be invariant probability measures for the process X_t, where $\pi = \pi(A)\pi_0 + \pi(A^c)\pi_1$.

Proof. Since A is the set of singularities of π, Obviously, $0 < \pi(A)$ and $\operatorname{Int} A = \emptyset$ with respect to $\operatorname{Supp}\pi$ as the topological set.

If $A = \overline{\operatorname{Supp}\pi}$, there is nothing to prove. Thus, let $A \neq \overline{\operatorname{Supp}\pi}$ implying that $0 < \pi(A) < 1$ (notice that if $\nu(E) = 1$ for a probability measure ν, then by the definition of support, $E = \overline{\operatorname{Supp}\nu}$).

If $\mathcal{A}$ is the infinitesimal generator corresponding to eq. 4-1, we know that π must be a solution of the partial differential equation $\mathcal{A}_y^*\nu(dy) = 0$, or equivalently, $\mathcal{A}_y^*\nu = 0$ in the distribution sense, which can be obtained from the Kolmogoroff forward equation. This equation can explicitly be written as

$$0 = \left(\frac{\partial}{\partial t} - \mathcal{A}_y^*\right) P(t, x, dy) \qquad \forall x, y \in M, \quad \forall t \geq o, \tag{A-1}$$

$$= \frac{\partial}{\partial t}P(t, x, dy) + \left(\frac{\partial}{\partial y}^T \left(P(t, x, dy)b(y)\right)\right) - \frac{1}{2}\operatorname{Tr}\left(\frac{\partial^2}{\partial y^2}\left(P(t, x, dy)G(y)G^T(y)\right)\right),$$

where clearly, $P(t, x, dy)$ can be replaced by $U_t\nu(dy)$ for every initial finite measure ν.

Now, regarding eq. (A-1), it is an standard argument in distribution theory or differential equation theory that if the coefficients of eq. 4-1 satisfy the specified conditions in the hypothesis, and the solution of $\mathcal{A}_y^* \nu = 0$, can be written as $\nu = \pi_0 + \pi_1$, where π_0, or its first or second differential, is singular with respect to Supp ν, then π_0 is also a solution of $\mathcal{A}_y^* \nu = 0$.

Of course, any time-independent solution of $\mathcal{A}_y^* \nu = 0$ is also a solution of $\left(\frac{\partial}{\partial t} - \mathcal{A}_y^*\right) \nu = 0$, while as explained before, $\left(\frac{\partial}{\partial t} - \mathcal{A}_y^*\right) U_t \nu = 0$ for any initial finite measure ν. Since the initial conditions for these two equations are the same; *i.e.* $U_0 \nu = \nu$, the solutions must be the same, too; *i.e.* $\nu = U_t \nu$ for all $t \geq 0$. The conclusion is that any finite solution of $\mathcal{A}_y^* \nu = 0$, and in particular π_0, must be an invariant measure, as well. Thus, with normalization, π_0 will be an invariant probability measure and by proposition 4-1, its support is invariant, which in this case is the set A.

Now, since $\pi(A) < 1$, $\pi(A^c) > 0$ and as it has already been proved, $\pi_1 = \frac{\mathbb{1}_{A^c}}{\pi(A^c)} \pi$ will also be an invariant probability measure and $\pi = \pi(A)\pi_0 + \pi(A^c)\pi_1$ is a non-extremal invariant probability measure. □

Appendix B. Proof of Theorem 5

(i) ⇒ *(ii)*:

Let Γ be a closed connected invariant set of the system. The stochastic stability of the system and corollary 1 yield an invariant probability measure π_x for every $x \in \Gamma$, where because of invariance of Γ, Supp $\pi \subset \Gamma$. If the first statement is true, since any submanifold with dimension lower than dim M has an empty interior, the implication is that Int $\Gamma \neq \emptyset$. Therefore, for every set C, $\partial C = (\overline{C} - C)$, which is a closed set with an empty interior, cannot be invariant.

Now, suppose that C is an invariant control set. By the above argument and lemma 1, C must be closed. Then by corollary 2, there exists a unique extremal invariant probability measure π such that $C = \overline{\text{Supp}\,\pi}$. Hence, with *(i)* being true, C must be a closed invariant set where Int $C \neq \emptyset$.

(ii) ⇒ *(iii)*:

By *(ii)* we know that there exist at most a countable number of closed invariant control sets $\{C_i,\ i \in \mathcal{N}\}$. Because of global stochastic stability, we have the following weak convergence

$$\frac{1}{t}\int_0^t P(s,x,\cdot)\,ds \longrightarrow \pi_x \qquad \text{as} \qquad t \to \infty, \tag{$B-1$}$$

for all $x \in M$, where by theorem 2, for some $i \in \mathcal{N}$, $\overline{\text{Supp}\,\pi_x} \supset C_i$. It is easy to see that $C_i \subset \overline{\Omega^+(x)}$ and since by *(ii)*, $\text{Int}\, C_i \neq \emptyset$, a simple limit set argument shows that $\text{Int}\, C_i$ is in the reachability set of x, $\Omega^+(x)$.

Therefore, we are able to divide M into countable (not necessarily disjoint) regions $\{\Gamma_i,\ i \in \mathcal{N}\}$ where for all $x \in \Gamma_i$, $\text{Int}\, C_i \subset \Omega^+(x)$. Obviously, each Γ_i is an accessibility set containing C_i itself. By continuity of the process, this means that Γ_i is a connected submanifold of M with a non-empty interior. Moreover, if for any $i \neq j$, $\Gamma_i \cap \Gamma_j \neq \emptyset$, then all points in Γ_i and Γ_j are also accessible from each other.

Equivalently, one can say that if there exists no Γ_i such that it is disjoint from all other similar sets, then the system will be weakly controllable in $\bigcup_{l \in \mathcal{N}} \Gamma_l = M$.

Now, suppose that there exists a Γ_i disjoint from every Γ_l, $i \neq l$. Since $\bigcup_{l \in \mathcal{N}} \Gamma_l = M$, for some integer j, there exists a point $x \in \Gamma_i$ (or Γ_j) such that for every neighborhood of x, N_x, $N_x \cap \Gamma_j\,(\Gamma_i) \neq \emptyset$. $\text{Int}\, C_i\,(C_j) \subset \Omega^+(x)$, hence by the Feller property and openness of $\text{Int}\, C_i\,(C_j)$, there is a neighborhood N_x such that for all $y \in N_x$, $\Omega^+(y) \cap \text{Int}\, C_i\,(C_j) \neq \emptyset$, that is to say $\Gamma_i \cap \Gamma_j \neq \emptyset$ which is a contradiction.

(iii) $\Rightarrow$ *(iv)* is true by theorem 4.

By the definition of support, *(iv)* $\Rightarrow$ *(i)* is quite obvious.

Appendix C. Proof of Proposition 3

Before proceeding with the proof of this proposition, we prove two auxiliary lemmas which we will use later in the proof:

Lemma 3. *Consider* $\text{Supp}\,\pi$ *where* π *is any invariant probability measure for the diffusion process* X_t, *a solution of eq. (4-1), and suppose that*

$$\inf\{\dim M';\ M' \text{ any submanifold of } M \text{ such that } \text{Supp}\,\pi \subset M'\} = r \leq n.$$

Then if $b(\cdot)$ *and* $G(\cdot)$, *the coefficients of eq. (4-1), are* $C^\infty(M)$ *functions, there exists* M_0, *a piecewise* C^∞ r*-dimensional submanifold of* M, *such that* $\text{Supp}\,\pi \subseteq M_0$.

Proof. The proof is simply an application of the support theorem for the C^∞ case. According to this theorem, by the smoothness assumption, for every two points connected by a trajectory in M, there exists a C^∞ (smooth) trajectory passing through them. This includes every two point in the support of an invariant probability measure. Thus, the support is a union of these C^∞ trajectories which should be part of a piecewise C^∞ submanifold. In other words, if one represents M_0 by the set $\{x \in M \mid U(x) = 0\}$, there is a neighborhood, with respect to M_0 as the topological set, inside Supp π on which the function U is infinitely differentiable. □

Lemma 4. *Consider the autonomous linear hybrid system of eq. (1-2) with $u_t = -Ky_t$ where K is a uniformly bounded C^∞ functional on y^t. Let Φ_t denote the indicator function of the Jump-Markov process.*

Then for every $\tau \geq 0$, $E[\Phi_\tau \mid y_\tau]$, the (nonlinear) least-square estimate of Φ_t with respect to the point observation of the output process y at time τ, is almost surely equal to p_τ, the unconditional probability distribution of the Jump-Markov process at time τ.

Proof. Let τ be fixed. If we take the observation process to be $\widetilde{y}_t = g_t y_t$, where $g_t = \mathbb{1}_{\{t \geq \tau\}}$, instead of y_t, then for every $\tau \geq 0$, $E[\Phi_\tau \mid y_\tau]$ may be regarded as the least square estimate of Φ_t with respect to the observation of $\widetilde{y}_t$ up to time τ.

However, because of the discontinuity of g_t, we are not able to directly apply the nonlinear filtering theorem (and equations) presented in [15] to this situation. Instead, we shall consider a class of parameter-dependent observation processes, $\widetilde{y}_\epsilon(t) = g_\epsilon(t) y_t$, where $g_\epsilon(t)$ is a $C^\infty(\mathcal{R}^+)$ function such that

$$g_\epsilon(t) = 0 \quad \text{for } t \leq \tau - \epsilon \quad \text{and} \quad g_\epsilon(t) = 1 \quad \text{for } t \geq \tau \qquad \forall \epsilon \in [0,1].$$

Obviously, $g_\epsilon \to g$ as $\epsilon \to 0$.

Since for every $\epsilon \in (0,1]$, g_ϵ is infinitely differentiable, using Ito's formula, we are able to obtain the following parameter-dependent SDE for the process $\widetilde{y}_\epsilon(t)$, for all $\epsilon \in (0,1]$:

$$d\widetilde{y}_\epsilon(t) = \left(g'_\epsilon(t) y_t + (A(\theta_t) - K_t B(\theta_t))\, \widetilde{y}_\epsilon(t)\right) dt + F\, dw_t.$$

By the smoothing property of the conditional expectations, we know that

$$E[g'_\epsilon(t) h_t {y_t}^T \mid \mathcal{F}_t^{\widetilde{y}_\epsilon}] = g'_\epsilon(t) E[h_t \mid \mathcal{F}_t^{\widetilde{y}_\epsilon}] {y_t}^T \qquad \forall h_t \text{ an } \mathcal{F}_t\text{-measurable function.}$$

Therefore, using the nonlinear filtering theorem in [15] and the above equality, we may derive the following differential equation for $E[\Phi_t^j \mid \mathcal{F}_t^{\widetilde{y}_\epsilon}]$ for all $\epsilon \in (0,1]$:

$$\begin{aligned} dE[\Phi_t^j \mid \mathcal{F}_t^{\widetilde{y}_\epsilon}] = \Pi^j E[\Phi_t \mid \mathcal{F}_t^{\widetilde{y}_\epsilon}]\, dt + E[\Phi_t^j \mid \mathcal{F}_t^{\widetilde{y}_\epsilon}] g_\epsilon(t) {y_t}^T (\widetilde{H}_{j\epsilon} - \overline{\widetilde{H}}_\epsilon)^T F^{-2} \times \\ \left(g_\epsilon(t) \left(A(\theta_t) - K_t B(\theta_t) - \overline{\widetilde{H}}_\epsilon\right) y_t\, dt + F\, dw_t\right), \qquad (C-1) \end{aligned}$$

where $\widetilde{H}_{j\epsilon} = A_j - E[K_t \mid \widetilde{\mathcal{F}}_t^{y_\epsilon}]B_j$ and $\widetilde{\overline{H}}_\epsilon = E[A(\theta_t) - K_t B(\theta_t) \mid \widetilde{\mathcal{F}}_t^{y_\epsilon}]$. Considering the original system equation, eq. (1-2), the smoothness of all the coefficient functions in y and the fact that for appropriate initial conditions, $E[\Phi_t \mid \widetilde{\mathcal{F}}_t^{y_\epsilon}]$ is always in the simplex S and uniformly bounded, one can see that the solution of the equation (C-1) is an almost surely continuous process for every $\epsilon \in (0,1]$. Let $\widetilde{p}_t$ denote this solution when $\epsilon = 0$. Obviously, $\widetilde{p}_t$ is also almost surely continuous (notice that for almost sure continuity of the solutions, it is not necessary for the coefficients to be continuous in time). It only remains to show that $\lim_{\epsilon \to 0} E[\Phi_t^j \mid \widetilde{\mathcal{F}}_t^{y_\epsilon}] = \widetilde{p}_t$. To show this, we will use theorem 5-5-2 in [7] . By using this theorem, we will have the following limit:

$$E\left\|\lim_{\epsilon \to 0} E[\Phi_\tau^j \mid \widetilde{\mathcal{F}}_\tau^{y_\epsilon}] - \widetilde{p}_\tau\right\|^2 = 0, \qquad (C-2)$$

where because of the boundedness of Φ_t, we have been able to change the order of the limit and the expectation operators.

Now, putting ϵ equal to zero in eq. (C-1), it is clear that for all $0 \leq t < \tau$,

$$d\widetilde{p}_t = \Pi^j \widetilde{p}_t \, dt.$$

Thus, if $\widetilde{p}_0 = E[\Phi_0] = p_0$, $\widetilde{p}_t = p_t$ for all $0 \leq t < \tau$. In particular, since $\widetilde{p}_t$ is almost surely continuous, $\widetilde{p}_\tau = p_\tau$ almost surely.

On the other hand, considering the integrability of Φ_t for all $t \geq 0$, one may change the order of the limit and the expectation operators for $E[\Phi_t \mid \widetilde{\mathcal{F}}_t^{y_\epsilon}]$ to obtain

$$\lim_{\epsilon \to 0} E[\Phi_t \mid \widetilde{\mathcal{F}}_t^{y_\epsilon}] = E[\Phi_t \mid \lim_{\epsilon \to 0} \mathcal{F}_t^{y_\epsilon}] = E[\Phi_t \mid y_t] \qquad \text{a.s.}$$

Assuming that the initial condition for all the processes is equal to p_0, the above equality for $t = \tau$ and eq. (C-2), together, imply that $E[\Phi_\tau \mid y_\tau] = \widetilde{p}_\tau = p_\tau$ almost surely, which will obviously be true for any arbitrary $\tau \geq 0$. □

Proof of Proposition 3. Since $K_{\hat{p}}$ is stochastically stabilizing and infinitely differentiable, the system is obviously a stochastically stable system with $C^\infty(M)$ coefficient functions for the corresponding SDE. Therefore, all the results previously obtained in section 4 are also applicable here.

If there is no closed submanifold of M, invariant with respect to the state process and of dimension lower than $n+N-1$, then obviously, for every invariant probability measure π, $\overline{\text{Supp}\,\pi}$ (a closed invariant subset of M) cannot be of any dimension smaller than $n+N-1$, either.

Suppose that there exists such a submanifold, M_0. By the definition of invariance, M_0 contains all the trajectories of the system originating at some point in M_0. By lemma 3, we know that M_0 can be a piecewise C^∞ submanifold and hence, with no loss of generality, it can be shown by the following set:

$$M_0 = M \cap \{x \in \mathcal{R}^{n+N-1} \mid U(x) = \bar{0}\}.$$

where U represents an r-dimensional piecewise C^∞ vector function with independent elements such that M_0 cannot be shown by any other similar vector function of higher dimension. Since M_0 is an invariant set for the state process x_t, this is equivalent to

$$U(y_t, \hat{p}_t^1, \hat{p}_t^2, \ldots, \hat{p}_t^{N-1}) = \bar{0} \qquad \forall t \geq 0, \tag{C-3}$$

By proposition 2, we know that every support of an invariant probability measure is at least n-dimensional, meaning that $r \leq N-1$. In addition, it will be proved that every element of U is a piecewise C^∞ function, independent of y:

Consider $U^k(y, \hat{p}^1, \hat{p}^2, \ldots, \hat{p}^{N-1})$, an arbitrary element of U. Since U^k is a piecewise C^∞ function for every k, there are open neighborhoods such that $\frac{\partial U^k}{\partial \hat{p}^j}$ exists. Also, there should exist some j and some open neighborhood where $\frac{\partial U^k}{\partial \hat{p}^j} \neq 0$, otherwise, U^k would be independent of $\hat{p}$, which is a contradiction by proposition 2. Without any loss of generality, we assume that this holds for $\hat{p}^1$. Hence, by the implicit function theorem, $\hat{p}^1$ can be expressed as a function of the other arguments of U^k, *i.e.*

$$\hat{p}_t^1 = v(y_t, \hat{p}_t^2, \ldots, \hat{p}_t^{N-1}) \qquad \forall t \geq 0. \tag{C-4}$$

By taking the conditional expectation of both sides with respect to y_t and using lemma 4, we get

$$p_t^1 = E[v(y_t, \hat{p}_t^2, \ldots, \hat{p}_t^{N-1}) \mid y_t] \qquad \forall t \geq 0 \qquad \text{a.s.,}$$

(notice that $E[\hat{p}_t \mid y_t] = E[E[\Phi_t \mid \mathcal{F}_t^y] \mid y_t] = E[\Phi_t \mid y_t]$ almost surely).

By proposition 2, we know that for every $t > 0$, y_t can be in any open neighborhood of $\mathcal{R}^n$ with some positive probability. On the other hand, $p_t = e^{\Pi t} p_0$. A comparison of these two clearly shows that the RHS of the above equation should be independent of y_t for all $t > 0$.

Since for every $t > 0$, y_t can take any value in $\mathcal{R}^n$, the conditional expectation of v with respect to y_t is independent of y_t for all $t > 0$, if and only if either v is independent of y_t or its conditional expectation is zero almost surely for all $t > 0$ and all initial conditions in M satisfying equation (C-4). Hence,

$$v(y_t, \hat{p}_t^2, \ldots, \hat{p}_t^{N-1}) = v_1(\hat{p}_t^2, \ldots, \hat{p}_t^{N-1}) + v_2(y_t, \hat{p}_t^2, \ldots, \hat{p}_t^{N-1}),$$

where $E[v_2(y_t, \hat{p}_t^2, \ldots, \hat{p}_t^{N-1}) \mid y_t] = 0$ almost surely for all $t > 0$ and all initial conditions in M satisfying eq. (C-4). But then, by almost sure continuity of the processes and continuity and uniform boundedness of v_1 and v_2 $(0 < u < 1)$,

$$\lim_{t \to 0} E[v_2(y_t, \hat{p}_t^2, \ldots, \hat{p}_t^{N-1}) \mid y_t] = 0 \qquad \text{a.s.}$$

The above limit implies that $v_2 = 0$ at every point in M which satisfies eq. (C-4), a contradiction unless $v_2 = 0$. This proves that $v = v_1$ and U^k are independent functions of y. Since U^k is an arbitrary element of U, the same is true for all the elements of U. Hence,

$$M_0 = \left(S \cap \{p \in \mathcal{R}^{N-1} \mid U(p) = \bar{0}\}\right) \times \mathcal{R}^n$$

and $U(\hat{p}_t^1, \hat{p}_t^2, \ldots, \hat{p}_t^{N-1}) = \bar{0}$ for all $t \geq 0$ and all initial conditions in M_0.

In this way, one can take $\{U = \bar{0}\}$ as a submanifold of $\mathcal{R}^{N-1}$ inside the simplex S, that is to say the cross-section of M_0 with any $y =$ constant plane, which will be denoted by S_0. Clearly, for all initial conditions in S_0, $dU(\hat{p}_t^1, \hat{p}_t^2, \ldots, \hat{p}_t^{N-1}) = \bar{0}$.

Consider again an arbitrary element of U, U^k. Using the Stratonovich-sense SDE's for $\hat{p}_t$, the differential equation for this element can be expanded into

$$\sum_{j=1}^{N-1} \frac{\partial U^k}{\partial \hat{p}_t^j} \left(\Pi^j \hat{p}_t^j + \frac{1}{2} \left(\hat{p}_t^j \mathrm{Tr}(Y_j) + \hat{p}_t^j {y_t}^T ({Y_j}^T F^{-2} Y_j + \sum_{i=1}^{N-1} \frac{\partial Y_j}{\partial \hat{p}_t^i} F^{-2} Y_i) y_t \right) \right) dt$$
$$+ \sum_{j=1}^{N-1} \frac{\partial U^k}{\partial \hat{p}_t^j} \hat{p}_t^j {y_t}^T {Y_j}^T F^{-1} \delta \nu_t = 0 \qquad \forall t \geq 0, \quad (C-5)$$

where $Y_j = H_j - \overline{H}_{\hat{p}}$.

Because of the independence of dt and $\delta\nu_t$, the above equality can be true if and only if the coefficients of dt and $\delta\nu_t$ are simultaneously zero. Since this should be true for y_t equal to every point in $\mathcal{R}^n$, the result is that all trajectories in M_0 satisfy the following equalities:

$$0 = \sum_{j=1}^{N-1} \frac{\partial U^k}{\partial \hat{p}_t^j} \Pi^j \hat{p}_t^j,$$

$$0 = \sum_{j=1}^{N-1} \frac{\partial U^k}{\partial \hat{p}_t^j} \hat{p}_t^j {Y_j}^T$$

and

$$0 = \sum_{j=1}^{N-1} \frac{\partial U^k}{\partial \hat{p}_t^j} \left({Y_j}^T F^{-2} Y_j + \sum_{i=1}^{N-1} \frac{\partial {Y_j}^T}{\partial \hat{p}_t^i} F^{-2} Y_i \right). \qquad (C-6)$$

(The reason that $\hat{p}_t^j \mathrm{Tr}(Y_j)$ does not appear in the first equality is that $\sum_{j=1}^{N-1} \frac{\partial U^k}{\partial \hat{p}_t^j} \hat{p}_t^j \mathrm{Tr}(Y_j) = 0$ as a result of the second inequality.)

The first equality of the eq.'s (C-6) can also be interpreted as the submanifold $\{U^k = 0\}$, a subset of S, consisting of a set of trajectories of the following ODE:

$$\dot{X} = \Pi' X. \qquad (C-7)$$

In the same manner, the second inequality implies that $U^k = 0$ consists of a set of trajectories of each of the following n^2 ODE's, where Y^{ik} denotes the i-th-row, k-th-column element of the matrix Y:

$$\dot{X} = \mathrm{Diag}\left((H_1 - \overline{H}_{\hat{p}})^{ik}, (H_2 - \overline{H}_{\hat{p}})^{ik}, \ldots, (H_{N-1} - \overline{H}_{\hat{p}})^{ik}\right) X \quad \forall i, k = 1, 2, \ldots, n. \qquad (C-7')$$

And similarly, for the last equality of eq.'s (C-6).

If the Jump-Markov process is irreducible, then for eq. (C-7), there will exist a unique asymptotically stable equilibrium point $p_\infty \in \mathrm{Int}\, S$. On the other hand, for the eq.'s $(C-7')$, the equilibrium points are located at the vertices of the simplex S, which may be stable or unstable. (Notice that in almost every case, Y_j, for each $j = 1, 2, \ldots, N-1$, cannot be equal to zero everywhere in S, otherwise the parameters of the system would be constant resulting in a time-invariant system.)

Therefore, comparing eq.'s (C-7) and $(C-7')$, it is clear that in almost every case, no set of trajectories of these equations, except for the simplex S, itself, can be identical.

Obviously, using the same argument for other elements of U will not improve the situation. This implies that U must be an identically zero vector function. Since we have assumed that $S_0 = S \cap \{p \in \mathcal{R}^{N-1} \mid U(p) = \overline{0}\}$, this is equivalent to saying that $S_0 = S$ or $M_0 = M$. However, in almost every case, S_0 is only a subset of $S \cap \{p \in \mathcal{R}^{N-1} \mid U(p) = \overline{0}\}$ with the same dimension. It is easy to see that in this case, exactly the same proof, which has been used so far, is applicable again and therefore the same result, that U is an identically zero vector function, will be obtained. The conclusion is that $\dim S_0 = \dim S = (N-1)$, or $\dim M_0 = \dim M = (n+N-1)$, and the proposition is proved.

Of course, if the Jump-Markov process were not irreducible, the above result would not be correct. In this case, some elements of p_∞ would be equal to zero which means that for some j, $p_j = 0$ would represent a common invariant plane for eq.'s (C-7) and $(C-7')$. In other words,

M_0 could be equal to $\{p \in \mathcal{R}^{N-1} \mid p_j = 0\} \times \mathcal{R}^n$, an $(n+N-2)$–dimensional submanifold of M.

□

Appendix D. Proof of Proposition 4

By corollary 3, theorem 5 and theorem 2, for every invariant probability measure, π, $\operatorname{Supp} \pi$ is an open set and $\overline{\operatorname{Supp} \pi}$ is a closed invariant closed set. Therefore, since all distinct invariant control sets are disjoint, the number of distinct closed invariant control sets, which by corollary 2 is equal to the number of distinct extremal invariant probability measures, is countable. Finiteness of this number is proved in the following way:

Let π_i and C_i for all $i \in \mathcal{N}$ denote the countable extremal invariant probability measures and the closure of their supports, respectively. We know that C_i is a closed invariant set and $C_i \cap C_j = \emptyset$ for all $i, j \in \mathcal{N}$ such that $i \neq j$.

By proposition 2, we know that each C_i is an unbounded set which has non-empty intersection with every $y =$ constant plane. Obviously every such intersection is a closed subset of the simplex S.

Consider the sequence of points (x_i) where $x_i \in C_i \cap \{(y, \hat{p}) \in M \mid y = 0\}$ for each $i \in \mathcal{N}$. The union of these intersections is a compact set, thus, there exists a subsequence of (x_i) converging to some limit point x inside this union.

$\overline{\Omega^+(x)}$ is obviously a closed invariant set which by stochastic stability of the system, contains the support of an invariant probability measure. Therefore, there exists an integer j such that $C_j \subset \overline{\Omega^+(x)}$ or $\operatorname{Int} C_j \subset \Omega^+(x)$.

Now, the Feller property of the process implies that for every $y \in \operatorname{Int} C_j$ and every neighborhood of y, N_y, there exists a neighborhood of x, N_x, such that for all $z \in N_x$, $\Omega^+(z) \cap N_y \neq \emptyset$.

Since x is the limit of a subsequence of (x_i), (x_{in}), there exists an integer $N > 0$ such that for all $n \geq N$, $x_{in} \in N_x$. On the other hand, there is a neighborhood of y inside C_j. This means that for all $n \geq N$, $C_j \cap \Omega^+(x_{in}) \neq \emptyset$ where $\Omega^+(x_{in}) \subset C_{in}$. Since every two distinct invariant control sets are disjoint, this is equivalent to saying that $C_j = C_{in}$ for all $n \geq N$, which proves that the number of distinct invariant control sets with non-empty interiors as well as the number of distinct extremal invariant probability measures is finite.

The second statement about the ergodicity property is only an application of theorem 3 to the stochastically stable system which is true for any number of extremal invariant probability measures provided that all of the supports are open sets.

Acknowledgement. Part of this work were done when the authors were visiting INRIA at Sophia-Antipolis, France and University of California at Berkeley, respectively. It has also been partially supported by the NSERC grant A1329. We wish to express our gratitude for the full support and help that we have received from our colleagues in McGill Research Center for Intelligent Machines at McGill University, specially to Prof. L. Gerencser.

References

[1] V. E. Beneš, "Finite Regular Invariant Measures for Feller Processes," *J. of Appl. Prob.*, vol. 5, 1968, pp. 203-209.

[2] P. E. Caines, and H. F. Chen, "Optimal Adaptive LQG Control for Systems with Finite State Process Parameters," *IEEE Trans. on Aut. Control*, vol. AC-30, no. 2, Feb. 1985, pp. 185-189.

[3] H. F. Chen, and P. E. Caines, "On The Adaptive Control of a Class of Systems with Random Parameters and Disturbances," *Automatica*, vol. 21, no. 6, June 1985, pp. 737-741.

[4] H. F. Chen, and P. E. Caines, "On The Adaptive Stabilization of Linear Stochastic Systems with Jump Process Parameters," *Proc. of The IEEE Conf. on Decision & Control, Tampa, Fl*, Dec. 1989.

[5] J. L. Doob, *Stochastic Processes*, John Wiley & Sons Inc., 1953.

[6] J. Ezzine, and A. H. Haddad, "On Largest Lyapunov Exponent Assignment and Almost Sure Stabilization of Hybrid Systems," *Proc. of Am. Control Conf.*, 1989, pp. 805-809.

[7] A. Friedman, *Stochastic Differential Equations and Applications, Vol. I*, Academic Press, 1975.

[8] R. Hermann, and A. J. Krener, "Nonlinear Controllability and Observability," *IEEE Trans. on Aut. Control*, no. 5, Oct. 1977, pp. 728-740.

[9] W. E. Hopkins, "Optimal Stabilization of Families of Linear Stochastic Differential Equations With Jump Coefficients and Multiplicative Noise," *SIAM J. of Control & Opt.*, vol. 25, no. 6, Nov. 1987, pp. 1587-1601.

[10] L. Hörmander, "Hypoelliptic Second Order Differential Equations," *Acta Mathematica*, vol. 119, Feb. 1968, pp. 147-171.

[11] K. Ichihara, and H. Kunita, "A Classification of The Second Order Degenerate Elliptic Operators and its Probabilistic Characterization," *Z. Wahrscheinlichkeits theorie verw. Gebiete*, vol. 30, 1974, pp. 235-254.

[12] G. Kallianpur, *Stochastic Filtering Theory*, Springer-Verlag, 1980.

[13] W. Kliemann, "Recurrence and Invariant Measures for Degenerate Diffusions," *The Annals of Prob.*, vol. 15, no. 2, 1987, pp. 690-707.

[14] N. N. Krasovskiĭ, and E. A. Lindskiĭ, "Analytical Design of Controllers in Systems with Random Attributes, parts I, II & III," *Automation And Remote Control*, vol. 2, 1962, pp. 1021-1025, 1141-1146, 1289-1294.

[15] R. S. Liptser, and A. N. Shiryaev, *Statistics of Random Processes, Vol. I & II*, Springer-Verlag, 1977.

[16] M. Mariton, "Controllability, Stability and Pole allocation for Jump Linear Systems," *Proc. of The IEEE Conf. on Decision and Control, Athens, Greece*, Dec. 1986, pp. 2193-2194.

[17] M. Mariton, "Stochastic Controllability of Linear Systems with Markovian Jumps," *Automatica*, vol. 23, 1987, pp. 783-785.

[18] S. P. Meyn, and P. E. Caines, "Asymptotic Behaviour of Stochastic Systems Possessing Markovian Realizations," *SIAM J. on Control & Optimization*. To appear.

[19] S. P. Meyn, and P. E. Caines, "A New Approach to Stochastic Adaptive Control," *IEEE Trans. on Aut. Control*, vol. AC-32, no. 3, Mar. 1987, pp. 220-226.

[20] S. H. Saperstone, *Semidynamical Systems in Infinite Dimensional Spaces*, Springer-Verlag, 1981.

[21] D. W. Stroock, and S. R. Varadhan, "On The Support of Diffusion Processes with Applications to the Strong Maximum Principle," *Proc. of the 6th Berkeley Symp. on Math. Statist. & Prob.*, 1972, pp. 333-368.

[22] D. D. Sworder, "Control of Systems Subject to Sudden Change in Character," *Proc. of The IEEE*, vol. 64, no. 8, Aug. 1976, pp. 1219-1225.

[23] D. D. Sworder, and R. O. Rogers, *An LQ Solution to a Control Problem Associated with a Solar Thermal Central Receiver*, IEEE Trans. on Aut. Control, Oct. 1983.

[24] W. M. Wonham, "Some Applications of Stochastic Differential Equations to Optimal Nonlinear Filtering," *SIAM J. of Control*, vol. 2, no. 3, 1965, pp. 347-369.

[25] W. M. Wonham, "Random Differential Equations in Control Theory," *Probabilistic Methods in Applied Mathematics, Vol. II*, edited by A. T. Bharucha-Reid, Academic Press, 1970, pp. 131-212.

IDENTIFICATION AND ADAPTIVE CONTROL FOR ARMAX SYSTEMS*

Han-Fu Chen Lei Guo Ji-Feng Zhang

Institute of Systems Science, Academia Sinica, Beijing 100080, P.R. China

* Work supported by the National Natural Science Foundation of China and by the TWAS RG MP 898-117.

IDENTIFICATION AND ADAPTIVE CONTROL FOR ARMAX SYSTEMS*

Han-Fu Chen *Lei Guo* *Ji-Feng Zhang*

Institute of Systems Science, Academia Sinica, Beijing 100080, P.R. China

1. INTRODUCTION

Consider the stochastic feedback control system called ARMAX process:

$$A(z)y_n = B(z)u_n + C(z)w_n, \quad n \geq 0; \quad y_n = w_n = 0, \ u_n = 0, \ n < 0, \tag{1.1}$$

where z denotes the shift-back operator and $A(z)$, $B(z)$ and $C(z)$ are polynomial matrices

$$A(z) = I + A_1 z + \cdots + A_p z^p, \quad p \geq 0, \tag{1.2}$$

$$B(z) = B_d z^d + B_{d+1} z^{d+1} + \cdots + B_q z^q, \quad q \geq 0, \ d \geq 1, \tag{1.3}$$

$$C(z) = I + C_1 z + \cdots + C_r z^r, \quad r \geq 0, \tag{1.4}$$

with unknown orders p, q, r, time-delay d and unknown matrix coefficient

$$\theta^{\tau} = [-A_1 \ \cdots - A_p \ \ B_d \ \cdots \ B_q \ \ C_1 \ \cdots \ C_r] \tag{1.5}$$

$\{y_n\}$ is the m-output and $\{u_n\}$ is the l-input, and $C(z)w_n$ represents the dynamic noise of the system.

Assume that $\{w_n, F_n\}$ is a martingale difference sequence with respect to a nondecreasing family $\{F_n\}$ of σ-algebras

$$E\{w_n | F_{n-1}\} = 0, \quad n \geq 0.$$

The control u_n depends upon the past measurement $\{y_0, \ldots, y_n, u_0, \ldots, u_{n-1}\}$, i.e. u_n is F_n-measurable. The feedback control nature of u_n makes the system different from that considered in time series analysis (Box and Jenkins, 1970; Hannan and Deistler, 1988).

In this paper we present only the results developed in our group on coefficient, order and time-delay estimation, stochastic adaptive control and time-varying systems, and hence many other interesting results in the field (see, e.g. Goodwin and Sin, 1984; Ljung and Soderstom, 1983; Caines 1988; Kumar and Varaiya, 1986; Chen, 1985, etc,) are not included.

2. COEFFICIENT ESTIMATION

In this section the system orders and time-delay are not estimated, and p, q, r as upper bounds of the system orders are assumed known. Also, the system time delay d is simply taken as 1 in this section.

2.1. Extended Least-Squares (ELS) Algorithm

One of commonly used methods for parameter estimation is the ELS algorithm which is the extension of the least squares method to the correlated noise and is defined as follows

$$\theta_{n+1}=\theta_n+a_nP_n\phi_n(y_{n+1}^\tau-\phi_n^\tau\theta_n), \tag{2.1}$$

$$P_{n+1}=P_n-a_nP_n\phi_n\phi_n^\tau P_n,\quad a_n=(1+\phi_n^\tau P_n\phi_n)^{-1}, \tag{2.2}$$

$$\phi_n^\tau=[y_n^\tau \ \cdots \ y_{n-p+1}^\tau \ \ u_n^\tau \ \cdots \ u_{n-q+1}^\tau \ \ y_n^\tau-\phi_{n-1}^\tau\theta_n \ \cdots \ y_{n-r+1}^\tau-\phi_{n-r}^\tau\theta_{n-r+1}] \tag{2.3}$$

with θ_0 arbitrarily chosen, $P_0=cI$, $c=mp+lq+mr$. It is easy to verify

$$P_{n+1}^{-1}=\frac{1}{c}I+\sum_{i=0}^{n}\phi_i\phi_i^\tau. \tag{2.4}$$

In the following we will denote the largest and the smallest eigenvalue of P_{n+1}^{-1} by $\lambda_{\max}(n)$ and $\lambda_{\min}(n)$, respectively. The conditions that will be used are as follows.

Condition 2.1. $\sup_{n\geq 0}E\{\|w_{n+1}\|^\beta|F_n\}\leq\sigma^2<\infty$, *a.s., for some* $\beta\geq 2$.

Condition 2.2. *The transfer function* $C(z)^{-1}-\frac{1}{2}I$ *is strictly positive real.*

The following Theorem is based on the work of Moore (1978), Solo (1979), Lai and Wei (1982, 1986), Chen (1982) and Chen and Guo (1986b).

Theorem 2.1. *If Conditions 2.1 and 2.2 hold, then*

$$\|\theta_n-\theta\|^2=O\left(\frac{\log\lambda_{\max}(n)(\log\log\lambda_{\max}(n))^{\delta(\beta-2)}}{\lambda_{\min}(n)}\right)\quad a.s., \tag{2.5}$$

where θ_n *is given by (2.1)-(2.3) and*

$$\delta(x)=\begin{cases}c, & x=0,\\ 0, & x\neq 0,\end{cases}\quad \text{for some } c>1,$$

We note that $\lambda_{\max}(n)$ and $\lambda_{\min}(n)$ are estimate-dependent, hence so is the right-hand side of (2.5). We now remove this dependence by using

$$\phi_n^{0\tau}=[y_n^\tau \ \cdots \ y_{n-p+1}^\tau \ \ u_n^\tau \ \cdots \ u_{n-q+1}^\tau \ \ w_n^\tau \ \cdots \ w_{n-r+1}^\tau],\quad n\geq 0,\quad \phi_i^0=0,\ i<0 \tag{2.6}$$

instead of ϕ_n.

Let $\lambda_{\max}^0(n)$ and $\lambda_{\min}^0(n)$ respectively denote the largest and the smallest eigenvalues of the matrix

$$\frac{1}{c}I+\sum_{i=0}^{n-1}\phi_i^0\phi_i^{0\tau}.$$

We have the following result (see, Chen and Guo, 1986b).

Theorem 2.2. *Under the conditions of Theorem 2.1, if*

$$\log\lambda_{\max}^0(n)(\log\log\lambda_{\max}^0(n))^{\delta(\beta-2)}=o(\lambda_{\min}^0(n)),\ a.s.,$$

then as $n\to\infty$,

$$||\theta_n-\theta||^2=O\left[\frac{\log\lambda^0_{max}(n)(\log\log\lambda^0_{max}(n))^{\delta(\beta-2)}}{\lambda^0_{min}(n)}\right] \quad a.s., \tag{2.7}$$

2.2. Stochastic Gradient (SG) Algorithm

We now introduce

$$\psi_n^\tau=[y_n^\tau \ \cdots \ y_{n-p+1}^\tau \ u_n^\tau \ \cdots \ u_{n-q+1}^\tau \ y_n^\tau-\psi_{n-1}^\tau\theta_{n-1} \ \cdots \ y_{n-r+1}^\tau-\psi_{n-r}^\tau\theta_{n-r}]$$

where for estimating w_n, the *a posteriori* estimate $y_n-\theta_{n-1}^\tau\psi_{n-1}$ rather than the *posteriori* estimate $y_n-\theta_n^\tau\phi_{n-1}$ as in (2.3) is used. The SG algorithm for estimating θ is defined as follows

$$\theta_{n+1}=\theta_n+\frac{\psi_n}{r_n}(y_{n+1}^\tau-\psi_n^\tau\theta_n), \tag{2.8}$$

$$r_n=1+\sum_{i=1}^n||\psi||^2, \quad r_0=1 \tag{2.9}$$

with θ_0 arbitrarily chosen.

Condition 2.3. $C(z)-\frac{1}{2}I$ *is strictly positive real.*

Theorem 2.3. *(Chen and Guo, 1985, 1987b) If Conditions 2.1 (with* $\beta=2$*) and 2.3 are satisfied,* $r_n\to\infty$ *and if there are* $\alpha\in[0,1/4)$ *and* N_0 *and* M *such that*

$$\frac{r_{n+1}}{r_n}\le M(\log r_n)^\alpha, \quad \text{and} \quad \frac{\mu_{max}(n)}{\mu_{min}(n)}\le M(\log r_n)^{1/4-\alpha} \tag{2.10}$$

for all $n\ge N_0$*, then* $\theta_n \underset{n\to\infty}{\to} \theta$*, a.s., where* θ_n *is defined by (2.8)-(2.9) and* $\mu_{max}(n)$ *and* $\mu_{min}(n)$ *respectively denote the maximum and minimum eigenvalues of* $\frac{1}{c}I+\sum_{i=1}^n\psi_i\psi_i^\tau$.

Corresponding to Theorem 2.2 we obtain the following results for SG algorithm.

Theorem 2.4. *(Chen and Guo, 1987b) Let* $r_n^0=1+\sum_{i=0}^n||\phi_i^0||^2$ *with* ϕ_i^0 *given by (2.6). If Conditions 2.1 (with* $\beta=2$*) and 2.3 hold,* $r_n^0\to\infty$ *and if there are* $\alpha\in[0,1/4)$*,* N_0 *and* M *(possibly depending on sampling point) such that for any* $n\ge N_0$,

$$\frac{r_{n+1}^0}{r_n^0}\le M(\log r_n^0)^\alpha, \quad \text{and} \quad \frac{\lambda^0_{max}(n)}{\lambda^0_{min}(n)}\le M(\log r_n^0)^{1/4-\alpha} \tag{2.11}$$

with $\lambda^0_{max}(n)$ *and* $\lambda^0_{min}(n)$ *defined by Theorem 2.2, then*

$$\theta_n \underset{n\to\infty}{\to} \theta, \ a.s..$$

Remark. For consistency of the parameter estimate θ_n the condition number $\mu_{max}(n)/\mu_{min}(n)$ is allowed to diverge at a rate of $(\log r_n)^{1/4}$, but there exists an example showing that the conclusion of $\theta_n\to\theta$ does not hold if $(\log r_n)^{1/4}$ changes to $(\log r_n)^{1+\delta}$, $\delta>0$ (Chen and Guo, 1986d).

2.3. Estimation With Attenuating Dither

From Theorems 2.1-2.4 we see that for consistency of parameter estimates some conditions are imposed on eigenvalues of the corresponding regression matrix (see (2.5), (2.7), (2.10)-(2.11)). These conditions are not easy to verify and are often not satisfied for adaptive control systems as shown by Chen and Guo (1987a). By attenuating excitation method (Chen and Guo, 1986b, 1987a-b), we shall replace the not easily verifiable conditions mentioned above by some easily verifiable ones. To be specific, let $\{\varepsilon_n\}$ be a sequence of l-dimensional i.i.d random vectors with continuous distribution and let it be independent of $\{w_n\}$ and such that

$$E\varepsilon_n=0,\quad E\varepsilon_n\varepsilon_n^{\tau}=\mu I,\quad \|\varepsilon_n\|\le \mu_0,\ a.s.,\quad \mu>0. \tag{2.12a}$$

Without loss of generality, assume that

$$F_n=\sigma\{w_i,\ \varepsilon_i,\ i\le n\},\quad F_n'=\sigma\{w_i,\ \varepsilon_{i-1},\ i\le n\}, \tag{2.12b}$$

Let F_n'-measurable u_n^0 be the desired control. The attenuating excited control u_n applied to the system is defined by

$$u_n=u_n^0+v_n, \tag{2.13}$$

where

$$v_n=\frac{\varepsilon_n}{n^{\varepsilon/2}},\quad \varepsilon\in[0,\frac{1}{2(t+1)}),\quad t=max(p,q,r)+mp-1 \tag{2.14}$$

The above "excitation" idea is motivated by the "continuously disturbed controller" introduced by Caines and Lafortune (1984), where the disturbance sequence v_n has constant variance rather than disminishing as here.

Condition 2.4. *There is a positive definite matrix $R>0$ (possibly depending on relizations) so that*

$$\lim_{n\to\infty}\frac{1}{n}\sum_{i=1}^{n}w_iw_i^{\tau}=R,\quad a.s.. \tag{2.15}$$

Condition 2.5. *$A(z)$, $B(z)$ and $C(z)$ have no left common factor and A_p is of full rank.*

Theorem 2.5. *(Chen and Guo, 1986b) Suppose that Conditions 2.1-2.2 and 2.4-2.5 are fullfiled and u_n is given by (2.13)-(2.14) and that as $n\to\infty$,*

$$\frac{1}{n}\sum_{i=1}^{n}(\|y_i\|^2+\|u_i\|^2)=O(n^b),\quad a.s., \tag{2.16}$$

with $b\in[0,\frac{1-2\varepsilon(t+1)}{2t+3})$. Then for θ_n given by ELS

$$\|\theta_n-\theta\|^2=O\left(\frac{(logn)(loglogn)^{\delta(\beta-2)}}{n^{\alpha}}\right)\quad a.s. \tag{2.17}$$

for any $\alpha\in(\frac{1}{2}(1+b),1-(t+1)(\varepsilon+b))$ with $\delta(x)$ is defined as Theorem 2.1.

A similar result is true for SG algorithm.

3. ALL-PARAMETER ESTIMATION

In this section we assume that true time-delay is d_0, and true orders for $A(z)$, $B(z)$, $C(z)$ are p_0, q_0, r_0 respectively, that is to say, in (1.2)-(1.4), $A_i=0$ for $i>p_0$, $B_j=0$ for $j>q_0$ or $j<d_0$ and $C_k=0$ for $k>r_0$.

Condition 3.1. *The true orders (p_0,q_0,r_0) belong to a known finite set M_0:*

$$M_0=\{(p,q,r):\ 0\le p\le p^*,\ 0\le q\le q^*,\ 0\le r\le r^*\},$$

and the true time-delay d_0 belongs to a known finite set M_d:

$$M_d=\{d:\ d^*\le d\le q^*\}.$$

Condition 3.2. *$A(z)$, $B(z)$ and $C(z)$ have no left common factor and C_{r_0} is of full rank.*

Condition 3.3. *There is a constant $b\ge 1$ such that*

$$\sum_{i=0}^{n}(\|y_i\|^2+\|u_i\|^2)=O(n^b),\quad a.s..$$

3.1. ELS Based Recursive Estimation for Order, Time-Delay and Coefficients

We now write down methods for estimating time-delay d_0, system orders p_0, q_0, r_0 and coefficient θ. To this end, by using the largest possible orders p^*,q^*,r^* and the smallest possible time-delay d^*, we first give the estimate $\hat{w}_n$ for the driven noise w_n:

$$\hat{w}_n=y_n-\bar{\theta}_n^\tau\bar{\phi}_{n-1},\quad n\ge 0;\quad \hat{w}_n=0,\ n<0, \tag{3.1a}$$

$$\bar{\theta}_{n+1}=\bar{\theta}_n+\bar{a}_n\bar{P}_n\bar{\phi}_n(y_{n+1}^\tau-\bar{\phi}_n^\tau\bar{\theta}_n), \tag{3.1b}$$

$$\bar{P}_{n+1}=\bar{P}_n-\bar{a}_n\bar{P}_n\bar{\phi}_n\bar{\phi}_n^\tau\bar{P}_n,\quad \bar{a}_n=(1+\bar{\phi}_n^\tau\bar{P}_n\bar{\phi}_n)^{-1}, \tag{3.1c}$$

$$\bar{\phi}_n^\tau=[y_n^\tau\ \cdots\ y_{n-p^*+1}^\tau\ u_{n-d^*+1}^\tau\ \cdots\ u_{n-q^*+1}^\tau\ \hat{w}_n^\tau\ \cdots\ \hat{w}_{n-r^*+1}^\tau],\ n\ge 0;\ \bar{\phi}_n^0=0,\ n<0 \tag{3.1d}$$

with initial value $\bar{P}_0=I$ and $\bar{\theta}_0$ being given arbitrarily.

For any $(p,q,r)\in M_0$, $d\in M_d$ and $\bar{\theta}_0$, denote

$$\theta^\tau(p,d,q,r)=[-A_1\ \cdots\ -A_p\ B_d\ \cdots\ B_q\ C_1\ \cdots\ C_r] \tag{3.2a}$$

$$\phi_n^\tau(p,d,q,r)=[y_n^\tau\ \cdots\ y_{n-p+1}^\tau\ u_{n-d+1}^\tau\ \cdots\ u_{n-q+1}^\tau\ \hat{w}_n^\tau\ \cdots\ \hat{w}_{n-r+1}^\tau] \tag{3.2b}$$

$$\theta(r)=\theta(p^*,d^*,q^*,r),\quad \phi_n(r)=\phi_n(p^*,d^*,q^*,r), \tag{3.2c}$$

with definition:

$$A_i=0,\ B_j=0,\ C_k=0,\quad \text{for any}\ \ i>p_0,\ j>q_0,\ j<d_0\ \text{ or }\ k>r_0. \tag{3.2d}$$

We now give the estimate $\theta_n(r)$ for $\theta(r)$ by using $\hat{w}_n$ obtained above:

$$\theta_{n+1}(r)=\theta_n(r)+a_n(r)P_n(r)\phi_n(r)(y_{n+1}^\tau-\phi_n^\tau(r)\theta_n(r)), \tag{3.3a}$$

$$P_{n+1}(r)=P_n(r)-a_n(r)P_n(r)\phi_n(r)\phi_n^\tau(r)P_n(r),\quad a_n(r)=(1+\phi_n^\tau(r)P_n(r)\phi_n(r))^{-1}, \tag{3.3b}$$

with $P_0(r)=I$ and $\theta_0(r)$ being given arbitrarily.

Write $\theta_n(r)$ in the following block form

$$\theta_n^\tau(r)=[-A_{1n}(r)\ \cdots\ -A_{p^*n}(r)\ B_{d^*n}(r)\ \cdots\ B_{q^*n}(r)\ C_{1n}(r)\ \cdots\ C_{rn}(n)] \tag{3.4}$$

for any $(p,q,r)\in M_0$ and $d\in M_d$, we take

$$\theta_n^\tau(p,d,q,r)=[-A_{1n}(r)\ \cdots\ -A_{pn}(r)\ B_{dn}(r)\ \cdots\ B_{qn}(r)$$

$$C_{1n}(r) \quad \cdots \quad C_{rn}(r)] \tag{3.5}$$

as the estimate of $\theta(p,d,q,r)$ defined by (3.2a).

For any $(p,q,r)\in M_0$ and $d\in M_d$, we introduce

$$\sigma_n(p,d,q,r)=\sum_{i=0}^{n-1}||y_{i+1}-\theta_n^\tau(p,d,q,r)\phi_i(p,d,q,r)||^2, \tag{3.6}$$

$$CIC_1(p,q,r)_n=\sigma_n(p,d^*,q,r)+(p+q+r)s_n, \tag{3.7}$$

with $s_n=(\log n)^2$. Then the estimates r_n, q_n and p_n for system orders r_0, q_0 and p_0 are given as follows:

$$r_n=\arg\min_{0\le r\le r^*} CIC_1(p^*,q^*,r)_n, \tag{3.8a}$$

$$q_n=\arg\min_{0\le q\le q^*} CIC_1(p^*,q,r_n)_n, \tag{3.8b}$$

$$p_n=\arg\min_{0\le p\le p^*} CIC_1(p,q_n,r_n)_n, \tag{3.8c}$$

Similarly, by introducing

$$CIC_2(d)_n=\sigma_n(p_n,d,q_n,r_n)-ds_n, \tag{3.9}$$

we estimate the time-delay by

$$d_n=\arg\min_{d^*\le d\le q_n} CIC_2(d)_n, \tag{3.10}$$

The criteria CICs are motivated by the work of Rissanen (1978, 1986), Hannan and Rissanen (1982), Akaike (1969) and Hannan and Deistler (1988), etc.

It is easy to see that $\sigma_n(p,d,q,r)$ can be calculated recursively as follows:

$$\sigma_{n+1}(p,d,q,r)=\sigma_n(p,d,q,r)+G_{n+1}(p,d,q,r). \tag{3.11a}$$

$$\begin{aligned} G_{n+1}(p,d,q,r)=&||y_{n+1}-\theta_n^\tau(p,d,q,r)\phi_n(p,d,q,r)||^2+tr\{(\theta_{n+1}(p,d,q,r)-\theta_n(p,d,q,r))^\tau \\ &\times(N_{n+1}(p,d,q,r)(\theta_{n+1}(p,d,q,r)+\theta_n(p,d,q,r)) \\ &-2H_{n+1}(p,d,q,r))\} \end{aligned} \tag{3.11b}$$

$$N_{n+1}(p,d,q,r)=N_n(p,d,q,r)+\phi_n(p,d,q,r)\phi_n^\tau(p,d,q,r), \quad N_0(p,d,q,r)=0, \tag{3.11c}$$

$$H_{n+1}(p,d,q,r)=H_n(p,d,q,r)+\phi_n(p,d,q,r)y_{n+1}^\tau, \quad H_0(p,d,q,r)=0, \tag{3.11d}$$

The following Theorems applicable to feedback and nonstationary systems are based on the work of Guo *et al* (1989) and Chen and Zhang (1989a).

Theorem 3.1. *If Conditions 2.1-2.2 and 3.1 hold, and*

$$\frac{\log s_n^0(\log\log s_n^0)^{\delta(\beta-2)}}{(\log n)^2}\xrightarrow[n\to\infty]{}0,\ a.s. \quad \text{and}\ \frac{(\log n)^2}{\lambda_{\min}^{0(n)}(r_0)}\xrightarrow[n\to\infty]{}0,\ a.s., \tag{3.12}$$

where

$$\lambda_{\min}^{0(n)}(r_0)=\lambda_{\min}\left[\sum_{i=0}^{n-1}\phi_i^0(r_0)\phi_i^{0\tau}(r_0)\right] \tag{3.13a}$$

$$s_n^0=1+\sum_{i=0}^{n-1}||\phi_i^0(r^*)||^2, \quad \phi_n^0(r)=\phi_n^0(p^*,d^*,q^*,r), \tag{3.13b}$$

$$\phi_n^0(p,d,q,r)=[y_n^\tau \cdots y_{n-p+1}^\tau \quad u_{n-d+1}^\tau \cdots u_{n-q+1}^\tau \quad w_n^\tau \cdots w_{n-r+1}^\tau]^\tau, \tag{3.13c}$$

then the order and time-delay estimates given by (3.8)-(3.10) are strongly consistent

$$(p_n, d_n, q_n, r_n) \underset{n\to\infty}{\longrightarrow} (p_0, d_0, q_0, r_0) \tag{3.14}$$

and the coefficient estimate given by (3.3)-(3.10) is strongly consistent with convergence rate

$$\|\tilde{\theta}_n\|^2 = O\left[\frac{\log s_n^0 (\log\log s_n^0)^{\delta(\beta-2)}}{\lambda_{\min}^{0(n)}(r_0)}\right] \ a.s., \tag{3.15}$$

where

$$\tilde{\theta}_n = \theta_n(p_n, d_n, q_n, r_n) - \theta(p_0, d_0, q_0, r_0), \tag{3.16}$$

for $(p_n, d_n, q_n, r_n) = (p_0, d_0, q_0, r_0)$.

Although condition (3.12) is a general one, but it is not easy to verify in practice. Now we give an easily verifiable one to replace it (see, Chen and Zhang, 1989a).

Theorem 3.2. *If Conditions 2.1-2.2, 2.4 and 3.1-3.3 hold,* u_n *is given by (2.12)-(2.14) but with* $b = 1+\delta$, $\delta \in [0, (1-2\varepsilon(t+1))/(2t+3)]$ *and* $t = (m+1)p^* + q^* + r^*$, *then the order and time-delay estimates given by (3.8)-(3.10) are strongly consistent*

$$(p_n, d_n, q_n, r_n) \underset{n\to\infty}{\longrightarrow} (p_0, d_0, q_0, r_0)$$

and the coefficient estimate given by (3.3)-(3.10) is strongly consistent with convergence rate

$$\|\tilde{\theta}_n\|^2 = O\left[\frac{(\log n)(\log\log n)^{\delta(\beta-2)}}{n^{1-(t+1)(\varepsilon+\delta)}}\right] \ a.s.,$$

where $\tilde{\theta}_n$ *is given by (3.16).*

3.2. Estimation Under Relaxed Conditions

In subsection 3.1, we have discussed the all-parameter (i.e., order, time-delay and coefficient) estimation problem under the requirement that upper bounds for system orders and a lower bound for time-delay are known, and the transfer function $C^{-1}(z) - \frac{1}{2}I$ is positive real. In this subsection we shall remove them by considering different algorithms. The following results are based on the work of Huang and Guo (1989), Guo and Huang (1989).

Let $\{h_n\}$ be a sequence of nondecreasing positive integers. Introduce the following regression vectors for any $n \ge 1$:

$$\Psi_t(h_n) = [y_t^\tau \ \cdots \ y_{t-h_n+1}^\tau \ \ u_t^\tau \ \cdots \ u_{t-h_n+1}^\tau]^\tau, \quad 0 \le t \le n, \tag{3.17}$$

The driven noise $\{w_t\}$ can be estimated by either one of the following two estimates.

(a). The **"honest estimate"** $\{\hat{w}_t(n),\ 1 \le t \le n\}$:

$$\hat{w}_t(n) = y_t - \hat{\alpha}_t^\tau(n)\Psi_{t-1}(h_n), \quad 1 \le t \le n, \tag{3.18a}$$

$$\hat{\alpha}_{t+1}(n) = \hat{\alpha}_t(n) + b_t(n)P_t(n)\Psi_t(h_n)(y_{t+1}^\tau - \Psi_t^\tau(h_n)\hat{\alpha}_t(n)), \tag{3.18b}$$

$$\hat{P}_{t+1}(n) = \hat{P}_t(n) - b_t(n)P_t(n)\Psi_t(h_n)\Psi_t^\tau(h_n)\hat{P}_t(n), \tag{3.18c}$$

$$b_t(n) = (1 + \Psi_t^\tau(h_n)P_t(n)\Psi_t(h_n))^{-1}, \tag{3.18d}$$

where the initial value $\hat{\alpha}_0(n) = 0$ and $P_0(n) = \beta I$, $\beta > 0$.

(b). The **"final estimate"** $\{\hat{\varepsilon}_t(n),\ 1 \le t \le n\}$:

$$[\hat{\varepsilon}_1(n)\ \hat{\varepsilon}_2(n)\ \cdots\ \hat{\varepsilon}_n(n)]^\tau = Y_n - \Phi_n(\Phi_n^\tau\Phi_n)^{-1}\Phi_n^\tau Y_n, \tag{3.19a}$$

$$Y_n=[y_1\ y_2\ \cdots\ y_n]^\tau, \tag{3.19b}$$

$$\Phi_n=[\Psi_0(h_n)\ \Psi_1(h_n)\ \cdots\ \Psi_{n-1}(h_n)]^\tau \tag{3.19c}$$

The "honest estimate" and the "final estimate" are so named because , as estimates for w_t, $\hat{w}_t(n)$ is $\sigma\{y_i,u_i,\ i\le t\}$-measurable and $\hat{\varepsilon}_t(n)$ is $\sigma\{y_i,u_i,\ i\le n\}$-measurable for any $t\in[1,n]$.

For any nonnegative integers p,d,q and r, denote

$$\phi_t(p,d,q,r)=[y_t^\tau\ \cdots\ y_{t-p+1}^\tau\ u_{t-d+1}^\tau\ \cdots\ u_{t-q+1}^\tau\ \hat{w}_t^\tau\ \cdots\ \hat{w}_{t-r+1}^\tau]^\tau, \tag{3.20a}$$

$$X_n(p,d,q,r)=[\phi_0(p,d,q,r)\ \cdots\ \phi_{n-1}(p,d,q,r)]^\tau, \tag{3.20b}$$

$$Z_n(p,d,q,r)=Y_n-X_n(p,d,q,r)(X_n^\tau(p,d,q,r)X_n(p,d,q,r))^{-1}X_n^\tau(p,d,q,r)Y_n, \tag{3.20c}$$

$$\hat{\sigma}_n(p,d,q,r)=tr(Z_n^\tau(p,d,q,r)Z_n(p,d,q,r)), \tag{3.20d}$$

where $\{\hat{w}_n\}$ and Y_n are defined by (3.18a) and (3.19b) respectively.

The following information criterion "CIC" for estimating the system orders is

$$CIC_3(p,q,r)_n=\hat{\sigma}_n(p,1,q,r)+(p+q+r)a_n^*, \tag{3.21}$$

where $\{a_n^*\}$ is a nondecreasing sequence of positive numbers specified later on.

(a). **Order estimation procedure.** For any $n\ge 1$,

$$\hat{m}(n)=\underset{0\le k\le [\log n]}{arg\ min}\ CIC_3(k,k,k)_n, \tag{3.22}$$

$$\hat{p}(n)=\underset{0\le p\le \hat{m}(n)}{arg\ min}\ CIC_3(p,\hat{m}(n),\hat{m}(n))_n, \tag{3.23a}$$

$$\hat{q}(n)=\underset{0\le q\le \hat{m}(n)}{arg\ min}\ CIC_3(\hat{p}(n),q,\hat{m}(n))_n, \tag{3.23b}$$

$$\hat{r}(n)=\underset{0\le r\le \hat{m}(n)}{arg\ min}\ CIC_3(\hat{p}(n),\hat{q}(n),r)_n, \tag{3.23c}$$

(b). **Time-delay estimation procedure.**

$$\hat{d}(n)=\underset{1\le d\le \hat{q}(n)}{arg\ min}\ CIC_4(d)_n, \tag{3.24}$$

where

$$CIC_4(d)_n=\hat{\sigma}_n(\hat{p}(n),d,\hat{q}(n),\hat{r}(n))-da_n^*, \tag{3.25}$$

(c). **Coefficient estimation procedure.** For any $n\ge 1$, the estimate $\hat{\theta}(n)$ for the unknown coefficient $\theta(p_0,d_0,q_0,r_0)$ (see (3.2a)) is defined by

$$\hat{\theta}(n)=\hat{\theta}_n(\hat{p}(n),\hat{d}(n),\hat{q}(n),\hat{r}(n)) \tag{3.26}$$

where $(\hat{p}(n),\hat{d}(n),\hat{q}(n),\hat{r}(n))$ is given by (3.23)-(3.24) and

$$\hat{\theta}_n(p,d,q,r)=(X_n^\tau(p,d,q,r)X_n(p,d,q,r))^{-1}X_n^\tau(p,d,q,r)Y_n, \tag{3.27}$$

with $X_n(p,d,q,r)$ and Y_n the same as those in (3.20c).

It is worth noting that in the above order and time-delay estimation procedures, the first step (3.22) corresponds to estimating the value of $m_0=\max\{p_0,q_0,r_0\}$. In the second step, the true orders p_0, q_0 and r_0 are estimated. Finally, in the third step, we give the estimate for time-delay d_0.

We now give the conditions that will be used later on.

Condition 3.4. *$\det C(z) \neq 0$, for any $|z| \leq 1$.*

Condition 3.5. *$\sup_t E\{\|w_{t+1}\|^2 | F_t\} < \infty$, $\|w_t\| = O(\phi(t))$, a.s., where the function $\phi(.)$ is positive, deterministic, nondecreasing and satisfies*

$$\sup_k \phi(e^{k+1})/\phi(e^k) < \infty .$$

Theorem 3.3. *(Huang and Guo, 1989) (i). If Conditions 3.3-3.5 hold, $h_n = O((\log n)^\alpha)$ with $\alpha > 1$, $\log n = o(h_n)$ and if in criteria (3.21) and (3.25), the sequence $\{a_n^*\}$ satisfies*

$$(h_n \log n + (\phi(n) \log\log n)^2)/a_n^* \to 0, \text{ a.s., as } n \to \infty$$

and

$$a_n^* / \lambda_{\min}(n) \to 0, \text{ a.s., as } n \to \infty, \tag{3.28}$$

where $\lambda_{\min}^0(n)$ is defined as follows

$$\lambda_{\min}^0(n) = \lambda_{\min}\left(\sum_{i=0}^{n-1} \phi_i^0 \phi_i^{0\tau}\right) \tag{3.29}$$

with

$$\phi_t^0 = [y_t^\tau \cdots y_{t-m_0+1} \ u_t^\tau \cdots u_{t-m_0+1} \ w_t^\tau \cdots w_{t-m_0+1}]^\tau, \quad m_0 = \max\{p_0, q_0, r_0\}.$$

Then for the estimation algorithm defined by (3.20)-(3.27), as $n \to \infty$,

$$\hat{m}(n) \to m_0, \text{ a.s.,} \tag{3.30}$$

$$(\hat{p}(n), \hat{d}(n), \hat{q}(n), \hat{r}(n)) \to (p_0, d_0, q_0, r_0), \text{ a.s.} \tag{3.31}$$

$$\|\tilde{\theta}_n\|^2 = O\left(\frac{h_n \log n}{\lambda_{\min}^0(n)}\right) + o\left(\frac{(\phi(n) \log\log n)^2}{\lambda_{\min}^0(n)}\right) \text{ a.s.,} \tag{3.32}$$

where

$$\tilde{\theta}(n) = \hat{\theta}(n) - \theta(p_0, d_0, q_0, r_0) \tag{3.33}$$

for $(\hat{p}(n), \hat{d}(n), \hat{q}(n), \hat{r}(n)) = (p_0, d_0, q_0, r_0)$.

(ii). If Condition 3.5 is replaced by

$$E\{\sup_t E\{\|w_{t+1}\|^2 | F_t\}\} < \infty,$$

$\{a_n^*\}$ satisfies (3.28) and

$$(h_n^2 \log n (\log\log n)^{2+\delta})/a_n^* \to 0, \text{ a.s., for some } \delta > 0,$$

where $h_n = [c(\log n)^\alpha]$ is the integer part of $c(\log n)^\alpha$ for some $\alpha > 1$ and $c > 0$. Then (3.30) and (3.31) also hold, and

$$\|\tilde{\theta}\|^2 = O\left(\frac{h_n^2 \log n (\log\log n)^{2+\delta}}{\lambda_{\min}^0(n)}\right) \text{ a.s.,}$$

where $\tilde{\theta}(n)$ is given by (3.33).

As one would have probably noted, a major feature of the theorem is that the criteria CIC_3 and CIC_4 depends on $\{a_n^*\}$, which in turn depends on the growth rate of $\lambda_{\min}^0(n)$ as exposed in (3.28). This is highly expected because there is no specific constraint on the input sequence $\{u_t\}$ except those in Condition 3.3, and the $\{u_t\}$ determines completely the "excitation extent": the growth rate of $\lambda_{\min}^0(n)$.

Let us now consider the "attenuating excitation" method as used in the above subsections. To be specific, let ε_n be given by (2.12), and take

$$\zeta_1 = 0, \quad \zeta_n = \frac{\varepsilon_n}{(\log n)^{1/2}}, \quad \text{for any } n \geq 2.$$

Theorem 3.4. *Suppose that in system (1.1), $A(z)$, $B(z)$ and $C(z)$ have no common left factor, A_{p_0}, B_{q_0}, C_{r_0} are of row full rank, the system input u_n is defined by*

$$u_n = u_n^0 + \zeta_n$$

with u_n^0 being F_n'-measurable, where F_n' is given by (2.12b). If Condition 2.4 holds, and

$$\limsup_{n \to \infty} \frac{1}{n} \sum_{i=0}^{n-1} (\|y_i\|^2 + \|u_i\|^2) < \infty, \text{ a.s..}$$

Then Theorem 3.1 still holds with $\lambda_{\min}^0(n)$ replaced by $\frac{n}{\log n}$ in (3.28) and (3.32)-(3.33).

4. ADAPTIVE CONTROL

4.1. Adaptive Tracking with Coefficient Estimation

In this section we define adaptive control based on the SG algorithm in order that the output y_n of the system follows a given bounded deterministic signal $\{y_n^*\}$. At time n the predicting value for y_{n+1} is $\theta_n^\tau \psi_n$. Hence it is natural to define adaptive control from the equation

$$\theta_n^\tau \psi_n = y_{n+1}^* \tag{4.1}$$

which leads to the global stability of the system and minimality of tracking error (Goodwin *et al*, 1981). However, under such a control the parameter estimate, in general, is inconsistent, unless y_n^* itself is sufficiently rich.

Define

$$Y_n^* = [y_n^{*\tau} \ \cdots \ y_{n-p-l(q-1)}^{*\tau}]^\tau$$

and denote by $\lambda_{\min}(A)$ the minimum eigenvalue of a matrix A.

Theorem 4.1. *(Chen and Guo, 1986c) Assume that Conditions 2.1 and 2.3-2.4 hold,* $m=l$, $d=1$, $p\geq 1$, $q\geq 1$, B_1 *and* $B_1^{-1}B_q$ *are of full rank,* $z^{-1}B(z)$ *is stable,* $A(z)$ *and* $B(z)$ *are left-coprime, the components of* $\{w_n\}$ *are mutually independent and have continuous distributions and finally*

$$\liminf_{n\to\infty} \frac{(\log n)^{1/4}}{n} \lambda_{\min}\left(\sum_{i=1}^{n} (Y_i^* Y_i^{*\tau})\right) \neq 0, \quad a.s.. \tag{4.2}$$

Let θ_n *be given by (2.8)-(2.9). Then the control* u_n *can be defined from (4.1) and which leads to*

$$\limsup_{n\to\infty} \frac{1}{n} \sum_{i=1}^{n} (\|y_i\|^2 + \|u_i\|^2) < \infty, \quad a.s.. \tag{4.3}$$

$$\lim_{n\to\infty} \frac{1}{n} \sum_{i=1}^{n} (y_i - y_i^*)(y_i - y_i^*)^\tau = R, \quad a.s., \tag{4.4}$$

and

$$\lim_{n\to\infty} \theta_n = \theta, \quad a.s.. \tag{4.5}$$

Condition (4.2) means that the reference signal is rich in a certain sense. Related results may be found in (Kumar and Praly, 1987). We now solve the problem for the general reference signals by applying attenuating excitation technique.

Theorem 4.2. *Suppose that Conditions 2.1, 2.3 and 2.5 are satisfied,* $m=l$, $d=1$, $B(z)z^{-1}$ *is stable, the initial value for estimating* B_1 *is non-degenerate. Then* u_n *can be defined by*

$$u_n = u_n^0 + v_n$$

and

$$u_n^0 = (B_{1n})^+ (y_{n+1}^* - \theta_n^\tau \psi_n + B_{1n} u_n)$$

where θ_n *is given by (2.8)-(2.9) and* v_n *by (2.18). This control leads to all conclusions (4.3)-(4.5) of Theorem 4.1.*

4.2. Adaptive Tracking with All-Parameter Estimation

In this section, we assume that $m=l$ and $q_0 \geq d_0 \geq 1$, and the true system orders and time-delay belong to the known sets M_0 and M_d respectively. The problem we want to deal with is to find an F_n-measurable system input u_n such that the following performance index is minimized,

$$\limsup_{n\to\infty}\frac{1}{n}\sum_{i=0}^{n}||A^0(z)y_i-B^0(z)y_i^*+Q(z)u_{i-1}||^2, \tag{4.6}$$

and all-parameter estimates are strongly consistent, where y_n^* is a given bounded reference signal, y_{n+1}^* is F_n-measurable, $A^0(z)$, $B^0(z)$ and $Q(z)$ are given by the designers. Obviously, the pure tracking is a special case of this index.

Suppose that $A^0(0)=I$ and let $G(z)$ and $\bar{N}(z)$ be defined by

$$(detC(z))I=(adjC(z))A(z)+zG(z), \tag{4.7}$$

$$A^0(z)=I+z\bar{N}(z). \tag{4.8}$$

Then from Chen and Zhang (1989c), we know that for any F_n-measurable input u_n, the system output y_n satisfies

$$\limsup_{n\to\infty}\frac{1}{n}\sum_{i=0}^{n}||A^0(z)y_i-B^0(z)y_i^*+Q(z)u_{i-1}||^2\geq \limsup_{n\to\infty}\frac{1}{n}\sum_{i=0}^{n}||w_i||^2. \tag{4.9}$$

Denote

$$G_0(z)=\sum_{i=0}^{g_0}G_{0i}z^i=\bar{N}(z)(adjC(z))A(z)+A^0(z)G(z), \tag{4.10a}$$

$$G_1(z)=\sum_{i=0}^{g_1}G_{1i}z^i=A_0(z)(adjC(z))B(z)z^{-1}-\bar{N}(z)(adjC(z))B(z)$$

$$+(detC(z))Q(z), \tag{4.10b}$$

$$G_2(z)=\sum_{i=0}^{g_2}G_{2i}z^i=-(detC(z))B^0(z), \tag{4.10c}$$

$$detC(z)=1+\sum_{i=0}^{mr^*}c_iz^i \tag{4.10d}$$

$$\bar{y}_n=A^0(z)y_i-B^0(z)y_i^*+Q(z)u_{i-1} \tag{4.10e}$$

where

$$g_i=deg(G_i(z)),\quad i=0,\ 1,\ 2,$$

Then we have

$$(detC(z))(\bar{y}-w_0)=G_n(z)y_{\ 1}+G_n(z)u_{\ 2}+G_n^*(z)y\ . \qquad (4$$

From the above expression it is easy to see that the optimal control can be obtained by

$$G_0(z)y_{n-1}+G_1(z)u_{n-1}+G_2(z)y_n^*=0.$$

With this fact in mind, we now estimate the parameter

$$\theta^{*\tau}=[G_{00}\,G_{01}\ \cdots\ G_{0g_0}\ G_{10}\,G_{11}\ \cdots\ G_{1g_1}$$

$$G_{20}\,G_{21}\ \cdots\ G_{2g_2}\ c_1\,c_2\ \cdots\ c_{mr^*}]$$

by the following recursive algorithm:

$$\theta_n^*=\theta_{n-1}^*+\frac{\phi_{n-1}^*}{r_{n-1}^*}(y_n^\tau-\phi_{n-1}^{*\tau}\theta_{n-1}^*), \tag{4.12a}$$

$$\phi_n^*=[y_n^\tau\ \cdots\ y_{n-g_0}^\tau\ \ u_n^\tau\ \cdots\ u_{n-g_1}^\tau\ \ y_{n+1}^{*\tau}\ \cdots\ y_{n+1-g_2}^{*\tau}$$

$$-\phi_{n-1}^{*\tau}\theta_{n-1}^{*} \cdots -\phi_{n-mr^*}^{*\tau}\theta_{n-mr^*}^{*}]^{\tau}. \tag{4.12b}$$

$$r_n^{*}=1+\sum_{i=0}^{n}\|\phi_i^{*}\|^2, \tag{4.12c}$$

Finally, we take the adaptive control u_n as follows

$$u_n=u_n^{(1)}+v_n, \tag{4.13a}$$

where v_n is given by (2.12) and (2.14) but with $t=(m+1)p^*+q^*+r^*$, and $u_n^{(1)}$ is given by

$$G_{10n}u_n^{(1)}=G_{10n}u_n-\theta_n^{*\tau}\phi_n^{*}, \tag{4.13b}$$

where

$$G_{10n}=[\theta_n^{*\tau}(mg_0+m+1) \cdots \theta_n^{*\tau}(mg_0+m+l)], \tag{4.13c}$$

and $\theta_n^{*\tau}(i)$ denotes the $i-th$ column of $\theta_n^{*\tau}$.

Theorem 4.3. *(Chen and Zhang, 1989a) If (i) Conditions 2.1 and 2.3 hold; (ii) $(detA^0(z))B(z)z^{-1}+A(z)(adjA^0(z))Q(z)$ and $A^0(z)$ are stable, then (4.13) leads to*

$$\limsup_{n\to\infty}\frac{1}{n}\sum_{i=0}^{n}(\|y_i\|^2+\|u_i\|^2)<\infty, \ a.s., \tag{4.14}$$

$$\limsup_{n\to\infty}\Big\|\frac{1}{n}\sum_{i=0}^{n}\bar{y}_i\bar{y}_i^{\tau}-\frac{1}{n}\sum_{i=0}^{n}w_iw_i^{\tau}\Big\|=0, \ a.s.. \tag{4.15}$$

Furthermore, if Conditions 2.1-2.4 and 3.2 hold, then in addition to (4.14)-(4.15), the order and time-delay estimates given by (3.8)-(3.10) are strongly consistent:

$$(p_n,d_n,q_n,r_n) \underset{n\to\infty}{\longrightarrow} (p_0,d_0,q_0,r_0)$$

and the coefficient estimate given by (3.3)-(3.10) is strongly consistent with convergence rate:

$$\|\tilde{\theta}_n\|^2=O\left(\frac{(\log n)(\log\log n)^{\delta(\beta-2)}}{n^{1-(t+1)\varepsilon}}\right), \ a.s.,$$

where $\tilde{\theta}_n$ is given by (3.16).

We note that for sovability of $u_n^{(1)}$ in (4.13b), the non-singularity of G_{10n} is necessary and this is guaranteed by continuity of the distribution of v_n and the non-singularity of $B_1+Q(0)$, which is true because of the stability assumption (i) in Theorem 4.3. For details, see Chen and Guo (1986c, 1987c).

In Theorem 4.3 the convergence rate for performance index is not given. For this we now combine the ELS and SG algorithms and apply the stochastic truncation method introduced by Chen and Guo (1987a-b) to design an optimal adaptive control for the following performance index

$$\limsup_{n\to\infty}\frac{1}{n}\sum_{i=0}^{n}\|A^0(z)y_i-B^0(z)y_i^{*}\|^2,$$

which is the performance (4.6) with the control term deleted and where $y_{n+q^*}^{*}$ is F_n-measurable.

Let us first indicate the minimum value of this performance index. Set

$$L(z)=\sum_{i=0}^{d_0-1}L_iz^i$$

and let $G(z)$ be the solution of the following Diophantine equation

$$(detC(z))I=L(z)(adjC(z))A(z)+G(z)z^{d_0},$$

$\bar{N}(z)$ and

$$\bar{F}(z)=\sum_{i=0}^{d_0-1}\bar{F}_i z^i$$

be defined by

$$A^0(z)L(z)=\bar{F}(z)+\bar{N}(z)z^{d_0}.$$

Then according to Chen and Zhang (1989c), the performance index has the following lower bound:

$$\lim_{n\to\infty}\sup\frac{1}{n}\sum_{i=0}^{n}\|A^0(z)y_i-B^0(z)y_i^*\|^2\geq \lim_{n\to\infty}\sup\frac{1}{n}\sum_{i=0}^{n}\|\bar{F}(z)w_i\|^2.$$

To obtain the desired results mentioned above, for r_n, q_n, d_n, p_n and $\theta_n(r_n)$ given by (3.3)-(3.10), we set

$$A_n(z)=I+\sum_{i=1}^{p_n}A_{in}(r_n)z^i,\quad B_n(z)=\sum_{i=d_n}^{q_n}B_{in}(r_n)z^i,\quad C_n(z)=I+\sum_{i=1}^{r_n}C_{in}(r_n)z^i. \tag{4.16}$$

Let

$$L_n(z)=\sum_{i=0}^{d_n-1}L_{in}z^i,$$

and $G_n(z)$ be the solution of the following Diophantine equation

$$(\det C_n(z))I=L_n(z)(\operatorname{adj}C_n(z))A_n(z)+G_n(z)z^{d_n}.$$

Futher, define $\bar{N}_n(z)$ and

$$\bar{F}_n(z)=\sum_{i=0}^{d_n-1}\bar{F}_{in}z^i$$

satisfying

$$A^0(z)L_n(z)=\bar{F}_n(z)+\bar{N}_n(z)z^{d_n}. \tag{4.18}$$

Let $u_n^{(2)}$ be the solution of the following equation

$$\begin{aligned}&B_{d_n n}u_n^{(2)}+(A^0(z)-I)B_{d_n n}u_n\\ &=(\det C_n(z))(B^0(z)y^*_{n+d_n}-\bar{N}_n(z)(y_n-\theta_n^\tau(p_n,d_n,q_n,r_n)\phi^\tau_{n-1}(p_n,d_n,q_n,r_n)))\\ &\quad -A^0(z)(G_n(z)y_n+(L_n(z)(\operatorname{adj}C_n(z))B_n(z))z^{-d_n}u_n-B_{d_n n}u_n).\end{aligned} \tag{4.19}$$

Introduce the stopping times $\{\tau_k\}$ and $\{\sigma_k\}$:

$$0=\tau_1<\sigma_1<\tau_2<\sigma_2<\ \dots, \tag{4.20a}$$

$$\sigma_k=\sup\{\mu>\tau_k:\ \sum_{i=\tau_k}^{j-1}\|y_i\|^2\leq(j-1)^{1+\delta/2}+\|y_{\tau_k}\|^2,\quad j\in(\tau_k,\mu]\}, \tag{4.20b}$$

$$\begin{aligned}\tau_{k+1}=\inf\{\mu>\sigma_k:\ &\sum_{i=\sigma_k}^{\mu}\|y_i\|^2\leq\frac{\mu\log\mu}{2^k},\quad \sum_{i=\tau_k}^{\sigma_k-1}\|y_i\|^2\leq\frac{\mu\log\mu}{2^k},\\ &\sum_{i=\sigma_k}^{\mu}\|u_i\|^2\leq\frac{\mu\log\mu}{2^k},\quad \sum_{i=\tau_k}^{\sigma_k-1}\|u_i\|^2\leq\frac{\mu\log\mu}{2^k}\},\end{aligned} \tag{4.20c}$$

where δ is defined by Theorem 3.2.

Denote

$$\Lambda=\{j:\ \|u_j^{(2)}\|^2\leq j^{1+\delta}\},$$

$$u_n^0=\begin{cases}u_n^{(2)}, & \text{if } n\in[\tau_k,\sigma_k)\cap\Lambda \text{ for some } k,\\ 0, & \text{if } n\in[\tau_k,\sigma_k)\cap\Lambda^c \text{ for some } k,\\ u_n^{(1)}, & \text{if } n\in[\sigma_k,\tau_{k+1}) \text{ for some } k.\end{cases} \tag{4.21}$$

Theorem 4.4. *(Chen and Zhang, 1989a-b) Suppose that Conditions 2.1-2.4 and 3.1-3.2 hold,* $A^0(z)B(z)z-d_0$ *is stable, and that there is a polynomial matrix* $Q(z)$ *such that* $(\det A^0(z))B(z)z^{-1}+A(z)(\operatorname{adj}A^0(z))Q(z)$ *is stable. Then (4.12)-(4.13), (4.19)-(4.21) and (2.12)-(2.13) with* $t=(m+1)p^*+q^*+r^*$ *lead to*

$$\lim_{n\to\infty}\sup\frac{1}{n}\sum_{i=0}^{n}(\|y_i\|^2+\|u_i\|^2)<\infty;\quad a.s.,$$

$$\Big\|\frac{1}{n}\sum_{i=0}^{n}(A^0(z)y_i-B^0(z)y_i^*)(A^0(z)y_i-B^0(z)y_i^*)^\tau-\frac{1}{n}\sum_{i=0}^{n}(\bar F(z)w_i)(\bar F(z)w_i)^\tau\Big\|=O(n^{-\varepsilon/2}),\ a.s.,$$

$$(p_n,d_n,q_n,r_n)\underset{n\to\infty}{\longrightarrow}(p_0,d_0,q_0,r_0)$$

$$\|\tilde\theta_n\|^2=O\left[\frac{(\log n)(\log\log n)^{\delta(\beta-2)}}{n^{1-(t+1)\varepsilon}}\right],\ a.s.,$$

where $\tilde\theta_n$ *is given by (3.16).*

Notice that the sovability of $u_n^{(2)}$ from (4.19) is guaranteed by the continuity of the distribution of v_n and the non-singularity of B_{d_0}. For details, see Chen and Zhang (1989c).

4.3. Adaptive Linear Quadratic Control

In this section we shall consider an adaptive LQ problem for system (1.1) with orders, time-delay and coefficients all unknown. The loss function is

$$J(u)=\lim_{n\to\infty}\sup J_n(u),\tag{4.22}$$

where

$$J_n(u)=\frac{1}{n}\sum_{i=0}^{n-1}(y_i^\tau Q_1y_i+u_i^\tau Q_2u_i),\quad Q_1\ge 0,\ Q_2>0.\tag{4.23}$$

In this section we assume that

Condition 4.1. $\frac{1}{n}\sum_{i=0}^{n}w_iw_i^\tau=R+O(n^{-\rho}),\ a.s.,$ *for some* $\rho>0$ and $R\ge 0$.

We first write (1.1) in the state space form

$$x_{k+1}=Ax_k+Bu_k+Cw_{k+1},\quad x_0=0,\tag{4.24a}$$

$$y_k=C^\tau x_k,\tag{4.24b}$$

with

$$A=\begin{bmatrix}-A_1 & I & 0 & \cdots & 0\\ -A_2 & 0 & I & \cdots & 0\\ \vdots & & & & \vdots\\ -A_{h-1} & 0 & 0 & \cdots & I\\ -A_h & 0 & 0 & \cdots & 0\end{bmatrix}_{mh\times mh},\tag{4.25a}$$

$$B^\tau=[0\ \cdots\ 0\ B_{d_0}^\tau\ \cdots\ B_h^\tau]_{l\times mh},\tag{4.25b}$$

$$C^\tau=[I\ \cdots\ 0]_{m\times mh},\tag{4.25c}$$

where $h=\max(p_0,q_0,r_0+1)$.

From Chen and Guo (1986b) it is known that

$$\inf_{u\in U} J(u)=RC^{\tau}SC, \tag{4.26}$$

and the optimal control is

$$u_n=Lx_n, \tag{4.27}$$

where

$$U=\{u:\ \sum_{i=0}^{n}||u_i||^2=O(n),\ \ ||u_n||^2=o(n),\ a.s.,\ \ u_n\in F_n\}, \tag{4.28}$$

$$L=-(B^{\tau}SB+Q_2)^{-1}B^{\tau}SA, \tag{4.29}$$

S satisfies

$$S=A^{\tau}SA-A^{\tau}SB(B^{\tau}SB+Q_2)^{-1}B^{\tau}SA+CQ_1C^{\tau} \tag{4.30}$$

for which there is a unique positive definite S in the class of non-negative definite matrices if (A,B,D) is controllable and observable for some D fullfiling $D^{\tau}D=CQ_1C^{\tau}$.

Based on the estimate p_n,d_n,q_n,r_n and $\theta_n(p_n,d_n,q_n,r_n)$ given by (3.3)-(3.10), we estimate A,B,C,S and x_n by $A(n)$, $B(n)$, $C(n)$, $S(n)$ and $\hat{x}_n$ respectively as follows:

$$A(n)=\begin{bmatrix} -A_{1n}(r_n) & I & 0 & \cdots & 0 \\ -A_{2n}(r_n) & 0 & I & \cdots & 0 \\ \vdots & & & & \vdots \\ -A_{h_n-1n}(r_n) & 0 & 0 & \cdots & I \\ -A_{h_n}(r_n) & 0 & 0 & \cdots & 0 \end{bmatrix}_{mh_n\times mh_n}, \tag{4.31a}$$

$$B^{\tau}(n)=[0\ \cdots\ 0\ B^{\tau}_{d_n n}(r_n)\ \cdots\ B^{\tau}_{h_n}(r_n)]_{l\times mh_n}, \tag{4.31b}$$

$$C^{\tau}(n)=[I\ \cdots\ 0]_{m\times mh_n}, \tag{4.31c}$$

where $h_n=max(p_n,q_n,r_n+1)$.

$$S(n)=A^{\tau}(n)S'(n-1)A(n)-A^{\tau}(n)S'(n-1)B(n)(B^{\tau}(n)S'(n-1)B(n)+Q_2)^{-1}$$
$$\times B^{\tau}(n)S'(n-1)A(n)+C(n)Q_1C^{\tau}(n). \tag{4.32}$$

Here $S(0)=0$, $S'(n-1)$ is a square matrix of dimension $mh_n\times mh_n$:

$$S'(n-1)=\begin{cases}\begin{bmatrix} S(n-1) & 0 \\ 0 & 0\end{bmatrix}, & \text{if } h_{n-1}< h_n, \\ M^{\tau}(n)S(n-1)M(n), & \text{if } h_{n-1}\geq h_n,\end{cases}$$

$$M^{\tau}(n)=[I\ \ 0]_{mh_n\times mh_n}$$

and finally,

$$\hat{x}_{n+1}=A(n)\hat{x}_n'+B(n)u_n+C(n)(y_{n+1}-C^{\tau}(n)A(n)\hat{x}_n'-C^{\tau}(n)B(n)u_n), \tag{4.33a}$$

$$\hat{x}_0=y_0=0, \tag{4.33b}$$

where $\hat{x}_n'$ is of dimension of mh_n and is defined by

$$\hat{x}_n'=\begin{cases}[\hat{x}_n^{\tau}\ \ 0]^{\tau}, & \text{if } h_{n-1}< h_n, \\ M^{\tau}(n)\hat{x}_n, & \text{if } h_{n-1}\geq h_n.\end{cases} \tag{3.34}$$

We now have the estimate L_n for the optimal gain L given by (4.29).

$$L_n=-(B^\tau(n)S(n)B(n)+Q_2)^{-1}B^\tau(n)S(n)A(n). \tag{4.35}$$

However, we cannot directly take $L_n x_n$ as the desired control u_n^0, because $L_n x_n$ may grow too fast so that condition 3.3 is not satisfied. Therefore, we define

$$L_n^0=\begin{cases} L_n, & \text{if } n\in[\tau_k,\sigma_k) \text{ for some } k, \\ 0, & \text{if } n\in[\sigma_k,\tau_{k+1}) \text{ for some } k, \end{cases} \tag{4.36}$$

$$u_n^0=L_n^0\hat{x}_n', \tag{4.37}$$

where stopping times $\{\tau_k\}$ and $\{\sigma_k\}$ are defined by

$$1=\tau_1<\sigma_1<\tau_2<\sigma_2<\cdots, \tag{4.38a}$$

$$\sigma_k=sup\{\tau>\tau_k:\ \sum_{i=\tau_k}^{j-1}||L_i\hat{x}_i'||^2\le(j-1)^{1+\delta}+||L_{\tau_k}\hat{x}_{\tau_k}'||^2,\ j\in(\tau_k,\tau]\}, \tag{4.38b}$$

$$\tau_{k+1}=\{\tau>\sigma_k:\ \sum_{i=\tau_k}^{\sigma_k-1}||L_i\hat{x}_i'||^2\le\frac{\tau^{1+\delta}}{2^k},\ \sum_{i=1}^{\tau}||\hat{x}_i'||^2\le\tau^{1+\delta/2},$$

$$||L_\tau\hat{x}_\tau'||^2\le\tau^{1+delat}\}. \tag{4.38c}$$

The following Theorem is based on the work of Chen and Guo (1987b) and Chen and Zhang (1989b).

Theorem 4.5. *If Conditions 2.1 and 4.1 hold, $A(z)$ is stable, (A,B,D) is controllable and observable for some D satisfying $D^\tau D=CQ_1C^\tau, p_n, d_n, q_n, r_n$ and $\theta_n(p_n,d_n,q_n,r_n)$ are given by (3.3)-(3.10), then u_n defined by (4.31)-(4.38) and (2.13) with $t=(m+1)p^*+q^*+r^*$ is optimal in the following sense that*

$$(p_n,d_n,q_n,r_n)\underset{n\to\infty}{\longrightarrow}(p_0,d_0,q_0,r_0),\ a.s.,$$

$$||\tilde{\theta}_n||^2=O\left[\frac{(logn)(loglogn)^{\delta(\beta-2)}}{n^{1-(t+1)\varepsilon}}\right],\ a.s.$$

and

$$J_n(u)=RC^\tau SC+O(n^{-min(\rho,\varepsilon)}),\ a.s.,$$

where $\tilde{\theta}_n$ is given by (3.16).

4.4. Robust Adaptive Control

In practice a real system may not be exactly described by (1.1) but with some unmodelled dynamics η_k. Let us consider the following model:

$$y_{k+1}=\theta^\tau\phi_k+v_{k+1}+\eta_k, \tag{4.39}$$

where

$$\theta^\tau=[a_1\ \cdots\ a_p\ b_1\ \cdots\ b_q]$$

$$\phi_n^\tau=[y_n\ \cdots\ y_{n-p+1}\ u_n\ \cdots\ u_{n-q+1}]$$

and $\{v_{k+1}\}$ and $\{\eta_k\}$ are, respectively, the system noise and unmodelled dynamics.

For the random noise $\{v_k\}$ we assume that $\{v_k,F_k\}$ is an adapted sequence and for some integer $r\ge 0$ and constants $\varepsilon>0$, $M>0$,

$$E[exp(\varepsilon||v_{k+1}||^2|F_{k-r})]\le exp(M),\quad k\ge 0. \tag{4.40}$$

For the unmodelled dynamics $\{\eta_k\}$ we assume that it has the following known but time-varying upper bound (see,e.g.,Ioannou and Kokotovic, 1984; Ioannou and Sun, 1988; Ioannou and Tsakalis, 1985; Kreiselmeier and Anderson, 1985 and Middleton *et al*, 1988):

$$|\eta_k| \leq \varepsilon^* m_k, \quad m_k = \gamma m_{k-1} + ||\phi_k||, \quad m_0 > 0, \quad k \geq 0, \tag{4.41}$$

where $\varepsilon^* > 0$, $\gamma \in (0,1)$.

Assume that

$$B(z) = b_1 + b_2 z + \cdots + b_q z^{q-1}, \quad \textit{is stable.} \tag{4.42}$$

Without loss of generality, assume that $b_1 > 0$.

Let us define

$$D = \{x = (x_1 \ \cdots \ x_{p+q}) \in R^{p+q} : |x_i| \leq L, \ 1 \leq i \leq p+q, \ x_{p+1} \geq b_0\}. \tag{4.43}$$

The estimate $\hat{\theta}_n$ for θ is given by

$$\hat{\theta}_{k+1} = \pi_D \{\hat{\theta}_k + \frac{\phi_k}{c + (m_k)^2}(y_{k+1} - \phi_k^\tau \hat{\theta}_k)\}, \tag{4.44}$$

where $\pi_D(x)$ is the projecting from x to D, and $c > 0$ is a constant. The following results can be proven along the lines of the argument in (Guo, 1989).

Theorem 4.6. *For system described by (4.39)-(4.42) if u_k is chosen from*

$$\hat{\theta}_k^\tau \phi_k = 0,$$

where $\hat{\theta}_k$ is defined by (4.33)-(4.44) with L and c (b_0) suitably large (small), then

$$\sup_k E[|y_k|^\beta + |u_k|^\beta] < \infty,$$

for some $\beta > 2$.

We remark that the above theorem does not need the stability of $A(z)$ and the positive realness of a certain noise model. Related robust adaptive control results, giving precise analysis for the control performance, was given in (Chen and Guo, 1988) under more conditions on the modelled part of the system.

5. TIME-VARYING SYSTEMS

5.1. Parameter Estimation

Let us consider the following time-varying regression model

$$y_k=\phi_k^\tau\theta_k+v_k \tag{5.1}$$

where y_k and v_k are the scalar observation and noise respectively, and ϕ_k and θ_k are respectively the p-dimensional stochastic vector and the unknown time-varying parameter. Denote the parameter variation at time k by w_k:

$$w_k=\theta_k-\theta_{k-1}, \quad k\geq 1, \tag{5.2}$$

For estimating the unknown parameter θ_k, we introduce the following Kalman filter based adaptive estimator:

$$\hat{\theta}_{k+1}=\hat{\theta}_k+\frac{P_k\phi_k}{R+\phi_k^\tau P_k\phi_k}(y_k-\phi_k^\tau\hat{\theta}_k), \tag{5.3a}$$

$$P_{k+1}=P_k-\frac{P_k\phi_k\phi_k^\tau P_k}{R+\phi_k^\tau P_k\phi_k}+Q, \tag{5.3b}$$

where P_0, $R>0$ and $Q>0$ as well as $\hat{\theta}_0$ are deterministic and can be arbitrarily chosen (here Q and R may be regarded as the *a priori* estimates for the variances of v_k and w_k respectively).

Most of the work done in the area of adaptive filtering was concerned with the least mean square (LMS) algorithm, in which the adaptive gain is $\mu\phi_k$ rather than $\dfrac{P_k\phi_k}{R+\phi_k^\tau P_k\phi_k}$ as in (5.3) (Benveniste and Ruget, 1982; Macchi, 1986 and Solo, 1988).

We now introduce the assumptions of this subsection.

Condition 5.1. *$\{v_k, w_k\}$ is a random or deterministic process satisfying*

$$\sigma_\beta=\sup_k E[\|v_k\|^\beta+\|w_k\|^\beta]<\infty, \ \textit{for some}\ \beta>0, \tag{5.4a}$$

$$\mu_\beta=\limsup_{n\to\infty}\frac{1}{n}\sum_{k=0}^{n-1}(\|v_k\|^\beta+\|w_k\|^\beta)<\infty, \ a.s., \ \textit{for some}\ \beta>0, \tag{5.4b}$$

Condition 5.2. *$\{\phi_k, F_k\}$ is an adapted sequence (i.e. ϕ_k is F_k-measurable for any k, where F_k is a sequence of non-decreasing σ-algebras), and there exists an integer $h>0$ such that*

$$E\left\{\sum_{k=m+1}^{m+h}\frac{\phi_k\phi_k^\tau}{1+\|\phi_k\|^2}\,|F_m\right\}\geq\frac{1}{\alpha_m}I, \ a.s., \quad m\geq 1, \tag{5.5a}$$

where $\{\alpha_m, F_m, m\geq 1\}$ is a positive adapted stochastic sequence satisfying

$$\alpha_{m+1}\leq a\alpha_m+\eta_{m+1}, \quad m\geq 0, \quad E\alpha_0<\infty, \tag{5.5b}$$

where $\{\eta_m, F_m\}$ is an adapted nonnegative sequence, and

$$\eta_m\in F_m, \quad \sup_m E[\eta_{m+1}^{1+\delta}|F_m]\leq M, \ a.s. \tag{5.5c}$$

and where $a\in[0,1)$, $\delta>0$, $M<\infty$ are deterministic constants.

We remark that the above assumptions are relaxations of those made in (Guo, 1990). To understand the generality of our condition 5.2, we present the following examples.

Example 5.1. Let $\{\phi_k\}$ be a ϕ-mixing process, that is, there is a deterministic sequence $\{\phi(n),\ n\geq 0\}$, such that

(i). $\phi(n)\to 0$, *as* $n\to\infty$;

(ii). $\sup_{\substack{A\in F^{\infty}_{s+n}\\ B\in F^{s}_{0}}} |P(A|B)-P(A)|\leq \phi(n),\quad s\geq 0,\quad n\geq 0,$

where, for any nonnegative integers s and n, $F^{\infty}_{s+n}=\sigma\{\phi_k,\ s+n\leq k<\infty\}$ and $F^{s}_{0}=\sigma\{\phi_k, 0\leq k<s\}$.

Suppose that

$$\inf_k \lambda_{\min}(E[\phi_k\phi_k^{\tau}])>0 \quad\text{and}\quad \sup_k E\|\phi_k\|^4<\infty .$$

Then Condition 5.2 holds for $F_m=F^{m}_{0}$.

Example 5.2. Let $\{\phi_k\}$ be generated by a linear model

$$\phi_k=F\phi_{k-1}+G\zeta_k,\quad k\geq 1,$$

where F is a stable matrix, (F,G) is controllable, and $\{\phi_0,\ \zeta_k,\ k\geq 1\}$ is an independent stochastic vector sequence with the following properties:

(i). $E\phi_0=0,\quad E\|\phi_0\|<\infty$;

(ii). $E\zeta_k=0,\ k\geq 0;\quad \inf_{k\geq 0} E\zeta_k\zeta_k^{\tau}>0;\quad \sup_{k\geq 0} E\|\zeta_k\|^{4(1+\delta)}<\infty$, *for some* $\delta\in(0,1)$.

Then Condition 5.2 holds.

The following Theorems are based on those in (Guo, 1990).

Theorem 5.1. *If in (5.1),* $\theta_k\equiv\theta,\ v_k\equiv 0,\ \{\phi_k\}$ *satisfies Condition 5.2, and for some deterministic constant* $\beta>0,\ E[\|\theta_0\|^{\beta}]<\infty$, *then for any* $\alpha\in(0,\beta)$, *as* $n\to\infty$,

(i). $E\|\hat{\theta}_n-\theta\|^{\alpha}\to 0$, *exponentially fast, and*

(ii). $\hat{\theta}_n-\theta\to 0$, *a.s., exponentially fast,*

where $\{\hat{\theta}_n\}$ *is given by (5.3).*

Theorem 5.2. *If in (5.1),* $\theta_k\equiv\theta,\ v_k\equiv 0,\ \{\phi_k\}$ *satisfies Condition 5.2, and for some deterministic constants* $\bar{\beta}_0>0,\ \bar{\varepsilon}_0>0,\ E\exp[\bar{\varepsilon}_0\|\theta_0\|^{\bar{\beta}_0}]<\infty$, *then for any* $\varepsilon\in(0,\bar{\varepsilon}_0)$, *as* $n\to\infty$,

(i). $E\exp[\varepsilon\|\hat{\theta}_n-\theta\|^{\bar{\beta}_0}]\to 1$, *exponentially fast, and*

(ii). $\exp[\varepsilon\|\hat{\theta}_n-\theta\|^{\bar{\beta}_0}]\to 1$, *a.s., exponentially fast,*

where $\{\hat{\theta}_n\}$ *is given by (5.3).*

Theorem 5.3. *Suppose that there is a deterministic constant* $\beta>0$ *such that* $E\|\theta_0\|^{\beta}<\infty$, *then under Conditions 5.1-5.2, for any* $\alpha\in(0,\beta)$, *there exist two deterministic nonnegative constants* C *and* D *independent of* σ_β *and* μ_β *such that*

(i). $\limsup_{n\to\infty} E\|\hat{\theta}_n-\theta_n\|^{\alpha}\leq C(\sigma_\beta)^{\frac{\alpha}{\beta}}$, *and*

(ii). $\limsup_{n\to\infty} \frac{1}{n}\sum_{i=0}^{n-1}\|\hat{\theta}_i-\theta_i\|^{\alpha}\leq D(\mu_\beta)^{\frac{\alpha}{\beta}}$, *a.s.*,

where $\{\hat{\theta}_n\}$ *is given by (5.3).*

5.2. Adaptive Control

In this subsection, we consider adaptive control of the following two class of time-varying models:

Model 5.1. (random parameter model).

$$y_{k+1}=a_1(k)y_k+\cdots+a_p(k)y_{k-p+1}+u_k+v_{k+1},\ k\geq 0;\ y_k=u_k=v_k=0,\ k<0, \tag{5.6}$$

where y_k, u_k and v_k are the scalar output, input and random noise process respectively, and $a_i(k)$, $1\leq i\leq p$, are the unknown random time-varying parameters.

Model 5.2. (deterministic parameter model).

$$y_{k+1}=a_1(k)y_k+\cdots+a_s(k)y_{k-s+1}+b_1(k)u_k+\cdots+b_t(k)u_{k-t+1}+v_{k+1},\ k\geq 0; \tag{5.7a}$$

$$y_k=u_k=v_k=0,\ k<0, \tag{5.7b}$$

where $a_i(k)$, $1\leq i\leq s$ and $b_j(k)$, $1\leq j\leq t$ are the unknown deterministic time-varying parameters.

Note that both Model 5.1 and Model 5.2 can be rewritten in the following regression form:

$$z_{k+1}=\phi_k^\tau\theta_k+v_{k+1}, \tag{5.8}$$

where for (5.6), $z_{k+1}=y_{k+1}-u_k$,

$$\phi_k=[y_k\ \cdots\ y_{k-p+1}]^\tau,\quad \theta_k=[a_1(k)\ \cdots\ a_p(k)]^\tau, \tag{5.9}$$

while for (5.7), $z_{k+1}=y_{k+1}$, and

$$\phi_k=[y_k\ \cdots\ y_{k-s+1}\ u_k\ \cdots\ u_{k-t+1}]^\tau, \tag{5.10a}$$

$$\theta_k=[a_1(k)\ \cdots\ a_s(k)\ b_1(k)\ \cdots\ b_t(k)]^\tau, \tag{5.10b}$$

Let us now introduce the assumptions on the random noise sequence $\{v_k\}$.

Condition 5.3. *(noise assumption) $\{v_k, F_k\}$ is an adapted sequence where $\{F_k\}$ is a non-decreasing family of σ-algebras, and for some integer $r\geq 0$ and deterministic positive constants ε_0 and M_v,*

$$E[\exp[\varepsilon_0\|v_{k+1}\|^2]|F_{k-r}]\leq \exp[M_v],\ a.s.,\quad k\geq 0. \tag{5.11}$$

A major difference between deterministic systems considered in (e.g., Ioannou and Tsakalis, 1985; Middleton and Goodwin, 1988) and the stochastic systems studied here is that the system noise may not be uniformly bounded. Let us now give an example where the noise sequence $\{v_k\}$ is unbounded almost surely.

Example 5.3. Let $\{v_k\}$ be the following time-varying moving average process:

$$v_k=e_k+c_1(k)e_{k-1}+\cdots+c_r(k)e_{k-r},\quad k\geq 0,$$

with deterministic coefficients $\{c_i(k)\}$ satisfying

$$\sum_{i=0}^{r}|c_i(k)|^2\leq c<\infty,\quad k\geq 0,\quad (c_0(k)=1).$$

Assuming that $\{e_k\}$ is a Gaussian white noise sequence with variance $\sigma^2>0$. Then

$$\limsup_{k\to\infty}\frac{|v_k|}{(2\log k)^{1/2}}\geq\sigma,\quad a.s.$$

and the noise assumption (5.11) holds for any

$$\varepsilon_0<\frac{1}{2c\sigma^2},\quad M_v\geq\frac{\varepsilon_0 c\sigma^2(r+1)}{1-2c\varepsilon_0\sigma^2}.$$

Condition 5.4. *(random parameter assumption)* $\{\theta_k, F_k\}$ *defined in (5.9) is an adapted sequence which satisfies,*

$$E[exp[M_0||\theta_{k+1}||^2]|F_{k-m}] \le exp[M_\theta], \ a.s., \quad k \ge 0. \tag{5.12}$$

$$E[exp[M_0||w_{k+1}||^2]|F_{k-m}] \le exp[\delta_0], \ a.s., \quad k \ge 0. \tag{5.13}$$

where w_{k+1} *is the parameter variation process given by (5.2) and where* $m \ge 0$ *is an integer and* M_0, M_θ *and* $\delta_0 < 1$ *are positive deterministic constants. (Actually, in the theorems to follow,* $M_0(\delta_0)$ *is assumed to be appropriately large (small), the precise lower (upper) bound can be found in (Guo, 1989)).*

We now give some discussions on this condition. The condition (5.12) means that the random parameter process $\{\theta_k\}$ is bounded in an average sense, and not necessarily bounded in sample path, while (5.13) means that the parameters are slowly varying in an average sense, and again, the variation is not necessarily small in sample path. In particular, these conditions do not rule out occasional but possibly large jumps of the parameter process. Let us give a concrete example.

Example 5.4. Let the unknown parameter θ_k be a constant vector plus a p-dimensional moving average process:

$$\theta_k = \theta + \varepsilon_k + D_1\varepsilon_{k-1} + \cdots + D_{m-1}\varepsilon_{k-m+1}, \quad k \ge 0,$$

where D_i, $1 \le i \le m-1$ are deterministic matrices, and $\{\varepsilon_k\}$ is a Gaussian white noise sequence with covariance matrix $(\sigma_\varepsilon)^2 I$. Then for any $\sigma_\varepsilon > 0$,

$$\limsup_{k \to \infty} ||\theta_k|| = \infty, \ a.s., \qquad \limsup_{k \to \infty} ||\theta_k - \theta_{k-1}|| = \infty, \ a.s..$$

Furthermore, if σ_ε is suitably small, then the above random parameter assumptions hold.

We now describe the estimation algorithm. Let $L > 0$ and $c > 0$ be two constants (which will be specified later on). We take D as the following bounded domain:

$$D = \{x = (x_1 \ \cdots \ x_p) \in R^p: |x_i| \le L, \ 1 \le i \le p\} \tag{5.14}$$

and $\pi_D(x)$ is the projecting from x to D.

The estimate $\hat{\theta}_k$ for the unknown process θ_k is given by the following projected version of the gradient algorithm:

$$\hat{\theta}_{k+1} = \pi_D\{\hat{\theta}_k + \frac{\phi_k}{c + ||\phi_k||^2}(y_{k+1} - u_k - \phi_k^\tau \hat{\theta}_k)\}, \tag{5.15}$$

with arbitrary initial condition $\hat{\theta}_0 \in D$, where $c > 0$ is a constant and ϕ_k is defined as in (5.9).

The certainty equivalent minimum variance adaptive control law is

$$u_k = -\phi_k^\tau \hat{\theta}_k. \tag{5.16}$$

The stability result is the following.

Theorem 5.4. *(Guo, 1989) For the random parameter model (5.6), if the noise assumption (5.11) and the parameter assumptions (5.12)-(5.13) hold, and if in the estimation algorithm (5.14)-(5.15), L and c are taken suitably large, then under the adaptive control law (5.16), the closed-loop system is stable in the sense that*

$$\limsup_{n \to \infty} E[|y_n|^\beta + |u_n|^\beta] < \infty$$

and

$$\limsup_{N \to \infty} \frac{1}{N} \sum_{n=0}^{N} (|y_n|^\beta + |u_n|^\beta) < \infty,$$

where $\beta > 2$ *is a constant depending on* M_0, δ_0, L *and* c.

Condition 5.5. *(deterministic parameter assumption)*

(i). There is a positive constant $b_1 > 0$ *such that*

$$b_1(k) \geq b_1, \quad k \geq 0, \tag{5.17}$$

and the model (5.7) is uniformly stably invertible in the sense that there are two constants $A > 0$, $\rho \in (0,1)$ *such that*

$$|u_k|^2 \leq A \sum_{i=0}^{k+1} \rho^{k+1-i}(|y_i|^2 + |u_i|^2), \quad k. \tag{5.18}$$

(ii). The parameter is slowly varying in the sense that

$$\|\theta_k\| \leq M_1, \quad \|\theta_{k+1} - \theta_k\| \leq \delta_1, \quad k \geq 0, \tag{5.19}$$

where M_1 *and* δ_1 *are some constants (Guo, 1989).*

We remark that since $\{\theta_k\}$ is bounded, the assumption (5.18) is implied by uniformly asymptotic stability of the following time-varying polynomial:

$$B_k(z) = b_1(k) + b_2(k)z + \cdots + b_t(k)z^{t-1}, \tag{5.20}$$

which in the constant parameter case is the standard minimum phase condition.

Let us introduce the following bounded domain:

$$D = \{x = (x_1 \cdots x_{s+t}) \in R^{s+t}: |x_i| \leq L, \ 1 \leq i \leq s+t, \ x_{s+1} \geq b_1\}. \tag{5.21}$$

The estimation algorithm is also a projected gradient one:

$$\hat{\theta}_{k+1} = \pi_D \{\hat{\theta}_k + \frac{\phi_k}{c + \|\phi_k\|^2}(y_{k+1} - \phi_k^\tau \hat{\theta}_k)\}, \tag{5.22}$$

with arbitrary initial condition $\hat{\theta}_0 \in D$, where $c > 0$ is a constant and ϕ_k is defined as in (5.10).

The certainty equivalent minimum variance adaptive control law is

$$\phi_k^\tau \hat{\theta}_k = 0. \tag{5.23}$$

The stability result is the following.

Theorem 5.5. *(Guo, 1989) For the deterministic parameter model (5.7), if the noise assumption (5.11) and the parameter assumptions (5.17)-(5.20) hold, and if in the estimation algorithm (5.21)-(5.22), L and c are taken suitably large, then under the adaptive control law (5.23), the closed-loop system is stable in the sense that*

$$\limsup_{n \to \infty} E[|y_n|^\beta + |u_n|^\beta] < \infty$$

and

$$\limsup_{N \to \infty} \frac{1}{N} \sum_{n=0}^{N} (|y_n|^\beta + |u_n|^\beta) < \infty,$$

where $\beta > 2$ *is a constant depending on* M_0, δ_0, L *and* c.

We remark that further results including the convergence of the performance $E[u_k^2 + y_k^2]$, could be obtained by applying the Markov chain ergodic theory (Meyn and Caines, 1987 and Meyn and Guo, 1989a). Moreover, generalizations of the parameter model to include e.g. ARMA parameters are also possible (Meyn and Guo, 1989b).

REFERENCES

Akaike H. (1969), Fitting autoregressive models for prediction, *Ann. Inst. Statist. Math.*, **21**, 243-247.

Benveniste A. and G. Ruget (1982), A measure of the tracking capability of recursive stochastic algorithms with constant gains, *IEEE Trans. on Automatic Control*, **AC-27**, 639-649.

Box G.E.P. and G.M. Jenkins (1970), *Time Series Analysis, Forecasting and Control*, Holden-Day, San Francisco.

Caines P.E. (1988), *Linear Stochastic Systems*, John Wiley and Sons.

Caines P.E. and S. Lafortune (1984), Adaptive control with recursive identification for stochastic linear systems, *IEEE Trans. on Automatic Control*, **AC-29**, 312-321.

Chen H.F. (1982), Strong consistency and convergence rate of least squares identification, *Scientia Sinica (Series A)*, **25**, 771-784.

Chen H.F. (1985), *Recursive Estimation and Control for Stochastic Systems*, John Wiley, New York.

Chen H.F. and L. Guo (1985), Strong consistency of parameter estimates for discrete-time stochastic systems, *J. Sys. & Math. Scis.*, **5**, 81-93.

Chen H.F. and L. Guo (1986a), Optimal adaptive control with quadratic index, *Int. J. Control*, **43**, 869-881.

Chen H.F. and L. Guo (1986b), Convergence rate of least-squares identification and adaptive control for stochastic systems, *Int. J. Control*, **44**, 1459-1476.

Chen H.F. and L. Guo (1986c), Strong consistency of parameter estimates in optimal stochastic adaptive tracking systems, *Scientia Sinica (Series A)*, **29**, 1145-1156.

Chen H.F. and L. Guo (1986d), The limit of stochastic gradient algorithm for identifying systems excited not persistently, *Kexue Tongbao (Science Bulletin)*, 6-9.

Chen H.F. and L. Guo (1987a), Asymptotically optimal adaptive control with consistent parameter estimates, *SIAM J. on Control and Optim.*, **25**, 558-575.

Chen H.F. and L. Guo (1987b), Optimal adaptive control and consistent parameter estimates for ARMAX model with quadratic cost, *SIAM J. on Control and Optim.*, **25**, 845-867.

Chen H.F. and L. Guo (1987c), Adaptive control with recursive identification for stochastic linear systems, *Advances in Control and Dynamic Systems*, **26**, 277-327.

Chen H.F. and L. Guo (1988), A robust stochastic adaptive controller, *IEEE Trans. on Automatic Control*, **AC-33**, 1035-1043.

Chen H.F. and J.F. Zhang (1989a), Identification and adaptive control for systems with unknown orders, time-delay and coefficients (for correlated noise case), *J. Sys. Sci. & Math. Scis.*, **9**, 355-363.

Chen H.F. and J.F. Zhang (1989b), Identification and adaptive control for systems with unknown orders, time-delay and coefficients (for uncorrelated noise case), *IEEE Trans. on Automatic Control*, to appear.

Chen H.F. and J.F. Zhang (1989c), Convergence rate in stochastic adaptive tracking, *Int. J. Control*, **49**, 1915-1935.

Goodwin G.C., P.J. Ramadge and P.E. Caines (1981), Discrete time stochastic adaptive control, *SIAM J. on Control and Optim.*, **19**, 829-853.

Goodwin G.C. and K.S. Sin (1984), *Adaptive Filtering, Prediction and Control*, Prentice-Hall, INC., Englewood Cliffs, New Jersey.

Guo L. (1989), On adaptive stabilization of time-varying stochastic systems, *SIAM J. on Control and Optim.*, to appear.

Guo L. (1990), Estiamting time-varying parameters by Kalman filter based algorithm: stability and convergence, *IEEE Trans on Automatic Control*, **AC-35**, No.2.

Guo L., Chen H.F. and J.F. Zhnag (1989), Consistent order estimation for linear stochastic feedback control systems (CARMA model), *Automatica*, **25**, 147-151.

Guo L. and D.W. Huang (1989), Least-squares identification for ARMAX models without the positive real condition, *IEEE Trans. on Automatic Control*, **AC-34**, No.10.

Hannan E.J. and M. Deistler (1988), *The Statistical Theory of Linear Systems*, John Wiley and Sons, New York.

Hannan E.J. and J. Rissanen (1982), Recursive estimation of ARMA order, *Biometrika*, **69**, 81-97.

Ioannou P. and P.V. Kokotovic (1984), Robust redesign of adaptive control, *IEEE Trans. on Automatic Control*, **AC-29**, 202-211.

Ioannou P. and J. Sun (1988), Theory and design of robust direct and indirect adaptive control, *Int. J. Control*, **47**, 775-813.

Ioannou P. and K. Tsakalis (1985), Robust discrete-time adaptive control, in *Adaptive and Learning Systems: Theory and Applications* (Ed. K.S. Narendra), 73-85, Plenum Press, New York.

Kreisselmeier G. and B.D.O. Anderson (1986), Robust model reference adaptive control, *IEEE Trans. on Automatic Control*, **AC-31**, No.2.

Kumar P.R. and L. Praly (1987), Self-tuning trackers, *SIAM J. on Control and Optim.*, **25**, 1053-1071.

Kumar P.R. and P.P. Varaiya (1986), *Stochastic Systems: Estimation; Identification, Adaptive Control*, Prentice-Hall, Englewood Cliffs, NJ.

Lai T.L. and C.Z. Wei (1982), Least squares estimation in stochastic regression models with application to identification and control of dynamic systems, *The Annals of Statistics*, **10**, 154-166.

Lai T.L. and C.Z. Wei (1986), Extended least squares and their application to adaptive control and prediction in linear systems, *IEEE Trans. on Automatic Control*, **AC-31**, 898-904.

Ljung L. and T. Soderstrom (1983), *Theory and Practice of Recursive Identification*, MIT Press, Cambridge.

Macchi Q. (1986), Optimization of adaptive identification for time varying filters, *IEEE Trans. on Automatic Control*, **AC-31**, 283-287.

Meyn S.P. and P.E. Caines (1987), A new approach to stochastic adaptive control, *IEEE Trans. on Automatic Control*, **AC-32**, 220-226.

Meyn S.P. and L. Guo (1989a), Stability, performance and convergence of an adaptive algorithm applied to a random varying system, submitted for publication.

Meyn S.P. and L.Guo (1989b), Adaptive control of time varying stochastic systems, submitted for publication

Middleton R.H. and G.C. Goodwin (1988), Adaptive control of time-varying linear systems, *IEEE Trans. on Automatic Control*, **AC-33**, 150-157.

Middleton R.H., G.C. Goodwin, D.J. Hill and D.Q. Mayne (1988), Design issues in adaptive control, *IEEE Trans. on Automatic Control*, **AC-33**, 50-58.

Moore J.B. (1978), On strong consistency of least squares identification algorithm, *Automatica*, **14**, 505-509.

Rissanen J. (1978), Modelling by shortest data description, *Automatica*, **14**, 467-471.

Rissanen J. (1986), Order estimation by accumulated prediction errors, Essays in Time Series and Allied Process (Eds. J. Gani and M.B. Priestly), *J. Appl. Probab.*, **Special Vol.23A**, 55-61.

Solo V. (1979), The convergence of AML, *IEEE Trans. on Automatic Control*, **AC-24**, 958-963.

Solo V. (1988), The limit behavior of LMS, *IEEE Trans. on Acoustics, Speech and Signal Processing*, to appear.

SOME METHODS FOR THE ADAPTIVE CONTROL OF CONTINUOUS TIME LINEAR STOCHASTIC SYSTEMS*

T. E. Duncan and B. Pasik-Duncan
Department of Mathematics
University of Kansas
Lawrence, Kansas 66045

1. INTRODUCTION

Stochastic adaptive control has become an important area of activity in control theory. This activity has developed especially during the past twenty years or so because many mathematical models of physical phenomena are stochastic systems which contain unknown parameters and for which control is required. With the maturity of this area of research there are a number of books and general surveys (e.g. [2, 4, 21, 22, 25, 26]). In this paper, some techniques are described for the solution of a family of related problems in stochastic adaptive control. Initially the model for the adaptive control problem is broadly described. The models are assumed to evolve in continuous time rather than discrete time because this assumption is natural for many models and it is important for the study of discrete time models when the sampling rates are large and for the analysis of numerical round-off errors. The stochastic systems are described by linear stochastic differential equations where the unknown parameters occur affinely in the drift term. It is assumed that there are observations of the complete state. The families of linear systems that are considered include finite dimensional systems, delay time systems, and infinite dimensional systems that are described by infinitesimal generators of C_0 semigroups.

The general approach to adaptive control that is described here exhibits a splitting or separation of the problems of identification of the unknown parameters and adaptive control. Maximum likelihood (or equivalently least squares) estimates are used for the identification of the unknown constant parameters. These estimates are given recursively and are shown to be strongly consistent. An approach to some periodic, time varying parameters is given using exponential discounting of the data. In this case asymptotic properties are described as the discount rate approaches zero. The adaptive control is usually constructed by the so-called certainty equivalence principle, that is, the optimal stationary controls are computed replacing the unknown true parameter values by the current estimates of these values. Since the optimal stationary controls can be shown to be continuous functions of the unknown parameters, the self tuning property is verified. It is shown that the family of average

* Research partially supported by NSF Grant ECS-8718026.

costs using the controls from the certainty equivalence principle converges to the optimal average cost. This verifies the self optimizing property. To give some measure for this rate of convergence the asymptotic distribution of some quadratic functionals of the state is given. If the optimal stationary controls are not used in the adaptive control then it is shown that some anomalies can occur. Specifically using variance in the asymptotic distribution as a measure of performance it is shown that this variance is less for a quadratic functional that is formed from a control based on estimates of the parameters than the same quadratic functional that is formed from the corresponding control based on the true parameter values.

2. FINITE DIMENSIONAL LINEAR SYSTEMS

The model for the adaptive control of finite dimensional linear stochastic systems used here is

$$dX(t) = \left(F_0X(t) + \sum_{i=1}^{p} \alpha_i F_i X(t) + BU(t)\right) dt + dW(t) \tag{1}$$

where $X(t) \in \mathbb{R}^n$, $U(t) \in \mathbb{R}^m$, $F_i \in L(\mathbb{R}^n, \mathbb{R}^n)$ $i = 0, 1, \ldots, p$, $B \in L(\mathbb{R}^m, \mathbb{R}^n)$, $(W(t), t \geq 0)$ is a standard $\mathbb{R}^n$-valued Wiener process and $X(0) \equiv a \in \mathbb{R}^n$. The following assumptions are made for the adaptive control problem

(A1) $\alpha_i \in I_i \subset \mathbb{R}$ is an unknown parameter where I_i is a bounded, open interval for $i = 1, 2, \ldots, p$.

(A2) The pair $(F_0 + \sum_{i=1}^{p} \alpha_i F_i, B)$ is reachable for each $\alpha_i \in I_i$ for $i = 1, 2, \ldots, p$.

(A3) The family $(F_i, i = 1, 2, \ldots, p)$ is linearly independent.

The probability space $(\Omega, \mathcal{F}, P)$ can be chosen so that Ω is the Fréchet space of $\mathbb{R}^n$-valued continuous functions on $\mathbb{R}_+ = [0, \infty)$ with the seminorms of local uniform convergence, P is the Wiener measure on Ω and $\mathcal{F}$ is the P-completion of the Borel σ-algebra of Ω.

The cost functional $C(t)$ for the control problem is

$$C(t) = \int_0^t \langle QX(s), X(s)\rangle + \langle PU(s), U(s)\rangle \, ds \tag{2}$$

where $Q \in L(\mathbb{R}^n, \mathbb{R}^n)$ and $P \in L(\mathbb{R}^m, \mathbb{R}^m)$ are symmetric and positive definite.

The control policies $(U(t),\ t \geq 0)$ that are considered are linear feedback controls, that is,

$$U(t) = K(t)X(t) \tag{3}$$

where $(K(t),\ t \geq 0)$ is an $L(\mathbb{R}^n, \mathbb{R}^m)$-valued process that is adapted to the sub σ-algebras $(\sigma(X(s),\ s \leq t),\ t \geq 0)$ and is uniformly bounded (almost surely).

Let $\delta > 0$ be fixed. It is assumed that the $L(\mathbb{R}^n, \mathbb{R}^m)$-valued process $(K(t),\ t \geq 0)$ has the property that $K(t)$ is adapted to $\sigma(X(s),\ s \leq t - \delta)$ for each $t \geq \delta$ and it is assumed that $(K(t),\ t \in [0, \delta))$ is a deterministic, uniformly bounded $L(\mathbb{R}^n, \mathbb{R}^m)$-valued function. For the adaptive control problem this measurability is accomplished by computing $K(t)$ from the parameter estimates at time $t - \delta$ which is quite natural for applications. If $(K(t),\ t \geq 0)$ satisfies this measurability condition and the boundedness condition, then it is easy to verify that there is one and only one strong solution of the stochastic differential equation

$$dX(t) = (A(\alpha) + BK(t))X(t)dt + dW(t) \tag{4}$$

where

$$A(\alpha) = F_0 + \sum_{i=1}^{p} \alpha_i F_i. \tag{5}$$

Fix the control law $(K(t),\ t \geq 0)$ and let $(X(t),\ t \geq 0)$ be the solution of (4). Define $A(t)$ and $\tilde{A}(t)$ in $L(\mathbb{R}^p, \mathbb{R}^p)$ as

$$A(t) = (a_{ij}(t)) \tag{6}$$

and

$$\tilde{A}(t) = (\tilde{a}_{ij}(t)) \tag{7}$$

where

$$a_{ij}(t) = \int_0^t \langle F_i X(s), F_j X(s)\rangle \, ds$$

and

$$\tilde{a}_{ij}(t) = \frac{a_{ij}(t)}{a_{ii}(t)}$$

for $i, j \in \{1, \ldots, p\}$.

For the computation of the maximum likelihood estimates (or equivalently the least squares estimates) the following is used.

(A4) For each admissible adaptive control law $(\tilde{A}(t),\ t \geq 0)$ in (7) satisfies

$$\liminf_{t\to\infty} |\det \tilde{A}(t)| > 0 \quad \text{a.s.}$$

However it can be shown that without assumption (A4), det $A(t) \neq 0$ a.s. for each $t > 0$. One can easily construct examples that satisfy (A4). For example, let $(f^i_k, k = 1, 2, ..., n)$ be the columns of F_i for $i = 1, 2, ..., p$. If $<f^i_k, f^j_l> = 0$ for all $i \neq j$ and $k, l \in \{1, 2, ..., n\}$ then (A4) is trivially satisfied. More generally well known inequalities can be used to verify (A4). For a specific example let $n = 3$, $p = 2$, $F_1 = E_{11} + E_{22}$ and $F_2 = E_{12} + E_{31}$.

In this approach to adaptive control it is initially shown that the maximum likelihood (or least squares) estimates are strongly consistent for a large family of control policies.

Theorem 1. Let $(K(t), t \geq 0)$ be a feedback control policy that is adapted to $(\sigma(X(s), s \leq t - \delta), t \geq \delta)$ where $\delta > 0$ is fixed and that is uniformly bounded almost surely. Assume that ((A1) - (A4)) are satisfied. Then the family of maximum likelihood estimates $(\hat{\alpha}(t), t \geq 0)$ is strongly consistent, that is,

$$P_{\alpha_0}(\lim_{t\to\infty} \hat{\alpha}(t) = \alpha_0) = 1$$

where α_0 is the true parameter vector.

The proof of the result is given in [16].

The estimate $\hat{\alpha}(t)$ is found by solving the following family of linear equations

$$\tilde{A}(t)\hat{\alpha}(t) = \tilde{A}(t)\alpha_0 + \tilde{b}(t) \tag{8}$$

where $\tilde{b}(t) = (\tilde{b}_1(t), ..., \tilde{b}_p(t))'$ and

$$\tilde{b}_j(t) = \frac{1}{a_{jj}(t)} \int_0^t <F_jX(s), dW(s)> ds.$$

For practical applications of identification it is important to describe the estimator recursively. This can be done for the estimator in Theorem 1. The family of maximum likelihood estimates $(\hat{\alpha}(t), t \geq 0)$ satisfies the linear stochastic differential equation (whose coefficients depend on $(X(t), t \geq 0)$)

$$d\hat{\alpha}(t) = A^{-1}(t) <\tilde{F}X(t), dX(t) - A(\hat{\alpha}(t))X(t)\, dt - BU(t)\, dt> \tag{9}$$

where $<\tilde{F}X, Y>$ is the vector whose i^{th} component is $<F_iX, Y>$.

For some adaptive control problems it may be difficult to verify (A4) so it is natural to obtain other sufficient conditions for consistency of least squares estimators.

Let ℓ be a nonnegative definite, symmetric matrix. The least squares estimate of α_0 based on $(X(t), t \leq T)$ and $(U(t), t \leq T)$, denoted $\alpha^*(T)$, is the minimum of the formal quadratic form

$$\int_0^T <\dot{X}(t) - A(\alpha)X(t) - U(t), \ell(\dot{X}(t) - A(\alpha)X(t) - U(t))> dt.$$

This minimization is accomplished by considering the following well defined expression

$$\int_0^t <\ell(A(\alpha)X(t) + U(t)), A(\alpha)X(t) + U(t)> dt$$
$$- 2\int_0^t <\ell(A(\alpha)X(t) + U(t)), dX(t)>.$$

The minimum is found by equating the partial derivatives of this expression with respect to α^i $i = 1, ..., p$ to 0 to obtain the system of linear equations

$$\sum_{j=1}^{p} \int_0^T <\ell F_i X(t), F_j X(t)> dt\, \alpha^{*j}(T) = \int_0^T <\ell F_i X(t), dX(t) - F_0 X(t)dt - U(t)\, dt>. \tag{10}$$

If (1) is used then (10) can be rewritten as

$$\sum_{j=1}^{p} \int_0^T <\ell F_i X(t), F_j X(t)> dt(\alpha^{*j}(T) - \alpha_0^j) = \int_0^T <\ell F_i X(t), dW(t)>. \tag{11}$$

The least squares estimates can be shown to satisfy a stochastic differential equation in analogy with (9). This stochastic differential equation is

$$d\alpha^*(t) = A_\ell^{-1}(t)<\tilde{F}_\ell X(t), dX(t) - A(\alpha^*(t))X(t)\, dt - BU(t)\, dt> \tag{12}$$

where

$$A_\ell(t) = \left(\int_0^t <\ell F_i X(s), F_j X(s)> ds\right)$$

and $<\tilde{F}_\ell X, Y>$ is the vector whose i^{th} component is $<\ell F_i X, Y>$.

If $\ell = I$ then the least squares estimate is the same as the maximum likelihood estimate in Theorem 1.

A statement of the consistency of the family of least squares estimates under some stability conditions is given now. Its proof is given in [24].

Theorem 2. Assume that the family $(\sqrt{\ell}\, F_i,\ i = 1, ..., p)$ is linearly independent. If

$$\lim_{t\to\infty} \frac{|X(t)|^2}{t} = 0 \qquad (13)$$

in probability (respectively, almost surely) and

$$\frac{1}{T}\int_0^T |X(t)|^2 + |U(t)|^2\, dt \qquad (14)$$

is bounded in probability (respectively, almost surely) then

$$\lim_{T\to\infty} \alpha^*(T) = \alpha_0$$

in probability (respectively, almost surely), where $\alpha^*(T)$ is the least squares estimate at time T and α_0 is the true parameter vector.

To find the feedback control law that gives the optimal average cost for the stochastic system (1) or the optimal infinite time cost (2) for the associated deterministic system

$$\frac{dx}{dt} = Ax + Bu$$

it is sufficient to solve the algebraic Riccati equation

$$0 = \Pi A + A'\Pi - \Pi R \Pi + Q \qquad (15)$$

where $R = BP^{-1}B'$. The optimal feedback control is $Kx = -P^{-1}B'\Pi x$.

The adaptive control law uses the certainty equivalence principle, that is, the algebraic Riccati equation is solved using $\hat{\alpha}(t - \delta)$ instead of the true parameter vector to obtain the feedback gain $K(t)$. For this adaptive control law it is necessary to obtain a suitable stability property for (3). Let V be the unique positive semidefinite solution of the algebraic Riccati equation (15) using the true parameter vector α_0 and let $k_0 = -P^{-1}B'V$. Since $A_0 + Bk_0$ is stable there is a symmetric, positive definite matrix $\tilde{V}$ such that

$$(A_0 + Bk_0)'\tilde{V} + \tilde{V}(A_0 + Bk_0) + I < 0. \qquad (16)$$

Since the solution $\tilde{V}$ of the inequality (16) is a continuous function of the parameter vector α and the feedback gain K, it follows that the inequality (16) is satisfied for α sufficiently close to α_0 and K sufficiently close to k_0.

Proposition 1. Assume that ((A1) - (A3)) are satisfied. Let $(X(t), t \geq 0)$ be the solution of (3) using the feedback control law $(K(t), t \geq 0)$ such that

$$\lim_{t \to \infty} K(t) = k_0 \quad \text{a.s.}$$

Then the process $(X(t), t \geq 0)$ satisfies

$$\lim_{t \to \infty} \frac{1}{t} \langle X(t), X(t) \rangle = 0 \quad \text{a.s.} \tag{17}$$

$$\limsup_{t \to \infty} \frac{1}{t} \int_0^t \langle X(s), X(s) \rangle \, ds \leq c \quad \text{a.s.} \tag{18}$$

where $c \in \mathbb{R}_+$.

The proof of this result is given in [16].

The unknown parameters take values in bounded subsets of the real line and the solution of the Riccati equation (15) is a continuous function of the parameters. By truncation of the (maximum likelihood) estimates, if necessary, to the bounded subsets of the parameter values and the continuity of the algebraic Riccati equation with respect to the parameters, it can be assumed that the feedback gains determined from the algebraic Riccati equation, using the strongly consistent (maximum likelihood) estimates as the parameter values, are uniformly bounded and converge (almost surely) to the optimal feedback gain k_0. By Proposition 1 the process $(X(t), t \geq 0)$ using the certainty equivalence control law satisfies the stability conditions (17 - 18). Furthermore it can be shown that the certainty equivalence control law is self optimizing, that is, the family of average costs converges (almost surely) to the optimal average cost [16].

Theorem 3. Assume that ((A1) - (A4)) are satisfied. Let $(K(t), t \geq \delta)$ be the family of feedback gains determined from the algebraic Riccati equation (15) using the (truncated) maximum likelihood estimates $(\hat{\alpha}(t - \delta), t \geq \delta)$. Then

$$\lim_{t \to \infty} K(t) = k_0 \quad \text{a.s.}$$

and

$$\lim_{t \to \infty} \frac{1}{t} C(t) = \operatorname{tr} V \quad \text{a.s.} \tag{19}$$

where V is the unique, symmetric positive semidefinite solution of (15) using the true parameter vector α_0, $C(t)$ is the cost at time t given in (3) using $(K(t), t \geq 0)$ and tr is the trace.

For some applications of adaptive control it is not sufficient to know that the family of average costs converges to the optimal average cost but it is important to estimate some probabilities associated with this convergence. Thus it is important to obtain the asymptotic distribution of the cost functional (3) as $t \to +\infty$ [13].

Theorem 4. Let $(C(t), t \geq 0)$ be the family of costs (3) associated with the feedback gains $(K(t), t \geq 0)$ given in Theorem 3. Then

$$\frac{C(t) - \operatorname{tr} V}{t^{1/2}} \to Z$$

where Z is $N(0, \operatorname{tr} \Gamma)$, the convergence is in distribution, V is the unique, positive semidefinite solution of (15) with the true parameter vector α_0 and Γ is the unique positive semidefinite solution of

$$\Gamma(A(\alpha_0) + Bk_0) + (A(\alpha_0) + Bk_0)'\Gamma + 4V^2 = 0 \tag{20}$$

where $k_0 = - P^{-1}B'V$.

In some applications of stochastic adaptive control it is desirable to use linear feedback control laws that differ from the optimal stationary control law. If the linear stochastic system is stable with a particular control law using the true parameter values then the limit of the average costs exists and can be easily described. Furthermore the asymptotic distribution of the suitably normalized cost functional is normal. Likewise with a stability condition (17 - 18) the average costs using the same control law based on the strongly consistent family of estimates of the unknown parameters converge to the same constant as the control law based on the true parameter values and the asymptotic distribution of the suitably normalized cost functional is again normal. In both cases if the average costs are translated by the limiting constant then the asymptotic distributions are normal with zero mean. Thus the variances that are obtained in the two cases can be used as a measure of dispersion or risk for the two control laws. It is shown that there are feedback control laws for a stable system for which the asymptotic distribution of the average costs for the control law using the maximum likelihood estimates of the unknown parameters has smaller variance than the asymptotic distribution for the control law using the true values of the parameters. Thus the former control law has less risk than the latter. It seems unusual that the control law based on the estimates of the unknown parameters performs better in this asymptotic distribution sense than the control policy based on the true parameter values.

Let V be the unique, positive definite, symmetric solution of the equation

$$(A(\alpha_0) + Bk_0)'V + V(A(\alpha_0) + Bk_0) + I = 0$$

where $k_0 = K(\alpha_0)$ and α_0 is the true parameter vector. Define $\theta = tr((Q + k_0'k_0)V)$. Let P be the unique positive semidefinite solution of the equation

$$P(A(\alpha_0) + Bk_0) + (A(\alpha_0) + Bk_0)'P + Q + k_0'k_0 = 0.$$

To describe the results the following quantities are defined:

$$\begin{aligned}
g_{ij} &= tr(F_i'F_jV) \\
G &= (g_{ij}) \\
p_i &= tr(PF_iV) \\
p &= (p_1, \ldots, p_p)' \\
D_ik &= \frac{\partial k}{\partial \alpha^i} \\
r_i &= 2tr((k_0' + PB)(D_ik)V) \\
r &= (r_1, \ldots, r_p)' \\
h &= 4tr(PVP)
\end{aligned}$$

for $i,j \in \{1, 2, \ldots, p\}$.

Proposition 2. Let $K(\alpha) \in L(\mathbb{R}^n, \mathbb{R}^m)$ for each parameter vector α be twice continuously differentiable at the true parameter vector α_0 and let the control be

$$U(t) = K(\hat{\alpha}(t - \delta))X(t)$$

where $(X(t), t \geq 0)$ is the solution of (3), $\delta > 0$ is fixed and $(\hat{\alpha}(t), t \geq 0)$ is the family of maximum likelihood estimates. Assume that $A + Bk_0$ is stable and that the family $(\hat{\alpha}(t), t \geq 0)$ is strongly consistent. Then the family of random variables

$$([\frac{C(t)}{t} - \theta]\sqrt{t}, t > 0)$$

converges in distribution to $N(0, \Delta)$ where

$$\Delta = h + 2(p + r)'G^{-1}r. \tag{21}$$

A proof of this result is given in [23].

The variance (21) for the asymptotic normal distribution of the average costs provides a measure of dispersion or risk associated with the limiting average cost θ. The following result shows that there are feedback control laws based on the estimates of the unknown parameters that have less risk than the corresponding control laws based on the true parameter values.

Proposition 3. Let $(X^c(t), t \geq 0)$ be the process that satisfies the stochastic differential equation

$$dX^c(t) = (A(\alpha_0) + cBK(\alpha_0))\, X^c(t)\, dt + dW(t)$$

and $(\hat{X}^c(t), t \geq 0)$ be the process that satisfies the stochastic differential equation

$$d\hat{X}^c(t) = (A(\alpha_0) + cBK(\hat{\alpha}(\tau - \delta))\, \hat{X}^c(t)\, dt + dW(t)$$

where $c \in \mathbb{R}$. Assume that there is a neighborhood N of $0 \in \mathbb{R}$ where $A(\alpha_0) + cBK(\alpha_0)$ is stable for each $c \in N$ and that $(\hat{\alpha}(t), t \geq 0)$ is strongly consistent and $K \in C^2$. Then there exist values of c such that $(\hat{X}^c(t), t \geq 0)$ is more risk averting than $(X^c(t), t \geq 0)$, that is, $\Delta(c) < h$.

The proof of Proposition 3 is given in [13].

For many applications of adaptive control it may not be clear if the unknown parameters are constants or slowly varying functions. To determine parameter variations it is necessary to "forget" the past of the state. A survey of some methods is given in [20]. A natural approach is to use exponential discounting of the past information which has been studied for some practical applications [1,2]. If it is unclear whether the unknown parameters are constants or time varying, then an estimator must compromise between the conflicting goals of accuracy in the case of constant parameters and response to changes in the parameters in the case of time varying parameters.

The least squares estimate at time T given the infinite past history, $\alpha^*(T)$, is obtained by minimizing the formal expression

$$\int_{-\infty}^{T} e^{\lambda t} \langle \ell(\dot{X}(t) - A(a)X(t) - BU(t)),\, \dot{X}(t) - A(a)X(t) - BU(t)\rangle\, dt.$$

In analogy with the minimization without discounting the estimate $\alpha^*(T)$ is obtained by the solution of the system of linear equations

$$\sum_{j=1}^{p} \int_{-\infty}^{T} e^{\lambda t} \langle \ell F_i X(t), F_j X(t)\rangle\, dt\, \alpha^{*j}(T) = \int_{-\infty}^{T} e^{\lambda t} \langle \ell F_i X(t), dX(t) - F_0 X(t) dt - BU(t)\, dt\rangle$$

for $i = 1, 2, \ldots, p$.

To formulate the problem and to put some restrictions on the problem to obtain the asymptotic properties of the family of estimates the following assumptions are used.

(A5) Let $\mathcal{A}$ be the set of possible values for the parameter vector and let $\mathcal{K}$ be the set of values of the feedback gains. It is assumed that $\mathcal{A} \subset \mathbb{R}^p$ and $\mathcal{K} \subset L(\mathbb{R}^n,$

$\mathbb{R}^m$) are closed and bounded. There is one quadratic Lyapunov function that ensures stability uniformly for $A(a) + Bk(a)$ for $a \in \mathcal{A}$.

(A6) The family of matrices $(\sqrt{\ell}\, F_i,\ i = 1, 2, \ldots, p)$ is linearly independent where $\sqrt{\ell}$ is the symmetric square root of ℓ.

(A7) The true parameter vector evaluated at time $t \in \mathbb{R}$ is

$$\alpha(t) = a(\lambda t).$$

The function $a(\cdot)$ is a periodic, piecewise continuously differentiable mapping of $\mathbb{R}$ into $\mathcal{A}$ with period $\tau > 0$.

(A8) The feedback gain $k(\cdot)$ has the form

$$k(a) = k_0 + \varepsilon j(a) \tag{22}$$

where $\varepsilon > 0$ is a small parameter and $j(\cdot)$ is bounded and Lipschitz continuous. For $\varepsilon > 0$ and $\lambda > 0$ sufficiently small there is a periodic state of $(X(t), \alpha^*(t), Y(t);\ t \in \mathbb{R})$ such that $E[\exp(p|X(0)|^2)]$ is bounded for some $p > 0$ where

$$Y(T) = \int_{-\infty}^{T} e^{\lambda(t - T)} Q(t)\, dt$$

and

$$Q(t) = (\langle \ell F_i X(t), F_j X(t) \rangle).$$

In the following propositions the asymptotic behaviors of $\alpha^*(t)$, $k(\alpha^*(t))$ and a quadratic cost functional as $\lambda \downarrow 0$ are given. The proofs of the results are given in [14].

Proposition 4. Assume that ((A5) - (A8)) are satisfied. Let

$$\bar{a}(y) = \left(\int_{-\infty}^{y} e^s \theta(s,k_0)\, ds \right)^{-1} \int_{-\infty}^{y} e^s \theta(s,k_0) a(s)\, ds$$

where

$$\theta(y,k) = (\mathrm{tr}(v(a(y), k) F_i' \ell F_j))$$
$$(A(a(t)) + Bk) v(a(t), k) + v(a(t),k)(A(a(t)) + Bk)' + I = 0.$$

Then for $\lambda \downarrow 0$

$$(\alpha^*(T/\lambda) - \bar{a}(T))/\sqrt{\lambda}$$

has asymptotically a normal distribution with zero mean and covariance matrix

$$V(T) = \bar{\theta}^{-1}(T) \int_{-\infty}^{T} e^{2s}\Delta(T,s)\, ds\, \bar{\theta}^{-1}(T)$$

where

$$\bar{\theta}(T) = \int_{-\infty}^{T} e^{s}\theta(s,k_0)\, ds$$

$$\Delta(T,s) = (tr(b_i'(T,s)b_j(T,s)v(a(s), k_0)))$$

$$b_i(T,s) = 2\sum_{j=1}^{p} (a^j(s) - \bar{a}^j(T))\, \omega_{ij}(s) + \ell F_i$$

$$\omega_{ij}(s)(A(a(s)) + Bk_0) + (A(a(s)) + Bk_0)'\omega_{ij}(s) + F_i'\ell F_j = 0.$$

Proposition 5. Assume that ((A5) - (A8)) are satisfied. There is an $\bar{\varepsilon} > 0$ such that for $\varepsilon \in (0, \bar{\varepsilon})$ in (22) the following are satisfied:

i) There is a unique periodic solution $\bar{a}(\cdot)$ of

$$\bar{a}(y) = \left(\int_{-\infty}^{y} e^{s}\theta(s,\, k(\bar{a}(s)))\, ds \right)^{-1} \int_{-\infty}^{y} e^{s}\theta(s,\, k(\bar{a}(s)))a(s)\, ds$$

for $y \in \mathbb{R}$ and

ii) For $\bar{k}(y) = k(\bar{a}(y))$ and the discount rate $\lambda > 0$ sufficiently small

$$E|k(\alpha^*(T)) - \bar{k}(\lambda T)|^2 \le C\lambda$$

where $C \in \mathbb{R}_+$.

Proposition 6. Let $C(\lambda)$ be the average cost over one period, that is,

$$C(\lambda) = \frac{\lambda}{\tau} \int_0^{\tau/\lambda} \langle QX(t), X(t)\rangle + \langle U(t), U(t)\rangle\, dt$$

and define $\gamma(\cdot)$ as

$$\gamma(y) = tr(v(a(y), \bar{k}(y))(Q + \bar{k}(y)\bar{k}'(y))).$$

Assume that ((A5) - (A8)) are satisfied. There is an $\bar{\varepsilon} > 0$ such that for $\varepsilon \in (0, \bar{\varepsilon})$ in (22)

$$E\left|C(\lambda) - \frac{1}{\tau}\int_0^\tau \gamma(y)\,dy\right| \le c\sqrt{\lambda}$$

where $c \in \mathbb{R}_+$.

3. DELAY TIME SYSTEMS

Many physical systems are naturally modelled by linear delay time stochastic differential equations. Since some parameters of one of these systems are often unknown and it is desired to control the system, there is the problem of stochastic adaptive control. The approach to the adaptive control problem for these systems is analogous to the approach for finite dimensional linear systems. A family of maximum likelihood (or equivalently least squares) estimates is shown to be strongly consistent for a family of controls. Using the certainty equivalence principle for the construction of the adaptive control, this control is shown to be self optimizing.

A deterministic, (discrete) delay time, linear system is described by a delay differential equation of the form

$$\frac{dx}{dt} = \sum_{i=1}^{k} A_i x(t - \tau_i) + Bu(t) \tag{23}$$

where $\tau_i \ge 0$, $A_i \in L(\mathbb{R}^n, \mathbb{R}^n)$ for $i = 1, 2, \ldots, k$, $u \in \mathbb{R}^m$, $B \in L(\mathbb{R}^m, \mathbb{R}^n)$, $t \ge 0$ and the initial condition is the pair $(h(0), h(s))$ where $s \in [-\tau, 0]$ and $\tau = \max_i (\tau_i)$.

With the above initial condition and for applications to control it is useful to describe the solution of (23) in the function space $M^2(-\tau, 0; \mathbb{R}^n)$ [10]. The space $M^2(-\tau, 0; \mathbb{R}^n)$ is the Hilbert space of equivalence classes of $\mathbb{R}^n$-valued Lebesgue measurable functions on $[-\tau, 0]$ with the inner product

$$\langle f,g\rangle_{M^2} = \langle f(0), g(0)\rangle + \int_{-\tau}^{0} \langle f(s), g(s)\rangle\, ds.$$

The space M^2 is isometrically isomorphic to $\mathbb{R}^n \times L^2(-\tau, 0; \mathbb{R}^n)$. With this isomorphism $\Lambda \in L(M^2, M^2)$ can be represented as $\Lambda = (\Lambda_{ij})$ where $\Lambda_{ij} \in L(X_i, X_j)$ $i = 1, 2$ and $X_1 = \mathbb{R}^n$ and $X_2 = L^2(-\tau, 0; \mathbb{R}^n)$. This decomposition is used subsequently.

Equation (23) can be described as an evolution equation in M^2. The state at time t is

$$\tilde{x}(t; h, u) = (\tilde{x}(t; h, u)^1, \tilde{x}(t; h, u)^2) \in M^2$$

where

$$\tilde{x}(t;\, h,\, u)^1 = x(t;\, h,\, u)$$

$$\tilde{x}(t;\, h,\, u)^2 = \begin{cases} x(t+\theta;\, h,\, u) & \text{for } t+\theta \geq 0 \\ h(t+\theta) & \text{for } t+\theta \leq 0 \end{cases}$$

and $\theta \leq 0$. Let $\tilde{A} \in L(M^2, M^2)$ and $B \in L(\mathbb{R}^n, M^2)$ be defined as

$$\begin{aligned} \tilde{A}_1 h &= \sum_{i=1}^{k} A_i h(-\tau_j) \\ \tilde{A}_2 h(\theta) &= \frac{dh}{d\theta} \\ [\tilde{A}h]^i &= \tilde{A}_i h \quad \text{for } i = 1, 2 \\ \tilde{B}u &= (Bu, 0). \end{aligned}$$

Then the controlled evolution equation is

$$\begin{aligned} \frac{d\tilde{x}}{dt} &= \tilde{A}\tilde{x} + \tilde{B}u \\ \tilde{x}(0) &= h. \end{aligned} \tag{24}$$

Now consider the linear, discrete delay stochastic differential equation for the adaptive control problem

$$dX(t) = \sum_{i=1}^{k} F_0^i X(t-\tau_i)dt + \sum_{i=1}^{k}\sum_{j=1}^{n_i} \alpha_{ij} F_j^i X(t-\tau_i)dt + BU(t)dt + dW(t) \tag{25}$$

where $F_j^i \in L(\mathbb{R}^n, \mathbb{R}^n)$, $B \in L(\mathbb{R}^m, \mathbb{R}^n)$, $(W(t),\, t \geq 0)$ is a standard n-dimensional Wiener process and $\tilde{X}(0) = h$ is a fixed function.

The probability space $(\Omega, \mathcal{F}, P)$ can be chosen to be the same as for the adaptive control of finite dimensional linear systems.

The uncontrolled system (23) is said to be L^2-stable if

$$\lim_{t\to\infty} \int_0^t |x(s;\, h)|^2\, ds < \infty$$

for all $h \in M^2$. If (23) is L^2-stabilizable then the infinite time control problem (that is $t = +\infty$ in (2)) is well posed [10] and the optimal feedback control is $u(t) = -\tilde{P}^{-1}\tilde{B}^*\pi\tilde{x}(t)$ where π is the unique positive semidefinite self adjoint operator that satisfies the stationary Riccati equation

$$0 = \pi\tilde{A} + \tilde{A}^*\pi - \pi\tilde{R}\pi + Q. \tag{26}$$

Let $(\rho_1, ..., \rho_\ell)$ be the smallest family of positive numbers such that τ_i is an $\mathbb{N}$-sum of $(\rho_1, ..., \rho_\ell)$ for $i = 1, 2, ..., k$. Define the operator σ_i as $\sigma_i X(t) = X(t - \rho_i)$. Then the linear system (23) can be considered as a system over a ring of polynomials $\mathbb{R}[\sigma_1, ..., \sigma_\ell]$. Let $\bar{A}$ be the nxn matrix over $\mathbb{R}[\sigma_1, ..., \sigma_\ell]$. If $(\bar{A}, B)$ satisfies an algebraic reachability condition [11] then it can be shown [12] that

$$\frac{dx}{dt} = \bar{A}x + Bu \tag{27}$$

is L^2-stabilizable.

The following assumptions are used subsequently.

(A9) $\alpha_{ij} \in I_{ij} \subset \mathbb{R}$ is an unknown parameter where I_{ij} is a bounded, open interval for $i \in \{1, ..., k\}$ and $j \in \{1, ..., n_i\}$.

(A10) The pair $(\bar{A}, B)$ satisfies the algebraic reachability condition for all $\alpha_{ij} \in I_{ij}$.

(A11) For each $i \in \{1, 2, ..., k\}$ the family of linear transformations $(F^i_j, j = 1, 2, ..., n_i)$ is linearly independent.

Let $(X(t), t \geq 0)$ be the solution of (25) with the feedback control law $(K(t), t \geq 0)$. Define $A(t)$ and $\tilde{A}(t)$ in $L(\mathbb{R}^{\Sigma n_i}, \mathbb{R}^{\Sigma n_i})$ as

$$A(t) = (a_{ij}(t)) \tag{28}$$

$$\tilde{A}(t) = (\tilde{a}_{ij}(t)) \tag{29}$$

where

$$a_{mn}(t) = \int_0^t \langle F^i_j X(s - \tau_i), F^p_\ell X(s - \tau_p)\rangle \, ds$$

and $m = \sum_{r<i} n_r + j$ and $n = \sum_{r<p} n_r + \ell$.

$$\tilde{a}_{mn}(t) = \frac{a_{mn}(t)}{a_{mm}(t)}$$

(A12) For each admissible adaptive control law $(\tilde{A}(t), t \geq 0)$ in (29) satisfies

$$\liminf_{t \to \infty} |\det \tilde{A}(t)| > 0 \qquad \text{a.s.}$$

The following result provides strong consistency of the family of maximum likelihood estimates [15].

Theorem 5. Assume that ((A8) - (A12)) are satisfied. Let $(K(t), t \geq 0)$ be an $L(M^2, M^2)$ -valued control law such that for each $t \geq 0$ the integral operators that define $K(t)$ have kernels that are piecewise continuous functions and the family $(K(t), t \geq 0)$ is uniformly bounded (almost surely) in the Hilbert-Schmidt norm. Then the family of maximum likelihood estimates $(\hat{\alpha}(t), t \geq 0)$ is strongly consistent, that is,

$$P_{\alpha_0}(\lim_{t\to\infty} \hat{\alpha}(t) = \alpha_0) = 1.$$

The maximum likelihood estimate $\hat{\beta}(t)$ is the solution of the family of linear equations

$$\tilde{A}(t)\hat{\beta}(t) = \tilde{A}(t)\beta + \gamma(t) \tag{30}$$

where β is a vector formed from α_{ij} by letting $\beta_{f(i,j)} = \alpha_{ij}$, $f(i,j) = \sum_{k=1}^{i-1} n_k + j$ and

$$\gamma_i(t) = \int_0^t <F_\ell^j X(s - \tau_j), dW(s)> \tag{31}$$

and $i = \sum_{k<j} n_k + \ell$. A recursive equation can be given for $(\hat{\alpha}(t), t \geq 0)$ in analogy with (9).

Since for some adaptive control problems it may be difficult to verify (A12), it is natural to obtain other sufficient conditions for consistency of least squares estimators.

The least squares estimator $\alpha^*(t)$ of the unknown parameters (α_{ij}) based on $(X(t), t \leq T)$ and $(U(t), t \leq T)$ is the minimizer of the formal expression

$$\int_0^T <\dot{\tilde{X}}(t) - \tilde{A}(\alpha)\tilde{X}(t) - BU(t), \dot{\tilde{X}}(t) - \tilde{A}(\alpha)\tilde{X}(t) - BU(t)> dt.$$

A statement of the consistency of the family of least squares estimates under some stability conditions is given now.

Theorem 6. Let ((A9) - (A11)) be satisfied. If

$$\frac{1}{T}\int_0^T <X(t), X(t)> + <U(t), U(t)> dt$$

is bounded in probability (respectively almost surely) for $T > 0$ and

$$\lim_{T\to\infty} \frac{<X(T), X(T)>}{T} = 0$$

in probability (respectively almost surely), then

$$\lim_{T\to\infty} \dot{\alpha}^*(T) = \alpha_0$$

in probability (respectively almost surely) where α_0 is the true parameter vector.

The verification of this result is given in [27].

The adaptive control law uses the certainty equivalence principle, that is, the stationary Riccati equation (26) is solved using $\hat{\alpha}(t - \tau)$ (or $\alpha^*(t - \tau)$) instead of the true parameter vector to obtain the feedback gain $K(t)$. For the adaptive control law it is necessary to obtain a suitable stability property for (25). Let V be the unique positive semidefinite solution of the stationary Riccati equation (26) using the true parameter vector α_0 and let $k_0 = - \tilde{P}^{-1}\tilde{B}^*V$. Using the implications of stability of the linear delay time equation (27) it follows that there is a symmetric, positive semidefinite operator $\tilde{V} \in L(M^2, M^2)$ such that

$$\tilde{I} + (\tilde{A}(\alpha_0) + \tilde{B}k_0)^*\tilde{V} + V(\tilde{A}(\alpha_0) + \tilde{B}k_0) < 0.$$

It can be shown that this inequality is a continuous function of the unknown parameters α and the feedback gain K so that the inequality is satisfied for α sufficiently close to α_0 and K sufficiently close to k_0.

Proposition 7. Assume that ((A9) - (A11)) are satisfied. Let $(X(t), t \geq 0)$ be the solution of (25) using the feedback control law $(K(t), t \geq 0)$ such that for each $t \geq 0$ the integral operators that define $K(t)$ have kernels that are piecewise continuous functions, $(K(t), t \geq 0)$ is uniformly bounded (almost surely) in the Hilbert-Schmidt norm and

$$\lim_{t\to\infty} K(t) = k_0 \quad \text{a.s.}$$

where the topology is induced from the Hilbert-Schmidt norm, $k_0 = - \tilde{P}^{-1}\tilde{B}^*V$ and B is the solution of (26) using α_0. Then the process $(X(t), t \geq 0)$ satisfies

$$\lim_{t\to\infty} \frac{1}{t} <X(t), X(t)> = 0 \quad \text{a.s.} \tag{32}$$

$$\limsup_{t\to\infty} \frac{1}{t} \int_0^t <X(s), X(s)> ds \leq c \quad \text{a.s.} \tag{33}$$

where $c \in \mathbb{R}_+$.

The proof of this result is given in [15].

The unknown parameters take values in bounded subsets of the real line and it can be shown that the solution of the stationary Riccati equation is a continuous function of the parameters. By truncation of the estimates, if necessary, to the bounded subsets of the parameter values and the continuity of the stationary Riccati equation with respect to the unknown parameters, it can be assumed that the feedback gains determined from the stationary Riccati equation using the estimates as the parameter values are uniformly bounded in Hilbert-Schmidt norm and converge (almost surely) to the optimal feedback gain k_0. By Proposition 7 the process $(X(t), t \geq 0)$ using the certainty equivalence control law satisfies the stability conditions (32 - 33). Furthermore it can be shown that the certainty equivalence control law is self optimizing, that is, the family of average costs converges (almost surely) to the optimal average cost [15].

Theorem 7. Assume that ((A9) - (A12)) are satisfied. Let $(K(t), t \geq \tau)$ be the family of feedback gains determined from the stationary Riccati equation (26) using the (truncated) maximum likelihood (or least squares) estimates $(\hat{\alpha}(t - \tau), t \geq \tau)$. Then

$$\lim_{t \to \infty} K(t) = k_0 \qquad \text{a.s.}$$

where the topology is induced from the Hilbert-Schmidt norm and

$$\lim_{t \to \infty} \frac{1}{t} C(t) = \operatorname{tr} V \qquad \text{a.s.} \tag{34}$$

where V is the unique, symmetric positive semidefinite solution of (26) using the true parameter vector α_0, $C(t)$ is the cost at time t given in (3) using $(K(t), t \geq 0)$ and tr is the operator trace.

4. INFINITE DIMENSIONAL SYSTEMS

Infinite dimensional linear systems can describe many linear delay time differential equations and various families of linear partial differential equations. The infinite dimensional systems that are considered here are linear stochastic evolution systems where the infinitesimal generators generate strongly continuous semigroups. A semigroup $(G(t), t \geq 0)$ of bounded linear operators on a Hilbert space H is called a strongly continuous semigroup of bounded linear operators if

$$\lim_{t \downarrow 0} G(t)x = x \tag{35}$$

for each $x \in H$ in the strong topology. A strongly continuous semigroup of bounded linear operators on H is called a C_0 semigroup. If $(G(t), t \geq 0)$ is a C_0 semigroup then there are real numbers $\omega > 0$ and $M \geq 1$ such that

$$|G(t)| \leq Me^{\omega t}$$

for $t \in [0, \infty)$. If A is the infinitesimal generator of a C_0 semigroup then the domain of A, $\mathcal{D}(A)$, is dense in H and A is a closed, linear operator (e.g. [28]).

The model for this adaptive control problem is described by the following infinite dimensional stochastic differential equation

$$dX(t) = A(\alpha)X(t)\,dt + BU(t)\,dt + dW(t) \tag{36}$$

where

$$A(\alpha) = F_0 + \sum_{i=1}^{p} \alpha_i F_i \tag{37}$$

$X(t) \in H$, H is a real, separable, infinite dimensional Hilbert space, $X(0) = X_0$ is an element of H, $(W(t), t \geq 0)$ is an H-valued Wiener process such that $W(1)$ has the nuclear covariance Q_w that is positive, $A(\alpha)$ is the generator of a C_0 semigroup, $B \in L(H_1, H)$ where H_1 is a real, separable Hilbert space and $t \geq 0$.

The following assumptions are selectively used.

(A13) The semigroup with the infinitesimal generator $A(\alpha)$ is exponentially stable uniformly for $\alpha \in \mathcal{K}$, that is, there is an $M_0 > 0$ and $\omega > 0$ such that

$$|e^{tA(\alpha)}| \leq M_0 e^{-\omega t}$$

for all $\alpha \in \mathcal{K}$ where $\mathcal{K}$ is compact and $\mathcal{A} = I_1 \times I_2 \times \cdots \times I_p \subset \mathcal{K}$.

(A14) For each $\alpha \in \mathcal{K}$, $A(\alpha)$ is the infinitesimal generator of a C_0 semigroup and $\bigcap_{\alpha \in \mathcal{A}} \mathcal{D}(A(\alpha))$ is dense in H.

(A15) The map $\alpha \mapsto e^{tA(\alpha)}$ is continuous for $t > 0$ and $\alpha \in \mathcal{K}$ in the uniform operator topology.

(A16) $\bigcap_{i=0}^{p} \mathcal{D}(F_i^*)$ is dense in H.

(A17) There is a finite dimensional projection $\tilde{P}$ with range in $\bigcap_{i=0}^{p} \mathcal{D}(F_i^*)$ such that the family $(\tilde{P}F_i, i = 1, 2, \ldots, p)$ is linearly independent on $\bigcap_{i=0}^{p} \mathcal{D}(F_i^*)$.

The probability space $(\Omega, \mathcal{F}, P)$ can be chosen such that Ω is the Fréchet space of H-valued continuous functions on $[0,\infty)$ with the seminorms of local uniform convergence, P is the Wiener measure on Ω for the process $(W(t), t \geq 0)$ such that $E[W(1)W(1)'] = Q_w$ and $\mathcal{F}$ is the P-completion of the Borel σ-algebra on Ω.

The notion of solution of the stochastic differential equation (36) is the mild solution, that is, the solution of the integral equation

$$X(t) = G(t)X(0) + \int_0^t G(t - s)BU(s)\, ds + \int_0^t G(t - s)dW(s) \tag{37}$$

where $(G(t), t \geq 0)$ is the semigroup $(e^{tA}, t \geq 0)$ and the dependence on α has been suppressed. The H-valued stochastic integral

$$\int_0^t G(t - s)\, dW(s)$$

has a continuous modification by [7]. For (square) integrable controls it is known that there is one and only one mild solution of (36).

It has been shown [5] that the mild solution is equivalent to the following two other notions of solution

i) $X(t) = X(0) + A\int_0^t X(s)\, ds + W(t)$

and

ii) for each $y \in \mathcal{D}(A^*)$

$$\langle y, X(t)\rangle = \langle y, X(0)\rangle + \int_0^t \langle A^*y, X(s)\rangle\, ds + \langle y, W(t)\rangle.$$

Consider the following deterministic optimal control problem in H. The system is

$$\frac{dx}{dt} = Ax(t) + Bu(t) \tag{38}$$

where $x(0) = x_0$, $t \in [0,T]$, A is the infinitesimal generator of a C_0 semigroup, $u(t) \in H_1$, $x(0) \in \mathcal{D}(A)$ and $B \in L(H_1, H)$. The cost functional that is to be minimized is

$$C(T) = \int_0^T \langle Qx(s), x(s)\rangle + \langle Pu(s), u(s)\rangle\, ds \tag{39}$$

over all $u \in L^2(0, T; H_1)$ where $Q \in L(H, H)$ is bounded, symmetric positive definite and $P \in L(H_1, H_1)$ is bounded and symmetric such that $\langle Px,x\rangle \geq c|x|^2$ where $c > 0$ is a fixed constant.

Let $L^+(H, H)$ be the cone of bounded, linear symmetric, nonnegative operators from H to H and let $C_S([0,T], L^+(H, H))$ be the family of all maps $S: [0, T] \to L^+(H, H)$ such that $S(\cdot)x$ is continuous for each $x \in H$. Consider the Riccati differential equation

$$\frac{dV}{dt} = A^*V + VA - VRV + Q \tag{40}$$

where $V(0) = V_0$, $t \in [0,T]$ and $R = BP^{-1}B^*$. A mild solution of (40) is a function $V \in C_S([0,T], L^+(H, H))$ such that

$$V(t)x = G^*(t)V(0)G(t)x + \int_0^t G^*(t - s)[Q - V(s)RV(s)]G(t - s)\, x\, ds \tag{41}$$

for all $x \in H$ where $G(t) = e^{tA}$. The optimal feedback control is

$$u^*(s, x(s)) = - P^{-1}B^*V(T - s)\, x(s)$$

If (A, B) is stabilizable and the conditions for a mild solution to (41) are satisfied then the stationary Riccati equation

$$A^*V + VA - VRV + Q = 0 \tag{42}$$

has one and only one nonnegative solution ([3, 29]). Furthermore this solution is the strong limit as $T \to \infty$ of the solutions of the Riccati differential equations.

For adaptive control the control laws $(U(t), t \geq 0)$ that are considered are linear feedback controls, that is,

$$U(t) = K(t)\, X(t) \tag{43}$$

where $(K(t), t \geq 0)$ is an $L(H, H_1)$-valued process that is adapted to the sub-σ-algebras $(\sigma(X(s), s \leq t), t \geq 0)$ and is uniformly bounded (almost surely).

Let $\delta > 0$ be fixed. It is assumed that the $L(H, H_1)$-valued process $(K(t), t \geq 0)$ has the property that $K(t)$ is adapted to $\sigma(X(u), u \leq t - \delta)$ for each $t \geq \delta$ and it is assumed that $(K(t), t \in [0, \delta))$ is a uniformly bounded, deterministic, operator-valued function. For the adaptive control problem this measurability is accomplished by computing $K(t)$ from the parameter estimates at time $t - \delta$ which is quite natural for applications. If $(K(t), t \geq 0)$ satisfies the above measurability condition and the boundedness condition, then it is easy to verify that there is one and only one mild solution of the stochastic differential equation

$$dX(t) = (A(\alpha) + BK(t))X(t)\,dt + dW(t). \tag{44}$$

Let $(X(t),\ t \geq 0)$ be the mild solution of (44). For notational convenience the dependence on the control has been suppressed. Define $A(t)$ and $\tilde{A}(t)$ in $L(\mathbb{R}^p, \mathbb{R}^p)$ as

$$A(t) = (a_{ij}(t)) \tag{45}$$

and

$$\tilde{A}(t) = (\tilde{a}_{ij}(t)) \tag{46}$$

where

$$a_{ij}(t) = \int_0^t \langle \tilde{P}F_iX(s), \tilde{P}F_jX(s)\rangle\, ds$$

and

$$\tilde{a}_{ij}(t) = \frac{a_{ij}(t)}{a_{ii}(t)}$$

for $i,j \in \{1, 2, ..., p\}$ and $\tilde{P}$ is defined in (A17).

For the computation of the least squares estimates the following is used.

(A18) For each admissible adaptive control law, $(\tilde{A}(t),\ t > 0)$ in (46) satisfies

$$\liminf_{t\to\infty} |\det \tilde{A}(t)| > 0 \quad \text{a.s.}$$

However, it can be shown that without the assumption (A18), $\det A(t) \neq 0$ a.s. for each $t > 0$. One can easily construct examples that satisfy (A18) in analogy to those given for (A4) as well as examples from systems described by partial differential equations [19].

Theorem 8. Let $(K(t),\ t \geq 0)$ be a feedback control law such that $K(t)$ is $\sigma(X(s),\ s \leq t - \delta)$ measurable for each $t \geq \delta$ and

$$\sup_{t\geq 0} |K(t)| \leq M \qquad \text{a.s.}$$

where $\delta > 0$ is fixed, $M \in \mathbb{R}_+$ and $|\cdot|$ is the operator norm. Assume that (A1), (A13), (A14) and ((A16) - (A18)) are satisfied. Then the family of least squares estimates $(\hat{\alpha}(t),\ t \geq 0)$ using $\tilde{P}$ is strongly consistent, that is,

$$P_{\alpha_0}(\lim_{t\to\infty} \hat{\alpha}(t) = \alpha_0) = 1.$$

The least squares estimate $\hat{\alpha}(t)$ in the above theorem is the minimizer of the formal expression

$$\int_0^t |\tilde{P}\dot{X}(s) - \tilde{P}A(\alpha)X(s) - \tilde{P}K(s)X(s)|^2 \, ds.$$

The estimates $(\hat{\alpha}(t), t \geq 0)$ can be computed recursively, that is, they satisfy the following stochastic differential equation

$$d\hat{\alpha}(t) = A^{-1}(t)<\tilde{F}X(t), d\tilde{P}X(t) - \tilde{P}A(\hat{\alpha}(t))X(t)\, dt - \tilde{P}BU(t)\, dt> \tag{47}$$

where $<\tilde{F}x, y>$ is the vector whose i^{th} component is $<\tilde{P}F_i x, y>$.

For the self tuning aspect of the adaptive control problem (39, 44) it is important to verify that the solution of the Riccati equations (40, 42) depends continuously on the parameters. This property apparently had not been verified for the infinite dimensional problem that is considered here.

Proposition 8. Assume that ((A13) - (A15)) are satisfied. Then the mild solution $(V(\alpha), \alpha \in \mathcal{K})$ of the stationary Riccati equation (42) is continuous in the uniform operator topology.

Subsequent to the verification of Proposition 8 in [19] this uniform operator continuity of solution of the stationary Riccati equation in Hilbert space with respect to parameters was obtained under only the conditions of uniform boundedness of the optimal costs, uniform detectability and uniform operator continuity of the semigroup [6].

The adaptive control law uses the certainty equivalence principle, that is, the stationary Riccati equation (42) is solved using $\hat{\alpha}(t - \delta)$ instead of the true parameter vector to obtain the feedback gain $K(t)$. For this adaptive control law it is necessary to obtain a suitable stability property for (44). Let V be the unique, positive semidefinite mild solution of the stationary Riccati equation (42) using the true parameter vector α_0 and let $k_0 = - P^{-1}B^*V$. The stability of the semigroup having infinitesimal generator $A_0 + Bk_0$ is equivalent to the existence of a nonnegative self adjoint operator $\tilde{V}$ such that

$$2<(A_0 + Bk_0)x, \tilde{V}x> = - <\tilde{Q}x, x>$$

for each $x \in \mathcal{D}(A_0)$ where Q is positive and symmetric [8] and $A_0 = A(\alpha_0)$. Since the stationary Riccati equation (42) is a continuous function of the parameters in the uniform operator topology, $A + BK$ is stable for α sufficiently close to α_0 and K is sufficiently close to k_0.

Proposition 9. Assume that (A1) and ((A13) - (A17)) are satisfied. Let $(X(t), t \geq 0)$ be the mild solution of

$$dX(t) = AX(t)dt + BK(t)X(t)dt + dW(t).$$

Assume that

$$\lim_{t \to \infty} K(t) = k_0 \quad \text{a.s.}$$

in the uniform operator topology where $k_0 = - P^{-1}B^*V$ and V is the unique mild solution of (42) using α_0. Then the process $(X(t), t \geq 0)$ satisfies

$$\lim_{t \to \infty} \frac{1}{t} <X(t), X(t)> = 0 \text{ a.s.} \tag{48}$$

and

$$\limsup_{t \to \infty} \frac{1}{t} \int_0^t <X(s), X(s)> ds \leq c \qquad \text{a.s.} \tag{49}$$

where $c \in \mathbb{R}_+$.

The unknown parameters take values in bounded subsets of the real line and the solution of the stationary Riccati equation (42) is a continuous function of the parameters in the uniform operator topology by Proposition 8. By truncation of the least squares estimates, if necessary, to the bounded subsets of the parameter values and the continuity of the Riccati equation with respect to parameters, it can be assumed that the feedback gains determined from the stationary Riccati equation, using the least squares estimates as the parameter values, are uniformly bounded in the operator norm and converge (almost surely) to the optimal feedback gain k_0. By Proposition 9 the process $(X(t), t \geq 0)$ using the certainty equivalence control law satisfies the stability conditions (48 - 49). Furthermore, it can be shown that the certainty equivalence control law is self optimizing, that is, the family of average costs converges (almost surely) to the optimal average cost.

Theorem 9. Assume that (A1) and ((A13)- (A18)) are satisfied. Let $(K(t), t \geq 0)$ be the family of feedback gains determined from the stationary Riccati equation (42) using the (truncated) least squares estimates $(\hat{\alpha}(t - \delta), t \geq \delta)$. Then

$$\lim_{t \to \infty} K(t) = k_0 \quad \text{a.s.}$$

in the uniform operator topology and

$$\lim_{t \to \infty} \frac{1}{t} C(t) = tr(VQ_w) \quad \text{a.s.} \tag{50}$$

where V is the unique bounded, nonnegative symmetric solution of the stationary Riccati equation (42) using the true parameter vector α_0, $C(t)$ is the cost at time t given in (39) using $(K(t), t \geq 0)$ and Q_w is the covariance of $W(1)$.

The previous results on the adaptive control of infinite dimensional systems are described and verified in [18, 19].

REFERENCES

[1] A. Aloneftis, Stochastic Adaptive Control, Lecture Notes in Control and Inf. Sc. 98, Springer-Verlag, Berlin 1987.

[2] K. J. Åström and B. Wittenmark, Adaptive Control, Addison-Wesley, Reading 1989.

[3] V. Barbu and G. Da Prato, Hamilton-Jacobi Equations in Hilbert Spaces, Res. Notes in Math. 86(1983), Pitman, Boston-London.

[4] P. E. Caines, Linear Stochastic Systems, John Wiley, New York, 1988.

[5] A. Chojnowska-Michalik, Stochastic differential equations in Hilbert spaces, Banach Center Pub. 5(1979), 53-74.

[6] A. Chojnowska-Michalik, T. E. Duncan, and B. Pasik-Duncan, Uniform operator continuity of the stationary Riccati equation in Hilbert space, to appear in J. Appl. Math. Optim.

[7] G. Da Prato, S. Kwapien, and J. Zabczyk, Regularity of solutions of linear stochastic equations in Hilbert spaces, Stochastics 23(1987), 1-23.

[8] R. Datko, Extending a theorem of A. M. Lyapunov to Hilbert space, J. Math. Anal. Appl. 32(1970), 610-616.

[9] M. C. Delfour and S. K. Mitter, Controllability, observability and optimal feedback control of affine hereditary differential systems, SIAM J. Control 10(1972), 298-328.

[10] M. C. Delfour, C. McCalla, and S. K. Mitter, Stability and the infinite-time quadratic cost problem for linear hereditary differential systems, SIAM J. Control 13(1975), 48-88.

[11] T. E. Duncan, On the steady state filtering problem for linear pure delay time systems, Séminaires IRIA, 1979, 25-42.

[12] T. E. Duncan, Estimation for linear pure delay systems, Proc. XVI Semester on Mathematical Control Theory, Banach Center, Warsaw 1980, 199-227.

[13] T. E. Duncan, P. Mandl, and B. Pasik-Duncan, Some quadratic functionals and self-tuning control, Proc. 27th IEEE Conf. on Decision and Control, 1219-1222.

[14] T. E. Duncan, P. Mandl, and B. Pasik-Duncan, On exponentially discounted adaptive control, Kybernetika, 26(1990), 361-372.

[15] T. E. Duncan and B. Pasik-Duncan, Adaptive control of linear delay time systems, Stochastics, 24(1988), 45-74.

[16] T. E. Duncan and B. Pasik-Duncan, Adaptive control of continuous time linear stochastic systems, Mathematics of Control, Signals and Systems 3(1990), 45-60.

[17] T. E. Duncan and B. Pasik-Duncan, The rate of convergence and the asymptotic destributions for some adaptive control problems, Proc. 26th Conf. on Decision and Control, 1666-1669.

[18] T. E. Duncan and B. Pasik-Duncan, Some aspects of the adaptive control of stochastic evolution equations, Proc. 28th Conf. on Decision and Control, 1989, 732-735, to be reprinted in Advances in Adaptive Control (K. Narendra, R. Ortega, and P. Dorato eds.) IEEE Press.

[19] T. E. Duncan, B. Goldys, and B. Pasik-Duncan, Adaptive control of linear stochastic evolution systems, to appear in Stochastics.

[20] L. Gerencser, Some new results in the theory of recursive identification, Proc. 28th Conf. on Decision and Control, 1989, 242-248

[21] P. R. Kumar, A survey of some results in stochastic adaptive control, SIAM J. Control Optim. 23(1985), 329-380.

[22] P. R. Kumar and P. Varaiya, Stochastic Systems: Estimation, Identification and Adaptive Control, Prentice Hall, Engelwood Cliffs, 1986.

[23] P. Mandl, Some connections between statistics and control theory, Mathematical Statistics and Probability Theory, Vol. B (P. Bauer, et. al., eds.) Reidel, 1987.

[24] P. Mandl, T. E. Duncan, and B. Pasik-Duncan, On the consistency of a least squares identification procedure, Kybernetika 24(1987), 3, 296-307.

[25] B. Pasik-Duncan, On adaptive control, Central School of Planning and Statistics Publishers, Warsaw, 1986.

[26] B. Pasik-Duncan, On adaptive control of continuous time linear stochastic systems, Lecture Notes in Control and Info. Sci. 136(1989), 328-343.

[27] B. Pasik-Duncan, On the consistency of a least squares identification procedure in continuous time linear stochastic systems with delays, to appear in Stochastics.

[28] A. Pazy, Semigroups of Linear Operators and Applications to Partial Differential Equations, Appl. Math. Sci. 44(1983), Springer, New York-Berlin.

[29] J. Zabczyk, Remarks on the algebraic Riccati equations in Hilbert space, Appl. Math. Optim. 2(1976), 251-258.

STRONG APPROXIMATION RESULTS IN ESTIMATION AND ADAPTIVE CONTROL

László Gerencsér
Department of Electrical Engineering, McGill University,
Montréal, Québec, Canada.
On leave from Computer and Automation Institute
of the Hungarian Academy of Sciences,
Budapest, Hungary.

1. INTRODUCTION

Every science ... must be not merely
a mirror of the facts, but a
genuine construction.

J. Ortega y Gasset

The purpose of this paper is to present relatively unknown results which establish a connection between the asymptotic theory of a fairly general class of estimator processes and of existing asymptotic theory for martingales, mixingales and L-mixing processes. The novelty of our approach is the widespread use of strong approximation techniques, by which we essentially get a representation of the estimator error processes in a very convinient form. A partial objective of this research has been to develop a mathematical theory which is tailored to the needs of estimation and adaptive control of linear stochastic systems.

When trying to organize the material the question we can ask is this: what should the central concepts and propositions of such a theory be? A possible answer is that it is the concept of L-mixing processes which has the advantage over other candidates that it is invariant under most of the operations we usually perform in estimation. The germs of this concept can be found in [53], [55], and [64], but it has been much further developed in [18]. The central propositions of this theory are moment inequalities for single integrals with respect to L-mixing processes, (c.f. [18]) a certain simple extension of Kolmogorov's theorem on the almost sure continuity of stochastic processes (c.f. [47]) and Móricz's maximal inequality and its extension (c.f. [59]). Some of the important and more or less direct corollaries of these fundamental results will be presented here. On a secondary level the central concepts of this theory are: a general estimation scheme proposed by [54] and in [13], the associated theory of random differential equations, a set of new stochastic averaging techniques, and certain extensions of the concept of predictive stochastic complexity.

The "leitmotive" of this research is that the best service mathematicians can offer to engineers is to create elegant mathematics which has some relevance to existing "practical problems", and which can provide some insight for further progress. The results that are presented here can be modified and extended in a number of ways. Our purpose is to present the focal points of this theory only.

The proposed theory is not the only thery of estimation and adaptive control. There has been at least three other schools of thought. Martingale theory and stochastic Lyapunov functions has been systematically used in [6], [39], [48], [49], [58]. The stochastic regression theory has been developed and used in [14], [52]. Finally some forms of pathwise analysis has been carried out in [9]. The solution of these various approaches and their relative strength is still to be investigated.

2. L-MIXING PROCESSES

We start the technical discussion with some conventions. The set of real numbers will be denoted by $\mathbb{R}$, the p-dimensional Euclidean space will be denoted by $\mathbb{R}^p$. Let $D \subset \mathbb{R}^p$ be compact domain and let a stochastic process $(x_n(\theta))$ be defined on $\mathbb{Z} \times D$, where $\mathbb{Z}$ denotes the set of natural numbers.

Definition 2.1. We say that $(x_n(\theta))$ is M-bounded if for all $1 \leq q < \infty$

$$M_q(x) = \sup_{\substack{n \geq 0 \\ \theta \in D}} \mathrm{E}^{1/q} |x_n(\theta)|^q < \infty.$$

If $(x_n(\theta))$ is M-bounded we shall also write $x_n(\theta) = O_M(1)$. Similarly if c_n is a positive sequence we write $x_n(\theta) = O_M(c_n)$ if $x_n(\theta)/c_n = O_M(1)$.

The definition extends to the case, when x_n does not depend on a parameter, or if (x_n) degenerates to a single random variable. We say that a sequence of random variables x_n tends to a random variables x in the M-sense if for all $q \geq 1$ we have

$$\lim_{n \to \infty} \mathrm{E}^{1/q} |x_n - x|^q = 0.$$

Similarly we can define differentiation in the M-sense.

Let $(\mathcal{F}_n), n \geq 0$ be a family of monotone increasing σ-algebras, and $(\mathcal{F}_n^+), n \geq 0$ be a monotone decreasing family of σ-algebras. We assume that for all $n \geq 0$, $\mathcal{F}_n$ and $\mathcal{F}_n^+$ are independent. For $n \leq 0$ $\mathcal{F}_n^+ = \mathcal{F}_0^+$. A typical example is provided by the σ-algebras

$$\mathcal{F}_n = \sigma\{e_i : i \leq n\} \qquad\qquad \mathcal{F}_n^+ = \sigma\{e_i : i > n\}$$

where (e_i) is an i.i.d. sequence of random variables.

Definition 2.2. A stochastic process $(x_n(\theta)), n \geq 0$ is L-mixing with respect to $(\mathcal{F}_n, \mathcal{F}_n^+)$ uniformly in θ if it is $\mathcal{F}_n$-adapted, M-bounded and with τ being a positive integer

$$\gamma_q(\tau, x) = \gamma_q(\tau) = \sup_{\substack{n \geq \tau \\ \theta \in D}} \mathrm{E}^{1/q} |x_n(\theta) - \mathrm{E}(x_n(\theta) | \mathcal{F}_{n-\tau}^+)|^q$$

we have for any $1 \leq q < \infty$.

$$\Gamma_q = \Gamma_q(x) = \sum_{\tau=1}^{\infty} \gamma_q(\tau) < \infty.$$

If we consider a single stochastic process (x_n) which does not depend on a parameter θ we can still use the above definition but the phrase "uniformly in θ" will be omitted.

Example. Discrete time stationary Gaussian ARMA processes are L-mixing. (This can be seen using a state space representation).

A fundamental technique mentioned earlier is a moment inequality which combined with Móricz's result [59] gives the following (c.f. Theorem 1.1 in [18] and for earlier results [57] and [71]):

Theorem 2.1. Let $(x_n), n \geq 0$ be an L-mixing process with $\mathrm{E}x_n = 0$ for all n and let (f_n) be a deterministic sequence. Then we have for all $1 \leq m < \infty$

$$\mathrm{E}^{1/2m} \sup_{1\le N'\le N} |\sum_{n=1}^{N'} f_n x_n|^{2m} \le C_m \Big(\sum_{n=1}^{N} f_n^2\Big)^{1/2} M_{2m}^{1/2}(x)\Gamma_{2m}^{1/2}(x)$$

where C_m depends only on m.

Corollary 2.2 Let (x_n) be as above. Then

$$\frac{1}{N}\sum_{n=1}^{N} x_n = O_M(N^{-1/2}).$$

Corollary 2.3 Let (x_n) be as above and let $0 < \lambda < 1$, then

$$\sum_{n=1}^{N}(1-\lambda)^{N-n}\lambda x_n = O_M(\lambda^{1/2}).$$

Define the process $\Delta x/\Delta\theta$ as $|x_n(\theta+h)-x_n(\theta)|/|h|$, defined for $n\ge 0$, $\theta\ne\theta+h\in D$.

Definition 2.3. The stochastic process $x_n(\theta)$ is M-Lipschitz-continuous in θ if the process $\Delta x/\Delta\theta$ is M-bounded, i.e. if for all $1\le q<\infty$ we have

$$M_q(\Delta x/\Delta\theta) = \sup_{\substack{n\ge 0\\ \theta\ne\theta+h\in D}} \mathrm{E}^{1/q}|x_n(\theta+h)-x_n(\theta)|^q/|h| < \infty.$$

Example If $(x_n(\theta))$ is absolutely continuous with respect to θ a.s. and the gradient process $(x_n(\theta))$ is M-bounded, then $(x_n(\theta))$ is M-Lipschitz-continuous.

Let us consider the case when $(x_n(\theta))$ is a stochastic process which is measurable, separable, M-bounded and M-Lipschitz continuous in θ for $\theta\in D$. By Kolmogorov's theorem (c.f. I+H) the realizations of $(x_n(\theta))$ are continuous in θ with probability 1, hence we can define for almost all ω

$$x_n^* = \max_{\theta\in D_0}|x_n(\theta)|$$

where $D_0\subset \mathrm{int}D$ is a compact domain.

A useful result which is often used is the following, which among others implies the validity of a uniform strong law of large numbers:

Theorem 2.4 Let $x=(x_n(\theta))$ be a separable stochastic process. Assume that x and $\Delta x/\Delta\theta$ are L-mixing uniformly in θ, for $\theta\in D$. Let D_0 be as above. Then

$$\max_{\theta\in D_0}|\frac{1}{N}\sum_{n=1}^{N} x_n(\theta)| = O_M(N^{-1/2}).$$

3. THE GENERAL ESTIMATION SCHEME

The general estimation scheme which was proposed in [54] and [13] can be described in its simplest form as follows: Let (e_n) be a second order stationary process which is L-mixing with respect to a pair of families of σ-algebras $(\mathcal{F}_n, \mathcal{F}_n^+)$ and let us define another second order stationary process $(\overline{x}_n(\theta))$ by the state space equation:

$$\overline{x}_{n+1}(\theta) = A(\theta)\overline{x}_n(\theta) + B(\theta)e_n \tag{3.1}$$

with $\theta \in D \subset \mathbb{R}^p$, $\overline{x}(\theta) \in \mathbb{R}^m$. (In practice the initial condition $x_0(\theta)$ is assumed to be M-bounded and $\mathcal{F}_0$-measurable, but otherwise arbitrary random variable.)

An important assumption is that the process $\overline{x}_n(\theta)$ is computable in the sense that for each fixed θ, $\overline{x}_n(\theta)$ is the output of a system which is obtained by coupling the physical system with a known, artificial system.

Assume that the matrix $A(\theta)$ is stable for all $\theta \in D$ and that the functions $(A(\theta))$, and $(B(\theta))$ are continuously differentiable in a neighborhood of D_0. Let Q be a quadratic function from $\mathbb{R}^r$ to $\mathbb{R}^p$, and define

$$G(\theta) = \mathrm{E}Q(\overline{x}_n(\theta)).$$

We will say that the function $G(\theta)$ is computable. $G(\theta)$ is well-defined in D and it has continuous and bounded partial derivatives up to second order. The estimation problem can now be formulated as follows: solve the nonlinear algebraic equation

$$G(\theta) = 0. \tag{3.2}$$

The solution will be denoted by θ^*. We assume that (3.2) has a unique solution in D, and that $G_\theta(\theta^*)$ is nonsingular.

Remark. The above scheme is the simplest possible and can be generalized in many different ways. E.g. G may depend directly on θ, i.e. we may have functions of the form $G(x,\theta)$. This is the situation in multivariable identification. As an extreme case it may even happen that x does not depend on θ at all, as in the case of the adaptive filtering problems. Or there may be some equality constraints imposed on θ. However the analysis of these extensions does not differ in any significant degree from the analysis of the simplest case. The potential of Ljung's scheme in estimation and adaptive control have been described in details in [56]. We mention a few examples below for further reference.

Example 1. The ARMA estimation problem: Let (y_n) be a wide sense stationary ARMA (p^*, q^*) process satisfying the difference equation

$$A^* y = C^* e$$

where A^*, C^* are polynomials of the shift operator of degree p^* and q^* respectively. Let $D \subset \mathbb{R}^{p^*+q^*}$ denote a large set of θ's such that the corresponding polynomials $A(z^{-1})$ and $C(z^{-1})$ are stable and relative prime. We shall also use the notation $D_{p^*,q^*} = \{(p,q): \ p = p^*, q \geq q^* \quad \text{or} \quad q = q^*, p \geq p^*\}$. Let D^* and D_0 be a compact domain such that $\theta^* \in D^* \subset \text{int } D_0$ and $D_0 \subset D$. Here int D_0 denotes the interior of D_0. Let us take an arbitrary $\theta \in D_0$ and define an estimated prediction error process $\varepsilon = (\varepsilon_n(\theta)), n \geq 0$ by the equation

$$\varepsilon = (A/C)y$$

The asymptotic cost function is defined by

$$W(\theta) = \frac{1}{2} \mathrm{E} \varepsilon_n^2(\theta).$$

and its gradient is then obtained by

$$W_\theta(\theta) = \mathrm{E} \varepsilon_{\theta n}(\theta) \varepsilon_n(\theta).$$

It is easy to see that $W_\theta(\theta)$ is computable and $W_\theta(\theta^*) = 0$. We shall later use the notation $R^* = W_{\theta\theta}(\theta^*)$.

Example 2.. The general multivariable linear stochastic system: This is given by

$$y = H(\theta^*)e$$

where $H(\theta) = (H((\theta, e^{i\lambda}))$ is an $m \times m$ rational stable and inverse stable transfer function with leading term I defined on a set D_θ of p-dimensional parametervectors θ. Also we assume that $H(\theta)$ is a smooth (i.e. C^∞) function of θ. Let $e = (e_n)$ be an m-dimensional wide sense stationary process satisfying

$$\mathrm{E}(e_n | \mathcal{F}_{n-1}) = 0 \qquad E(e_n e_n^T | \mathcal{F}_{n-1}) = \Lambda^*$$

with $\Lambda^* > 0$ (i.e. Λ^* is positive definite).

To estimate θ^* and Λ^* let $\theta \in D_\theta, \Lambda \in D_\Lambda$ and define the process $\varepsilon(\theta)$ by

$$\varepsilon(\theta) = H^{-1}(\theta) y$$

and define the asymptotic cost function (assuming stationary) by

$$W(\theta,\Lambda) = \frac{1}{2}\mathrm{E}(\varepsilon_n^T(\theta)\Lambda^{-1}\varepsilon_n(\theta) + \frac{1}{2}\log\ \det\Lambda. \tag{3.3}$$

Its partial derivatives with respect to θ and Λ^{-1} are given by

$$W_\theta(\theta,\Lambda) = \mathrm{E}\varepsilon_{\theta n}^T(\theta)\Lambda^{-1}\varepsilon_n(\theta) \qquad W_{\Lambda^{-1}}(\theta,\Lambda) = \frac{1}{2}(-\mathrm{E}\varepsilon_n(\theta)\varepsilon_n^T(\theta) + \Lambda)$$

It is easy to see that $W_\theta(\theta,\Lambda)$ and $W_{\Lambda^{-1}}(\theta,\Lambda)$ are computable and

$$W_\theta(\theta^*,\Lambda^*) = 0 \qquad W_{\Lambda^{-1}}(\theta^*,\Lambda^*) = 0. \tag{3.4}$$

Example 3.. The performance enhancing adaptive control problem (c.f. [23]). Let us consider a linear stochastic control system described by

$$y^o(\theta^*) = H^{ou}(\theta^*)u + H^{oe}(\theta^*)e$$

where u is a control-signal. Associated with this system is a feedback law $u = -K(\theta^*)y$ which is obtained by solving e.g. a linear quadratic regulator problem. Thus the function K is a fixed and known function of the system parameters. For each test value θ we can compute an estimated input noise. Assume that the control signal is

$$u = -K(\theta^*) + v$$

where v is a "dither", which is assumed to be a wide sense stationary process with spectral density which bounded away from zero. Also v is orthogonal to e. For each test value θ we can compute an estimated input noise. For this purpose consider the closed loop system

$$y^c(\theta,\theta^*) = H^{ou}(\theta^*)(-K(\theta)y^c(\theta,\theta^*) + v) + H^{oe}(\theta^*)e$$

and write it as

$$y^c(\theta,\theta^*) = H^{cu}(\theta,\theta^*)v + H^{ce}(\theta^*)e.$$

Then estimate e by ε^c which is computed as

$$\varepsilon^c(\theta,\theta^*) = (H^{ce}(\theta))^{-1}(y^c(\theta,\theta^*) - H^{cu}(\theta,\theta)v).$$

Now we can in principle proceed as in the case of an open loop identification problem. Pretending $\Lambda = I$ we consider the criterion function (assuming stationarity)

$$G(\theta,\theta^*) = \frac{1}{2}\mathrm{E}\frac{\partial}{\partial\theta}|\varepsilon_n^c(\theta,\theta^*)|^2.$$

and minimize it with respect to θ.

At this point we face an interesting problem which shows the limitations of Ljung's scheme. Namely the gradient-process $(\partial/\partial\theta)\varepsilon_n^c(\theta,\theta^*)$ is not computable (c.f. [23]). However as it was pointed out in the quoted paper, that there exists a computable process $z_n(\theta,\theta^*)$ such that $z_n(\theta^*,\theta^*) = (\partial/\partial\theta)\varepsilon_n^c(\theta,\theta^*)|_{\theta=\theta^*}$. Hence the problem of performance enhacing adaptive control can also be put in the general framework of Ljung's scheme.

There are estimation problems which do not fit so smoothly into the general framework that has just been described. As an example we mention the identification problem of open loop control systems with noisy measurement on the input. (c.f. [12])

4. OFF-LINE METHODS. THE FIRST REPRESENTATION THEOREM

Although Ljung's scheme was originally designated to provide a general framework for recursive estimation methods, we shall start with off-line estimation methods based on Ljung's scheme. It should be noted though that the off-line estimation problem does not make sense in the case of an adaptive control problem, since then the tuning can not be performed without actually repeating the experiment with the physical system with exactly the same input noise sequence. Obviously this can not be done.

The problem now is to estimate the parameter θ^* from the observed data $Q(\overline{x}_n(\theta))$. Let us define the $\mathbb{R}^p$-valued empirical function

$$U_N(\theta) = \sum_{n=1}^{N} Q(\overline{x}_n(\theta)).$$

Then the off-line estimation $\widehat{\theta}_N$ of θ^* will be defined as "the solution" to the equation

$$U_N(\theta) = 0. \tag{4.1}$$

More exactly, we define $\widehat{\theta}_N$ as the solution to (4.1) if a unique solution exists, and otherwise we define $\widehat{\theta}_N$ arbitrarily, subject to the condition that $\widehat{\theta}_N$ on the whole must be a D-valued random variable. A useful, but not very well-known lemma is the following.

Lemma 4.1. For any $d > 0$ the equation (4.1) has a unique solution in D such that it is also in the sphere $\{|\theta - \theta^*| < d\}$ with probability at least $1 - O(N^{-s})$ for any $s > 0$ where the constant in the error term $O(N^{-s}) = CN^{-s}$ depends only on d and s.

A careful examination of a well-known linearization technique gives the following basic strong approximation theorem:

Theorem 4.2 (The first representation theorem) We have

$$\widehat{\theta}_N - \theta^* = -G_\theta^{-1}(\theta^*)\frac{1}{N}\sum_{n=1}^{N} Q(\overline{x}_n(\theta^*)) + O_M(N^{-1}). \tag{4.2}$$

This theorem translates all reasonable question on the behaviour of the estimator process into an analogous question on L-mixing processes which in some cases are even martingale-difference processes (c.f. the examples above). Thus e.g. central limit theorem (c.f. [55]), laws of iterated logarithms, invariance principles (c.f. e.g. [45]) can all be derived from the corresponding results for L-mixing processes.

In addition new asymptotic results can also be derived which could not be obtained by other methods. E.g. we get a positive answer to Rissanen's conjecture on the validity of the so-called tail condition for Gaussian linear stochastic systems (Example 2.), which can be formulated in a milder form as follows:

Theorem 4.3. Let $\widehat{\theta}_N$ denote the off-line estimator for Example 2.. Assume that (e_n) is a Gaussian white-noise sequence. Then for any $c > 0$ we have the inequality

$$\sum_{n=1}^{\infty} P(N^{1/2}|\widehat{\theta}_N - \theta^*| > c\log N) < \infty. \tag{4.3}$$

Similarly the validity of the tail condition can be derived for $\widehat{\Lambda}_N - \Lambda^*$. Using this result the Rissanen-Shannon inequality (c.f. [62]) is applicable to finite dimenstional stationary Gaussian systems. Thus a lower bound for the mean cumulated prediction error can be obtained, which reflects the cost of parameter uncertainty (c.f. [34] for the scalar case).

5. RECURSIVE METHODS. RATE OF CONVERGENCE

The recursive estimator associated with Ljung's scheme is defined as follows. Let $\widehat{\widehat{\theta}}_0 = \xi$ be an initial estimate of θ^*, and then set

$$\widehat{\widehat{\theta}}_{n+1} = \widehat{\widehat{\theta}}_n + \frac{1}{n+1}Q(x_{n+1}) \tag{5.1}$$

$$x_{n+1} = A(\widehat{\widehat{\theta}}_n)x_n + B(\widehat{\widehat{\theta}}_n)e_n, \qquad x_0 = 0. \tag{5.2}$$

It has been shown in [56], that most of the well-known recursive estimation methods can be described by this scheme. However an important practical and theoretical aspect has not been very well addressed, namely the problem of keeping the estimator in a prescribed bounded "domain of stability". It is well-known that the algorithm given above may not converge without some form of resetting. It turns out that the analysis of the effect of resetting is the most difficult part of the convergence proof. (c.f. [22])

Let $D_0 \subset \text{int}D$ denote a compact domain such that $\theta^* \in \text{int}D_0$. To keep $(\widehat{\widehat{\theta}}_n)$ in D_0 we use a resetting mechanism: if $\widehat{\widehat{\theta}}_{n+1}$ leaves D_0 we redefine it to be $\widehat{\widehat{\theta}}_0$. To formalize this procedure let $\widehat{\widehat{\theta}}_{n+1-}$ denote the tentative value of θ computed at time $n+1$ by (5.1) and let $\widehat{\widehat{\theta}}_{n+1}$ be the actual value . The actual value will be the tentative value if this latter does not leave D_0, otherwise it will be reset to the initial value $\widehat{\widehat{\theta}}_0$. Let B_{n+1} denote the event that a resetting had to be done at time $n+1$, i.e. $B_{n+1} = \{\omega : \widehat{\widehat{\theta}}_{n+1-} \notin D_0\}$. Then (5.1) will be replaced by

$$\widehat{\widehat{\theta}}_{n+1} = \widehat{\widehat{\theta}}_n + (1 - \chi_{B_{n+1}})\frac{1}{n+1}Q(x_{n+1}) + \chi_{B_{n+1}}(\widehat{\widehat{\theta}}_0 - \widehat{\widehat{\theta}}_n) \qquad (5.1)'$$

To ensure the exponential stability of (5.2) we assume that the family of $r \times r$ matrices $A(\theta), \theta \in D \subset \mathbb{R}^p$ are jointly stable in the following sense: there exists a positive definite $n \times n$ matrix V such that $A^T(\theta)VA(\theta) \leq \lambda V$ with some $0 < \lambda < 1$ for all $\theta \in D$.

The analysis of the recursive estimation scheme is based on a sophisticated averaging technique in the result of which we finally can conclude that the estimator sequence $\widehat{\widehat{\theta}}_n$ closely follows the solution trajectories of the so-called associated differential equation, assuming that the latter satisfies certain stability conditions (c.f. also [6], [7], [10], [16], [54], [67]).

The associated differential equation will be defined as

$$\dot{y}_t = \frac{1}{t}G(y_t) \qquad y_s = \xi \qquad (5.3)$$

Under the conditions above (5.3) has a unique solution in $[s, \infty)$ which we denote by $y(t, s, \xi)$. It is well-known, that $y(t, s, \xi)$ is a continuously differentiable function of (t, s, ξ).

Condition 5.1 For every $\xi \in D_0, t > s > 0$ $y(t, s, \xi) \in D$ is defined and we have with some $C_0, \alpha > 0$

$$\|(\partial/\partial\xi)y(t, s, \xi)\| \leq C_0(s/t)^\alpha. \qquad (5.4)$$

Here $\|\cdot\|$ denotes the operator norm of a matrix. For the recursive estimator we need additional assumptions on the noise:

Definition 5.1. Let $(u_n), n \geq 0$ be a real-valued stochastic process. We say that (u_n) is in class M^* if for some $\varepsilon > 0$

$$\sup_n \mathrm{E} \exp \varepsilon |u_n| < \infty.$$

Let us remember that we defined for any positive integer τ and $q \geq 1$

$$\gamma_q(\tau, e) = \sup_{n \geq \tau} \mathrm{E}^{1/q} |e_n - (e_n | \mathcal{F}^+_{n-\tau})|^q.$$

It has been shown in [20], that if (u_n) is an L-mixing process then $\gamma_q(\tau, u) \leq 16\Gamma_q(u)/\tau$ for all $q \geq 1$ and $\tau \geq 1$ integer. We shall need a slightly stronger condition to be satisfied by (e_n).

Definition 5.2. We say that a stochastic process is L^+-mixing if it is L-mixing and for all $\tau \geq 1$ integer and $q \geq 1$ with some $c > 0$ $\gamma_q(\tau, e) = O(\tau^{-1-c})$.

We shall assume in this section that (e_n) is a wide sense stationary process in the class M^* and that it is also L^+-mixing. A main result of [22] is the following theorem:

Theorem 5.1 Assume that the differential equation (5.3) satisfies the stability condition (5.4) with $\alpha > 1/2$ and $\widehat{\widehat{\theta}}_0$ a sufficiently good initial approximation. Then we have $\widehat{\widehat{\theta}}_n - \theta^* = O_M(n^{-1/2})$. Moreover the following stronger result also holds: let $1 < q < \infty$ and

$$\widetilde{\theta}^*_n = \sup_{n \leq m \leq qn} |\widehat{\widehat{\theta}}_m - \theta^*|$$

then $\widetilde{\theta}^*_n = O_M(n^{-\alpha+\varepsilon})$ for any $\varepsilon > 0$. For $\alpha < 1/2$ we have for any $\varepsilon > 0$ $\widetilde{\theta}^*_n = O_M(n^{-\alpha+\varepsilon})$.

One of the best known recursive estimation methods is the recursive prediction error (RPE) method for ARMA processes. (c.f. [65], [56], [8], or [66]). It is given by first considering the following recursive steps to generate new, tentative values of the parameter estimator:

$$\widehat{\widehat{\theta}}_{N-} = \widehat{\widehat{\theta}}_{N-1} - \frac{1}{N} \widehat{\widehat{R}}_{N-1}^{-1} \frac{\partial}{\partial \theta} \varepsilon_N \cdot \varepsilon_N. \tag{5.5}$$

$$\widehat{\widehat{R}}_{N-} = \widehat{\widehat{R}}_{N-1} + \frac{1}{N}((\frac{\partial}{\partial \theta}\varepsilon_N)(\frac{\partial}{\partial \theta}\varepsilon_N)^T - \widehat{\widehat{R}}_{N-1}) \tag{5.6}$$

These tentative values will be reset to their initial value if they violate certain boundedness conditions, i.e. if $\widehat{\widehat{\theta}}_N \notin D_\theta$ or $\widehat{\widehat{R}}_N \notin D_R$ where D_θ and D_R are prescribed compact domains. The associated ordinary differential equation is

$$\dot{\theta}_t = -R_t^{-1}\frac{\partial}{\partial\theta}W(\theta_t) \qquad \dot{R}_t = S(\theta_t) - R_t. \tag{5.7}$$

where $S(\theta) = \mathbf{E}\overline{\varepsilon}_\theta(\theta)\varepsilon_\theta^T(\theta)$. The right hand side of this ordinary differential equation is defined in $D \times \mathbf{R}^+(p \times p)$ where $\mathbf{R}^+(p \times p)$ denotes the set of symmetric positive definite $p \times p$ matrices. It is well known that (5.7) has a unique stationary point (θ^*, R^*) and that this equilibrium point is asymptotically stable. We can even take $\alpha = 1 - c$ with any $c > 0$. Thus Theorem 5.1 implies

$$\widehat{\widehat{\theta}}_N - \theta^* = O_M(N^{-1/2}) \quad \text{and} \quad \widehat{\widehat{R}}_N - R^* = O_M(N^{-1/2}) \tag{5.8}$$

Theorem 5.1 has recently been extended to the LMS algorithm of the theory of adaptive filtering. In that algorithm no resetting is used but the special structure of the problem can be succesfully exploited. (c.f. [33])

6. OFF-LINE VERSUS ON-LINE. THE SECOND REPRESENTATION THEOREM

The heuristic derivation of the RPE method was initially obtained by considering an approximate recursion for the solution of the likelihood equation.This background has since been almost forgotten, although it indicates that there should be an intimate relation between the nonrecursive or off-line and the recursive or on-line maximum-likelihood estimator. This conviction has also been formulated by Tusnády in the following conjecture: If $\widehat{\widehat{\theta}}_N$ denotes the recursive prediction error estimator of an ARMA process and $\widehat{\theta}_N$ denotes the off-line estimator, then $\widehat{\widehat{\theta}}_N - \widehat{\theta}_N = o(N^{-1/2})$ in a sense still to be specified. A main result of our theory of recursive estimation is the validity of this conjecture for a fairly general class of ARMA-processes (c.f. [19]).

Theorem 6.1. Under the conditions of the previous section we have for ARMA processes

$$\widehat{\widehat{\theta}}_N - \widehat{\theta}_N = O_M(\log N/N).$$

Since the off-line estimator has been very accurately characterized in Theorem 4.2 specializing it to the ARMA case we have a similar result for the on-line estimator, namely:

Theorem 6.2. Under the conditions above we have

$$\widehat{\widehat{\theta}}_N - \theta^* = -(R^*)^{-1}\frac{1}{N}\sum_{n=1}^{N}\bar{\varepsilon}_{\theta n}(\theta^*)e_n + O_M(\log N/N).$$

The proof of Theorem 6.1 is based on showing that

$$V_N(\widehat{\widehat{\theta}}_N) - V_{N-1}(\widehat{\widehat{\theta}}_N) = O_M(N^{-1}) \tag{6.1}$$

hence it follows that $V_N(\theta_N) = O_M(\log N)$, from which the theorem follows by a standard linearization technique.

The essential point in the proof of (6.1) is the fact that the recursive prediction error method is a quasi-Newton method with respect to θ. In general a recursive scheme will be called quasi-Newton if $G_\theta(\theta^*) = -I$. Note however that the recursive prediction error method for ARMA processes is not fully quasi-Newton in the latter sense since we have

$$G_{\theta,R}(\theta^*, R^*) = \begin{pmatrix} -I & 0 \\ X & -I \end{pmatrix}.$$

Thus the general result below does not imply Theorem 6.1.

Theorem 6.3 Let $\widehat{\theta}_N$ denote the off-line estimator for the general estimation scheme and let $\widehat{\widehat{\theta}}_N$ denote the corresponding recursive estimator. Assume that the estimation scheme is quasi-Newton. Then

$$\widehat{\widehat{\theta}}_N - \widehat{\theta}_N = O_M(\log N/N).$$

The proof of this beautiful theorem is up to a certain point analogous to the proof of Theorem 6.1. However in the general case the martingale structure of the problem is lost and we need to develop a new deep tool to complete the proof. This new tool is a representation theorem for recursive estimators which is a basic theorem of this theory of recursive estimation, providing a new, more comprehensive meaning to the terminology ODE method:

Theorem 6.4 (The second representatin theorem) Assume that D_0 is starlike. Then for some $\varepsilon > 0$

$$\widehat{\widehat{\theta}}_N - \theta^* = \sum_{n=1}^{N}\frac{\partial y}{\partial \xi}(N, n, \theta^*)\cdot\frac{1}{n}Q(\overline{x}_n(\theta^*)) + O_M(N^{-1/2-\varepsilon}).$$

Since the matrix $(\frac{\partial}{\partial \xi})y(N,n,0)$ is the sensitivity matrix which indicates the relative effect of a perturbation at time n onto the solution of (5.3) at time N, this representation theorem has a nice heuristic appeal. Also it can be used to derive some of the classical results of the theory of stochastic approximation for recursive estimators generated by Ljung's scheme. Also a slightly weaker form of Theorem 6.3 can be directly derived from Theorem 6.4.

7. ABOUT THE METHODS

The proof of Theorem 5.1 is carried out in the following major steps. First let us "freeze" the parameter at some value θ and consider the random field

$$H(n,\theta,\omega) = Q(\overline{x}_n(\theta))$$

where $\overline{x}_n(\theta)$ is defined by (3.2). Imbedding the discrete sequence $H(n,\theta,\omega)$ into a piece-wise constant but continuous time "sequence" we write (5.1)′ as a random differential equation of the form

$$\dot{\widehat{\theta}}_t = \frac{1}{t}\left(H(t,\widehat{\theta}_t,\omega) + \delta\widetilde{H}(t,\omega)\right) \tag{7.1}$$

where $H = (H(t,\theta,\omega))$ is a random field defined in $[1,\infty) \times D$.

Now we consider the "frozen parameter" problem i.e. we set $\delta\widetilde{H}(t,\omega) = 0$ for all t, and consider the differential equation

$$\dot{\theta}_t = \frac{1}{t}H(t,\theta_t,\omega). \tag{7.1$'$}$$

Let $s \geq 1$, and $s \leq \sigma < qs$. Let $\tau(\sigma)$ denote the first moment after σ, at which θ_t "hits" ∂D_0. Further, let $\overline{y}_t$ denote the solution of (5.3) with initial condition $\overline{y}_\sigma = \theta_\sigma$ i.e. $\overline{y}_t = y(t,\sigma,\theta_\sigma)$. A main technical tool used in the proof is the development of an upper bound for the increments $|\widehat{\theta}_t - \overline{y}_t|$.

Lemma 7.1 Let

$$I_\sigma(q) = \sup_{\sigma \leq t < \tau(\sigma)\wedge q\sigma} |\theta_t - \overline{y}_t|.$$

Then

$$I_s^*(q) \triangleq \sup_{s \leq \sigma \leq qs} I_\sigma(q) = O_M(s^{-1/2}).$$

The proof of Lemma 7.1 is based on the analysis of the random variable

$$\delta_s^*(q) = \sup_{\substack{s \leq \sigma \leq qs \\ \sigma \leq t \leq qs \\ \theta \in D_0}} \left| \int_\sigma^t \frac{1}{r} \overline{H}(r, y(r, \sigma, \theta), \omega) dr \right|. \tag{7.2}$$

It has to be shown that $\delta_s^*(q) = O_M(s^{-1/2})$. The main difficulty of the proof is handling supremum over σ over a set of dilating intervals. Assuming the validity of Lemma 7.1 we can estimate the probability of resetting. Let C_s denote the event θ_t hits ∂D_0 in the interval $[s, qs)$. Then it can be shown that we have for any $m \geq 1$

$$P(C_s) = O(s^{-m}). \tag{7.3}$$

Combining this inequality with the inequality given by Lemma 7.1 we get

Lemma 7.2 Let K denote the diameter of D_0. Then we have

$$\sup_{s \leq t \leq qs} |\theta_t - \overline{y}_t| \leq I_s^*(q) + K\chi_{C_s} = O_M(s^{-1/2}). \tag{7.4}$$

Now it is easy to see that the "frozen parameter" analysis is not affected by adding a perturbation term $\delta\widetilde{H}(t, \omega)$ to $H(t, \widehat{\widehat{\theta}}_t, \omega)$ in (7.1)' provided the following condition holds (where now $\widehat{\widehat{\theta}}_t$ denotes the solution of (7.1)):

Condition 7.1. If $\tau(\sigma)$ denotes the first moment after σ when $\widehat{\widehat{\theta}}_t$ hits ∂D_0 and $q > 1$ is a fixed real number then for some $\varepsilon > 0$

$$\sup_{s \leq \sigma \leq qs} \int_\sigma^{\tau(\sigma) \wedge q\sigma} \frac{1}{r} |\delta\widetilde{H}(r, \omega)| dr = O_M(s^{-1/2-\varepsilon}). \tag{7.6}$$

Unfortunately in the analysis of Ljung's scheme the validity of this condition cannot a priori be derived, and its verification requires some sophisticated techniques of the theory of extremal processes. (c.f. GA)

8. STOCHASTIC COMPLEXITY

A predictive form of stochastic complexity has been introduced in [62] and by now has become one of the most important concept in the statistical analysis of linear stochstic systems. The concept of predictive stochastic complexity is a particularly useful concept in the context of dynamical systems since it is inherently "on-line" unlike other criteria such as AIC or BIC, (c.f. [1]) which are inherently off-line. For a more extended survey see [35]. It was originally meant to be a device for solving model selection problems but has since outgrown of this framework. Thus e.g. in adaptive filtering the effect of parameter uncertainty on prediction is a central issue. Similarly in adaptive control the effect of parameter uncertainty on performance will likely to be a central issue in the years to come.

To put the problem into a general setting let us consider a real-valued quadratic function $S = S(x)$ and consider the index of ideal performance $S(\overline{x}_n(\theta^*))$ at time n. We assume that

$$W(\theta) = \mathrm{E}S(\overline{x}_n(\theta^*))$$

is minimized at $\theta = \theta^*$ and that

$$T^* = \frac{\partial^2}{\partial\theta^2} W(\theta) \Big|_{\theta=\theta^*}$$

is positive definite.

For example we may consider a self-tuning regulator for an ARMAX system, where the function $G(\theta)$ consists of cross-correlations, and the function $S(x)$ is the square of the output (c.f [3]). We can now ask the classical question in the theory of stochastic complexity: how much do we lose in performance if we substitute θ^* by its estimate $\widehat{\theta}_{n-1}$?

Assume that (e_n) is stationary and consider the random variable

$$\eta_N = \frac{1}{N^{1/2}} \sum_{n=1}^{N} Q(\overline{x}_n(\theta^*)), \tag{8.1}$$

which has a normal limit distribution, say $\mathcal{N}(0, P^*)$, when N tends to infinity. (This follows from a general CLT for L-mixing processes proved in L+ C.) Moreover, it follows from the cited works that with $(R^*)^{-1} = G_\theta^{-1} P^* (G_\theta^T)^{-1}$ we have $N^{-1/2}(\widehat{\theta}_N - \theta^*) \xrightarrow{\mathcal{L}} \mathcal{N}(0, (R^*)^{-1})$. Now the arguments of [20] yield the following result. This theorem is a considerable extension of Davisson's formula. (c.f. [11])

Theorem 8.1 Under the conditions of Section 4 we have

$$\mathrm{E}(S(\overline{x}_n(\widehat{\theta}_{n-1})) - S(\overline{x}_n(\theta^*))) = \frac{1}{N}\mathrm{Tr}(T^*(R^*)^{-1})(1 + o(1)).$$

This theorem is a considerable extension of Davisson's formula (c.f. [11]). A much more difficult task is to establish the pathwise properties of the cost of parameter uncertainty, but again the arguments of [20] can be generalized. However in this case the martingale structure is lost and we have to rely on the fairly sophisticated result given in Theorem 8.2. Thus we get

Theorem 8.2 We have almost surely

$$\lim_{N\to\infty} \sum_{n=1}^{N} S(\overline{x}_n(\widehat{\theta}_{n-1})) - S(\overline{x}_n(\theta^*))/\log N = \mathrm{Tr}T^*(R^*)^{-1}.$$

Although this theorem is a substantial generalization of the result presented in Gerencsér [20], the methods of proof have been laid down in that earlier paper. A standard objection which can be raised against this result is that $S_n(\overline{x}_n(\widehat{\theta}_{n-1}))$ is very difficult to compute. A natural alternative would be to use $S_n(x_n))$ instead, where x_n is generated by the recursive estimator scheme. Let $N^{1/2}(\widehat{\widehat{\theta}}_N - \theta^*) \to \mathcal{N}(0,(R^{**})^{-1})$. (Note that in general $R^{**} \neq R^*$). Then we have the following result.

Theorem 8.3. Under the conditons of Section 5 we have

$$\lim_{N\to\infty} \sum_{n=1}^{N} (S(x_n) - S(x_n(\theta^*))/\log N = \mathrm{Tr}T^*(R^{**})^{-1} \quad \text{a.s.}$$

As a special case we mention the earlier result for the ARMA case: This result extends the results of [42], [43], [50], [60], [61].

Theorem 8.4. Let us consider a minimal ARMA(p^*,q^*) process and let ε_n be the prediction error generated by the recursive prediction error method, assuming that the model we fit is ARMA(p,q) with $(p,q) \in D_{p^*,q^*}$. Then under the conditions of Section 5 we have

$$\lim_{N\to\infty} \sum_{n=1}^{N} (\varepsilon_n^2 - e_n^2)/\sigma^2 \log N = (p+q) \qquad \text{a.s.}$$

A detailed exposition of the history of this group of problems and the potentials of Theorem 4.4 is given in [35].

9. ESTIMATION IN CONTINUOUS-TIME

Physical systems are governed by differential equations and therefore are inherently continuous-time systems. However in digital control we use sampled data and very often the resulting discrete time system is considered the "true sytem". This approach can cause certain problems. E.g. recent research in microrobotics demonstrated that the traditional approach does not yield the required high accuracy (c.f. [46]). Another example for the breakdown of the traditional approach is the problem of model selections. Assuming that the continuous-time data are generated by a continuous-time AR(p) process, the parameter estimation is fairly straightforward with today's techniques at least in the off-line sense. However if we discretize our system then the AR stucture is lost and the resulting discrete time system is an ARMA system. Thus the order determination becomes much more difficult. These two examples show that a profound statistical analysis of continuous-time systems is needed. Ultimately we shall end up using sampled data just like in the traditional approach. But the interpretation and "parametrization" of these data will be radically different.

First we summarize the continuous-time counterpart of some of the basic concepts used before. Let $D \subset \mathbb{R}^p$ be an open domain and let the stochastic process $(x_t(\theta))$ be defined on the parameter set $\mathbb{R}^+ \times D$, we here $\mathbb{R}^+ = \{t : t \geq 0\}$. We say that $(x_t(\theta))$ is M-bounded if for all $1 \leq q < \infty$

$$M_q(x) = \sup_{\substack{t \geq 0 \\ \theta \in D}} \mathrm{E}^{1/q} |x_t(\theta)|^q < \infty.$$

Let a probability space $(\Omega, \mathcal{F}, P)$ be given together with a pair of families of σ-algebras $(\mathcal{F}_t, \mathcal{F}_t^+)$ such that (i) $\mathcal{F}_t \subset \mathcal{F}$ is monotone increasing (ii) $\mathcal{F}_t^+ \subset \mathcal{F}$ is monotone decreasing and $\mathcal{F}_t^+$ is right continuous in t i.e. $\mathcal{F}_s^+ = \sigma\{\bigcup_{0<\varepsilon} \mathcal{F}_{s+\varepsilon}^+\}$ (iii) $\mathcal{F}_t$ and $\mathcal{F}_t^+$ are independent for all t. For $s < 0$ we set $\mathcal{F}_s^+ = \mathcal{F}_0^+$.

Definition 9.1 A stochastic process $(x_t(\theta)), t \geq 0, \theta \in D$ is L-mixing with respect to $(\mathcal{F}_t, \mathcal{F}_t^+)$ uniformly in $\theta \in D$ if it is $\mathcal{F}_t$ progressively measurable, M-bounded and with

$$\gamma_q(\tau, x) = \gamma_q(\tau) = \sup_{\substack{t \geq \tau \\ \theta \in D}} \mathrm{E}^{1/q} |x_t(\theta) - \mathrm{E}(x_t(\theta)|\mathcal{F}_{t-\tau}^+)|^q \qquad \tau \geq 0$$

we have

$$\Gamma_q = \Gamma_q(x) = \int_0^\infty \gamma_q(\tau) d\tau < \infty.$$

The continuous-time version of Ljung's scheme can be described as follows. Let $(w_t, \mathcal{F}_t)$ be an $\mathbb{R}$-valued standard Wiener-process defined over some probability space $(\Omega, \mathcal{F}, P)$ and let $(\overline{x}_t(\theta))$ be an stationary process defined by

$$d\overline{x}_t(\theta) = A(\theta)\overline{x}_t(\theta)dt + B(\theta)dw_t. \tag{9.1}$$

(In practice the initial condition $\overline{x}_0(\theta)$ is an arbitrary M-bounded and $\mathcal{F}_0$-measurable random variable).

Assume that the matrices $A(\theta), \theta \in D$, where D is some compact domain, are stable. It is also assumed that A, B are twice continuously differentiable in θ for $\theta \in \text{int}D$. Let $Q(x)$ and $S(x)$ be quadratic fuctions of x of appropriate dimension and assume that the available data for fixed θ are given by the following stochastic differential:

$$Q(\overline{x}_t(\theta))dt + S(\overline{x}_t(\theta))dw_t.$$

and set

$$G(\theta) = \mathrm{E}Q(\overline{x}_n(\theta)).$$

It is easy to see that $G(\theta)$ is well-defined and is twice continuously differentiable in int$D$.

As an example let us consider a linear time-invariant stochastic system given by the state space equations for $-\infty < t < \infty$

$$d\overline{x}_t(\theta^*) = A(\theta^*)\overline{x}_t(\theta^*)dt + K(\theta^*)dw_t \tag{9.2}$$

$$d\overline{y}_t(\theta^*) = C(\theta^*)\overline{x}_t(\theta^*)dt + dw_t \tag{9.3}$$

where dw_t is a standard Guassian white noise process in $\mathbb{R}^m$. The $\mathbb{R}^m$-valued process $(\overline{y}_t(\theta^*))$ is called the output process, while the $\mathbb{R}^n$-valued process $(\overline{x}_t(\theta^*))$ is the state-vector-process.

Assume that the matrices $A(\theta^*), K(\theta^*), C(\theta^*)$ are defined for $\theta^* \in D \subset \mathbb{R}^p$ where D is an open domain and they are C^∞ function of θ^*. Moreover $A(\theta^*)$ and $A(\theta^*) - K(\theta^*)C(\theta^*)$ are strictly stable for $\theta^* \in D$. It follows that $\overline{y}_t$ is a (second order) stationary process the innovation process of which dw_t.

To estimate θ^* on the basis of the observation process $\overline{y}_t$ we proceed as follows: fix a $\theta \in D$ and invert the system (9.2), (9.3) assuming that $\theta = \theta^*$ to get dw_t. The inverse system is given by the equations:

$$d\overline{x}_t(\theta,\theta^*) = (A(\theta) - K(\theta)C(\theta))\overline{x}_t(\theta,\theta^*)dt + K(\theta)d\overline{y}_t(\theta^*). \tag{9.4}$$

$$d\overline{\epsilon}_t(\theta,\theta^*) = d\overline{y}_t(\theta^*) - C(\theta)\overline{x}_t(\theta,\theta^*)dt. \tag{9.5}$$

In practice we set zero initial state, i.e. $\overline{x}_0(\theta,\theta^*) = 0$.

The cost function $V_T(\theta,\theta^*)$ can be written as (c.f. [2], [36], [51])

$$V_T(\theta,\theta^*) = \int_0^T \frac{1}{2}\frac{1}{dt}(|d\overline{\epsilon}_t(\theta,\theta^*)|^2 - |dw_t|^2). \tag{9.6}$$

If we work out the expression following the integral sign, it is easily seen that it is actually a one-form of the infinitesimal generators dt and dw_t. Differentiation with respect to t will be denoted by $\cdot$, and differentiation with respect to θ will be denoted by the subscript θ. Thus we get

$$V_{\theta T}(\theta, \theta^*) = \int_0^T \dot{\overline{\varepsilon}}_\theta(\theta, \theta^*) d\overline{\varepsilon}_t(\theta, \theta^*).$$

We set $R^* = \mathrm{E} V_{\theta\theta T}(\theta^*, \theta^*)/T$ and assume that R^* is nonsingular.

As an example for the validity of the "uniqueness" condition (c.f. text after (3.2)), assume that in the system equation the unknown parameters are the entries of K, and the system matrices A and C are known. In this case $\mathrm{E} V_\theta(\theta) = 0$ has a single solution in a reasonable domain (c.f. [68]). Now the continuous-time version fo Theorem 4.2 can be stated as follows:

Theorem 9.1. Under the conditions above we have

$$\widehat{\theta}_T - \theta^* = -G_\theta^{-1}(\theta^*) \frac{1}{T} \int_0^T (Q(\overline{x}_t(\theta^*))dt + S(\overline{x}_t(\theta^*))dw_t) + O_M(T^{-1})$$

The proof is a continuous-time extension of [21] and is given in [36] for the case of the identification problem of multivariable linear systems when we get the following:

Theorem 9.2. Under the conditions above we have with

$$\widehat{\theta}_T - \theta^* \;=\; -(R^*)^{-1} \frac{1}{T} \int_0^T \dot{\overline{\varepsilon}}_{\theta t}(\theta^*, \theta^*) dw_t \;+\; O_M(T^{-1}).$$

Using this theorem and another result on the strong approximation of multidimensional integrals (c.f. [24]) we get the following result in (c.f. [25]).

Theorem 9.3. Let $\widehat{\theta}_T$ be the off-line estimator of Theorem 9.2, and let

$$\overline{\widehat{\theta}}_T = \frac{1}{T} \int_0^T \widehat{\theta}_t dt.$$

Assume that the underlying probability space is sufficiently rich. Then under the conditions above we have for every $\varepsilon > 0$

$$\overline{\widehat{\theta}}_T - \widehat{\theta}_T = \frac{1}{T} (R^*)^{-1/2} \overline{w}_T + O_M(T^{-3/5+\varepsilon})$$

where $(\overline{w}_T)$ is a standard Wiener-process in $\mathbb{R}^p$.

One of the important techniques in the analysis of continuous-time systems is the following result (c.f. [18] and also [4])

Lemma 9.4. Let $u_t(\theta)$ be a progressively measurable measurable $\mathbf{R}^p$-valued M-bounded stochastic process. Assume that $\Delta u/\Delta\theta$ is also M-bounded. Let

$$x_t(\theta) = \frac{1}{\sqrt{t}} \int_0^t u_s(\theta) dw_s$$

Then

$$x_t^* = \sup_{\theta \epsilon D_0} |x_t(\theta)| = O_M(1).$$

The stochastic complexity issue can also be raised in the context of continuous-time systems. The continuous-time version of Theorem 8.2 is the following (c.f. [5], [29], [31])

Theorem 9.5 Under the conditions above we have with $p = \dim\theta$:

$$\lim_{T\to\infty} \int_0^T \frac{1}{dt}(|d\varepsilon_t(\widehat{\theta}_t, \theta^*)|^2 - |dw_t|^2)/\log T = p$$

almost surely.

Let us now turn to the harder problem of recursive estimation. The continuous-time recursive estimator is defined by the system of stochastic differential equations

$$dx_t = \frac{1}{t}(Q(\phi_t)dt + S(\phi_t)dw_t) \qquad x_1 = \xi \tag{9.7}$$

$$d\phi_t = A(x_t)\phi_t dt + B(x_t)dw_t \qquad \phi_1 = 0. \tag{9.8}$$

The "frozen parameter" counterpart can then be defined as

$$d\overline{x}_t = \frac{1}{t}(Q(\overline{\phi}_t(\overline{x}_t))dt + S(\overline{\phi}_t(\overline{x}_t))dw_t). \tag{9.9}$$

Unfortunately the analysis of the continuous-time recursion has not yet been completed. A complete analysis of an analogous algorithm has been given in [69] (c.f. also [32]). However the methods that have been applied for the proof of Theorem 5.1 can be applied to analyze the associated "frozen system".

10. FIXED GAIN ESTIMATION

Much of what has been said can be objected by the practising control engineer on the ground that in a real adaptive control problem we should in one way or another forget past data. This objection is justified but we should add: the analysis of more orthodox problems may reveal (and has revealed) many important structures which may be useful in the analysis of more realistic problems. This will be apparent from the theorems below.

In this section the systems are still time-invariant but we use an estimator with fixed gain. In the off-line case the terminology "fixed gain" indicates that each new observation gets the same weight in the recursive computation of the cost function

The off-line estimator of θ^*, say $\widehat{\theta}_n^\lambda$ is obtained as "the solution" of the nonlinear algebraic equation

$$U_N(\theta) \triangleq \sum_{n=1}^{N}(1-\lambda)^{N-n}\lambda Q(\overline{x}_n(\theta)) = 0 \tag{10.1}$$

where $0 < \lambda < 1$ is the forgetting factor. (In the limiting case $\lambda \searrow 0$ the empircal "index of performance" $U_N(\theta)$ becomes the arithmetic mean of the terms $Q(\overline{x}_n(\theta))$, i.e. we get the usual procedure of off-line estimation for time-invariant systems). By "the solution of (10.1)" we mean a D_θ-valued random variable $\widehat{\theta}_N$ which is equal to the unique solution of (10.1) in D_θ if a D_θ-valued unique solution exists. Such a random variable exists by the measurable selection theorem.

Lemma 10.1. For any $d > 0$ and $s > 0$ the equation (10.1) has a unique solution in D with the property that it is also in the sphere $\{|\theta - \theta^*| < d\}$ with probability at least $1 - O(\lambda^s)$ where the constant in the error term $O(\lambda^s) = C\lambda^s$ depends only on d and s.

The extension of Theorem 4.2 is then the following: (c.f. [27])

Theorem 10.2 (The third representation theorem) We have with for small λ's

$$\theta_N^\lambda - \theta^* = -G_\theta^{-1}(\theta^*)\sum_{n=1}^{N}(1-\lambda)^{N-n}\lambda Q(\overline{x}_n(\theta^*)) + O_M(\lambda).$$

It is easy to see that the first term on the right hand side is of order $O_M(\lambda^{1/2})$. Also it is easy to see that this dominant term is an L-mixing process, so we may expect that $\theta_n^\lambda - \theta^*$ itself will be an L-mixing process. However, since θ_n^λ is not uniquely defined on a set of nonvanishing probability, some precaution is needed. It can be shown that the solutions $\widehat{\theta}_N^\lambda$ are uniquely defined, except for a set

$$B_N = \{\omega : \delta_N \geq c > 0\} \tag{10.2}$$

where (δ_N) is an appropriately defined L-mixing process, and $\delta_N = O_M(\lambda^{1/2})$. However, $\chi_{B_N}(\omega) = \chi_{x>c}(\delta_n)$, where $\chi_{B_N}(\omega)$ is the characteristic function of B_N, is not necessarily L-mixing.

Definition 10.1 A stochastic process (x_n), $n \geq 0$ is L_0-mixing with respect to $(\mathcal{F}_n, \mathcal{F}_n^+)$, if it is L-mixing and for any $q \geq 1, m > 1, \gamma_q(\tau, x) = O(\tau^{-m})$.

A very useful device in the analysis of the exceptional sets is the following:

Lemma 10.3 Let (x_n), $n \geq 0$ be an L_0-mixing process, and let $I \subset \mathbb{R}$ be a fixed nonempty open interval. Then there exist a sequence of real numbers $\delta_n \in I$ such that the process $y_n = \chi_{x>\delta_n}(x_n)$ is L_0-mixing,

Using the above result we get the following theorem:

Theorem 10.4 If (e_n) is L_0-mixing then there exists an L_0 -mixing version of $\widehat{\theta}_N^\lambda$ say $\widehat{\theta}_N^{\lambda 0}$ and for any other solution $\widehat{\theta}_N^\lambda$ we have $|\widehat{\theta}_N^\lambda - \widehat{\theta}_N^{\lambda 0}| \leq r_N$ where r_N is an L_0-mixing process and $r_N = O(\lambda^m)$ for any $m \geq 1$.

This characterization of the estimator process is very useful in establishing pathwise properties of certain predictive stochastic complexities. (c.f. below).

Now we turn to the problem of recursive estimation. The recursive fixed gain estimator of θ^* is described by the following equations

$$\widehat{\widehat{\theta}}_{n+1}^\lambda = \widehat{\widehat{\theta}}_n^\lambda + \lambda Q(x_{n+1}) \tag{10.3}$$

$$x_{n+1} = A(\widehat{\widehat{\theta}}_n^\lambda)x_n + B(\widehat{\widehat{\theta}}_n^\lambda)e_n, \qquad x_0 = 0. \tag{10.4}$$

Although this scheme has been used for quite a while, the "asymptotic properties" have apparently not been investigated. Our interest in this method has been raised by our research on change point detection methods based on stochastic complexity. (c.f. B+G and [30]). The presently best available result is a fixed gain counterpart of Theorem 6.4. In this theorem $y^\lambda(t, s, \xi)$ denotes the general solution of the associated ODE which is defined in this case as

$$\dot{y}_t = \lambda G(y_t).$$

Theorem 10.5 (The fourth representation theorem) Let (e_n) be a bounded process, i.e. let $M_\infty(e) < \infty$ and such that $\Gamma_\infty(e) < \infty$. Then if $M_\infty(e)$ is sufficiently small then no resetting is needed and

$$\widehat{\theta}_N^\lambda - \theta^* = \sum_{n=1}^{N} \frac{\partial y^\lambda}{\partial \xi}(N, n, \theta^*)\lambda Q(\overline{x}_n(\theta^*)) + r_N$$

where (r_N) is majorated by an L-mixing process r_N^* such that $r_N^* = O_M(\lambda)$.

Note that the first term on the right hand side is an L-mixing process of the order of $O_M(\lambda^{1/2})$ as can be seen by using the moment inequality given in Theorem 2.1. Thus we also get $\widehat{\theta}_N^\lambda - \theta^* = O_M(\lambda^{1/2})$.

Let us now consider some of the complexity issues related to fixed gain estimation. First we consider the effect of parameter uncertainty on prediction (c.f. [28]).

Theorem 10.6 Assume that the conditions of Theorem 10.2 are satisfied. If (e_n) is L_0-mixing and $(p,q) \in D_{p^*,q^*}$ then for a suitable version of $\widehat{\theta}_N^\lambda$ the process $u_n = \varepsilon_n^2(\theta_{n-1}^\lambda) - e_n^2$ is L-mixing. Moreover for any solution $\widehat{\theta}_N^\lambda$

$$\overline{\lim}_{N\to\infty} |(\lambda N)^{-1} \sum_{n=1}^{N} (\varepsilon_n^2(\widehat{\theta}_{n-1}^\lambda) - e_n^2) - \sigma^2 \frac{p+q}{2}| \leq C\lambda^{1/2}$$

almost surely, where C is a nonrandom constant.

A startling aspect of fixed gain estimators is that they provide a more sensitive criterion for the detection of overestimation than the standard estimators of the previous section: the "badness" of the estimator increases the "badness" of overestimation. Indeed, let us define for $(p,q) \in D_{p^*,q^*}$

$$C_N^\lambda(p,q) = \sum_{n=1}^{N} \varepsilon_n^2(\theta_{n-1}^\lambda).$$

Then Theorem 10.6 implies the following

Theorem 10.7 Let (p,q) and say $q = q^*$ and hence $p \geq p^*$ in D_{p^*,q^*}. Then almost surely

$$\overline{\lim}_{N\to\infty} \left|(C_N^\lambda(p+1,q) - C_N^\lambda(p,q))/\lambda N - \frac{\sigma^2}{2}\right| \leq C\lambda^{1/2}.$$

This result has been further exploited in [31].

In the general case let $S(x)$ be a performance index given in Section 8 and assume that (e_n) is stationary. Assuming that the processes have been started at $n = -\infty$ let

$$\delta\theta = \sum_{n=-\infty}^{0} \frac{\partial y^{\lambda}}{\partial \xi}(0, n, \theta^*)\lambda Q(\bar{x}_n(\theta^*))$$

and let $(R^{**})^{-1} = \mathrm{E}\delta\theta \cdot \delta\theta^T/\lambda$.

Theorem 10.8 Assume that the conditions of Theorem 10.5 are satisfied and (e_n) is stationary. Then

$$\overline{\lim}_{N\to\infty}|\frac{1}{N}\sum_{n=1}^{N}(S(x_n) - S(\bar{x}_n(\theta^*)) - \lambda \mathrm{Tr} T^*(R^{**})^{-1}| \leq C\lambda^{1/2}$$

where C is a nonrandom constant.

11. TIME VARYING SYSTEMS. FURTHER REPRESENTATION THEOREMS

The above setup with minor modifications is also suitable to describe time-varying systems. For this purpose let us incorporate a "true system-parameter" ψ^* in our general estimation scheme and write it as

$$\bar{\bar{x}}_{n+1}(\theta, \psi^*) = A(\theta, \psi^*)\bar{\bar{x}}_n(\theta, \psi^*) + B(\theta, \psi^*)e_n \tag{11.1}$$

The dimension and the meaning of θ and ψ may be very different. E.g. θ may denote some controller parameters of a direct adaptive controller as in the Åström-Wittenmark regulator. The corresponding nonlinear equation is

$$G(\theta, \psi^*) = 0$$

The solution will be denoted by $\theta^*(\psi^*)$, to indicate the dependence on ψ^*. It is assumed that the mapping $\psi^* \to \theta^*(\psi^*)$ is Lipschitz-continuous from D_ψ to D_θ, where the latters are compact domains.

A time-varying system is obtained if the systems parameter ψ^* in (11.1) is replaced by ψ_n^*. The system is then described by the following equation:

$$\bar{x}_{n+1}(\theta) = A(\theta, \psi^*)\bar{x}_n(\theta, \psi_n^*) + B(\theta, \psi_n^*)e_n. \tag{11.2}$$

We assume that $\psi_n^* \in D_\psi$. The time-varying system is said to be slowly time-varying if

$$\sup_{n\geq 0} |\psi_{n+1}^* - \psi_n^*| \triangleq \dot{S} < \infty \tag{11.3}$$

and consequently $\sup_{n\geq 0} |\theta_{n+1}^* - \theta_n^*| \leq C\dot{S}$ with some constant C. The quantity $\dot{S}$ is called the rate of change. It is well known that if $\dot{S}$ is sufficiently small then (11.2) is "exponentially stable". The above description of slowly time-varying systems is a special case of a more general definition in [72].

The off-line estimation of $\theta_N^* = \theta^*(\psi_N^*)$ is obtained by considering the function

$$U_N(\theta) = \sum_{n=1}^{N} (1-\lambda)^{N-n} \lambda Q(\overline{x}_n(\theta)),$$

with some $0 < \lambda < 1$ and then getting the estimator $\widehat{\theta}_N^\lambda$ by solving $U_N(\theta) = 0$. The following theorem is an extension of Theorem 10.2 (c.f. also [17], [18]) :

Theorem 11.1 Under the conditions above we have for a suitable version of $\widehat{\theta}_N^\lambda$

$$\widehat{\theta}_N^\lambda - \theta_N^* = \delta_{rN} + \delta_{dN} + o_M(\alpha^N)$$

where (δ_{rN}) is an L-mixing process such that $\delta_r = O_M(\lambda^{1/2})$ and (δ_{dN}) is a deterministic process satisfying $\delta_d = O(\dot{S}/\lambda)$, and $0 < \alpha < 1$.

This decomposition theorem captures and to some extent separates the effects of noise and time variation. The optimizing value of λ is easily found to be $\lambda = O(\dot{S}^{2/3})$ which yields the tracking error $O_M(\dot{S}^{1/3})$. Thus we get a "law of the cubic root", which states that the tracking error can be made proportional to the cubic root of the rate of change. It can be shown that this is the best we can hope for.

The recursive estimator of θ_n^* is defined by a fixed gained estimation scheme. Assume that an initial value $\widehat{\widehat{\theta}}_0^\lambda \in D_\theta$ is given .Then the estimator $\widehat{\widehat{\theta}}_n^\lambda$ of the time-varying parameter θ_n^* is generated recursively as follows:

$$x_{n+1} = A(\widehat{\widehat{\theta}}_n^\lambda, \psi_n^*) x_n + B(\widehat{\widehat{\theta}}_n^\lambda, \psi_n^*) e_n \tag{11.4}$$

$$\widehat{\widehat{\theta}}_{n+1}^\lambda = \widehat{\widehat{\theta}}_n^\lambda + \lambda Q(x_{n+1}). \tag{11.5}$$

The associated ODE is now defined as

$$\dot{y}_t = \lambda G(y_t, \psi_t^*)$$

where $\psi_t^* = \psi_n^*$ for $n \le t < n+1$. Let the general solution of (11.5) be denoted by $y^\lambda(t,s,\xi)$. We present a theorem which captures the effect of the "noise" in a way analogous to Theorem 10.5.

Theorem 11.2 Under the conditions of Theorem 10.5 $\widehat{\widehat{\theta}}_N^\lambda$ will not leave D_0 and we have

$$\widehat{\widehat{\theta}}_N^\lambda - y_N = \sum_{n=1}^{N} \frac{\partial y^\lambda}{\partial \xi}(N,n,y_n)\lambda Q(\overline{x}_n(y_n)) + r_N$$

where $|r_N|$ is majorated by a L-mixing process r_N^* such that $r_N^* = O_M(\lambda)$.

Now it is easy to see that $|y_n - \theta_n^*| \le C\dot{S}/\lambda$, and hence we get a decomposition of the tracking error similar to the one given in Theorem 11.1, and the corresponding law of the cubic root holds.

As an illustration let us consider the adaptive control problem described in Example 3. Let the true system parameter be ψ^* and let us write the feedback law for the time-invariant system in the form

$$u = -K(\theta^*, \psi^*)y + v$$

where $\theta^* = \theta^*(\psi^*)$. In the time-varying case we use the "receding horizon controller" defined by

$$u_n^* = -(K(\theta^*, \psi_n^*)y^*)_n + v_n$$

where $\theta_n^* = \theta^*(\psi_n^*)$ and to evaluate the first term we use an apriori fixed state space representation to the system with transfer function $K(\theta^*, \psi^*)$. Let the corresponding output process be denoted by (y_n^*) and let $S(y^*)$ be a real-valued quadratic function of y^*, which is thought to be an index of performance. The "optimal" index of performance within the time-interval $[1,N]$ is then defined as $\sum_{n=1}^N S(y_n)$. The actual output is obtained when applying the feedback

$$u_n = (-K(\widehat{\widehat{\theta}}_n^\lambda, \psi_n^*)y)_n + v_n$$

where $\widehat{\widehat{\theta}}_n$ is the recursive fixed gain estimator for θ_n^*. The actual index of performance is related to the optimal index of performance as follows:

Theorem 11.3. Under the conditions of Theorem 11.2 we have for $c_1\dot{S}^{2/3} < \lambda < c_2\dot{S}^{2/3}$

$$\overline{\lim}_{N\to\infty}\left|\frac{1}{N}\sum_{n=1}^{N}(S(y_n) - S(y_n^*))\right| = C\dot{S}^{1/3}.$$

where C is a deterministic constant.

12. CONCLUSION. AN OPEN PROBLEM

The objective of the research that has been surveyed above was to develop user-friendly mathematical devices for the understanding of the behavior of estimators that are associated with the general estimation scheme described in Section 3. This objective has been achieved in a number of instances as shown by the four basic representation theorem and their extensions. Since an important class of stochastic adaptive control problems can be described by the general estimation scheme, the results have a nontrivial impact on the theory of stochastic adaptive control, too. These results also enabled as to answer the important question of how parameter uncertainty affects performance.

In spite of all this progress one closely related problem of stochastic adaptive control is still unsolved. Namely, some practical control systems may be infinite dimensional hence the general estimation scheme cannot be set up. We may then consider to have a controller parameter θ, a scalar-valued quadratic function $P(x)$ and the average performance defined as

$$F(\theta) = \mathrm{E}P(\overline{x}(\theta)).$$

Then the problem to be solved is

$$\min_{\theta} F(\theta).$$

In the general case the gradient of F is not computable and as far as we know no systematic work has been done to solve this problem.

12. ACKNOWLEDGEMENTS

This research was supported in part by the Natural Sciences and Engineering Research Council under Grant 01329. The author wishes to thank Andrew Heunis for discussing some parts of this paper and to Zsuzsanna Vágó, Jimmy Baikovicius, and Mindle Levitt for their considerable amount of work in the preparation of this paper.

REFERENCES

[1] H. Akaike, *Statistical predictor identification*, Ann. Inst. Statis. Math., 22 (1970), pp.202-217.

[2] Arató, M., *Linear Stochastic Systems with Constant Coefficients*. Lecture Notes in Control and Information Science, Springer, (1984).

[3] K.J. Åström, and B. Wittenmark, *On self-tuning regulators*. Automatica, 9 (1973), pp.185-199.

[4] A.Bagchi and V.Borkar, *Parameter identification in infinite-dimensional linear systems*,Stochastics,12 (1984),201-213.

[5] J. Baikovicius and L. Gerencsér, *Change point detection in a stochastic complexity framework*. To appear in the Proc of the 29-th IEEE CDC, Vol 6, (1990), 3554-3555.

[6] A. Benveniste, M. Metivier, and P. Priouret, *Algorithms adaptatifs et approximations stochastiques*, Masson, Paris, (1987).

[7] A.N. Borodin, *A stochastic approximation procedure in the case of weakly dependent observations*, Theory Probab. Appl, 24 (1979), 34-52.

[8] P.E. Caines, *Linear Stochastic Systems*, Wiley, (1988).

[9] H.F. Chen,*Recursive Estimation and Control for Stochastic Systems*, John Wiley & Sons, New York, (1985).

[10] M.H.A. Davis and R.B. Vinter, *Stochastics Modelling and Control*, Chapman and Hall, New York, (1985).

[11] L.D. Davisson, *Prediction error of stationary Gaussian time series of unknown covariance*, IEEE Trans. Inform. Theory, IT-19 (1965), pp.783-795.

[12] M. Deistler and B.D.O. Anderson, *Linear dynamic errors-in-variable models. Some structure theory*. J. of Econometrics, 41 (1989), pp.39-63,.

[13] D.P. Djereveckii and A.L. Fradko, *Applied theory of discrete adaptive control systems*, (In Russian), Nauka, Moscow (1981).

[14] T.E. Duncan, and B. Pasik-Duncan, *Adaptive Control of Continuous-Time Linear Stochastic Systems*, to appear in Mathematics of Control, Signals and Systems (1991).

[15] J. Galambos, *The Asymptotic Theory of Extreme Order Statistics*, John Wiley & Sons, New York, (1978).

[16] S. Geman, *Some averaging and stability results for random differential equations*, SIAM J. Appl. Math., 36 (1979) 87-105.

[17] L. Gerencsér, *Parameter Tracking of Time-Varying Continuous-Time Linear Stochastic Systems*, In Modelling, Identification and Robust Control (eds. Ch. I. Byrnes and A. Lindquist), North Holland, (1986), 581-595.

[18] L. Gerencsér, *On a class of mixing processes*. Stochastics, 26 (1989a), pp.165-191.

[19] L. Gerencsér, *Strong approximation of the recursive prediction error estimator of the parameters of an ARMA process*. McGill Research Centre for Intelligent Machines, TR-CIM 89-8 Submitted for publication.

[20] L. Gerencsér, *On Rissanen's predictive stochastic complexity for stationary ARMA processes.* McGill Research Centre for Intelligent Machines, TR-CIM-89-5. Submitted for publication.

[21] L. Gerencsér,*On the martingale approximation of the estimation error of ARMA parameters.* Systems and Control Letters, 15, (1990), 417-423.

[22] L. Gerencsér, *Rate of convergence of recursive estimators.* Submitted for publication.

[23] L. Gerencsér, *Closed loop parameter identifiability and adaptive control of a linear stochastic system.* Systems and Control Letters, 15, (1990), 411-416.

[24] L. Gerencsér, *Strong approximation of vector-valued stochastic integrals.* To appear in Statistics and Probability Letters (1991).

[25] L. Gerencsér, *Strong approximation theorems for estimator processes in continuous time,* Proc. of Limit Theorem in Probability and Statistics (eds. I. Berkes, E. Csáki and I. Révész), North Holland, (1991), to appear.

[26] L.Gerencsér, *Almost sure exponential stability of random linear differential equations.*To appear in Stochastics, (1991)

[27] L. Gerencsér, *Fixed gain estimation of ARMA parameters.* Submitted for publication.

[28] L. Gerencsér, *Predictive stochastic complexity associated with fixed gain estimators.* Submitted for publication.

[29] L. Gerencsér, *Predictive stochastic complexity for continuous-time systems.* Submitted for publication (1991).

[30] L. Gerencsér, and J. Baikovicius, *Change and point detection with stochastic complexity.* To appear in Proc. of 9-th IFAC/IFORS Symposium on Identification and System Parameter Estimation, Budapest, (1991), Pergamon Press, Oxford.

[31] L. Gerencsér, and J. Baikovicius, *Model selection, stochastic complexity, and badness amplification*, Submitted to the 30th IEEE Conference on Decision and Control, (1991).

[32] L. Gerencsér, I. Gyöngy, and Gy. Michaletzky, *Continuous time Recursive Maximum Likelihood Method. A New Approach to Ljung's Scheme*, Proc. of the 9th. IFAC World Congress, Budapest, 2, (eds. L. Ljung and K.J. Åström), Pergamon Press, Oxford, (1984), pp. 75-77.

[33] L. Gerencsér and A. Heunis, *Rate of convergence for the LMS algorithm*, Manuscript (1991).

[34] L. Gerencsér and J. Rissanen, *A prediction bound for Gaussian ARMA processes.* Proc. of the *25-th CDC*, Athens, Vol. 3 (1986), pp.1487-1490.

[35] L. Gerencsér and J. Rissanen, *Asymptotics of predictive stochastic complexity.* To appear in "New Directions in Time Series Analysis", Proc. of the 1990 IMA Workshop (eds. E. Parzen, D. Brillinger, M. Rosenblatt, M. Taqqu, J. Geweke and P.E. Caines,). Springer, (1991).

[36] L. Gerencsér and Zs. Vágó *A strong approximation theorem for estimators in continuous time*,Submitted for publication.

[37] L. Gerencsér, and Zs. Vágó, From fine asymptotics to model selection. To appear in "Identification of Continuous-Time Systems" (ed. N.K. Sinha), Kluwer Academic Publishers, Dordrecht, The Netherlands, (1991).

[38] L. Gertler and Cs. Bányász, A recursive maximum likelihood estimator, *Automatica*, (1971).

[39] G.C. Goodwin and K.S. Sin, *Adaptive Filtering, Prediction and Control*, Prentice-Hall Englewood Cliffs, NJ (1984).

[40] E.J. Hannan, *The convergence of some time-series recursions*, Ann. Stat., 4, (1976) 1258 - 1270.

[41] E.J. Hannan, and M. Deistler, *The statistical theory of linear systems.* Wiley (1988).

[42] E.J. Hannan, A.J. McDougall and D.S. Poskitt, *Recursive estimation of autoregressions.* J. Roy, Stat. Soc. Ser B. 51 (1990).

[43] E.M. Hemerley and M.A.H. Davis, *Strong consistency of the PLS criterion for order determination of autoregressive processes*, Ann. Stat., 17, (1989) 941 - 946.

44] A. Heunis, *Rates of convergence for an adaptive filtering algorithm driven by stationary finite-memory data.* Submitted for publication.

[45] A. Heunis, *Asymptotic Properties of Prediction Error Estimators in Approximate System Identification.* Stochastics,24 (1988),pp.1-43.

[46] I. Hunter, S. Lafontaine and J. Hollerbach, *Microrobotics and the study of muscle: special problems in control, system identification and modelling*, To appear in: Proc. of the 9th IFAC/IFORS Symposium on Identification and System Parameter Estimation, Budapest, (1991)., Pergamon Press, Oxford.

[47] I.A. Ibragimov and R.Z. Khasminskii,*Statistical Estimation. Asymptotic Theory*,Springer Verlag ,Berlin (1981).

[48] H.J. Kushner, *Approximation and Weak Convergence Methods for Random Processes*, MIT Press, (1984).

[49] H.J. Kushner and D.S. Clark, *Stochastic approximation methods for constrained and unconstrained optimization*, Springer, (1978).

[50] L. Kavalieris, *The estimation of the order of an autoregression using recursive residuals and cross-validation.* J. of Time Series analysis, 10 (1989), pp.271-281.

[51] Yu. A. Kutoyants, *Estimation of stochastic processes.* Publishing House of the Armenian Academy of Sciences, Yerevan (in Russian, 1980) and Heldermann, Berlin (1989).

[52] T.Z. Lai and C.Z. Wei, *Least squares estimates in stochastic regression models with application to identification and control of dynamic systems*, Ann. of Stat., 10 (1982) 154 - 165.

[53] L. Ljung *On Consistency and Identifiability*, Mathematical Programming Study, 5, (1976) 169-190.

[54] L. Ljung, *Analysis of recursive stochastic algorithms*, IEEE Trans. Aut. Cont., AC-22 (1977), pp.551-575.

[55] L. Ljung and P. Caines, *Asymptotic normality of prediction error estimation for approximate system models*, Stochastics, 31 (1979), pp.29-46.

[57] D.L. McLeish, *A maximal inequality and dependent strong laws*,The Annals of Probability, 3 (1975), pp.829-839.

[58] J.B. Moore, *On strong consistency of least squares identification algorithms*, Automatica, 14 (1978), pp.505 - 509.

[59] F. Móricz, *Moment Inequalities and the strong laws of large numbers*, Z. Wahrscheinlichkeitstheorie u. verw. Gebiete, 35 (1974), pp.299-314.

[60] J. Rissanen, *Modelling by shortest data description*, Automatica, 14 (1978), pp.465-471.

[61] J. Rissanen, *Order estimation by accumulated prediction errors*, Essays in Time Series and Allied Processes (eds. J. Gani, M.B. Priestley), pp.55-61 (1984).

[62] J. Rissanen, *Stochastic complexity and modeling*, Annals of Statistics 14 (1986), pp.1080-1100.

[63] J. Rissanen, *Stochastic complexity in statistical inquiry.* World Scientific Publisher (1989).

[64] J. Rissanen, and P.E. Caines, *The strong consistency of maximum likelihood estimators for ARMA processes.* Ann Statist. 7 (1979) 297 - 315.

[65] T. Söderström, *An On-line Algorithm for Approximate Maximum-Likelihood Identification of Linear Dynamic Systems* Report 7308 , (1973) Department of Automatic Control, Lund Institute of Technology, Lund, Sweden.

[66] T. Söderström and P. Stoica, *System identification* (1989), Prentice Hall.

[67] V. Solo, *The second order properties of a time series recursion*, Ann. Stat. 9 (1981) 307-317.

[68] Zs. Vágó and L. Gerencsér, *Uniqueness of the Maximum-Likelihood Estimates of the Kalman-gain Matrix of a State Space Model*, Proc. of the IFAC/IFORS, Conference on *Dynamic Modelling of National Economics*, Budapest, (1985).

[69] D.G. DeWolf and D.M.Wiberg An ordinary differential equation technique for continuous time parameter estimation, *IEEE Trans. Aut. Cont.*, submitted.

[70] G. Yin, and Y.M. Zhu, *On robustness of the Robbins-Monro method for parallel processing*, Systems and Control Letters, 12 (1989), pp. 77-86.

[71] K. Yoshihara, *Moment inequalities for mixing sequences*, Kodai Math. J. 1 (1978) 316-328.

[72] G. Zames, and L.Y. Wang, Local-Global Double Algebras for Slow H^∞ Adaptation: Part I–Inversion and Stability, *IEEE Trans. Aut. Cont.*, 36, 2, (1991), pp. 130-142.

STOCHASTIC ADAPTIVE CONTROL: RESULTS AND PERSPECTIVE

G.C. Goodwin
Dept. of Electrical & Computer Engineering University of Newcastle, NSW, 2308, Australia.

M. Gevers
Louvain University, Batiment Maxwell, B–1348, Louvain–La–Neuve, Belgium.

D.Q. Mayne
Dept. of Electrical Engineering & Computer Science,
University of California, Davis, CA 95616, USA

V. Wertz
Louvain University, Batiment Maxwell, B–1348, Louvain–La–Neuve, Belgium.

1. INTRODUCTION

A key question asked in stochastic adaptive control theory is whether or not it is possible to stabilize a linear system which is perturbed by unmeasured random disturbances and whose model is only partially known.

One approach to this problem is to use the, so called, Certainty Equivalence principle. In this strategy, one divides the problem into two parts. Firstly, one uses a gradient type estimation algorithm to obtain an on–line estimate of the model of the system and disturbances. Then one acts as if the estimated model were the correct model when designing the feedback control law. We will argue later that this strategy is limited since no attempt is made to quantify the accuracy of the estimated model. Hence it is not possible to use robust control design methods which, inter–alia, require a measure of fidelity of the model. For the moment however, we will adopt this approach as a working strategy.

Clearly, one would wish to say something about the performance of this method. For example, a minimal requirement would be to show that if the system is ideal (no nasty nonlinearities or high frequency unmodelled dynamics) then the design objectives are asymptotically achieved. This turns out to be a surprisingly difficult problem. The reason for this is that the full system (plant + estimator + feedback law) is a nonlinear, time varying, stochastic system and such systems are notoriously difficult to analyze. Thus the associated theory presents difficult and challenging mathematical problems.

To make progress on the problem, simplified versions have been studied. One line of attack has been to remove the random disturbances from the specification whilst retaining the format of the parameter estimation algorithm. The latter step is important since otherwise one could be led to trivial solutions in which the model parameters are estimated in infinitesimal time using open loop measurements.

The simplified deterministic certainty equivalence adaptive control problem still presents a significant mathematical challenge and it was not until 1978 [1] that global stability was established for a moderately general case. This first result covered the deterministic continuous time case. Soon after, simpler methods of analysis were found for the deterministic discrete time case [2]. These were subsequently extended to the corresponding continuous time problem [3]–[4].

These solutions suggest that the stochastic case might not be intractable. Indeed, as early as 1973, it had been shown [5] that, provided the estimated parameters converged to something (not necessarily their true values), that a certainty equivalence form of the stochastic minimum variance control law would asymptotically achieve the design objectives. However, parameter convergence remained an open question. By side–stepping the stability issue it was shown via an associated ODE approach [6] that a positive real condition was a necessary condition for convergence. However, the theory was still incomplete.

Finally in 1980, using Martingale Convergence theory, it was shown in [7] that for the discrete time stochastic adaptive control problem, a certainty equivalence minimum variance control law was capable of stabilizing a partially known system and of asymptotically achieving the design objective.

A remaining open question was the corresponding continuous time stochastic adaptive control problem. One might legitimately ask if this problem is worth studying given that all realistic controllers are certain to be implemented digitally. However, apart from the issue of intellectual completeness, there is also the issue that most existing discrete algorithms do not have a meaningful limit as the sampling rate is increased [8]. This suggests that the methods will have numerical and conceptual difficulties when used with moderately fast sampling rates. This is indeed the case as explored in detail in [8]. This problem is fundamental and applies to the discrete stochastic adaptive control laws studied in [5], [6], [7]. These turn out to depend on high order differences of the plant output and these approach high order derivatives as the sampling rate increases! Thus there is both theoretical and practical motivation to study the continuous time stochastic problem. In turn, this will have consequences for the discrete case since the continuous algorithms are likely to show how a sensible limit can be maintained as the sampling rate increases.

An initial objective of this chapter is to review the essential features of continuous time certainty equivalence stochastic adaptive control as originally presented in [9], [10]. These results specialize to the simpler cases of deterministic systems [1] to [4] and lead to new insights for the discrete time stochastic case [5] to [7].

Subsequently we will point to the limitations of the current theory and suggest some open research problems.

2. PRELIMINARIES

We assume that the system can be exactly modelled by a finite dimensional linear stochastic differential equation expressed in innovations form as:

$$dx_t = Ax_t dt + Bu_t dt + Kd\omega_t \tag{2.1}$$

$$dz_t = Cx_t dt + d\omega_t \tag{2.2}$$

where ω_t is a Wiener process with incremental covariance $\sigma^2 dt$, x_t is a state–vector of dimension n, u_t is a scalar control input and z_t is the integral of the output of the system. The matrices A, B, K, C contain unknown but fixed parameters. We shall denote by $\mathcal{F}_t$ the increasing σ– fields generated by $\{\omega_s,\ 0 \leq s \leq t\}$ and the unknown initial conditions x_0, and we shall assume that $\| x_0 \|^2$ is bounded.

We do not necessarily assume that (2.1) is stable. Our control objective is firstly to guarantee stability of the closed loop system and subsequently to ensure that the system output has certain desirable properties (e.g. follows a given reference input signal with minimal mean square error). We will approach this problem using the Centainty Equivalence Principle described earlier, i.e. we will first attempt to estimate the unknown parameters in A, B, K and C and then, assuming these estimates are correct, we will design a suitable feedback control law.

Towards this end, we will first express the model (2.1), (2.2) in a slightly different form which turns out to be more convenient for parameter estimation. Without loss of generality, we assume that the above model is in observer form where

$$A = \begin{bmatrix} -a_{n-1} & 1 & & \\ -a_{n-2} & & \ddots & \\ \vdots & & & 1 \\ -a_0 & & & 0 \end{bmatrix}; \quad B = \begin{bmatrix} b_{n-1} \\ \vdots \\ \vdots \\ b_0 \end{bmatrix}; \quad K = \begin{bmatrix} k_{n-1} \\ \vdots \\ \vdots \\ k_0 \end{bmatrix} \tag{2.3}$$

$$C = [1 \quad 0 \quad \cdots \quad 0] \tag{2.4}$$

To avoid differentiations in the parameter estimator it is helpful to re–express this model in fractional form [4]. We therefore re–parameterize the system as follows. Let

$$E(\varrho) = \varrho^n + e_{n-1}\varrho^{n-1} + \ldots + e_0 \tag{2.5}$$

$$G^T = [g_{n-1}, \ \ldots, \ g_0] \ \text{with} \ g_i = e_i - a_i; \quad i = 0, \ \ldots, \ n-1 \tag{2.6}$$

and

$$E = \begin{bmatrix} -e_{n-1} & 1 & & \\ \vdots & & \ddots & \\ \vdots & & & 1 \\ -e_0 & & 0 & 0 \end{bmatrix} \tag{2.7}$$

where the coefficients are arbitrary subject to $E(\varrho)$ having all its zeros in the open left half plane and $K \neq G$.

Then, by adding and subtracting $Ex_t dt$ to the right hand side, equations (2.1), (2.2) can be rewritten as

$$dx_t = Ex_t dt + Bu_t \, dt + (K - G) d\omega_t + G dy_t \tag{2.8}$$

$$dz_t = Cx_t \, dt + d\omega_t \tag{2.9}$$

Using superposition, equations (2.8), (2.9) can also be expressed as

$$d\phi_t^1 = E\phi_t^1 dt + G dy_t \tag{2.10}$$

$$d\phi_t^2 = E\phi_t^2 dt + B u_t dt \tag{2.11}$$

$$d\phi_t^3 = E\phi_t^3 dt + (K - G) d\omega_t \tag{2.12}$$

$$dz_t = C\left[\phi_t^1 dt + \phi_t^2 dt + \phi_t^3 dt\right] + d\omega_t \tag{2.13}$$

Since z_t is a scalar, we have that $C(sI - E)^{-1}B = B^T\left(sI - E^T\right)^{-1}C^T$ etc. Hence (2.10) to (2.13) can be rewritten as

$$d\phi_t^z = E^T\phi_t^z dt + C^T dz_t \tag{2.14}$$

$$d\phi_t^u = E^T\phi_t^u dt + C^T u_t dt \tag{2.15}$$

$$d\phi_t^\omega = E^T\phi_t^\omega dt + C^T d\omega_t \tag{2.16}$$

$$dz_t = \left[G^T\phi_t^z + B^T\phi_t^u + (K - G)^T\phi_t^\omega\right]dt + d\omega_t \tag{2.17}$$

In (2.14)–(2.16) we choose $\phi_0^z = 0$, $\phi_0^u = 0$ and ϕ_0^ω such that

$$(K - G)^T\phi_0^\omega = Cx_0 \tag{2.18}$$

Equation (2.17) is in the form of a linear regression, i.e.

$$dz_t = \phi_t^T\theta dt + d\omega_t \tag{2.19}$$

where

$$\phi_t^T = \left[(\phi_t^z)^T,\ (\phi_t^u)^T,\ (\phi_t^\omega)^T\right] \tag{2.20}$$

$$\theta^T = \left[G^T,\ B^T,\ (K - G)^T\right] = \left[G^T,\ B^T,\ F^T\right] \tag{2.21}$$

Note that ϕ_t does not depend on the unknown system parameters since E is known. Thus equations (2.14) to (2.16) simply represent a state space form of the usual regression vector as in [11] and [4]. To make the comparison with the regression vector formulations more complete, we note that with $\varrho \triangleq \frac{d}{dt}$ we have, with some abuse of notation and ignoring initial conditions

$$(\phi_t^u)^T = \left[\frac{\varrho^{n-1}}{E(\varrho)}u, \; \ldots, \; \frac{1}{E(\varrho)}u\right] \tag{2.22}$$

and similarly for ϕ_t^z, ϕ_t^ω.

3. PSEUDO LINEAR REGRESSION

Equation (2.19) suggests that the unknown parameter vector θ could be easily estimated by some form of least squares. However, the model is not quite in a form which is suitable for parameter estimation. This is because the component ϕ_t^ω depends on the unmeasured noise source ω_t. Thus, following a standard strategy [11] we define the predicted output by a pseudo regression in which $d\omega_t$ is replaced by the prediction error. Thus we define

$$d\hat{z}_t = \psi_t^T \hat{\theta}_t dt \tag{3.1}$$

where $\hat{\theta}_t$ is an on–line estimate of θ_t obtained from a stochastic gradient algorithm of the form:

$$d\hat{\theta}_t = \frac{\psi_t}{r_t}\left(dz_t - \psi_t^T \hat{\theta}_t dt\right) \tag{3.2}$$

In equation (3.1), r_t is a normalization variable which will be specified in Section 5.

Also, in (3.1) we have

$$\psi_t^T = \left[(\psi_t^z)^T, \; (\psi_t^u)^T, \; (\psi_t^e)^T\right] \tag{3.3}$$

$$d\psi_t^z = E^T\psi_t^z dt + C^T dz_t \tag{3.4}$$

$$d\psi_t^u = E^T\psi_t^u dt + C^T u_t dt \tag{3.5}$$

$$d\psi_t^e = E^T\psi_t^e dt + C^T de_t \tag{3.6}$$

$$de_t = dz_t - d\hat{z}_t = dz_t - \psi_t^T\hat{\theta}_t dt \tag{3.7}$$

It is assumed that $\psi_0 = 0$ and that $\hat{\theta}_0$ is $\mathcal{F}_0$ – measurable. With this choice of initial conditions $\psi_t^z = \phi_t^z$ and $\psi_t^u = \phi_t^u$ for $t \geq 0$.

The effect of a using pseudo regression is to disturb the direction of the gradient vector ϕ_t.

Heuristically one might expect that it would be necessary for the angle between the true gradient ϕ_t and the pseudo gradient ψ_t to be accute on average. In turn, this requires that the filter $E(\varrho)$ used in defining ψ_t be a reasonable approximation to the optimal Kalman Filter for the system which is implicitly used in defining ϕ_t. To make this clearer, let us consider the difference, η_t, between the true system quantity $\phi_t^T\theta$ and the estimated quantity $\psi_t^T\hat{\theta}_t$:

$$\eta_t = \phi_t^T\theta - \psi_t^T\hat{\theta}_t \tag{3.8}$$

From (2.19), (3.7), (3.8) we have

$$de_t - d\omega_t = \eta_t dt \tag{3.9}$$

Notice that η_t is the "deterministic part" of the prediction error.

Now let

$$\gamma_t \triangleq \psi_t^e - \phi_t^\omega \tag{3.10}$$

Then from (3.6), (3.8), (2.16) we have

$$d\gamma_t = E^T\gamma_t dt + C^T\eta_t dt, \quad \text{with} \quad \gamma_0 = -\phi_0^\omega \tag{3.11}$$

Noting that

$$\psi_0^z = \phi_0^z = 0 \quad \text{and} \quad \psi_0^u = \phi_0^u = 0 \tag{3.12}$$

then

$$(K-G)^T\gamma_t = -(\phi_t - \psi_t)^T\theta \tag{3.13}$$

In particular:

$$(K-G)^T\gamma_0 = -Cx_0 \tag{3.14}$$

Notice that γ_0 cannot be made zero if x_0 is unknown, but we will assume that $\| \gamma_0 \|^2$ is bounded; this is consistent with our assumption on x_0 and (3.14).

Denoting

$$\tilde{\theta}_t \triangleq \hat{\theta}_t - \theta \tag{3.15}$$

we then have from (3.13), (3.11) that

$$\eta_t = -(K-G)^T\gamma_t - \psi_t^T\tilde{\theta}_t \tag{3.16}$$

where

$$d\gamma_t = (A-KC)^T\gamma dt + C^T\left(-\psi_t^T\tilde{\theta}_t\right)dt \tag{3.17}$$

We thus see that η_t is related to $-\psi_t^T\tilde{\theta}_t$ by the following transfer function equation

$$D(\varrho)\eta_t = E(\varrho)\left[-\psi_t^T\tilde{\theta}_t\right] \tag{3.18}$$

where $D(\varrho)$ is the characteristic polynomial of the optimal Kalman filter for the system, i.e.

$$D(\varrho) = \varrho^n + (k_{n-1} + a_{n-1})\varrho^{n-1} + \dots (k_0 + a_0) = \det(\varrho I - A + KC) \tag{3.19}$$

and where $E(\varrho)$ is as in (2.5).

We will see later that the required "accute angle condition" between the regression vectors is captured by assuming that the transfer function $\frac{D}{E}$ in (3.18) satisfies the following restriction:

Assumption 1:

$D(\varrho)$ is strictly Hurwitz and the filter $E(\varrho)$ is chosen such that

1) $\operatorname{Re}\sigma_i(E(\varrho)) \leq -a < 0, \quad i = 1, \ldots, n$ where $\sigma_i(E)$

2) $\frac{D}{E}$ is input strictly passive, i.e. $\exists \epsilon > 0$ and $K > 0$ such that

$$\forall T > 0, \quad \int_0^T y_\tau u_\tau d\tau \geq \epsilon \int_0^T u_\tau^2 d\tau - K \| \gamma_0 \|^2 \tag{3.20}$$

where y_t is the output of the filter D/E driven by u_t.

Noting that $d\theta = 0$ and using the definition of e_t and η_t given in (3.7), (3.9), we observe from (3.15) and (3.2) that $\tilde{\theta}_t$ is the solution of the following stochastic differential equation

$$d\tilde{\theta}_t = \frac{\psi_t}{r_t}\eta_t dt + \frac{\psi_t}{r_t} d\omega_t \tag{3.21}$$

In the sequel we will require that $\tilde{\theta}_t$ be bounded. We guarantee this by introducing a projection scheme as described below. We first introduce the following assumption.

Assumption 2:

There exists a known parameter value θ_c and a positive number R_1 such that the true value θ lies inside $\mathcal{C}_1$ where

$$\mathcal{C}_1 = \{\Theta : \| \Theta - \theta_c \| \leq R_1\}$$

▽▽▽

Let R_2 be another positive number larger than R_1 and define another convex set $\mathcal{C}_2$ analogous to the one given in Assumption $\mathcal{C}_1$. We then modify the parameter estimator to ensure that $\| \hat{\theta}_t - \theta_c \| < R_2$ for all t. We do this by using the following projection scheme:

Parameter estimator with projection

Let τ be a time for which the solution of (3.2) is such that $\| \theta_\tau - \theta_c \| = R_2$. Denote the corresponding value of $\hat{\theta}_\tau$ by $\hat{\theta}_{\tau_-}$. At time τ, the estimate $\hat{\theta}_\tau$ is then defined as

$$\hat{\theta}_\tau \triangleq \theta_c + \frac{R_1}{R_2}\left(\hat{\theta}_{\tau_-} - \theta_c\right) \tag{3.22}$$

For $t \geq \tau$ equation (3.2) is then integrated with initial condition $\hat{\theta}_\tau$ defined by (3.23). This makes for $\hat{\theta}_t$ right continuous at the projection times and ensures $\hat{\theta}_t \in \mathcal{C}_2$ for all t.

4. FEEDBACK CONTROL

We consider a general class of control laws in state feedback form:

$$u_t = -\left[\hat{\ell}_{n-1}, \ \ldots, \ \hat{\ell}_0\right] \psi_t^u - [\hat{p}_{n-1}, \ \ldots, \ \hat{p}_0]\psi_t^z + z^* \tag{4.1}$$

where $\hat{\ell}_{n-1}, \ \ldots, \ \hat{\ell}_0, \ \hat{p}_{n-1}, \ \ldots, \ \hat{p}_0$ are Lipschitz functions of the estimated parameter vector $\hat{\theta}$. This is equivalent to the feedback law

$$\hat{Q}(\varrho)u_t = -\hat{P}(\varrho)z_t + E(\varrho)z_t^* \tag{4.2}$$

where

$$\hat{Q}(\varrho) = E(\varrho) + \hat{L}(\varrho); \ \hat{L}(\varrho) = \hat{\ell}_{n-1}\varrho^{n-1} + \ldots + \hat{\ell}_0; \ \hat{P}(\varrho) = \hat{p}_{n-1}\varrho^{n-1} + \ldots + \hat{p}_0 \tag{4.3}$$

Note that the control law transfer function is $\frac{-\hat{P}}{\hat{Q}}$ which is strictly proper. Also z_t^* denotes a bounded reference signal. For the moment, we make no further assumptions about $\hat{L},\ \hat{P}$.

From the model (2.1), (2.2), the general controller (4.1), the definition of ψ_t ((3.3) to (3.6)), and the definition of the errors ((3.7) to (3.9)) we can write:

$$d\psi_t = A_t\psi_t dt + B_1(\eta_t dt + d\omega_t) + B_2 z_t^* dt \tag{4.4}$$

where

$$A_t = \left[\begin{array}{ccc|ccc|ccc}
-\hat{a}_{n-1} & \cdots & -\hat{a}_0 & \hat{b}_{n-1} & \cdots & \hat{b}_0 & \hat{f}_{n-1} & \cdots & \hat{f}_0 \\
 & & 0 & & & & & & \\
 & I_{n-1} & \vdots & & & & & & \\
\cdots & & 0 & \cdots & & \cdots & \cdots & & \cdots \\
\hline
-\hat{p}_{n-1} & \cdots & -\hat{a}_0 & -\hat{q}_{n-1} & \cdots & -\hat{q}_0 & 0 & \cdots & 0 \\
 & & & & & 0 & & & \\
 & & & & I_{n-1} & \vdots & & & \\
\cdots & & \cdots & \cdots & & 0 & \cdots & & \cdots \\
\hline
0 & \cdots & 0 & 0 & \cdots & 0 & -\hat{e}_{n-1} & \cdots & -\hat{e}_0 \\
 & & & & & & & & 0 \\
 & & & & & & & I_{n-1} & \vdots \\
\cdots & & \cdots & \cdots & & \cdots & \cdots & & 0
\end{array}\right] \tag{4.5}$$

$$B_1^T = [10\ \ldots\ 010\ \ldots\ 0] \tag{4.6}$$

$$B_2^T = [0\ \ldots\ 010\ \ldots\ 0] \tag{4.7}$$

where B_1^T has 1's in the 1st and $(2n+1)^{st}$ positions and B_2^T has 1 in the $(n+1)^{st}$ position.

A key point about (4.5) is that A_t is a Lipschitz function of $\tilde{\theta}_t$ provided $\hat{\ell}_{n-1},\ \ldots,\ \hat{\ell}_0,\ \hat{p}_{n-1},\ \ldots,\ \hat{p}_0$ are Lipschitz in $\hat{\theta}$.

5. PROPERTIES OF THE ESTIMATION ALGORITHM

An important first step is to establish existence and uniqueness of the solution of the full set of equations describing the system and estimation algorithm. Without this step, one cannot proceed – indeed one can easily show $1+1 = 3$ if one is not careful [12]. Combining equations (3.15), (3.21) and (4.4), the full set of equations is

$$d\psi_t = \left[A_t\left(\bar{\theta}_t\right)\psi_t - B_1\psi_t^T\tilde{\theta}_t - B_1 \int_0^t h_{t-\tau}\psi_\tau^T\tilde{\theta}_\tau d\tau \right] dt$$

$$+ B_1 d\omega_t + B_2 y_t^* dt$$

$$d\tilde{\theta}_t = \left[-\frac{\psi_t\psi_t^T}{r_t(\psi.)}\tilde{\theta}_t - \int_0^t h_{t-\tau}\frac{\psi_t\psi_\tau^T}{r_t(\psi.)}\tilde{\theta}_\tau d\tau \right] dt$$

$$+ \frac{\psi_t}{r_t(\psi.)} d\omega_t \qquad (5.1)$$

where $A_t\left(\bar{\theta}_t\right) \equiv A_t\left(\hat{\theta}_t\right)$ and h_t is the impulse response of the strictly proper part of the transfer function E/D in (3.18).

Preliminary inspection of (5.1) indicates that existence and uniqueness of the solution will depend on being able to bound the coefficient $\left(\frac{\psi_t\psi_t^T}{r_t}\right)$. Thus we define the normalization variable r_t as follows:

$$r_t \triangleq \sup_{0 \le \tau \le t} \psi_\tau^T\psi_\tau + \int_0^t \psi_\tau^T\psi_\tau d\tau + c_0; \quad c_0 > 0 \qquad (5.2)$$

where c_0 is any positive deterministic number.

We then have the following result

Lemma 5.1: The composite set of equations (5.1) has a unique solution with continuous sample paths a.s. up to the random time T of the first explosion. (That is T is the first time that either a component of ψ or $\tilde{\theta}$ becomes infinite or $T = \infty$.

Proof: We first note from (4.5) that $A_t\left(\tilde{\theta}_t\right)$ is Lipschitz in $\tilde{\theta}_t$ due to the assumed form of the dependence of $\hat{\ell}_{n-1}, \ldots, \hat{\ell}_0, \hat{p}_{n-1}, \ldots, \hat{p}_0$ on $\hat{\theta}$. This implies that the coefficient vectors multiplying dt and $d\omega_t$ in (5.1) are locally Lipschitz with respect to the supremum norm on sample paths [9]. (That is given a compact set in the space of $\psi_t, \tilde{\theta}_t$, the functions are Lipschitz with constant depending on the choice of the set).

The result then follows from Theorems (14.18) and (14.20) of [13].

∇∇∇

Note that Lemma 5.1 does not use the projection of the parameter estimates as described at the end of Section 3. When this additional facet of the algorithm is included, we can strengthen Lemma 5.1 as follows:

Lemma 5.2: With the addition of the projection scheme (3.22) to equation (5.1) then $\hat{\theta}_t$ remains in $\mathcal{C}_2$ for all t and the composite set of equations (5.1) has a unique solution a.s. with sample paths $\left(\psi_t, \tilde{\theta}_t\right)$ which are continuous except at the projection times.

Proof: Up to the time of the first projection, θ_t is bounded and thus (5.1) is a linear time varying equation with bounded coefficients and hence ψ_t cannot become unbounded in a finite time. Hence the first projection occurs strictly before the explosion time T of Lemma 5.1 (unless both T and the first projection time are infinite).

After projection, we can apply Lemma 5.1 again and repeat the same argument. Hence due to the linearity of (5.1) for given $\psi_\cdot$, $\tilde{\theta}_t$ exists, is unique and it is bounded by the projection. Then (5.1) is a linear equation with bounded coefficients, and hence ψ_t exists a.s. for all t.

∇∇∇

Lemma 5.2 provides the basic existence and uniqueness result necessary to establish the following result giving properties of the parameter estimator. This result does not depend on a–priori boundedness of the system states. (Note, however, that $\tilde{\theta}_t$ and $\frac{\|\psi_t\|^2}{r_t}$ are bounded by the structure of the algorithm.)

Theorem 5.1: For the general class of feedback control laws described in Section 4 and under Assumptions 1 and 2, the following properties hold for the model ((2.1), (2.2)) and the estimator (3.2), (5.2) with the projection scheme (3.22):

$$\text{(i)} \quad \limsup_{t \to \infty} \int_0^t \frac{\eta_\tau^2}{r_\tau} d\tau \leq K_1 < \infty \qquad \text{a.s.} \tag{5.3}$$

where K_1 is a random variable (realization dependent)

$$\text{(ii) For all finite } \Delta, \lim_{t \to \infty} \sup_{0 \leq T \leq \Delta} \| \hat{\theta}_{t+T} - \hat{\theta}_t \| = 0 \qquad \text{a.s.} \tag{5.4}$$

(iii) There exists a finite random time t_R beyond which no further parameter projections occur.

<u>Proof</u>: (a) Starting from (3.21) and using Ito's rule (see e.g. [14]), we have that, between projections:

$$\begin{aligned} d\left(\tilde{\theta}_t^T \tilde{\theta}_t\right) &= 2\tilde{\theta}_t^T \frac{\psi_t}{r_t}(\eta_t dt + d\omega_t) + \sigma^2 \frac{\| \psi_t \|^2}{r_t^2} dt \\ &= -\frac{2}{r_t}\left(-\tilde{\theta}_t^T \psi_t \eta_t - \epsilon \eta_t^2\right) dt - 2\epsilon \frac{\eta_t^2}{r_t} dt + 2\tilde{\theta}_t^T \frac{\psi_t}{r_t} d\omega_t + \sigma^2 \frac{\| \psi_t \|^2}{r_t^2} dt \end{aligned} \tag{5.5}$$

for some $\epsilon > 0$.

Note that from (3.22) and using Assumption 2, at the times of projection we have $\| \tilde{\theta}_t \|^2 \leq \| \tilde{\theta}_{t-} \|^2 - (R_2 - R_1)^2$.

Defining $\xi_t \triangleq -\tilde{\theta}_t^T \psi_t \eta_t - \epsilon \eta_t^2$ where ϵ is as in Assumption 1, integrating (5.5) and accounting for projections yields

$$\begin{aligned} \tilde{\theta}_t^T \tilde{\theta}_t \leq \tilde{\theta}_s^T \tilde{\theta}_s - 2\int_s^t \frac{\xi_\lambda}{r_\lambda} d\lambda - 2\epsilon \int_s^t \frac{\eta_\lambda^2}{r_\lambda} d\lambda + 2\int_s^t \tilde{\theta}_\lambda^T \frac{\psi_\lambda}{r_\lambda} d\omega_\lambda \\ + \int_s^t \sigma^2 \frac{\| \psi_\lambda \|^2}{r_\lambda^2} d\lambda - N_{t,s}(R_2 - R_1)^2 \end{aligned} \tag{5.6}$$

where $N_{t,s}$ is the number of times that projections occur between times s and t.

Consider now the integral $\int_0^t \frac{\| \psi_\lambda \|^2}{r_\lambda^2} d\lambda$. We have, using (5.2),

$$\int_0^t \frac{\| \psi_\lambda \|^2}{r_\lambda^2} d\lambda \le \int_0^t \frac{dr_\lambda}{r_\lambda^2} = \frac{1}{r_0} - \frac{1}{r_t} \le \frac{1}{r_0} = \frac{1}{c_0} \tag{5.7}$$

(This operation makes sense due to Bonnet's and Du Bois–Reymond's formulae [15] which allow the integral to be considered as a Riemann integral.)

We now define $\mathcal{H}_t$ as the solution of the Ito integral

$$d\mathcal{H}_t = \frac{2\tilde{\theta}_t^T \psi_t}{r_t} d\omega_t \, ; \qquad \mathcal{H}_0 = \frac{\sigma^2}{c_0} + \frac{2K \| \gamma_0 \|^2}{r_0} + \tilde{\theta}_0^T \tilde{\theta}_0 \tag{5.8}$$

This integral makes sense thanks to (5.7) and Lemma 5.2. Since c_0, r_0, γ_0 and $\tilde{\theta}_0$ are $\mathcal{F}_0$ – measurable, $(\mathcal{H}_t, \ \mathcal{F}_t)$ is a Martingale. Moreover it follows from (5.6) and (5.7) that $\mathcal{H}_t$ satisfies

$$\mathcal{H}_t \ge \tilde{\theta}_t^T \tilde{\theta}_t + 2 \int_0^t \frac{\xi_\lambda}{r_\lambda} d\lambda + 2\epsilon \int_0^t \frac{\eta_\lambda^2}{r_\lambda} d\lambda + \frac{\sigma^2}{c_0} - \int_0^t \sigma^2 \frac{\| \psi_\lambda \|^2}{r_\lambda^2} d\lambda + \frac{2K \| \gamma_0 \|^2}{r_0}$$

$$+ N_{t,0} [R_2 - R_1]^2 \tag{5.9}$$

The second term on the right hand side o f (5.8) can be easily shown [9] to be positive using (3.20). Hence using (5.7) it follows that $\mathcal{H}_t$ is positive. Thus $(\mathcal{H}_t, \ \mathcal{F}_t)$ is a positive Martingale and hence

$$\lim_{t \to \infty} \mathcal{H}_t = \mathcal{H} < \infty \qquad \text{a.s.} \tag{5.10}$$

Using (5.10) and noting (5.7) and Assumption 1 we conclude that (5.3) holds for some finite random variable K_1.

This establishes (i). Also, since $R_2 > R_1$, then from (5.9), (5.10) $N_{t,0}$ is bounded a.s. Hence (iii) follows.

(b) Result (ii) follows in a straightforward way from (i) and (ii) using the Schwartz inequality. Details are given in [9].

∇∇∇

Comment 5.1

A key feature of our estimator is our choice of r_t (see (5.2)). It guarantees that $\frac{\| \psi_t \|^2}{r_t} \leq 1 \ \ \forall t$ and this is crucial in establishing Lemma 5.1, 5.2 and Theorem 5.1. If we were to define $r_t = \int_0^t \psi_\tau^T \psi_\tau d\tau$ which is the direct analogue of the discrete case [7], then proofs would require the assumption that $\frac{\| \psi_t \|^2}{r_t}$ is almost surely bounded [16]. This assumption is rather unrealistic given that ψ_t contains signals driven by white noise.

6. LINKING ESTIMATOR PROPERTIES TO CLOSED LOOP PROPERTIES

In establishing the estimator properties given in Theorem 5.1 we made only oblique appeal to the form of the certainty equivalence feedback law. In particular, it was not assumed that the law would stabilize the system if the true parameters were known. Thus, to make further progress we will need to strengthen the assumption on the feedback law made in Section 4.

The proof given below depends upon the fact that the certainty equivalence control law stabilizes the frozen estimated model. This is true of a wide class of algorithms. However, to be specific, we will illustrate the analysis procedure by considering two cases; namely adaptive pole assignment and adaptive model reference control. In this section we will study the pole assignment algorithm. The model reference problem will be taken up later. In the pole assignment case, the polynomials $\hat{L}$ and $\hat{P}$ defining the control law are computed from the estimated $\hat{A}(\hat{\theta}_t)$ and $\hat{B}(\hat{\theta}_t)$ as follows.

Let

$$\hat{\theta} = \left[\hat{g}_{n-1}, \ \ldots, \ \hat{g}_0, \ \hat{b}_{n-1}, \ \ldots, \ \hat{b}_0, \ \hat{f}_{n-1}, \ \ldots, \ \hat{f}_0\right]$$

and define (c.f. (2.6), (2.21))

$$\hat{a}_i = e_i - \hat{g}_i; \quad i = 0, \ldots, n-1.$$

$$\hat{A}(\varrho) = \varrho^n + \hat{a}_{n-1}\varrho^{n-1} + \ldots \hat{a}_0$$

$$\hat{B}(\varrho) = \hat{b}_{n-1}\varrho^{n-1} + \ldots + \hat{b}_0$$

$$\hat{F}(\varrho) = \hat{f}_{n-1}\varrho^{n-1} + \ldots + \hat{f}_0$$

Then, for a given possibly time varying A^* of degree 2n, solve the following equation for $\hat{Q}$ and $\hat{P}$

$$\hat{Q}\hat{A} + \hat{P}\hat{B} = A^* \quad \text{with} \quad \operatorname{Re}\lambda_i(A^*) \leq -\beta < 0, \quad i = 1, \ldots, 2n \tag{6.1}$$

where $\lambda_i(A^*)$ are the eigenvalues of the polynomial A^*.

Finally compute

$$\hat{L}(\varrho) = \hat{Q}(\varrho) - E(\varrho)$$

Equation (6.1) can also be written

$$M(\hat{\theta}) \begin{bmatrix} \hat{q}_0 \\ \cdot \\ \cdot \\ \hat{q}_{n-1} \\ \hat{p}_0 \\ \cdot \\ \cdot \\ \hat{p}_{n-1} \end{bmatrix} = \begin{bmatrix} a_0^* \\ \cdot \\ \cdot \\ \cdot \\ \cdot \\ \cdot \\ \cdot \\ a_{2n-1}^* \end{bmatrix} \tag{6.2}$$

The polynomial A^* is a design polynomial available to the user. The only restriction we require on the polynomial $A^*(t)$ is:

Assumption 3

$\forall t$, the polynomial $A^*(t)$ is continuous and bounded, has a uniform stability margin (i.e. $\mathrm{Re}\,\lambda_i(A^*(t)) \le -\beta < 0,\ i = 1, \ldots, 2n)$ and $\lim_{t\to\infty} \sup_{0\le T\le\Delta} \| a^*(t+T) - a^*(t) \| = 0$ for some $\Delta > 0$, where a^* is the vector of coefficients of A^*.

∇∇∇

From Section 4, we require that $\hat{L},\ \hat{P}$ be Lipschitz functions of $\hat{\theta}$ for all time. We note that the projection facility (3.22) ensures that $\hat{\theta} \in \mathcal{C}_2$ for all time. Then, the Lipschitz condition will be automatically satisfied provided all models corresponding to $\hat{\theta} \in \mathcal{C}_2$ are uniformly stabilizably.

This assumption will be eliminated in the case of model reference control of stably invertible systems studied later. For the case of pole assignment we introduce:

Assumption 4:

Assumption 2 is satisfied and there exists a known positive constant ϵ such that for all θ_t in $\mathcal{C}_2$,

$$\det M(\theta_t) \ge \epsilon \qquad \nabla\nabla\nabla$$

Subject to Assumption 4, the projection scheme ensures that $\det M(\theta_t) \ge \epsilon$ for all t. In view of (6.2) this will ensure that $\hat{L},\ \hat{P}$ are Lipschitz functions of $\hat{\theta}$ as required.

To link the estimated properties to the closed loop properties we will make use of the following Lemma.

Lemma 6.1 (Continuous Time Key Technical Lemma): Consider a realisation produced by the model (2.1)–(2.2). Suppose the estimator is such that property (i) of Theorem 5.1 holds and the controller is such that the following growth condition is satisfied:

$$\frac{r_t}{t} \le C + \frac{K_2}{t}\int_0^t \eta_\tau^2 d\tau \tag{6.3}$$

where r_t, η_t are defined before and C, K_2 are finite positive constants. Then

$$\text{(i)} \quad \lim_{t\to\infty} \sup \frac{r_t}{t} \le K_3 < \infty \tag{6.4}$$

$$\text{(ii)} \quad \lim_{t\to\infty} \frac{1}{t} \int_0^t \eta_\tau^2 d\tau = 0 \tag{6.5}$$

Proof: (i) From (6.3) and the nonnegativity of η_τ^2 and r_τ it follows that

$$\frac{r_t}{t} \le C + K_2 \int_0^t \frac{\eta_\tau^2}{r_\tau} \frac{r_\tau}{\tau} d\tau$$

The result then follows from the Bellman–Gronwall lemma (see e.g [18]) using (i) of Theorem 5.1.

(ii) Suppose first that $\lim_{t\to\infty} r_t = \infty$. Then, by the Kronecker lemma [7], [9]:

$$\lim_{t\to\infty} \frac{1}{r_t} \int_0^t \eta_\tau^2 d\tau = 0 \tag{6.6}$$

Hence, using (6.4) and (6.6)

$$\lim_{t\to\infty} \frac{1}{t} \int_0^t \eta_\tau^2 d\tau = \lim_{t\to\infty} \frac{r_t}{t} \frac{1}{r_t} \int_0^t \eta_\tau^2 d\tau = 0 \tag{6.7}$$

Alternatively, if $\lim_{t\to\infty} r_t \le K_4 < \infty$ then

$$\lim_{t\to\infty} \frac{1}{t} \int_0^t \eta_\tau^2 d\tau \le \lim_{t\to\infty} \frac{1}{t} \frac{K_4}{r_t} \int_0^t \eta_\tau^2 d\tau$$
$$\le \lim_{t\to\infty} \frac{K_4}{t} \int_0^t \frac{\eta_\tau^2}{r_\tau} d\tau = 0$$

using (i) of Theorem 5.1. The last inequality follows from the monotonicity of r_τ.

∇∇∇

Our key remaining task is to establish the growth condition (6.3). We will use the BIBO properties of the closed loop state space model (4.4). We first establish that the homogeneous part of (4.4) is exponentially stable.

Lemma 6.2: Consider the homogeneous time varying differential equation

$$\frac{d}{dt}\psi_t = A_t\psi_t$$

with A_t given by (4.5). Assume that the $\hat{a}_i$, $\hat{b}_i$, $\hat{f}_i$ are estimated using the parameter estimator of Section 3 including the projection scheme (3.22) and that Assumptions 1 to 4 hold. Then (6.1) is exponentially stable a.s..

Proof: We have shown in Theorem 5.1 that there exists a random time t_R beyond which no further projections occur. The result then follows using Lemma 3 of [17] since the pole assignment condition (6.1) ensures that the eigenvalues of each frozen matrix A_t are in the strict left half plane. Details are given in [9].

∇∇∇

We can now establish the following convergence result.

Theorem 6.1: Consider the system ((2.1), (2.2)), with the parameter vector θ satisfying Assumption 2, the parameter estimator of Section 3 with projection, an observer polynomial E satisfying Assumption 1, and an adaptive pole assignment control law of the form (4.2), (6.1) satisfying Assumptions 3 and 4. Then for arbitrary finite initial conditions and an arbitrary, piecewise continuous, uniformly bounded reference input z_t^*, the following results hold

(i) $\limsup_{t\to\infty} \frac{1}{t}\int_0^t \| \psi_\tau \|^2 \, d\tau < \infty$ a.s.

(ii) $\limsup_{t\to\infty} \frac{1}{t}\int_0^t u_\tau^2 d\tau < \infty$ a.s.

(iii) $\limsup_{t\to\infty} \frac{1}{t}\int_0^t |Cx_\tau|^2 d\tau < \infty$ a.s.

Proof: The proof uses a standard argument on robustness of bounded solutions of exponentially stable linear time varying systems. We use Lemma 6.2 to establish from (4.4) that the growth condition (6.3) holds. We then use Lemma 6.1 in an obvious way – see [9].

∇∇∇

7. MODEL REFERENCE STOCHASTIC ADAPTIVE CONTROL

The adaptive pole assignment result of the previous section established a form of closed loop stability ((i), (ii), (iii) of Theorem 6.1) but did not establish any form of asymptotic convergence to a desired tracking objective. We thus consider an alternative feedback law of the model reference type [3], [4]. An important preliminary observation from [3], [4] that the issue of plant relative degree is crucial to avoid differentiators in the control law. We will thus strengthen our assumptions to include knowledge of this index.

To relate our results more closely to those in the deterministic literature, particularly those results in [4] which are extended here to the stochastic case, we will use the following equivalent form of the model (2.1) (2.2):

$$\overline{A}(\varrho)y_t = \overline{B}(\varrho)u_t + \overline{C}(\varrho)\omega_t \tag{7.1}$$

where

$$\overline{A}(\varrho) \triangleq \varrho^n + a_{n-1}\varrho^{n-1} + \cdots + a_0 \tag{7.2}$$

$$\overline{B}(\varrho) \triangleq b_m\varrho^m + \cdots + b_0 \ , \ b_m > 0 \tag{7.3}$$

$$\overline{C}(\varrho) \triangleq \varrho^n + c_{n-1}\varrho^{n-1} + \cdots + c_0 \tag{7.4}$$

and ϱ denotes (d/dt). Clearly y_t and ω_t replace dz_t/dt and dw_t/dt respectively. The representation (7.1) is formal and should be regarded as shorthand for (2.1) and (2.2). The transcription from (2.1) and (2.2) to (7.1) and vice versa is simple if an appropriate canonical form is employed.

In (7.1), $\overline{C}(\varrho)$ is the Kalman observer polynomial and is monic and Hurwitz.

Assumptions 1 to 4 in the previous analysis are replaced by::

A1: The degrees n and m of the polynomials $\overline{A}$ and $\overline{B}$ are known.

A2: Upper and lower bounds are known for all parameters. Also the sign of b_m is known as well as a lower bound ($\overline{b}_{\min}$) on its magnitude.

A3: There are no unmodelled dynamics

A4: The reference model is exponentially stable and the plant is minimum phase (i.e. $\overline{B}$ is Hurwitz). The input to the reference model, ξ, is piecewise continuous and uniformly bounded.

We will use $\nu(A)$ to denote the degree of A and without loss of generality we will take b_m to be positive (b_m negative simply requires u to be redefined as $-u$).

In view of assumption A2, we can then define two convex regions $\mathcal{C}_1$ and $\mathcal{C}_2$ as follows:

$$\mathcal{C}_1 \triangleq \left\{\theta : \theta^i_{\text{min}} \le \theta_i \le \theta^i_{\text{max}}\right\} ; \; \theta^i_{\text{min}} = \overline{b}_{\text{min}} \text{ for } \theta_i \equiv b_m$$

$$\mathcal{C}_2 \triangleq \left\{\theta : \theta^i_{\text{min}} - b_{\text{min}} \le \theta_i \le \theta^i_{\text{max}} + b_{\text{min}}\right\} ; \; b_{\text{min}} = \frac{\overline{b}_{\text{min}}}{2}$$

Note that $\theta \in \mathcal{C}_1 \subset \mathcal{C}_2$ and that $b_m > b_{\text{min}}$ for $\theta \in \mathcal{C}_2$. Also, note that unlike Assumption 4 of Section 6, we do not now require any relative primeness assumptions inside $\mathcal{C}_2$.

To form a basis for the adaptive control algorithm it is first necessary to define the control objective in the non-adaptive case. Our basic aim is to have a 'minimum variance' type algorithm as in [2]. However, the simplistic stochastic discrete time objective of [2] does not have a continuous time counterpart since the direct analogue requires some of the closed loop poles to be placed at $-\infty$. Thus, we will instead, combine the essential idea from the stochastic discrete time case [2] of using an optimal Kalman observer, with the essential idea from the continuous time deterministic case [3], [4] of defining the feedback objective via a reference model. This gives a stochastic model reference control law.

The reference model is defined to be

$$y^* = \mathcal{M}\xi \tag{7.5}$$

where $\mathcal{M}$ is exponentially stable and proper. Since the relative degree of the closed loop system cannot be less than that of the plant, we further decompose $\mathcal{M}$ as

$$\mathcal{M} = M/E_2 \tag{7.6}$$

where M is any proper exponentially stable transfer function and E_2 is any monic Hurwitz polynomial with $\nu(E_2) = n - m$.

Since our aim is to achieve the closed loop transfer function given in (7.5), then the plant numerator $\overline{B}$ must be cancelled in forming the control law. Also we require that the "observer" be the optimal Kalman observer $\overline{C}$. Hence the closed loop characteristic polynomial is required to be of the form: $A^* = \overline{C}E_2\overline{B}$. Thus, if we consider linear feedback control laws of the form:

$$Lu = -Gy + Dr \tag{7.7}$$

where $\nu(L) = n$; $\nu(G) = n-1$ and D is an appropriate exponentially stable transfer function, then the desired closed loop characteristic polynomial is achieved if G and L satisfy the following pole–assignment identity

$$E_2CB = L\overline{A} + G\overline{B} \tag{7.8}$$

In (7.8) and subsequent equations we will suppress the explicit dependence on the operator ϱ when this is clear from the context.

Since $\overline{B}$ is a factor of two of the terms in (7.8), we can write $L = F\overline{B}$. Then cancelling $\overline{B}$ in (7.8) gives

$$E_2\overline{C} = F\overline{A} + G \tag{7.9}$$

where

E_2 is the denominator of the reference model and is any monic Hurwitz polynomial with $\nu(E_2) = n-m$;

$\overline{C}$ is the Kalman observer as in equation (7.1);

F is a monic polynomial; $\nu(F) = n-m$;

G is a (not–necessarily monic) polynomial; $\nu(G) = n-1$.

Note that (7.9) can always be solved for F and G, given E_2, $\overline{C}$ and $\overline{A}$ whether or not $\overline{A}$ and $\overline{B}$ are relatively prime.

We then have the following result:

Lemma 7.1: Consider the feedback control law (7.7) where $L = F\overline{B}$, $D = CM$ and F and G satisfy the identity (7.9). Then subject to assumption A4, the closed loop system is exponentially stable and (modulo exponentially decaying terms) y tracks the output of the reference model, y^*, with error $(F/E_2)\omega$ i.e.

$$y = (1/E_2)\overline{\xi} + (F/E_2)\omega \tag{7.10}$$

where $\overline{\xi} \triangleq M\xi$.

Proof: Immediate by substitution.

∇∇∇

In the next section we will construct a certainty equivalence adaptive form of the above control law.

8. DIRECT FORM OF THE ALGORITHM

To obtain a direct adaptive controller it is necessary to express (7.1) in a form such that the estimated parameters yield directly the control parameters. We proceed as follows:

To simplify the subsequent analysis we introduce the following notation. For every integer j, P^j is a row vector of polynomials (in ϱ) defined by:

$$P^j \triangleq (1, \varrho, \varrho^2, \ \dots \ , \varrho^j) \tag{8.1}$$

For any polynomial Γ of the form:

$$\Gamma(\varrho) \triangleq \gamma_p \varrho^p + \gamma_{p-1}\varrho^{p-1} + \ \dots \ + \gamma_0 \tag{8.2}$$

the coefficient vector $\mathrm{c}(\Gamma) \in \mathbb{R}^{p+1}$ is defined by

$$\mathrm{c}(\Gamma) \triangleq (\gamma_0, \gamma_1, \ \dots \ , \gamma_p) \tag{8.3}$$

Finally, for any signal z and any polynomial Γ, the signal z_Γ is defined by:

$$z_\Gamma \triangleq (1/\Gamma)z \tag{8.4}$$

In the sequel, we will define a number of regression vectors ψ, ϕ etc. depending on y, u in various ways. For convenience, the notation is summarized below.

<u>Notation for ϕ and ψ vectors:</u>

ψ_0, ϕ_0	full regression vectors containing ω (see (8.9), (8.10))
ψ, ϕ	full regression vectors with ω replaced by $\hat{\omega}$ (see (8.18), (8.19))
$\overline{\psi}, \overline{\phi}$	regression vectors with ω replaced by $\hat{\omega}$ and the last n–m terms set to zero (see (8.22), (9.3))
ψ_c, ϕ_c	regression vectors containing ω and the last n–m terms set to zero (see (8.13))

In all cases $\psi_* = (\phi_*/E_2)$

We have the following result:

Lemma 8.1:
(i) The model (7.1) can be rearranged into the following (non–minimal) form:

$$E_2 y = \left[\overline{G} y_{E_1} + \overline{F} G y_E + F\overline{F} B u_E - \overline{G} F \omega_E + (F - E_2)\omega\right] + E_2\omega \tag{8.5}$$

where E_1 is any Monic Hurwitz polynomial, $\nu(E_1) = n$; E_2 is as in (7.6); F and G satisfy (7.9) and $\overline{F}, \overline{G}$ satisfy

$$E \triangleq E_1 E_2 = \overline{F} C + \overline{G} \tag{8.6}$$

with $\overline{F}$ a monic polynomial; $\nu(\overline{F}) = n - m$; $\overline{G}$ a (not necessarily monic) polynomial; $\nu(\overline{G}) = n - 1$.

(ii) The model (8.5) can be expressed in the linear regression form (similar to (2.19)); i.e.

$$E_2 y = \phi_0^T \theta_0 + E_2 \omega \tag{8.7}$$

or

$$y = \psi_0^T \theta_0 + \omega \tag{8.8}$$

where

$$\phi_0^T \triangleq \left(P^{n-1} y_{E_1}, P^{2n-m-1} y_E, \overline{P}^{2n-m} u_E, P^{2n-m-1} \omega_E, P^{n-m-1} \omega \right) \tag{8.9}$$

$$\psi_0 \triangleq (1/E_2)\phi_0 \tag{8.10}$$

$$\theta_0^T \triangleq \left(\mathrm{e}(\overline{G}), \mathrm{e}(\overline{F}G), (\mathrm{e}(Q), b_m), \mathrm{e}(-\overline{G}F), \mathrm{e}(F - E_2) \right) \tag{8.11}$$

and

$$\overline{P}^{2n-m} \triangleq (P^{2n-m-1}, E) \tag{8.12}$$

so that $\overline{P}^{2n-m} u_E = (P^{2n-m-1} u_E, u)$.

It follows that θ_0, ϕ_0 and ψ_0 are p–dimensional vectors where $p \triangleq 8n - 4m + 1$ and $\phi_0^q = u$ and $\theta_0^q = b_m$ where $q \triangleq 5n - 2m + 1$.

(iii) The equivalent model (8.8) is in a form whereby the control law can be expressed as a direct function of θ_0. To do this, we define ϕ_c to be equal to ϕ_0 save that the last n–m (non causal) terms are set to zero:

$$\phi_c^T \triangleq (P^{n-1} y_{E_1}, P^{2n-m-1} y_E, \overline{P}^{2n-m} u_E, P^{2n-m-1} \omega_E, 0) \tag{8.13}$$

Then the stochastic model reference control law (7.7) is equivalent to

$$\phi_c^T \theta_0 = \overline{\xi} \tag{8.14}$$

which yields a closed loop response as in (7.10)

Proof: (i) Multiplying (7.1) by $F\overline{F}$ and substituting $E_2\overline{C} - G$ for $F\overline{A}$ and $E - \overline{G}$ for $\overline{FC}$ yields the modified system equation:

$$E_2\overline{F}\overline{C}y = \overline{F}Gy + F\overline{F}Bu + F(E - \overline{G})\omega \tag{8.15}$$

Substituting $E - \overline{G}$ for $\overline{F}C$ in the left hand side of (8.3) yields (8.5) after some simple algebra.

(ii) Noting that $\overline{F}F\overline{B} = b_m E + \overline{Q}$ where $\nu(\overline{Q}) = 2n - m - 1$, (8.5) may be expressed as (8.7).

(iii) From (8.13), (8.9) we have

$$\phi_0^T = E_2\psi_0^T = \phi_c + (0, 0, 0, 0, P^{n-m-1}\omega_{E_2}) \tag{8.16}$$

so that; from (8.5) and (8.7):

$$\begin{aligned} E_2 y &= E_2\left[\psi_0^T\theta_0 + \omega\right] = \phi_0^T\theta_0 + E_2\omega \\ &= \left[\overline{G}y_{E_1} + \overline{F}Gy_E + \overline{F}FBu_E - \overline{G}F\omega_E + F\omega\right] \\ &= \phi_c^T\theta_0 + F\omega \end{aligned} \tag{8.17}$$

Hence substituting the control law (8.14) into (8.17) immediately gives (7.10) i.e.

∇∇∇

The above result shows how the system model can be rearranged into a form in which the control law is a direct function of the model parameters. However, the regressor vector ψ_0 cannot be used for estimation since ω is not known. Therefore, as is in Section 3, we introduce a pseudo–regressor ψ obtained by replacing ω with $\hat{\omega}$. Thus we define

$$\psi \triangleq (1/E_2)\phi \tag{8.18}$$

where

$$\phi \triangleq \left(P^{n-1}y_{E_1}, P^{2n-m-1}y_E, \overline{P}^{2n-m}u_E, P^{2n-m-1}\hat{\omega}_E, P^{n-m-1}\hat{\omega}\right) \tag{8.19}$$

$$\hat{\omega} \triangleq y - \hat{y}, \tag{8.20}$$

$$\hat{y} \triangleq \psi^T \hat{\theta} = \hat{\overline{G}} y_E + (\overline{F}\hat{G}) y_{EE_2} + (F\hat{\overline{F}}B) u_{EE_2} - (\overline{G}\hat{F}) \hat{\omega}_{EE_2} + (F - \hat{}E_2) \hat{\omega}_{E_2} \tag{8.21}$$

and $\hat{\theta}$ is the current estimate of θ.

Similarly, we define the certainty equivalence stochastic model reference control law as follows: We first define $\overline{\phi}$ by replacing ω, θ_0 by $\hat{\omega}, \hat{\theta}$ respectively in the definition of ϕ_c (see (8.13)):

$$\overline{\phi}^T \triangleq \left(P^{n-1} y_{E_1}, P^{2n-m-1} y_E, \overline{P}^{2n-m} u_E, P^{2n-m-1} \hat{\omega}_E, 0 \right) \tag{8.22}$$

Then, the certainty equivalence control law is defined as in (8.14) but with ϕ_c, θ_0 replaced by $\overline{\phi}, \hat{\theta}$ respectively:

$$\overline{\phi}^T \hat{\theta} = \overline{\xi} \tag{8.23}$$

We then have the following result on certainty equivalence adaptive control:

Lemma 8.2:

(i) Using the regression vector ψ to define $\hat{y}$ as in (8.21) gives

$$y - \omega = \hat{y} + \eta = \psi_0^T \theta_0 \tag{8.24}$$

where

$$\eta \triangleq \hat{\omega} - \omega \tag{8.25}$$

is the "noise reduced prediction error".

(ii) η is related to the parameter error $\tilde{\theta}$ as follows:

$$(F\overline{F}\overline{C}/EE_2)\eta = -\psi^T \tilde{\theta} \tag{8.26}$$

where

$$\tilde{\theta} \triangleq \hat{\theta} - \theta_0 \tag{8.27}$$

(iii) Provided $\hat{\theta}$ is constrained so that $\hat{\theta} \in \mathfrak{C}_2$ then the control law (8.23) is well posed and involves proper functions of the input and output.

(iv) Use of the certainty equivalence control law (8.23) gives a closed loop system satisfying:

$$E_2 y = [\overline{\xi} + F\omega] + \left[(\overline{G}F/E)\eta - \overline{\phi}^T\tilde{\theta}\right] \tag{8.28}$$

$$E_2\overline{B}u = \overline{A}(\overline{\xi} - \overline{\phi}^T\tilde{\theta}) + [\overline{AG}F/E]\eta - G\omega \tag{8.29}$$

Proof: Straightforward by substitution – see [10].

∇∇∇

Comparing (7.10) with (8.28) we see that the adaptive controller, when compared with 'ideal' controller, introduces the extra errors $\overline{\phi}^T\tilde{\theta}$ and $(\overline{G}F/E)\eta$. One main concern will be to quantify these errors.

9. ESTIMATION ALGORITHM REVISITED

The estimation algorithm is exactly as in (3.2) with ψ_t etc. redefined as above. We also replace the projection scheme given at the end of Section 3 by:

Modified Projection Scheme:

Let τ be the time for which the solution of (3.2) is such that $\hat{\theta}_t$ lies on the boundary of $\mathfrak{C}_2$ (defined in section 7). Denote the corresponding value of $\hat{\theta}_\tau$ by $\hat{\theta}_{\tau-}$. At time τ, the estimate $\hat{\theta}_\tau$ is then defined as the point in $\mathfrak{C}_1$ closest to $\hat{\theta}_{\tau-}$. For $t \geq \tau$, equation (3.2) is then integrated with initial condition $\hat{\theta}_\tau$. This makes $\hat{\theta}_t$ right continuous at the projection times and ensures that $\hat{\theta}_t \in \mathfrak{C}_2$ for all t.

The initial value of $\hat{\theta}_0$ is chosen so that $\hat{\theta}_0 \in \mathcal{C}_1$.

For the case of stochastic model reference adaptive control we need to modify the normalization variable r_t to also bound the derivatives of $\overline{\psi}$. This will be needed in our subsequent analysis. We thus define r_t as follows: for all $j \in 1, 2 \ \dots \ , n-m$, μ_t^j is defined by

$$\mu_t^j \triangleq \sup_{\tau \in [0,t]} \left\{ \| \varrho^{j-1} \overline{\psi}_\tau \|^2 \right\} + C_0 \tag{9.1}$$

whereas μ_t^0 is given by:

$$\mu_t^0 \triangleq \sup_{\tau \in [0,t]} \left\{ \| \psi_\tau \|^2 \right\} + \int_0^t \| \psi_\tau \|^2 \, d\tau + C_0 \tag{9.2}$$

and $\overline{\psi}$ is defined to be $(1/E_2)\overline{\phi}$, i.e. using (8.26);

$$\overline{\psi}^T \triangleq (1/E_2) \left\{ P^{n-1} y_{E_1} \ ; \ p^{2n-m-1} y_E \ ; \ \overline{P}^{2n-m} u_E \ ; \ p^{2n-m-1} \hat{\omega}_E \ ; \ 0 \right\} \tag{9.3}$$

Also C_0 in (9.1) and (9.2) is an arbitrary positive constant. Finally, the normalization variable r_t is defined by:

$$r_t \triangleq \max \left\{ \mu_t^j | j \in \ \{0, 1, \ \dots \ , n-m\} \right\} \tag{9.4}$$

Clearly, r_t, is not less than μ_t^0 which we observe is the normalization employed in Section 3.

To establish convergence properties of the adaptive control law, we need the following additional assumption which is analogous to Assumption 1 of Section 3:

A5: The transfer function relating η to $-\psi^T \tilde{\theta}$, i.e., $(F\overline{F}C/EE_2)$ is strictly passive, i.e. there exists a $\beta > 0$ such that (ignoring the term due to initial conditions):

$$\int_0^t \bar{y}_s \bar{u}_s ds \geq \int_0^t \beta \bar{u}_s^2 ds$$

for all $t > 0$ and all $\bar{u}$ where $\bar{y} \triangleq (F\overline{FC}/EE_2)\bar{u}$.

We then immediately have the properties established in Lemma 5.1, Lemma 5.2 and Theorem 5.1.

10. CLOSED LOOP PROPERTIES

We will again employ Lemma 6.1. To do this we must first establish the growth condition given in (6.3). This is relatively straightforward using the inverse stability of the plant (assumption A4) – see [10] for details.

Finally, we obtain an extended form of theorem 6.1, namely:

Theorem 10.1: Consider the system (2.1) and (2.2) (equivalently, (7.1)), the parameter estimator of Sections 3 and 9, and the certainty equivalence controller of (8.23). Assume that assumptions A1 – A5 are satisfied, and that the reference input ξ is piecewise continuous and uniformly bounded. Then the algorithm converges in the sense that

(i) The regression vector is sample m.s. bounded almost surely; i.e.

$$\limsup_{t\to\infty}(1/t)\int_0^t \| \psi_s \|^2 ds < \infty \quad \text{a.s.} \tag{10.1}$$

(ii) The noise reduced output is sample m.s. bounded almost surely; i.e.

$$\limsup_{t\to\infty}(1/t)\int_0^t \| \overline{C}x_s \|^2 ds = \limsup_{t\to\infty}(1/t)\int_0^t (y_s - \omega_s)^2 ds < \infty \quad \text{a.s.} \tag{10.2}$$

(iii) The input is sample m.s. bounded almost surely; i.e.

$$\limsup_{t\to\infty}(1/t)\int_0^t u_s^2 ds < \infty \text{ a.s.} \tag{10.3}$$

(iv) The following asymptotic tracking performance is achieved:

$$\limsup_{t\to\infty}(1/t)\int_0^t (y_s - y_s^*)^2 ds = 0 \text{ a.s.} \tag{10.4}$$

where y_s^* is the desired reference model response together with a noise term, i.e.

$$y^* = (1/E_2)\xi + (\hat{F}/E_2)\omega \tag{10.5}$$

where $\hat{F}$ is the estimate of the polynomial F.

(Equations (10.4), (10.5) should be compared with (7.10) for the case of known parameters.) ❑

Proof: Basically as for the proof of theorem 6.1 – see [10] for details

∇∇∇

The above theorem establishes both stability ((10.1) to (10.3)) as well as asymptotic convergence to the desired tracking objective ((10.4), (10.5)).

A key assumption in the above analysis has been the requirement that the relative degree be known. We have recently shown [19] that this assumption can be removed by running parallel estimators corresponding to each possible relative degree and then using a switching strategy to decide which estimator to use for the purpose of defining the feedback control law.

11. PERSPECTIVE

The convergence analyses given above establish key existence results for stochastic adaptive control. Their significance is that they establish stability of the closed loop which is clearly an essential property. However, the form of stability is weak (L_2 functions can still be unbounded). Moreover, they fall well short of what is needed in practical applications. The difficulty is that the beauty of the theory lies in its critical dependence on the assumptions. Paradoxically this also equates to a practical weakness. Indeed, a key practical requirement is likely to be that one have insensitivity to assumptions but this is clearly contrary to the requirements of the current precisely "honed" theory.

An issue which is yet to be addressed satisfactorily is that of nasty effects in the plant (nonlinearities, unmodelled dynamics, nonstationary noise etc.) These effects need to be accounted for in the design of practical stochastic adaptive control algorithms.

An initial step has been taken in [20] where we show that it is possible to quantify the errors due to both noise and undermodelling in the estimated model with finite data. In principle this permits the following actions:

- use of robust control design methods which account for model inadequacy.
- optimal determination of model order. (Indeed, we argue in [20] that this will never be the true model order but a lower order model which gives an optimal trade–off between bias errors (due to undermodelling) and variance errors (due to noise)).

However, these are very preliminary results and much more work is needed to overcome the restrictions of the Certainty Equivalence approach. We suggest that a useful area for future research would be to study adaptive control algorithms which, inter alia, incorporate both an estimated nominal model together with a quantification of its accuracy accounting for noise, undermodelling, finite data and possible time variations. These algorithms will almost certainly look very different from those outlined above.

12. REFERENCES

[1] Feuer A. and S. Morse (1978). Adaptive control of single input single output linear systems. *IEEE Trans. Auto. Control*, AC–23, pp.557–570.

[2] Goodwin, G.C., P.J. Ramadge and P.E. Caines (1980). Discrete time multivariable adaptive control. *IEEE Trans. Auto. Control*, AC–25, pp.449–456.

[3] Morse, A.S. (1980). Global stability of parameter–adaptive control systems. *IEEE Trans. Auto. Control*, AC–25, pp.433–439.

[4] Goodwin, G.C. and D.Q. Mayne (1987). A parameter estimation perspective of continuous time model reference adaptive control. *Automatica*, 23, pp.57–70.

[5] Astrom, A.S. and B. Wittenmark (1973). On self tuning regulators. *Automatica*, 9, pp.195–199.

[6] Ljung, L. Analysis of recursive stochastic algorithms. *IEEE Trans. Auto. Control*, AC–22, pp.551–575.

[7] Goodwin, G.C., P.J. Ramadge and P.E. Caines (1981). Discrete time stochastic adaptive control. *SIAM J. Control and Optimization*, 19, pp.829–853.

[8] Middleton, R.H. and G.C. Goodwin (1990). Digital Control and Estimation: A Unified Approach. Prentice–Hall.

[9] Gevers, M., G.C. Goodwin and V. Wertz (1987). Continuous time stochastic adaptive control. To appear, *SIAM Journal on Control and Optimization*, 1991.

[10] Goodwin, G.C. and D.Q. Mayne (1991). Continuous time stochastic model reference adaptive control. To appear *IEEE Trans. Auto. Control*.

[11] Goodwin, G.C. and K.S. Sin (1984). Adaptive Filtering Prediction and Control, Prentice Hall.

[12] Morse, S. (1990). Private Correspondence.

[13] Jacod, J. (1979). Calcul Stochastique et Problemes de Martingales, 714, Lecture Notes in Mathematics, Springer Verlag, Berlin.

[14] Wong, E. (1971). Stochastic Processes in Information and Dynamical Systems. McGraw Hill.

[15] Deheuvels (1980). L'integrale. Presses Universitaires de France, Mathematiques.

[16] Chen, H.F. Recursive Estimation and Control for Stochastic Systems. Wiley, New York.

[17] Kreisselmeier, G. (1985). An approach to stable indirect adaptive control. *Automatica*, 21, 4, pp.425–433.

[18] Desoer, C.A. and M. Vidyasagar (1975). Feedback Systems: Input–Output Properties. Academic Press.

[19] Morse, A.S., D.Q. Mayne and G.C. Goodwin (1990). Identifier–based, switched–parameter algorithms for the adaptive stabilization of linear systems. *The Sixth Yale Workshop on Adaptive and Learning Systems.*

[20] Goodwin, G.C., M. Gevers and B. Ninness. Optimal model order selection and estimation of model uncertainty for identification with finite data. Submitted to *IEEE Trans. on Auto. Cont. Special Issue on System Identification for Control Design.*

INFORMATION BOUNDS, CERTAINTY EQUIVALENCE AND LEARNING IN ASYMPTOTICALLY EFFICIENT ADAPTIVE CONTROL OF TIME-INVARIANT STOCHASTIC SYSTEMS

Tze Leung Lai
Department of Statistics, Stanford University, Stanford, CA 94305

1. INTRODUCTION

The past two decades have witnessed important advances in the subject of adaptive control of stochastic systems with unknown but time-invariant parameters. A fundamental difficulty in this area is the apparent dilemma between the need for information to learn about the unknown system parameters and the control objective of driving the system outputs towards certain target values. This difficulty has been more or less completely resolved in several prototypical problems. Three such problems will be discussed herein. They are (i) the multi-armed bandit problem in §2, (ii) the multi-period control problem in stochastic regression models in §3, and (iii) adaptive control of an ARMAX system in §4.

In principle, given a prior distribution of the unknown system parameters and the joint probability distribution of the sequence of random variables that determine the stochastic system, one can formulate an adaptive control problem as a dynamic programming problem in which the "state" is the conditional distribution of the original system state and parameter vector given the past observations (cf. [1]). However, the dynamic programming equations are usually prohibitively difficult to handle, both computationally and analytically. Moreover, it is often not possible to specify a realistic probability law for all the random variables involved and a reasonable prior distribution for the unknown parameter vector.

Instead of the Bayesian approach, a much more practical and intuitively appealing alternative is the "certainty-equivalence" approach that first finds the optimal solution when the system parameters are known and then replaces the unknown parameters by their sample estimates at every stage. This approach incorporates sequential learning, in the form of updating the parameter estimates based on all the available data at every stage, into the control scheme that tries to mimic the optimal rule assuming known system parameters. The idea is particularly attractive when the optimal control law assuming known system parameters has a simple recursive form that can be implemented in real time, as is the case for the three classical problems to be discussed in the subsequent sections. If one can also develop real-time recursive algorithms for updating the parameter estimates which are used to replace the unknown parameters in the "fictitious" optimal rule, then the certainty-equivalence rule will indeed satisfy real-time computational constraints.

Instead of trying to solve a difficult optimization problem that represents ignorance of the actual parameter values in the form of a prior distribution as in the Bayesian approach, the certainty-equivalence approach combines a much simpler optimization problem assuming known parameter values with a recursive estimation problem to learn about the parameter values sequentially from the data. The performance of this approach, therefore, depends critically on how well the parameter values are estimated. In view of this, it is therefore highly desirable to estimate the parameters as efficiently as possible. However, there are also practical computational constraints here for on-line implementation. In the case of ARMAX systems to be discussed in §4, this poses formidable problems since the statistically efficient estimation procedures involve nonlinear estimating equations which have to be solved off-line and which also require storing all the data as they accumulate over time. §4 shows, however, tht it is possible to develop recursive estimators that are asymptotically as efficient as the off-line efficient estimators by a parallel implementation of a few well known recursive identification algorithms.

Clearly how well the parameter values can be estimated depends not only on the choice of efficient estimation procedures but also on how much information about the parrameters there is in the data. In statistical estimation theory, one usually summarizes the amount of information in the form of Kullback-Leibler information numbers or Fisher information matrices, and represents one's uncertainty about the estimates by using confidence intervals or Bayesian posterior distributions. These basic concepts provide important tools in assessing the amount of information about the unknown parameters in the data and the amount of uncertainty in the sequential parameter estimates for the certainty-equivalence approach. In particular, when the amount of information is inadequate and the parameter estimates are highly uncertain,, one should introduce "probing signals" to enhance the information content of the data instead of strictly following the certainty-equivalence rule which replaces the unknown parameters in the fictitious optimal rule by dubious estimates. On the other hand, when the standard errors of the parameter estimates are relatively small, the certainty-equivalence input should provide a good approximation to the optimal input that assumes known parameter values. This obvious idea plays an important role in the development of asymptotically efficient adaptive control rules that perturb the certainty-equivalence approach with probing signals in §§2-4.

A basic issue concerning this information enhancement of the certainty-equivalence approach is the maximal proportion of probing signals that can be introduced so that the resultant control rule may still attain full asymptotic efficiency. In this connection the notion of "asymptotic efficiency" of a control rule has also to be precisely defined. For each of the adaptive control problems considered in §§2-4, we address these issues by considering the fictitious situation which assumes knowledge of some crucial parameter(s) so that there is negligible conflict between estimation and control. The performance of asymptotically optimal solutions in this much more tractable situation then sets a lower bound that we can hope to achieve even when all the parameters are unknown, and asymptotic efficiency can therefore be defined as achieving such asymptotic performance. More importantly, the difference in performance between this fictitious situation assuming knowledge of some (but not all) parameters and the ideal situation in which all parameters are known provides an asymptotic lower bound on the amount of information about the unknown parameters needed for optimal control. By keeping the probing perturbations within these bounds, we show in §§2-4 that suitably chosen certainty-equivalence rules with suitable perturbations are indeed asymptotically efficient.

2. INFORMATION BOUNDS AND DYNAMIC ALLOCATION INDICES IN THE MULTI-ARMED BANDIT PROBLEM

Let Π_j, $j = 1, \cdots, k$, denote statistical populations (treatments, manufacting processes, etc.) specified, respectively, by univariate density functions $f(x; \theta_j)$ with respect to some nondegenerate measure ν, where $f(\cdot;\cdot)$ is known and the θ_j are unknown parameters belonging to some set Θ. Assume that $E_\theta|X| = \int_{-\infty}^{\infty} |x| f(x;\theta) d\nu(x) < \infty$ for all $\theta \in \Theta$. How should we sample $x_1, \cdots, x_N$ sequentially from the k populations in order to maximize, in some sense, the expected value of the sum $S_N = x_1 + \cdots + x_N$? This is the classical "multi-armed bandit problem", with specified horizon N, in the statistics and engineering literature. The name derives from an imagined slot machine with $k \geq 2$ arms. When an arm is pulled, the player wins a random reward. For each arm j, there is an unknown probability distribution Π_j of the reward, and the player's problem is to choose N pulls on the k arms so as to maximize the total expected reward. There is an apparent dilemma between the need to learn from all populations about their parameter values ("information") and the objective of sampling only from the best population ("control").

An *adaptive allocation rule* φ is a sequence of random variables $\varphi_1, \cdots, \varphi_N$ with values in the set $\{1, \cdots, k\}$ and such that the event $\{\varphi_i = j\}$, $j = 1, \cdots, k$, belongs to the σ-field $\mathcal{F}_{i-1}$ generated by the previous observations $\varphi_1, x_1, \cdots, \varphi_{i-1}, x_{i-1}$. Letting $\mu(\theta) = \int_{-\infty}^{\infty} x f(x;\theta) d\nu(x)$ and $\underset{\sim}{\theta} = (\theta_1, \cdots, \theta_k) \in \Theta^k$, it follows that for every $n \leq N$

$$E_{\underset{\sim}{\theta}} S_n = \sum_{i=1}^{n} \sum_{j=1}^{k} E_{\underset{\sim}{\theta}} \{ E_{\underset{\sim}{\theta}} (x_i I_{\{\varphi_i = j\}} | \mathcal{F}_{i-1}) \} = \sum_{j=1}^{k} \mu(\theta_j) E_{\underset{\sim}{\theta}} T_N(j), \tag{2.1}$$

where

$$T_n(j) = \sum_{i=1}^{n} I_{\{\varphi_i = j\}} \tag{2.2}$$

denotes the number of observations that φ samples from Π_j up to stage n. Hence, the objective of maximizing $E_{\underset{\sim}{\theta}} S_N$ is equivalent to that of minimizing the regret

$$R_N(\underset{\sim}{\theta}) = N\mu^*(\underset{\sim}{\theta}) - E_{\underset{\sim}{\theta}} S_N = \sum_{j:\mu(\theta_j) < \mu^*(\underset{\sim}{\theta})} (\mu^*(\underset{\sim}{\theta}) - \mu(\theta_j)) E_{\underset{\sim}{\theta}} T_N(j), \tag{2.3}$$

where $\mu^*(\underset{\sim}{\theta}) = \max_{1 \leq j \leq k} \mu(\theta_j)$. In particular, the usual Bayesian formulation of the multi-armed bandit problem, stated in the form of maximizing $\int E_{\underset{\sim}{\theta}} S_N dH(\underset{\sim}{\theta})$, can be restated in the more convenient form of minimizing the Bayes risk $\int R_N(\underset{\sim}{\theta}) dH(\underset{\sim}{\theta})$, where H is a prior distribution on Θ^k.

In principle, one can use dynamic programming to study the problem of minimizing $\int R_N(\underset{\sim}{\theta}) dH(\underset{\sim}{\theta})$. For the case where $k = 2$ and Θ has two elements, which we shall denote by a, b with $\mu(a) > \mu(b)$, Feldman [2] found by this approach that for the prior distribution which assigns probability p to the parameter vector $\theta = (a, b)$ and probability $1 - p$ to the vector (b, a), the allocation rule that chooses Π_1 or Π_2 at stage $i + 1$ according as $p_i \geq \frac{1}{2}$ or $p_i < \frac{1}{2}$ is Bayes, where p_i denotes the posterior probability in favor of the vector (a, b) at the end of stage i ($p_0 = p$). For the case of $k = 2$ Bernoulli populations, Fabius and van Zwet [3] and Berry [4] studied the dynamic programming equations analytically and obtained several interesting results about the Bayes rules with respect to general priors. Beyond the two-point priors considered by Feldman, Bayes rules are usually described only implicitly by the dynamic programming equations, whose numerical solution is too complicated for practical implementation when N is large.

Instead of considering the finite-horizon problem of maximizing the expected value of $S_N = \sum_1^N x_i$, one can study the discounted problem that maximizes the expected value of the series $\sum_1^\infty \beta^{i-1}x_i$, where $0 < \beta < 1$. Major advances in the discounted multi-armed bandit problem were made in the seventies by Gittins and Jones [5], Gittons [6] and Whittle [7]. Their results show, for general $k \geq 2$, that under the assumption of independent prior distributions G_j on θ_j $(j = 1, \cdots, k)$, the optimal solution to the Bayes problem of maximizing

$$\int \cdots \int E_{\underset{\sim}{\theta}}(\sum_{i=1}^{\infty} \beta^{i-1}x_i)dG_1(\theta_1)\cdots, dG_k(\theta_k) \tag{2.4}$$

is the "index rule" described below. To begin with, suppose that θ is a random variable having distribution G and that conditional on θ, $Y_1, Y_2, \cdots$ are i.i.d. with common density $f(x;\theta)$ with respect to ν. Let $G^{[r]}$ denote the conditional distribution of θ given $Y_1, \cdots, Y_r$. (Thus, $G_j^{[r]}$ is the posterior distribution of the parameter θ_j of population Π_j given r observations $Y_1, \cdots, Y_r$ from Π_j.) Set $G^{[0]} = G$. As is well known, the conditional distribution of $(\theta, Y_{r+1}, Y_{r+2}, \cdots)$ given $(Y_1, \cdots, Y_r)$ can be described by saying that θ has distribution $G^{[r]}$ and that $Y_{r+1}, Y_{r+2}, \cdots$ are i.i.d. with common density $f(x;\theta)$ given the value of θ. Therefore, for $m > r$,

$$E[Y_m|Y_1, \cdots, Y_r] = \int \mu(\theta)dG^{[r]}(\theta) = E[\mu(\theta)|Y_1, \cdots, Y_r]. \tag{2.5}$$

The *Gittins index* $M(G)$ of the distribution G on θ is defined as the infimum of the set of solutions M of the equation

$$\sup_{\tau \geq 0} E\left\{\sum_{i=0}^{\tau-1} \beta^i E[\mu(\theta)|Y_1, \cdots, Y_i] + M\sum_{i=\tau}^{\infty} \beta^i\right\} = M\sum_{i=0}^{\infty} \beta^i, \tag{2.6}$$

where $\sup_{\tau \geq 0}$ is over all stopping times τ. The *index rule* φ^* is to sample at stage $n+1$ from the population Π_{j^*} that has the largest Gittins index at the end of stage n, i.e.,

$$M(G_{j^*}^{[T_n(j^*)]}) = \max_{1 \leq j \leq n} M(G_j^{[T_n(j)]}). \tag{2.7}$$

For fixed j, the equation (2.6) defining the Gittins index of the posterior distribution $G = G_j^{[T_n(j)]}$ arises from the following (fictitious) optimization problem, cf. [6], [7]. Suppose that the largest mean, $\max_{i \neq j} \mu(\theta_i)$, of the remaining $k-1$ populations and its population label $i^* \neq j$ are both known. Since θ_j is unknown, we have only to decide whether $\mu(\theta_j)$ exceeds $M = \max_{i \neq j} \mu(\theta_i)$, in which case it is profitable to sample from Π_j, or $\mu(\theta_j) < M$, in which case we should switch to sampling from Π_{i^*}. Initially, in ignorance of θ_j, we should clearly sample from Π_j to learn its value and continue learning until there is enough evidence to reject the hypothesis $\mu(\theta_j) \geq M$, at which point we should switch to sampling from Π_{i^*} (which does not add any information about $\mu(\theta_j)$). The left hand side of (2.6) represents the expected discounted reward for such strategy with an optimal choice of the stopping rule. The right hand side of (2.6) represents the expected discounted reward of the strategy that samples from Π_{i^*} at all stages.

Using diffusion approximations to analyze the optimal stopping problem (2.6) in the case where $f(x;\theta)$ belongs to an exponential family of densities

$$f(x;\theta) = \exp(\theta x - \psi(\theta)) \tag{2.8}$$

with respect to some measure ν, Chang and Lai [8] found simple asymptotic approximations to Gittins indices as the discount factor β approaches 1. These asymptotic approximations turn out to be certain upper confidence bounds for the $\mu(\theta_j)$. Moreover, it is natural to expect that replacing $1-\beta$ by $1/N$ in these approximations should also lead to asymptotically optimal solutions to the finite-horizon problem of maximizing $\int \cdots \int E_{\underset{\sim}{\theta}} S_N dG_1(\theta_1)\cdots dG_k(\theta_k)$. Lai [9] has shown that this is indeed the case, assuming the G_i to put all its mass on an open interval $A \subset \Theta \triangleq \{\theta : \int e^{\theta x} d\nu(x) < \infty\}$ with endpoints $(-\infty \leq)\ a_1 < a_2 (\leq \infty)$ such that

$$\inf_{a_1 - r < \theta < a_2 + r} \psi''(\theta) > 0, \qquad \sup_{a_1 - r < \theta < a_2 - r} \psi''(\theta) < \infty \tag{2.9}$$

and ψ'' is uniformly continuous on $(a_1 - r, a_2 + r)$ for some $r > 0$.

In particular, if $Y_1, Y_2, \cdots$ are normally distributed, then Θ is the entire real line and we can take $A = \Theta$.

More generally, Lai [9] considers general prior distributions H on A^k (not necessarily assuming independent $\theta_1, \cdots, \theta_k$) and shows that index rules in which the dynamic allocation indices are certain upper confidence bounds for θ_j (noting that $\mu(\theta) = \psi'(\theta)$ is an increasing function for the exponential family (2.8)) provide asymptotically optimal solutions to the finite-horizon problem of maximizing $\int E_\theta S_N dH(\underset{\sim}{\theta})$. Moreover, these index rules can be regarded as natural perturbations of obvious certainty-equivalence rules to ensure that there is enough information to learn about all the population parameters, as will be explained below. For fixed j, based on the observations $Y_1, \cdots, Y_n$ from Π_j, the maximum likelihood estimator of θ_j, known to belong to A, is given by

$$\widehat{\theta}_{j,n} = \mu^{-1}(S_n/n), \ \text{ if } \mu(a_1) \leq S_n/n \leq \mu(a_2),$$

$$= a_2, \text{ if } S_n/n > \mu(a_2), \tag{2.10}$$
$$= a_1, \text{ if } S_n/n < \mu(a_1),$$

where $S_n = Y_1 + \cdots + Y_n$. Let $N > 1$ and let g be a nonnegative function on $(0, \infty)$ satisfying the following assumptions:

$$\sup_{t \geq a} g(t)/t < \infty, \quad \text{for all} \quad a > 0, \tag{2.11}$$

$$g(t) \sim \log t^{-1}, \quad \text{as} \quad t \to 0, \tag{2.12}$$

$$g(t) \geq \log t^{-1} + \xi \log\log t^{-1}, \quad \text{as} \quad t \to 0, \tag{2.13}$$

for some ξ. Based on the n observations $Y_1, \cdots, Y_n$, define the "upper confidence bound"

$$U_{j,n}(g, N) = \inf\{\theta \in A : \theta \geq \widehat{\theta}_{j,n} \text{ and } I(\widehat{\theta}_{j,n}, \theta) \geq n^{-1} g(n/N)\} \tag{2.14}$$

($\inf \emptyset = \infty$), where $I(\theta, \lambda)$ is the Kullback-Leibler information number given by

$$I(\theta, \lambda) = (\theta - \lambda)\psi'(\theta) - (\psi(\theta) - \psi(\lambda)). \tag{2.15}$$

A particular choice of g satisfying (2.11)-(2.13) derived in [9] from an approximating continuous-time optimal stopping problem for Brownian motion is $g_0(t) = h_0^2(t)/2t$, where

$$h_0(t) = (t^{-1} - 1)^{1/2}\{0.63883 - 0.40258(t^{-1} - 1)\}, \text{ if } 0.86 < t \leq 1, \tag{2.16}$$
$$= -0.5759t^2 + 0.2987t + 0.4034, \text{ if } 0.28 < t \leq 0.86,$$
$$= -1.58137t + 1.53343t^{1/2} + 0.073271, \text{ if } 0.01 < t \leq 0.28,$$
$$= \{t[2\log t^{-1} - \log\log t^{-1} - \log 16\pi + 0.99232\exp(-0.03812t^{-1/2})]\}^{1/2},$$
$$\text{if } 0 < t \leq 0.01.$$

These upper bounds $U_{j,n}(g, N)$, which we shall simply denote by $U_{j,n}$ for given N and g, will serve as dynamic allocation indices in the index rule below.

If the values $\theta_1, \cdots, \theta_k$ were known, the optimal rule would obviously be to sample from the population with the largest θ_j. In ignorance of $\theta_1, \cdots, \theta_k$, the certainty-equivalence approch is to estimate them at stage n with the estimators $\widehat{\theta}_{1,T_n(1)}, \cdots, \widehat{\theta}_{k,T_n(k)}$ and to sample at stage $n+1$ from the population with the largest $\widehat{\theta}_{j,T_n(j)}$. The difficulty with this "play-the-leader" rule is that we may have sampled too little from an apparently inferior population to get a reliable estimate of its parameter and may thereby miss the actually superior population. Instead of sampling at stage $n+1$ from the population with the largest $\widehat{\theta}_{j,T_n(j)}$, Lai [9] proposes the following simple modification:

$$\text{Sample at stage } n+1 \text{ from the population } \Pi_j \text{ with the largest upper confidence bound } U_{j,T_n(j)}, \tag{2.17}$$

where $U_{j,r}$ is defined in (2.14) and $n \geq k$. (During the first k stages, we sample once from each population.) To explain the heuristic idea behind this approach, we first note that the upper confidence bound $U_{j,r}$ inflates the estimator $\widehat{\theta}_{j,r}$ by an amount which decreases with the number r of observations already taken from the population. Thus, $U_{j,r}$ depends not only on the estimator $\widehat{\theta}_{j,r}$ but also on the sample size r, and comparing the k populations on the basis of $U_{j,T_n(j)}$ involves not only the parameter estimates but also the sample sizes of all populations.

EXAMPLE. Suppose that $Y_1, Y_2, \cdots$ are i.i.d. observations from a normal population Π_1 with mean θ and variance 1. Here $\mu(\theta) = \theta$, $I(\theta, \lambda) = (\theta - \lambda)^2/2$ and the confidence bound (2.14) reduces to

$$U_{1,n}(g, N) = S_n/n + (2n^{-1}g(n/N))^{1/2}. \tag{2.18}$$

For an example of g satisfying (2.11)-(2.13), consider a nonnegative continuous function on $(0,1]$ having the asymptotic expansion

$$g(t) = \log t^{-1} - \frac{1}{2}\log\log t^{-1} - \frac{1}{2}\log 16\pi + o(1), \quad \text{as } t \to 0, \tag{2.19}$$

The asymptotic expansion (2.19) first arose in the following one-armed bandit problem considered by Chernoff and Ray [10] and by Chernoff [11]. Suppose that an experimenter can choose at each stage $n(\leq N)$ between sampling from the normal population Π_1 with unknown mean θ and sampling from another normal population Π_2 with known mean 0. Assuming a normal prior on θ, the Bayes procedure (to maximize the expected sum of N observations) samples from Π_1 until stage

$$T^* = \inf\{n \leq N : n^{-1}\sum_{i=1}^{n} Y_i + a_{n,N} \leq 0\}, \tag{2.20}$$

and then takes the remaining $N - T^*$ observations from Π_2, where $a_{n,N}$ are positive constants. Writing

$$t = n/N, \quad w(t) = (Y_1 + \cdots + Y_n)/N^{1/2}, \quad \delta = \theta N^{1/2}, \tag{2.21}$$

and treating $0 < t \leq 1$ as a continuous variable for large N, we can approximate the Bayes stopping time (2.20) by $N\tau(h)$, where $\tau(h) = \inf\{t \in (0,1] : w(t)+h(t) \leq 0\}$ is the optimal stopping rule in the following continuous-time stopping problem: Assuming a flat prior for the drift coefficient δ of a Wiener process $w(t)$, find the stopping rule $\tau \leq 1$ to maximize $\int_{-\infty}^{\infty} E_\delta(\delta\tau)d\delta$. Using an asymptotic analysis of the free boundary problem associated with this continuous-time optimal stopping problem, Chernoff and Ray [10] found that as $t \downarrow 0$

$$h(t) = \{2t(\log t^{-1} - \frac{1}{2}\log\log t^{-1} - \frac{1}{2}\log 16\pi + o(1))\}^{1/2}.$$

Therefore, letting $h(t)/t^{1/2} = (2g(t))^{1/2}$, $g(t)$ satisfies the asymptotic expansion (2.19). From (2.21), it follows that the $a_{n,N}$ in (2.20) can be approximated by $(2n^{-1}g(n/N))^{1/2}$. With this approximation, the Bayes procedure for the one-armed bandit problem can be described in the form:

$$\begin{aligned} &\text{At stage } n+1 \text{ sample from } \Pi_1 \text{ or } \Pi_2 \text{ according as} \\ &U_{1,T_n(1)}(g,N) > 0 \text{ or } U_{1,T_n(1)}(g,N) \leq 0. \end{aligned} \tag{2.22}$$

The boundary (2.16) is obtained by approximating the optimal stopping boundary $h(t)$ that has been evaluated numerically, cf. [9].

For $j = 1,\cdots,k$, let $\underset{\sim}{\theta}^j = (\theta_1,\cdots,\theta_{j-1},\theta_{j+1},\cdots,\theta_k)$ and let $\theta_j^* = \max_{i\neq j}\theta_i$. For a prior distribution H of $\underset{\sim}{\theta}$, let H_j denote the marginal distribution of the $(k-1)$-dimensional vector $\underset{\sim}{\theta}^j$ and let $H^{(j)}(\theta|\underset{\sim}{\theta}^j) = P\{\theta_j \leq \theta|\underset{\sim}{\theta}^j\}$ denote the conditional distribution function of θ_j given $\underset{\sim}{\theta}^j$. Suppose that the prior distribution H on A^k staisfies the following four conditions for some $\rho > 0$ and every $j = 1,\cdots,k$:

(a) $\int |\theta_j| dH(\underset{\sim}{\theta}) < \infty$;

(b) for every fixed $\underset{\sim}{\theta}^j \in A^{k-1}$, $H^{(j)}(\theta|\underset{\sim}{\theta}^j)$ has a positive continuous derivative $h_j(\theta;\underset{\sim}{\theta}^j)$ for $\theta \in (\theta_j^* - \rho, \theta_j^* + \rho) \cap A$;

(c) $\int_{A^{k-1}} \sup_{\theta\in(\theta_j^*-\rho,\theta_j^*]\cap A} h_j(\theta;\underset{\sim}{\theta}^j)dH_j(\underset{\sim}{\theta}^j) < \infty$,

(d) $h_j(\theta;\underset{\sim}{\theta}^j)/h_j(\theta_j^*,\underset{\sim}{\theta}^j) \to 1$ as $\theta \to \theta_j^*$, uniformly in $\underset{\sim}{\theta}^j \in B^{k-1}$,

for every compact subset B of A. Under the mild assumptions (2.11)-(2.13) on g, Lai [9] has established that the index rule (2.17) has the following asymptotically minimal order of magnitude for the Bayes regret $\int_{A^k} R_N(\underset{\sim}{\theta})dH(\underset{\sim}{\theta})$:

$$\left\{\frac{1}{2}\sum_{j=1}^{k}\int_{A^{k-1}} h_j(\theta_j^*;\underset{\sim}{\theta}^j)dH_j(\underset{\sim}{\theta}^j)\right\}(\log N)^2 \sim \inf_{\varphi}\int_{A^k} R_N(\underset{\sim}{\theta})dH(\underset{\sim}{\theta}) \text{ as } N\to\infty, \tag{2.23}$$

where $\inf_\varphi$ is taken over all adaptive allocation rules φ and $R_N(\underset{\sim}{\theta})$ is defined in (2.3).

Replacing N by $1/(1-\beta)$ in the index rule (2.17) also gives an asymptotically optimal solution to the discounted Bayes problem of maximizing $\int_{A^k} E_{\underset{\sim}{\theta}}(\sum_{i=1}^{\infty} \beta^{i-1} x_i) dH(\underset{\sim}{\theta})$ as $\beta \uparrow 1$, cf. [8]. The index rule (2.17) also provides an asymptotically optimal solution to the following version of the multi-armed bandit problem first considered by Robbins [12] without assuming a finite horizon N or discount factor β: Find an adaptive allocation rule such that $n^{-1} E_{\underset{\sim}{\theta}} S_n$ approaches $\mu^*(\underset{\sim}{\theta}) = \max_{1 \le i \le k} \mu(\theta_i)$ as rapidly as possible as $n \to \infty$ for all $\underset{\sim}{\theta} \in \Theta^k$. In particular, for the case $k = 2$, Robbins [12] introduced the following class of simple adaptive allocation rules satisfying

$$\lim_{n \to \infty} n^{-1} E_{\underset{\sim}{\theta}} S_n = \mu^*(\underset{\sim}{\theta}) \quad \text{for all} \quad \underset{\sim}{\theta} \in \Theta^k. \tag{2.24}$$

Let $a_1 = 1 < a_2 < \cdots$ and $b_1 = 2 < b_2 < \cdots$ be two fixed, disjoint, increasing sequences of positive integers such that $a_n/n \to \infty$ and $b_n/n \to \infty$ as $n \to \infty$. At stage n, sample from Π_1 if $n \in \{a_1, a_2, \cdots\}$, sample from Π_2 if $n \in \{b_1, b_2, \cdots\}$, and if $n \notin \{a_1, a_2, \cdots, b_1, b_x, \cdots\}$, sample from Π_1 or Π_2 according as the mean of all previous observations from Π_1 exceeds or does not exceed the mean of all previous observations from Π_2. From the strong law of large numbers, it follows easily that this procedure attains (2.24).

The sequences a_n and b_n above are assumed to be prescribed in advance, and a natural question is how to choose them so that $n^{-1} E_{\underset{\sim}{\theta}} S_n$ approaches $\mu^*(\underset{\sim}{\theta})$ (or equivalently, $n^{-1} R_n(\underset{\sim}{\theta})$ approaches 0) as rapidly as possible. To answer this question, Lai and Robbins [13] first came up with the following asymptotic lower bound for the regret of adaptive allocation rules so that the phrase "as rapidly as possible" can be given a more precise meaning. It is shown in [13] that for every allocation rule satisfying $R_n(\underset{\sim}{\theta}) = o(n^a)$ for all $a > 0$ and $\underset{\sim}{\theta} \in \Theta^k$,

$$\liminf R_n(\underset{\sim}{\theta}) / \log n \ge \sum_{j:\mu(\theta_j) < \mu^*(\underset{\sim}{\theta})} (\mu^*(\underset{\sim}{\theta}) - \mu(\theta_j)) / I(\theta_j, \lambda) \tag{2.25}$$

for all $\theta_1, \cdots, \theta_k$, $\lambda \in \Theta$ such that $\mu(\lambda) > \mu^*(\underset{\sim}{\theta})$, where

$$I(\theta, \lambda) = \int_{-\infty}^{\infty} \{\log(f(x;\theta)/f(x;\lambda))\} f(x;\theta) d\nu(x). \tag{2.26}$$

For the exponential family of densities (2.8), consider the allocation rule $\varphi(g)$ that samples at stage $n + 1 (> k)$ from the population Π_j with the largest upper confidence bound

$U_{j,T_n(j)}(g,n)$, where $U_{j,r}(g,N)$ is defined in (2.14) and g is a nonnegative function on $(0,1]$ satisfying (2.11)-(2.13) in which $\xi > -\frac{1}{2}$. Suppose that the condition (2.9) is satisfied by $A = \Theta$. Then for the allocation rule $\varphi(g)$, we have at every fixed $\underset{\sim}{\theta} \in \Theta^k$,

$$R_n(\underset{\sim}{\theta}) \sim \left\{ \sum_{j:\theta_j < \theta^*} (\mu(\theta^*) - \mu(\theta_j))/I(\theta_j, \theta^*) \right\} \log n, \tag{2.27}$$

where $\theta^* = \max_{1 \le j \le k} \theta_j$, cf. Theorem 4 of [14]. Note that $\mu = \psi'$ is an increasing function and that (2.26) reduces to (2.15) in this case. Hence the index rule $\varphi(g)$ attains the asymptotic lower bound (2.25) for the regret. Such allocation rules are called "asymptotically efficient" by Lai and Robbins [13] who also show how to construct upper confidence bounds to form asymptotically efficient index rules for several other parametric families of distributions. Their construction has been extended from the i.i.d. to Markov sequences by Ananthraman, Varaiya and Walrand [15] and to the setting with switching costs by Agrawal, Hedge and Teneketzis [16].

The asymptotic lower bound (2.25) can be interpreted as (and follows from) that the minimal expected sample size (information) from an inferior population Π_j is at least as large as $(1 + o(1))(\log n)/I(\theta_j, \lambda)$ if the rule does not perform too badly at every $\underset{\sim}{\theta}$, cf. [13]. In other words, we need to sample at least that many observations on the average from the inferior population Π_j for the rule to have a regret $R_n(\underset{\sim}{\theta}) = o(n^a)$ for every fixed $\underset{\sim}{\theta}$ and $a > 0$. In particular, for the exponential family (2.8), this implies that for an inferior population Π_j we need approximately $(\log n)/I(\theta_j, \theta^*)$ observations from Π_j to be reasonably confident that it is not a contender. Of course, since θ_j and θ^* are unknown, we have to see if there is enough information to adequately estimate $I(\theta_j, \theta^*)$. Lai and Robbins [13] found that replacing the maximum likelihood estimate $\widehat{\theta}_{i,T_n(i)}$ for each population parameter by a perturbation ("upper confidence bound")

$$U_{i,T_n(i)}(g,n) = \inf\{\theta : \theta \ge \widehat{\theta}_{i,T_n(i)} \quad \text{and} \quad T_n(i) I(\widehat{\theta}_{i,T_n(i)}, \theta) \ge g(T_n(i)/n)\}$$

provides a simple way to get around these difficulties. Such upper confidence bounds turn out also to provide approximations to Gittins indices for the discounted Bayes problem and also to give asymptotically optimal solutions to the finite-horizon Bayes problem, as explained above.

3. LEAST SQUARES ADAPTIVE PREDICTORS AND MULTI-PERIOD ADAPTIVE CONTROL IN STOCHASTIC REGRESSION MODELS

The "multi-period control problem under uncertainty" in the econometrics literature is to choose successive inputs $u_1, \cdots, u_n$ in the linear regression model

$$y_i = \alpha + \beta u_i + \epsilon_i \quad (i = 1, \cdots, n), \tag{3.1}$$

where α and β are unknown parameters and the random disturbances $\epsilon_1, \epsilon_2, \cdots$ are i.i.d. with mean 0 and variance σ^2, so that the outputs $y_1, \cdots, y_n$ are as close as possible in some sense to a given target value y^*. A Bayesian formulation is often used so that the problem becomes that of minimizing

$$\int_{-\infty}^{\infty}\int_{-\infty}^{\infty} E_{\alpha,\beta}[\sum_{i=1}^{n}(y_i - y^*)^2]d\pi(\alpha,\beta) = n\sigma^2 + \int_{-\infty}^{\infty}\int_{-\infty}^{\infty} \beta^2 E_{\alpha,\beta}[\sum_{i=1}^{n}(u_i - \theta)^2]d\pi(\alpha,\beta), \tag{3.2}$$

where π is a prior distribution of the unknown parameters α, β and $\theta = (y^* - \alpha)/\beta$, cf. [17], [18], [19]. The ϵ_i are often assumed to be normally distributed, and (3.2) can in principle be minimized by using dynamic programming. However, because of the complexities in the numerical solution of the dynamic programming equations and the analytical difficulties in studying the Bayes rules, not much is known about their performance.

A first departure from the Bayesian approach was due to Aoki [20]. Assuming that the sign of β is known, say $\beta > 0$, he proposed the use of a stochastic approximation scheme of the form

$$u_{t+1} = u_t - a_t(y_t - y^*), \tag{3.3}$$

where $\{a_i\}$ is a sequence of positive constants such that $\sum_1^\infty a_i = \infty$ and $\sum_1^\infty a_i^2 < \infty$. This approach has the property that $u_t \to \theta$ a.s. (almost surely). Another non-Bayesian approach is the certainty-equivalence rule proposed by Anderson and Taylor [21]. If α and $\beta(\neq 0)$ are both known, the optimal choice of u_t is clearly at $\theta = (y^* - \alpha)/\beta$. Without assuming α and β to be known, suppose that bounds K_1, K_2 are known such that $K_1 < \theta < K_2$. Assuming the ϵ_i to be normally distributed, the maximum likelihood estimator of θ at stage $t \geq 2$ is

$$\widehat{\theta}_t = K_2 \wedge \{\widehat{\beta}_t^{-1}(y^* - \widehat{\alpha}_t) \vee K_1\}, \tag{3.4}$$

where $\widehat{\beta}_t = \{\sum_1^t(u_i - \bar{u}_t)y_i\}/\{\sum_1^t(u_i - \bar{u}_t)^2\}$ and $\widehat{\alpha}_t = \bar{y}_t - \widehat{\beta}_t\bar{u}_t$ are the least squares estimates of β and α, and $\bar{u}_t = t^{-1}\sum_1^t u_i$, $\wedge$ and $\vee$ denote minimum and maximum, respectively. The initial values u_1 and u_2 are distinct but otherwise arbitrary numbers between K_1 and K_2, and for $t \geq 2$, the certainty-equivalence rule sets $u_{t+1} = \widehat{\theta}_t$.

Despite its simplicity for implementation, the certainty-equivalence rule is difficult to analyze. A basic difficulty with this approach is that although $\widehat{\theta}_t$ may conceivably represent one's best current guess of θ, how good the guess is depends on how the inputs $u_1, \cdots, u_t$ are chosen. In particular, if the inputs $u_i = \widehat{\theta}_{i-1}, 1 \leq i \leq t$, tend to cluster around their mean $\bar{u}_t$, then there may not be enough information to give a reliable estimate $\widehat{\theta}_t$, even though $\widehat{\theta}_t$ may well be one's closest possible guess of θ at stage t. Lai and Robbins [22] subsequently exhibited an event Ω_0 with positive probability on which $\widehat{\beta}_t$ is of the wrong sign for all t and u_t does not converge to θ as $t \to \infty$. There is, therefore, again a dilemma between the desire to set the current input as close to θ as the data suggest ("control") and the need to set the inputs in a wide enough scatter so as to provide a reliable estimate of θ ("information"). To resolve this dilemma, we have first to find out how reliable an estimate is really needed and how much information (design scatter) is required to produce such an estimate. A bench mark that gives valuable clues to these questions is provided by studying the "fictitious" situation that assumes β to be known.

Suppose that $\beta \neq 0$ is known. Rewriting the regression model (3.1) as

$$(y_i - y^*)/\beta = u_i - \theta + \epsilon_i/\beta, \tag{3.5}$$

the maximum likelihood estimator of θ at stage t is

$$\theta_t^* = \bar{u}_t - \beta^{-1}(\bar{y}_t - y^*) = \theta - \beta^{-1}\bar{\epsilon}_t.$$

Thus, *irrespective of how the inputs x_i are chosen,*

$$\theta_t^* - \theta = -\beta^{-1}\bar{\epsilon}_t \sim N(0, \sigma^2/\beta^2). \tag{3.6}$$

Hence, in the present setting, there is *no conflict* between estimation and control. In particular, if we set u_{t+1} to be our best current guess θ_t^* of θ, i.e.,

$$u_{t+1} = \theta_t^* = \bar{u}_t - \beta^{-1}(\bar{y}_t - y^*), \quad t \geq 1, \tag{3.7}$$

and let u_1 = initial guess of θ, then Lai and Robbins [22] showed that $u_t \to \theta$ a.s. and

$$\beta^2 \sum_{t=1}^{n} (u_t - \theta)^2 / \log n \to \sigma^2 \text{ a.s.}, \tag{3.8}$$

$$\beta^2 E_\theta \sum_{t=1}^{n} (u_t - \theta)^2 = \sigma^2 \log n + O(1). \tag{3.9}$$

Let $R_n = \beta^2 \sum_{t=1}^{n} (u_t - \theta)^2$ denote the "regret" of an input sequence $\{u_t : t \leq n\}$, and note that $E_\theta\{\sum_{t=1}^{n} (y_i - y^*)^2\} = n\sigma^2 + E_\theta(R_n)$. Putting a prior distribution π on θ, the solution to the Bayes problem of minimizing $\int_{-\infty}^{\infty} E_\theta(R_n) d\pi(\theta)$ is simply

$$u_{t+1} = E(\theta | u_1, y_1, \cdots, u_t, y_t) = \text{posterior mean of } \theta; \tag{3.10}$$

in particular, if π is normal, then (3.8) and (3.9) still hold for (3.10), cf. [23].

In this auxiliary problem that assumes knowledge of β, we do not have to wrestle with the difficult issue of potential conflict between estimation and control. The auxiliary problem, which we can solve completely, sets a lower bound on the regret R_n that we hope to achieve even when β is unknown. Lai and Robbins [22], [24] proposed two simple methods to achieve this, under the assumption that the ϵ_i are i.i.d. random variables (not necessarily normal) with mean 0 and variance σ^2.

The first method, considered in [24], is adaptive stochastic approximation. The stochastic approximation scheme (3.3) was originally introduced by Robbins and Monro [25] to estimate sequentially the solution θ, assumed unique, of the equation $f(u) = y^*$, based on successive observations (u_i, y_i) from the regression model $y_i = f(u_i) + \epsilon_i$, where f is an unknown function satisfying certain assumptions. In the setting of (3.1), $f(u) = y^* + \beta(u - \theta)$ in which β and θ are unknown parameters. If $\beta(\neq 0)$ is known, then for the choice $a_t = (\beta t)^{-1}$ in (3.3), the stochastic approximation recursion (3.3) is equivalent to the recursion (3.7), for which (3.8) and (3.9) hold, cf. [24]. Without assuming β to be known, Lai and Robbins [24] suggested using $a_t = (b_t t)^{-1}$ in (3.3), where $b_t = b_t(u_1, y_1, \cdots, u_t, y_t)$ is an estimate of β based on the data already observed, and showed that it is possible to choose b_t such that $b_t \to \beta$ a.s., which in turn ensures that (3.8) and (3.9) still hold and that

$$\sqrt{t}(u_t - \theta) \text{ has a limiting normal } N(0, \sigma^2/\beta^2) \text{ distribution as } t \to \infty. \tag{3.11}$$

This approach was subsequently generalized by Wei [26] to the multivariate case in which α, y_i, u_i and ϵ_i in (3.1) are $k \times 1$ vectors and β is a $k \times k$ nonsingular matrix.

The second method, considered in [22], starts with the certainty-equivalence rule (3.7) that assumes β known and then uses a modified least squares estimate $\widehat{\beta}_t$ to replace β in (3.7), leading to the certainty-equivalence rule

$$u_{t+1} = \bar{u}_t - \widehat{\beta}_t^{-1}(\bar{y}_t - y^*), \quad t \geq 1. \tag{3.12}$$

Although this rule appears to be motivated by similar certainty-equivalence principles as in the Anderson-Taylor rule $u_{t+1} = K_2 \wedge \{\widehat{\beta}_t^{-1}(y^* - \widehat{\alpha}_t) \vee K_1\}$ in (3.4), there are important differences. While the Anderson-Taylor rule applies the certainty-equivalence principle to the regression model (3.1), the rule (3.12) works with the reparametrized model $y_i = y^* + \beta(u_i - \theta) + \epsilon_i$. Furthermore, while Anderson and Taylor assume known *a priori* bounds $K_1 < \theta < K_2$, it is assumed in [22] that upper and lower bounds of the same sign for β are known, say $0 < b < \beta < B$. Therefore [22] uses the following truncated version of the least squares estimator of β:

$$\widehat{\beta}_t = B \wedge \{[\sum_{i=1}^{t}(u_i - \bar{u}_t)y_i]/[\sum_{i=1}^{t}(u_i - \bar{u}_t)^2] \vee b\} \text{ if } \sum_{i=1}^{t}(u_i - \bar{u}_t)^2 \geq (2B^2)^{-1}s_t^2 \log t,$$
$$= \widehat{\beta}_{t-1} \text{ otherwise,} \tag{3.13}$$

where $s_t^2 = t^{-1}\sum_{i=1}^{t}(y_i - \bar{y}_t)^2$ and the initial value $\widehat{\beta}_1$ is any constant between b and B. With this choice of $\widehat{\beta}_t$, define the certainty-equivalence rule (3.12). It is shown in [22] and [23] that (3.8), (3.9) and (3.11) still hold for the adaptive control rule (3.12)-(3.13).

To extend the above least squares certainty-equivalence approach to the multivariate case in which α, y_i, u_i and ϵ_i in (3.1) are $k \times 1$ vectors and β is a $k \times k$ nonsingular matrix, Lai [27] again reparametrized (3.1) as $y_i = y^* + \beta(u_i - \theta) + \epsilon_i$ and assumed (by considering $y_i - y^*$ instead of y_i) that $y^* = 0$. In fact, [27] considers a more general stochastic regression model of the form

$$Y_i = B(U_i - CX_i) + \epsilon_i, \tag{3.14}$$

where

(i) Y_i is a $k \times 1$ vector of outputs at stage i;

(ii) C is a $k \times h$ matrix and B is a $k \times k$ nonsingular matrix of parameters;

(iii) ϵ_i is a $k \times 1$ vector of random disturbances such that $\{\epsilon_i\}$ is a martingale difference sequence with respect to an increasing sequence of σ-fields $\{\mathcal{F}_i\}$ (i.e., ϵ_i is $\mathcal{F}_i$-measurable and $E(\epsilon_i|\mathcal{F}_{i-1}) = 0$ for all i);

(iv) U_i is a $k \times 1$ vector of inputs depending on the current and past observations $X_1, U_1, Y_1, \cdots, X_{i-1}, U_{i-1}, Y_{i-1}, X_i$;

(v) X_i is an $\mathcal{F}_{i-1}$-measurable $h \times 1$ vector of regressors.

The random disturbances ϵ_i are assumed to satisfy the condition

$$\sup_i E(\|\epsilon_i\|^\gamma|\mathcal{F}_{i-1}) < \infty \text{ a.s. for some } \gamma > 2, \tag{3.15}$$

while the regressors X_i are assumed to satisfy the condition

$$X_n = O(1) \text{ and } \lambda_{\min}(\sum_{i=1}^n X_i X_i^T) \to \infty \text{ a.s.}, \tag{3.16}$$

where $\lambda_{\min}$ and $\lambda_{\max}$ denote, respectively, the minimum and maximum eigenvalues of a symmetric matrix. In particular, the case $h = 1 = X_i$ and $C = \theta$ reduces to the model $Y_i = B(U_i - \theta) + \epsilon_i$ considered above.

If B is known, then we can write (3.14) as

$$U_i - B^{-1}Y_i = CX_i - B^{-1}\epsilon_i. \tag{3.17}$$

The least squares estimate C_t of C based on $X_1, U_1, Y_1, \cdots, X_t, U_t, Y_t$ can be expressed in the following recursive form:

$$C_t = C_{t-1} + (U_t - B^{-1}Y_t - C_{t-1}X_t)X_t^T P_t, \tag{3.18a}$$

$$P_t^{-1} = P_{t-1}^{-1} + X_t X_t^T. \tag{3.18b}$$

Note that (3.18b) can be expressed in the computationally more convenient form

$$P_t = P_{t-1} - P_{t-1}X_t X_t^T P_{t-1}/(1 + X_t^T P_{t-1} X_t).$$

Now suppose that B is unknown but that there exist nonsingular $k \times k$ matrices B_i such that B_i is $\mathcal{F}_i$-measurable and

$$B_i \to B \quad \text{w.p.1.} \tag{3.19}$$

Replacing B by B_{t-1} in (3.18) leads to

$$C_t = C_{t-1} + (U_t - B_{t-1}^{-1}Y_t - C_{t-1}X_t)X_t^T P_t, \tag{3.20a}$$

$$P_t^{-1} = P_{t-1}^{-1} + X_t X_t^T. \tag{3.20b}$$

If B and C are known, then $E(Y_i|\mathcal{F}_{i-1}) = B(U_i - CX_i)$ is the minimum variance predictor of Y_i and the prediction error $Y_i - E(Y_i|\mathcal{F}_{i-1})$ is equal to ϵ_i. When B and C are unknown, replacing B by B_{i-1} and C by C_{i-1} leads to the adaptive predictor $B_{i-1}(U_i -$

$C_{i-1}X_i$) with prediction error $e_i = Y_i - B_{i-1}(U_i - C_{i-1}X_i)$. Under the assumptions (3.15), (3.16) and (3.19), the cumulative square difference between e_i and ϵ_i has been shown in [27] to satisfy the basic inequality

$$\begin{aligned}\sum_1^n \|e_i - \epsilon_i\|^2 + \mathrm{tr}\{B(C_n - C)P_n^{-1}(C_n - C)^T B^T\} \\ \leq (\sigma^2 + o(1)) \log \det(\sum_1^n X_i X_i^T) + o(\sum_1^n \|U_i - CX_i\|^2) \text{a.s.}\,,\end{aligned} \tag{3.21}$$

where $\sigma^2 = \limsup_{i\to\infty} E(\|\epsilon_i\|^2 | \mathcal{F}_{i-1})$.

The inequality (3.21) provides important clues for the construction of asymptotically efficient certainty-equivalence-type adaptive control laws for the regulation problem of choosing the inputs $U_1, U_2, \cdots$ sequentially so that $\sum_1^n \|Y_t\|^2$ is minimized in some sense in ignorance of the parameter matrices B and C. Minimizing $E(\sum_1^n \|Y_t\|^2)$ is equivalent to minimizing $E(R_n)$, where

$$R_n = \sum_1^n \|B(U_t - CX_t)\|^2 = \sum_1^n \|Y_t - \epsilon_t\|^2 \tag{3.22}$$

Clearly if C is known, then the $\mathcal{F}_{t-1}$-measurable input $U_t = CX_t$ (which is equivalent to $E[Y_t|\mathcal{F}_{t-1}] = 0$) minimizes $E\|Y_t\|^2$, and the output of this rule is ϵ_t. The certainty-equivalence rule $U_t = C_{t-1}X_t$ has output $Y_t = e_t$, and therefore its regret is $R_n = \sum_1^n \|e_i - \epsilon_i\|^2$, a bound for which is provided by (3.21).

To apply (3.21), we have to construct estimates B_i such that $B_i \to B$ a.s. A simple way of doing this is to apply the method of least squares to "well-excited" blocks of input-output data. To generate such data, we introduce white-noise probing inputs at certain prespecified times to ensure that there is enough information for estimating B. Let

$$\begin{aligned}\tau^* = \inf\Bigg\{ i : \lambda_{\min}\left[\sum_{t=1}^i \binom{X_t}{U_t}(X_t^T, U_t^T)\right] \geq \eta \quad \text{and} \\ \left[\sum_{t=1}^i Y_t(X_t^T, U_t^T)\right]\left[\sum_{t=1}^i \binom{X_t}{U_t}(X_t^T, U_t^T)\right]^{-1} \text{ is nonsingular}\Bigg\}.\end{aligned} \tag{3.23}$$

where η is a positive constant, and take integers $1 < \nu_1 < \nu_2 < \cdots$ Let

$$J = \{1, \cdots, \tau^*\} \cup \{\tau^* + \nu_1, \tau^* + \nu_2, \cdots\} \tag{3.24}$$

be the set of stages when probing inputs are to be introduced. Thus, at stage $t \in J$, the input

$$U_t \text{ is a random vector independent of } X_1, U_1, Y_1, \cdots, X_{t-1}, U_{t-1}, Y_{t-1}, X_t \text{ such that the components of } U_t \text{ are independently distributed with mean } 0, \text{ variance } c, \text{ and } \|U_t\| \leq M, \tag{3.25}$$

where c and M are positive constants. Note that $\tau^* < \infty$ a.s. by (3.14), (3.16), (3.25) and the fact that B is nonsingular. Rewriting (3.14) as $Y_i = AX_i + BU_i + \epsilon_i$, where $A = -BC$, let

$$(\widetilde{A}_\nu \vdots \widetilde{B}_\nu) = \Big\{\sum_{\substack{t\in J\\ t\leq\nu}} Y_t(X_t^T, U_t^T)\Big\}\Big\{\sum_{\substack{t\in J\\ t\leq\nu}} \binom{X_t}{U_t}(X_t^T, U_t^T)\Big\}^{-1}, \nu \geq \tau^*, \tag{3.26}$$

be the least squares estimate of $(A \vdots B)$ based only on observations at stages $t \in J$ with $t \leq \nu$. Then $\widehat{B}_\nu \to B$ a.s. In fact, letting m_ν denote the number of elements in the set $\{t \in J : t \leq \nu\}$ and letting $\widehat{B}_\nu(i,j)$ and $B(i,j)$ denote the (i,j) elements of $\widehat{B}_\nu$ and B, Lemma 1 of [27] shows that

$$\max_{1\leq i,j\leq k} |\widehat{B}_\nu(i,j) - B(i,j)| = O(m_\nu^{-1/2}(\log m_\nu)^{1/2}) \text{ a.s.} \tag{3.27}$$

Suppose that the ν_i in (3.24) are so chosen that $i = o(\log \nu_i)$; for example, take $\nu_i = \exp((1 + o(1))i^\delta)$ with $\delta > 1$. Then for $\nu_i \leq \nu < \nu_{i+1}$, $m_\nu = \tau^* + i = o(\log \nu)$. Therefore the number m_ν of probing inputs up to stage ν diverges to ∞ but within the order of $o(\log \nu)$ as $\nu \to \infty$, so the contribution of these probing inputs to the regret (3.22) is of the order $o(\log n)$ a.s. Define the input U_t at stage $t \in J$ by (3.25) ("probing input"), and at stage $t \notin J$ define $U_t = C_{t-1}X_t$ (certainty equivalence). For this certainty-equivalence rule with forced learning, $\sum_{i\in J, i\leq n} \|U_i - CX_i\|^2 = o(\log n)$ a.s., and it follows from (3.21) that

$$\sum_{i\notin J, i\leq n} \|B(U_i - CX_i)\|^2 = \sum_{i\notin J, i\leq n} \|Y_i - \epsilon_i\|^2 = \sum_{i\notin J, i\leq n} \|e_i - \epsilon_i\|^2$$
$$\leq (\sigma^2 + o(1))h \log n + o(\sum_{i=1}^n \|U_i - CX_i\|^2) \text{ a.s.},$$

noting that $\log\det(\sum_1^n X_iX_i^T) \leq h(1 + o(1)) \log n$ a.s. by (3.16). Hence

$$R_n = \sum_{i=1}^n \|B(U_i - CX_i)\|^2 \leq (\sigma^2 + o(1))h \log n \text{ a.s.} \tag{3.28}$$

Note that in the special case $h = 1 \equiv X_i$, the right hand side of (3.28) is $(1+o(1))\sigma^2 \log n$, which agrees with the asymptotically minimal order (3.8) for the regret for the model (3.1). In fact, for the multivariate model (3.14), a similar Bayesian argument assuming B to be known shows that the right hand side of (3.28) is also asymptotically minimal under certain additional assumptions, cf. [27].

4. ASYMPTOTICALLY EFFICIENT CERTAINTY-EQUIVALENCE-TYPE ALGORITHMS IN ADAPTIVE CONTROL OF ARMAX SYSTEMS

A widely used stochastic model in the time series and stochastic control literature is the ARMAX system (autoregressive moving average system with exogenous inputs) defined by the linear difference equation

$$A(q^{-1})y_n = B(q^{-1})u_{n-d} + C(q^{-1})\epsilon_n, \tag{4.1}$$

where $\{y_n\}, \{u_n\}$ and $\{\epsilon_n\}$ denote the output, input and disturbance sequences, respectively, $d \geq 1$ represents the delay and $A(q^{-1}) = 1 + a_1 q^{-1} + \cdots + a_p q^{-p}$, $B(q^{-1}) = b_1 + \cdots + b_k q^{-(k-1)}$, $C(q^{-1}) = 1 + c_1 q^{-1} + \cdots + c_h q^{-h}$ are scalar polynomials in the backward shift operator q^{-1}. Throughout the sequel we shall assume that $b_1 \neq 0$ and that

$$\begin{aligned}&\{\epsilon_n\} \text{ is a martingale difference sequence with respect to an increasing}\\&\text{sequence of } \sigma-\text{fields } \mathcal{F}_n \text{ such that } E(\epsilon_n^2|\mathcal{F}_{n-1}) = \sigma^2 \text{ (nonrandom)} > 0 \text{ for all } n.\end{aligned} \tag{4.2}$$

Because of its theoretical interest and practical importance, the problem of determining the inputs u_n, based on current and past observations $y_n, y_{n-1}, u_{n-1}, \cdots$ (i.e., u_n is $\mathcal{F}_n$-measurable), to keep the outputs y_{n+d} as close as possible to certain target values y^*_{n+d} when the system parameters are not known in advance but have to be estimated "on-line" (i.e., during the operation of the system) has been one of the major topics in the subject of stochastic adaptive control. Let $x_0 = (y_0, \cdots, y_{1-p}, u_0, \cdots, u_{2-d-k}, \epsilon_0, \cdots, \epsilon_{1-h})^T$ denote the "initial condition" of (1.1). The polynomial $B(z) = b_1 + \cdots + b_k z^{k-1}$ will be called *stable* if all its zeros lie outside the unit circle. We shall also call two or more polynomials *coprime* (or *relatively prime*) if their greatest common divisors have degree 0.

In principle, given a probability distribution of the random sequence $\{x_0, \epsilon_1, \epsilon_2, \cdots\}$ and a prior distribution π of the unknown parameter vector

$$\theta = (-a_1, \cdots, -a_p, b_1, \cdots, b_k, c_1, \cdots, c_h)^T,$$

we can use backward induction to solve the dynamic programming equations defining the inputs $u_1, \cdots, u_{N-d}$ that minimize

$$\int E_\theta\{ \sum_{i=d+1}^{N} (y_i - y_i^*)^2 \} d\pi(\theta), \tag{4.3}$$

for every given horizon N, where the y_i^* are nonrandom target values for the outputs. Despite the analytical and computational difficulties in the implementation of the Bayesian approach, Bayesian analysis of some very simple examples has provided important insights into the structure of optimal control rules. In particular, Feldbaum [28] and subsequent authors have shown that Bayes rules have the "dual control" function of both probing the system for information about its parameters and trying to drive the outputs towards their target values. Åström [29, p. 478] has provided an interesting numerical example to illustrate this dual control effect in the adaptive control problem (4.3) for the simple ARX model $y_n - \alpha y_{n-1} = \beta u_{n-1} + \epsilon_n$, with i.i.d. zero-mean normal ϵ_n, a normal prior distribution for β, known value $\alpha = 1$, and $N = 30$, $y_i^* \equiv 0$. The example shows that the Bayes rule takes relatively large and irregular control actions to probe the system when the Bayes estimate $\widehat{\beta}_t$ of β has poor precision, but when $\widehat{\beta}_t$ has high precision the Bayes rule is well approximated by the "certainty-equivalence" rule $\widehat{\beta}_t u_t = -y_t$.

Suppose that in Åström's example, instead of assuming α to be known and β to have a normal prior distribution, we assume $\beta \neq 0$ to be known and α to have a normal prior distribution. This problem turns out to be more tractable and can be used to derive lower bounds on the cost of Bayes rules. First note that for any input sequence $\{u_n\}$,

$$\int_{-\infty}^{\infty} E_\alpha\{\sum_{i=2}^{N}(y_i - \epsilon_i)^2\} d\pi(\alpha) = \sum_{i=1}^{N-1} E\{(\alpha y_i + \beta u_i)^2\}$$

$$\geq \sum_{i=1}^{N-1} E\{[\alpha - E(\alpha | y_1, u_1, \cdots, y_i, u_i)]^2 y_i^2\}. \tag{4.4}$$

In particular, for an input sequence $\{u_n\}$ that has the property

$$\lim_{N\to\infty} N^{-1} \sum_{i=2}^{N} (y_i - \epsilon_i)^2 = 0 \text{ a.s. } [P_\alpha] \text{ for every } \alpha, \tag{4.5}$$

it can be shown that

$$\sum_{i=1}^{n} \{\alpha - E(\alpha | y_1, u_1, \cdots, y_i, u_i)\}^2 y_i^2 \tag{4.6}$$

$$= (1 + o(1)) \sum_{i=1}^{n} \{\sum_{t=2}^{i} y_{t-1}\epsilon_t / \sum_{t=2}^{i} y_{t-1}^2\}^2 y_i^2 + O(1) \sim \sigma^2 \log n \text{ a.s. } [P_\alpha],$$

for every α. From (4.4) and (4.6), it follows by an application of Fatou's lemma and Fubini's theorem that

$$\int_{-\infty}^{\infty} E_\alpha\{\sum_{i=2}^{N}(y_i - \epsilon_i)^2\}d\pi(\alpha) \geq (1 + o(1))\sigma^2 \log N, \tag{4.7}$$

for all input sequences $\{u_n\}$ satisfying (4.6).

Lai [27] has extended the results (4.4), (4.6) and (4.7) above for the problem (4.3) with $y_i^* \equiv 0$ to general ARX models (4.1) (with $C(q^{-1}) = 1$) and unit delay $d = 1$. Specifically, assuming $b_1(\neq 0)$ to be known and putting a truncated normal prior distribution π on $\lambda = b_1^{-1}(-a_1, \cdots, -a_p, b_2, \cdots, b_k)^T$, Lai [27] showed that in analogy with (4.7),

$$\int E_\lambda\{\sum_{i=2}^{N}(y_i - \epsilon_i)^2\}d\pi(\lambda) \geq (1 + o(1))\sigma^2(p + k - 1)\log N, \tag{4.8}$$

for all input sequences $\{u_n\}$ satisfying (4.5) and the additional growth condition that $u_n^2 = O(n^\delta)$ a.s. for some $0 < \delta < 1$. The truncated normal prior distribution π in (4.8) is the restriction of a standard multivariate normal distribution to the λ-region defined by

$A(z)$ and $B(z)$ are stable, and the polynomials $a_1 z^{p-1} + \cdots + a_p$ and $z^{k-1}B(z^{-1})$ are relatively prime. (4.9)

In the case of unit delay $d = 1$, if all the parameters of the system (1.1) are known, then the optimal controller chooses the input u_t at stage t so that $E(y_{t+1}|\mathcal{F}_t) = y_{t+1}^*$, and its output at stage $t + 1$ is $y_{t+1}^* + \epsilon_{t+1}$. In view of this, Lai [27] defines the "regret" at stage N of an input sequence $\{u_n\}$ to be

$$R_N = \sum_{i=2}^{N}\{y_i - (y_i^* + \epsilon_i)\}^2 = \sum_{i=2}^{N}\{E(y_i|\mathcal{F}_{i-1}) - y_i^*\}^2. \tag{4.10}$$

Note that (4.8) above represents a lower bound for the expected regret in the regulation problem $y_i^* \equiv 0$ within a Bayesian framework. For general delay d, the definition (4.10) of "regret" can be extended to

$$R_N = \sum_{i=d+1}^{N}\{E(y_i|\mathcal{F}_{i-d}) - y_i^*\} = \sum_{i=d+1}^{N}(y_i - y_i^* - \eta_i)^2, \tag{4.11}$$

where $\eta_i = y_i - E(y_i|\mathcal{F}_{i-d})$. The regret R_N, which is 0 for the optimal controller assuming knowledge of all system parameters so that u_t is determined by $E(y_{t+d}|\mathcal{F}_t) = y_{t+d}^*$, can be regarded as the cumulative cost up to stage N due to lack of knowledge of the system parameters in an adaptive controller. Since $E\{\sum_{i=d+1}^{N}(y_i - y_i^*)^2\} = E(R_N) + E(\sum_{i=d+1}^{N}\eta_i^2)$, the problem of minimizing the total expected quadratic cost is equivalent to that of minimizing the expected value of the regret R_N.

An input seuence is called "self-optimizing" (or "globally convergent") if

$$R_n/n \to 0 \text{ a.s.} \tag{4.12}$$

Since the seminal paper of Åström and Wittenmark [30] on "self-tuning regulators", an active area of research has been to find self-optimizing control schemes for (4.1) that can be implemented in real time. To review the Åström-Wittenmark [30] idea of self-tuning regulators for the stochastic linear system (4.1), first consider the case of unit delay ($d = 1$) and white noise ($C(q^{-1}) = 1$). Replacing b_1 in (4.1) by a prior guess $b \neq 0$, they proposed to estimate the other parameters at stage t by $a_1^{(t)}, \cdots, a_p^{(t)}, \beta_2^{(t)}, \cdots, \beta_k^{(t)}$ that minimize

$$\sum_{i=1}^{t}(y_i + \alpha_1 y_{i-1} + \cdots + \alpha_p y_{i-p} - bu_{i-1} - \beta_2 u_{i-2} - \cdots - \beta_k u_{i-k})^2, \tag{4.13}$$

and to determine the input u_t by the certainty-equivalence rule

$$u_t = \{\alpha_1^{(t)} y_t + \cdots + \alpha_p^{(t)} y_{t-p+1} - \beta_2^{(t)} u_{t-1} - \cdots - \beta_k^{(t)} u_{t-k+1}\}/b. \tag{4.14}$$

They showed that if the estimates should converge as $t \to \infty$, then $\alpha_1^{(t)}/b, \ldots, \alpha_p^{(t)}/b, \beta_2^{(t)}/b, \cdots, \beta_k^{(t)}/b$ must necessarily converge to the coefficients $a_1/b_1, \cdots, a_p/b_1, b_2/b_1, \cdots, b_k/b_1$ in the optimal controller $u_t^* = \{a_1 y_t + \cdots + a_p y_{t-p+1} - b_2 u_{t-1} - \cdots - b_k u_{t-k+1}\}/b_1$ that assumes knowledge of the system parameters, and therefore the adaptive controller u_t "self-tunes" itself in the sense that its defining equation has asymptitically negligible difference from that of u_t^*. By reparametrizing the system (4.1) and using least squares or extended least squares to directly estimate the transformed parameters, they also suggested natural extensions of (4.14) to general delay and colored noise.

An open problem with the Åström-Wittenmark approach is whether with positive probability the parameter estimates may fail to converge. Instead of adhering to a prior guess b of b_1 in their rule, an obvious modification is to update this guess with the current and past data. Extending the ideas described in §3 for adaptive control of the stochastic regression model (3.14), Lai and Wei [31] studied this modification of the original Åström-wittenmark scheme (4.13)-(4.14). Specifically, under the assumption that $A(z)$ is stable, they introduced occasional blocks of white-noise probing inputs and used only the data from these stages of forced learning to estimate the system parameters, thus obtaining a strongly consistent estimate $b(t)$ of b_1. Replacing b by $b(t)$ in (4.13)-(4.14) whenever $b(t) \neq 0$ and the number of probing inputs up to stage t does not fall below some threshold K_t (with $K_t \to \infty$ and $K_t = o(\log t)$) leads to a certainty-equivalence controller, for the system (4.1) with $d = 1$ and $C(q^{-1}) = 1$, whose regret has the logarithmic order

$$R_n \leq (1 + o(1))\sigma^2(p + k - 1)\log n \text{ a.s.}, \tag{4.15}$$

cf. [31]. Note that the right hand side of (4.15) is the same as the lower bound (4.8) for the Bayes regret assuming knowledge of b_1. Lai [27] calls adaptive regulators having the property (4.15) "asymptotically efficient". As shown in [31], not only does $b(t)$ converge a.s. to b_1, but the least squares estimates $\alpha_1^{(t)}, \cdots, \alpha_p^{(t)}, \beta_2^{(t)}, \cdots, \beta_k^{(t)}$ determined by (4.13) with b replaced by $b(t)$ also converge a.s to $a_1, \cdots, a_p, b_2, \cdots, b_k$.

In the unit-delay, white noise case (i.e., $d = 1 = C(q^{-1})$), (4.1) can be written as a stochastic regression model

$$y_{n+1} = \theta^T \psi_n + \epsilon_{n+1}, \tag{4.16}$$

where $\theta = (-a_1, \cdots, -a_p, b_1, \cdots, b_k)^T$ and $\psi_n = (y_n, \cdots, y_{n-p+1}, u_n, \cdots, u_{n-k+1})^T$. An obvious certainty-equivalence rule in this case is to define u_t by the equation

$$\theta_t^T \psi_t = y_{t+1}^*, \tag{4.17}$$

where θ_t is the least squares estimator which has the recursive representation

$$\theta_t = \theta_{t-1} + P_{t-1}\psi_{t-1}(y_t - \theta_{t-1}^T \psi_{t-1}), \tag{4.18a}$$

$$P_t = P_{t-1} - P_{t-1}\psi_t \psi_t^T P_{t-1}/(1 + \psi_t^T P_{t-1}\psi_t). \tag{4.18b}$$

A first difficulty with this straightforward approach is that (4.17) need not be well defined since the coefficient $b_1^{(t)}$ of u_t in $\theta_t^T \psi_t$ may be 0 unless some continuity assumptions are made on the distribution of $\{\epsilon_t\}$. For the case $y_i^* \equiv 0$, Åström and Wittenmark [30] circumvented this difficulty by reparametrizing (4.16) as

$$y_{n+1} = b_1(u_n - \lambda^T X_n) + \epsilon_{n+1}, \quad \text{where} \quad X_n = (y_n, \cdots, y_{n-p+1}, u_{n-1}, \cdots, u_{n-k+1})^T,$$

and replacing the unknown b_1 by $b \neq 0$ while using least squares criterion (4.13) to estimate λ. A second difficulty with (4.17) or with (4.14) lies in the convergence analysis of the least squares estimator of θ or λ. A well known condition that ensures consistency of θ_t is the "persistent excitation" property:

$$n^{-1}\sum_{i=1}^{n}\psi_i\psi_i^T \text{ converges a.s. to a positive definite matrix.} \tag{4.19}$$

However, for the regulation problem $y_i^* \equiv 0$, the self-optimizing property (4.12) precludes persistent excitation for $\{\psi_i\}$, although it does not preclude persistent excitation for $\{X_i\}$, as has been noted by Kumar [1, pp. 372-373] and Lai and Wei [32, pp. 230 and 251].

For the unit-delay colored-noise cae, (4.1) can still be written in the form of a stochastic regression model (4.16) with

$$\psi_n = (y_n, \cdots, y_{n-p+1}, u_n, \cdots, u_{n-k+1}, \epsilon_n, \cdots, \epsilon_{n-h+1})^T. \tag{4.20}$$

However, the regressor ψ_n contains unobservable components $\epsilon_n, \cdots, \epsilon_{n-h+1}$. Replacing the unobservable ϵ_i by their estimates $\widehat{\epsilon}_i$ in the recursion (4.18) leads to the "extended least squares" algorithm of the form

$$\theta_t = \theta_{t-1} + P_{t-1}\phi_{t-1}(y_t - \theta_{t-1}^T\phi_{t-1}), \quad P_t^{-1} = P_{t-1}^{-1} + \phi_t\phi_t^T, \tag{4.21}$$

where

$$\phi_t = (y_t, \cdots, y_{t-p+1}, u_t, \cdots, u_{t-k+1}, \widehat{\epsilon}_t, \cdots, \widehat{\epsilon}_{t-h+1})^T. \tag{4.22}$$

The estimates $\widehat{\epsilon}_i$ of ϵ_i in (4.22) are given either by the residuals $\widehat{\epsilon}_i = y_i - \theta_i^T\phi_{i-1}$, in which case (4.22) is called the AML algorithm, or by the prediction errors $\widehat{\epsilon}_i = y_i - \theta_{i-1}^T\phi_{i-1}$, in which case (4.22) is called the RML1 algorithm, cf. [33]. Assuming that $\mathrm{Re}(1/C(e^{it}) - 1/2) > 0$ for all $t \in [-\pi, \pi]$, Solo [33] proved the strong consistency of θ_t under the persistent excitation condition (4.19).

Because of the difficulties in satisfying the persistent excitation condition for consistency of least squares or extended least squares estimates in a closed loop system where the inputs are determined by a certainty-equivalence rule, it was natural that the first major breakthrough in the development of self-optimizing controllers made use of a different recursive identification scheme which does not involve matrix-type analysis and conditions like the persistent excitation property (4.19). Instead of the matrix gain $P_{t-1} = (P_{t-1}^{-1})^{-1}$ in (4.18) or (4.21), Goodwin, Ramadge and Caines [34] used the scalar gain $1/\mathrm{tr}(P_{t-1}^{-1}) = 1/(\mathrm{tr}P_0^{-1} + \sum_1^{t-1}\|\phi_i\|^2)$, which circumvents the difficulties of matrix analysis. Moreover, instead of convergence analysis of the recursive identification algorithm, they proceeded directly to prove that the certainty-equivalence control rule associated with this scalar-gain algorithm is self-optimizing in the sense of (4.12). An important first step in their approach is to reparametrize (4.1) in the following prediction form. By the division algorithm, there exist polynomials $F(z) = 1 + f_1 z + \cdots + f_{d-1} z^{d-1}$ and $G(z) = g_1 + \cdots + g_{p(d)} z^{p(d)-1}$ with $p(d) = p \vee (h - d + 1)$ such that

$$C(z) = A(z)F(z) + z^d G(z), \tag{4.23}$$

and therefore (4.1) can be rewritten in the form

$$C(q^{-1})\{y_{n+d} - F(q^{-1})\epsilon_{n+d}\} = G(q^{-1})y_n + B(q^{-1})F(q^{-1})u_n, \tag{4.24}$$

cf. [35]. Hence, in the case of known system parameters, the optimal d-step ahead predictor $\widetilde{y}_{n+d} \stackrel{\Delta}{=} E(y_{n+d}|\mathcal{F}_n)$ is given by

$$\widetilde{y}_{n+d} + c_1 \widetilde{y}_{n+d-1} + \cdots + c_h \widetilde{y}_{n+d-h} = G(q^{-1})y_n + (BF)(q^{-1})u_n, \tag{4.25}$$

where $(BF)(z) = B(z)F(z) = b_1 + (bf)_2 z + \cdots + (bf)_{k+d-1} z^{k+d-2}$, and its prediction error is

$$\eta_{n+d} \stackrel{\Delta}{=} y_{n+d} - \widetilde{y}_{n+d} = F(q^{-1})\epsilon_{n+d}. \tag{4.26}$$

Let

$$\widetilde{\theta} = (g_1, \cdots, g_{p(d)}, b_1, (bf)_2, \cdots, (bf)_{k+d-1}, -c_1, \cdots, -c_h)^T. \tag{4.27}$$

Since $y_{n+d} = \widetilde{y}_{n+d} + \eta_{n+d}$, we obtain from (4.25) the prediction form of (4.1):

$$\begin{aligned} y_{n+d} &= \widetilde{\theta}^T \widetilde{\psi}_n + \eta_{n+d}, \quad \text{where} \\ \widetilde{\psi}_n &= (y_n, \cdots, y_{n-p(d)+1}, u_n, \cdots, u_{n-k-d+2}, \widetilde{y}_{n+d-1}, \cdots, \widetilde{y}_{n+d-1})^T. \end{aligned} \tag{4.28}$$

First consider the case of unit delay $d = 1$. Here $F(z) = 1$, $zG(z) = C(z) - A(z)$, $p(d) = p \vee h$, and $\eta_t = \epsilon_t$. In the case of known system parameters, the optimal controller chooses u_t so that $\widetilde{y}_{t+1} = y^*_{t+1}$. Without assuming knowledge of the parameters, Goodwin, Ramadge and Caines [34] therefore replaced the $\widetilde{y}_i$ in $\widetilde{\psi}_n$ by the target value y^*_i and the unknown parameter vector $\widetilde{\theta}$ in (4.27) by a stochastic gradient estimate of the form

$$\theta^*_t = \theta^*_{t-1} + (a/r^*_{t-1})\phi^*_{t-1}(y_t - \theta^{*T}_{t-1}\phi^*_{t-1}), \quad r^*_t = r^*_{t-1} + \|\phi^*_t\|^2, \tag{4.29}$$

where

$$\phi_n^* = (y_n, \cdots, y_{n+1-p\vee h}, u_n, \cdots, u_{n+1-k}, y_n^*, \cdots, y_{n+1-h}^*)^T \quad \text{for} \quad n \geq h. \tag{4.30}$$

They used the certainty-equiavlence rule that defines the input u_t for $t \geq h$ by the equation

$$\theta_t^{*T} \phi_t^* = y_{t+1}^*. \tag{4.31}$$

To ensure that the component $b_{t,1}$ of θ_t^* is nonzero a.s. so that u_t is well defined by (4.31), they assumed that $(x_0, \epsilon_1, \cdots, \epsilon_n)$ is absolutely continuous with respect to Lebesgue measure for every $n \geq 1$. Under certain additional stability and positive real assumptions on $B(z)$ and $C(z)$ and boundedness assumptions on $\{y_i^*\}$, they used a martingale argument to establish the self-optimizing property (4.12) for the certainty-equivalence rule (4.31). For general delay, they made use of an "interlacing" technique to extend this control scheme.

As mentioned earlier, a long-standing difficulty with certainty-equiavlence rules based on standard recursive identification methods like least squares or extended least squares has been that the persistent excitation conditions commonly used in their convergence analysis may fail to hold for self-optimizing systems. For the least squares method in the white-noise, unit-delay case, Lai and Wei [36] succeeded in establishing its strong consistency under a much weaker excitation condition. Specifically, for a general stochastic regression model (4.16) in which $\{\epsilon_n, \mathcal{F}_n, n \geq 1\}$ is a martingale difference sequence such that $\sup_n E(|\epsilon_n|^\alpha | \mathcal{F}_{n-1}) < \infty$ a.s. for some $\alpha > 2$ and ψ_n is $\mathcal{F}_n$-measurable, they showed that the least squares estimate θ_n defined recursively by (4.18) converges a.s. to θ if

$$\lambda_{\min}(\sum_1^n \psi_i \psi_i^T) \to \infty \quad \text{and} \quad \log \lambda_{\max}(\sum_1^n \psi_i \psi_i^T) = o(\lambda_{\min}(\sum_1^n \psi_i \psi_i^T)) \text{ a.s.} \tag{4.32}$$

They also gave an example of a stochastic regression model in which $\log \lambda_{\max}(\sum_1^n \psi_i \psi_i^T)$ is of the same order of magnitude as $\lambda_{\min}(\sum_1^n \psi_i \psi_i^T)$ and θ_n is inconsistent. For the unit-delay case $d = 1$ in (4.1), under the stability and positive real assumption $\min_{|t| \leq \pi} \mathrm{Re}\{1/C(e^{it}) - \frac{1}{2}\} > 0$, Lai and Wei [37] showed that (4.32) is again sufficient to ensure the strong consistency of the AML algorithm (4.21)-(4.22). The relaxation of the excitation condition from the persistent type (4.19) to the weak type (4.32) suggests new possibilities in using the AML algorithm (which reduces to least squares in the white noise case $C(q^{-1}) = 1$) for consistent parameter estimation in an adaptive control environment. For example, for systems with bounded sample mean square inputs and outputs (i.e., $n^{-1} \sum_1^n (y_i^2 +$

$u_i^2) = O(1)$ a.s.), $\log \lambda_{\max}(\sum_1^n \psi_i \psi_i^T)$ is of the order $O(\log n)$, and therefore the excitation condition (4.32) is satisfied if $\lambda_{\min}(\sum_1^n \psi_i \psi_i^T)/\log n \to \infty$ a.s., which can typically be achieved by introducing white-noise probing inputs occasionally so that the total variance V_n of these probing inputs up to stage n is of the order $(\log n)(\log\log n)$. However, if we want to keep the regret R_n to be no larger than $O(\log n)$, then we cannot afford to have $V_n/\log n \to \infty$ but must have at most $V_n = O(\log n)$. Although this is in conflict with the excitation condition (4.32) for consistency of the AML algorithm θ_n, the convergence analysis of the least squares algorithm in [36] and of the AML algorithm [37] is fortunately versatile enough to provide two other useful ideas to handle this delicate situation.

The first idea is to obtain bounds for $\theta_n - \theta$, and the convergence analysis of [37] in fact shows that

$$\|\theta_n - \theta\| = O(\{\log \lambda_{\max}(P_n^{-1})\}^{1/2}/\lambda_{\min}^{1/2}(P_n^{-1})) \text{ a.s.} \tag{4.33}$$

The second idea is to obtain bounds for the cumulative squared difference $\sum_1^n(\theta^T\psi_i - \theta_i^T\phi_i)^2$ between the optimal one-step ahead predictor $\theta^T\psi_i$ of the response y_{i+1} (that assumes knowledge of the parameter vector θ and initial condition x_0) and the adaptive predictor $\theta_i^T\phi_i$ (that does not assume knowledge of θ and x_0). This cumulative squared difference is of particular interest in adaptive control applications since the optimal contoller assuming knowledge of the system parameters is given by $\theta^T\psi_t = y_{t+1}^*$, while the certainty-equivalence rule based on the adaptive predictor is given by $\theta_t^T\phi_t = y_{t+1}^*$. It is shown in [37] that for every $0 < \delta < 1$,

$$\sum_{i=1}^n (\theta^T\phi_i - \theta_i^T\phi_i)^2 I_{\{\phi_i^T P_i \phi_i \le \delta\}} = O(\log \lambda_{\max}(P_n^{-1})) \text{ a.s.} \tag{4.34}$$

Making use of (4.33) and (4.34), Lai and Wei [37] proved not only the self-optimizing property (4.12) but also the much stronger conclusion

$$R_n = O(\log n) \text{ a.s.} \tag{4.35}$$

for a modification of the certainty-equivalence rule based on the AML algorithm, under the assumptions of boundedness of the target values y_i^*, stability of the open-loop plant (i.e., $A(z)$ and $B(z)$ are stable), and positive real assumptions for $C(z)$. A basic ingredient of this modification is a simple criterion to decide whether information is inadequate for approximating the unobservable $\theta^T\psi_t$ by $\theta_t^T\phi_t$. When the data show inadequate information for such approximation, instead of adhering to the certainty-equivalence formula

$\theta_t^T \phi_t = y_{t+1}^*$ to determine the output u_t, [37] proposes to introduce a block of white-noise perturbations to improve the information content of the design, in such a way that the number of these perturbations up to stage n is kept within the order $O(\log n)$.

Despite the much stronger results (4.15) and (4.35) on least squares and extended least squares certainty-equivalence rules with forced learning in [31] and [36] than the self-optimizing property (4.12) for the stochastic gradient certainty-equivalence rule (4.31), the requirement that $A(z)$ be stable in [31] and [36] imposes a serious limitation, particularly in light of the fact that (4.31) can be used to stabilize a system that would be unstable in the open loop. Another limitation in the methods and results of [31] and [36] is that they are applicable only to the unit delay case. Lai and Ying [38] recently removed these limitations by a parallel implementation of stochastic gradient and recursive maximum likelihood certainty-equivalence rules and by using occasional white-noise perturbations of the target values instead of the simple white-noise inputs in [31] and [36]. For the case of general delay d and without assuming $A(z)$ to be stable, this approach gives the following logarithmic order for the regret:

$$R_n \leq (1 + o(1))(1 + f_1^2 + \cdots + f_{d-1}^2)\sigma^2(p(d) + h + k + d - 1)(2d - 1)\log n \text{ a.s.} \qquad (4.36)$$

Moreover, if $\log(1 + \sum_{i=1}^n y_i^{*2}) = o(\log n)$, as in the regulation problem ($y_i^* \equiv 0$), we can further strengthen (4.36) into

$$R_n \leq (1 + o(1))(1 + f_1^2 + \cdots + f_{d-1}^2)\sigma^2(p(d) + d + k - 2)(2d - 1)\log n \text{ a.s.} \qquad (4.37)$$

Note that in the case $d = 1$ and $C(q^{-1}) = 1$ (so that $h = 0$), (4.37) reduces to (4.15).

Although the extended least squares recursive identification (4.21) has been called "approximate maximum likelihood" (AML) or "recursive maximum likelihood of type 1" (RML1), its statistical properties are actually unrelated to those of the off-line maximum likelihood estimator even when the ϵ_i are normal. Linear approximations to the nonlinear estimating equations defining the off-line maximum likelihood estimator in some neighborhood of the true parameter lead to the RML2 algorithm, cf. [39]. For adaptive control in the case of general delay d, Lai and Ying [38] use an extension of the RML2 algorithm to estimate the parameter $\widetilde{\theta}$ in the regression model (4.28) which is the prediction form of the model (4.1). This extension also includes monitoring by a consistent recursive estimator $\widetilde{\theta}_{m_j}$ constructed by the method of moments at stages $m_1 < m_2 < \cdots$ that will be described later. Letting $\mathbf{I}_j$ be a cube in $\mathbb{R}^{p(d)+k+d-1+h}$ with center at $\widetilde{\theta}_{m_j}$ so that

$$P\{\widetilde{\theta} \in \mathbf{I}_j \text{ for all large } j\} = 1, \lim_{j\to\infty} (\text{width of } \mathbf{I}_j) = 0, \qquad (4.38)$$

as will be specified later, the monitored recursive maximum likelihood algorithm $\theta_{n,M}$ is constrained to lie inside the "confidence interval" $\mathbf{I}_j$ for $m_j \le n < m_{j+1}$. The projection which we use to constrain $\theta_{n,M}$ is taken with respect to the norm induced by the positive definite matrix P_{n-d}^{-1} defined in (4.40d) below, instead of the usual Euclidean norm. For $x \in \mathbb{R}^{p(d)+k+d-1+h}$ and $m_j \le n < m_{j+1}$, let $\pi_n(x)$ denote the unique solution of the quadratic programming problem

$$(\pi_n(x)-x)^T P_{n-d}^{-1}(\pi_n(x)-x) = \min_{y\in \mathbf{I}_j}\{(y-x)^T P_{n-d}^{-1}(y-x)\}, \tag{4.39}$$

i.e., $\pi_n(x)$ is the projection of x into $\mathbf{I}_j$ with repect to the norm induced by P_{n-d}^{-1}. The choice of a cube for the confidence region $\mathbf{I}_j$ implies linear constraints for the quadratic programming problem (4.39), which can be handled by simple computational methods. Define $\theta_{n,M} = (\widehat{g}_{n,1},\cdots,\widehat{g}_{n,p(d)},\widehat{B}_{n,1},(\widehat{bf})_{n,2},\cdots,(\widehat{bf})_{n,k+d-1},-\widehat{c}_{n,1},\cdots,-\widehat{c}_{n,h})^T$ for $n > m_1$ by the recursion

$$\theta_{n,M} = \pi_n(\theta_{n-1,M} + P_{n-d}\xi_{n-d}(y_n - \widehat{y}_{n,M})), \tag{4.40a}$$

$$\xi_n + \widehat{c}_{n,1}\xi_{n-1} + \cdots + \widehat{c}_{n,h}\xi_{n-h} = \phi_{n,M}, \quad \text{where} \tag{4.40b}$$

$$\phi_{n,M} = (y_n,\cdots,y_{n-p(d)+1},u_n,\cdots,u_{n-k-d+2},\widehat{y}_{n+d-1,M},\cdots,\widehat{y}_{n+d-h,M})^T,$$

$$\widehat{y}_{n+d,M} = \theta_{n,M}^T\phi_{n,M}, \tag{4.40c}$$

$$P_n^{-1} = P_{n-1}^{-1} + \xi_n\xi_n^T + I/n, \tag{4.40d}$$

where P_{m_1} is a positive definite matrix and $\theta_{m_1,M}$ represents an initial guess of $\widetilde{\theta}$ (e.g., set $\theta_{m_1,M} = \widetilde{\theta}_{m_1}$), and I denotes the identity matrix.

Suppose that $C(z)$ is stable and that the ϵ_n satisfy assumption (4.2) and the boundedness condition $\sup_n |\epsilon_n| < \infty$ a.s. Defining ξ_n and P_n as in (4.40), assume that

$$\xi_n^T P_n \xi_n \to 0, \sum_{i=1}^{n}\|\xi_i\|^2 \le n^{1+o(1)}, \sum_{j:m_j\le n}\sum_{r=1}^{h+d-1}(\|\xi_{m_j}-r\|^2 + \|\xi_{m_j-r}\|) = O(\log n) \text{ a.s.} \tag{4.41}$$

Let $\sigma_d^2 = \sigma^2(1+f_1^2+\cdots+f_{d-1}^2)$. Then as shown in [38],

$$\sum_{t=1}^{n}\{E(y_{t+d}|\mathcal{F}_t) - \widehat{y}_{t+d,M}\}^2 \le (1+o(1))\sigma_d^2(p(d)+h+k+d-1)(2d-1)\log n \text{ a.s.} \tag{4.42}$$

Note that the right hand side of (4.42) is the same as the logarithmic order of magnitude (4.36) of the regret for adaptive controllers that we want to achieve. Therefore the certainty-equivalence rule based on the monitored recursive maximum likelihood algorithm

$$\theta_{n,M}^T\phi_{n,M} = y_{n+d}^* \tag{4.43}$$

has the desired order of magnitude (4.36) for its regret if it can be shown that (4.41) holds and if consistent auxiliary estimates can be developed to define the monitoring sets $\mathbf{I}_j$. This is the basic idea behind the asymptotically efficient adaptive control scheme in [38].

With probability 1, since $\theta_{n,M} \to \widetilde{\theta}$ and $b_1 \neq 0$, (4.43) is well defined for all large n. To define strongly consistent estimators $\widetilde{\theta}_{m_j}$, Lai and Ying [38] introduce blocks of well-excited input-output data from which the $\widetilde{\theta}_{m_j}$ are to be constructed. Since $A(z)$ may be unstable, they modify an idea due to Caines and Lafortune [40] to construct such blocks of well excited data. Without assuming $A(z)$ to be stable but assuming that $B(z)$ is stable and $\min_{|t|\le\pi} \mathrm{Re}\{C(e^{it}) - a/2\} > 0$, Caines and Lafortune [40] made use of the stochastic gradient certainty-equivalence rule (4.31) to stabilize the system but introduced white-noise perturbations w_t into the target values y^*_{t+1} of (4.31) to excite the system. Thus, they replaced the Goodwin-Ramadge-Caines scheme (4.31) for the unit-delay case by

$$\theta_t^{*T}\phi_t^* = y^*_{t+1} + w_t, \quad \text{where}$$

$$\phi_t^* = (y_t, \cdots, y_{t+1-p\vee h}, u_t, \cdots, u_{t+1-k}, y_t^* + w_{t-1}, \cdots, y^*_{t+1-h} + w_{t-h})^T,$$

and w_t are independent random variables such that $Ew_t^2 = v > 0$ and $\sup_t Ew_t^4 < \infty$. Lai and Ying [38] propose to apply the stochastic gradient scheme only to broken blocks of successive observations for which white-noise dither signals are used as target values in the corresponding certainty-equivalence equation defining the inputs. Specifically, assuming that $\min_{|t|\le\pi} \mathrm{Re}\{C(e^{it}) - (d-\frac{1}{2})a\} > 0$ for some $a > 0$ and that $\sup_n |\epsilon_n| < \infty$ a.s., they define the inputs u_n at stages $n \in J = \bigcup_{i=1}^{\infty}\{n_i, n_i+1, \cdots, m_i - 1\}$ (representing the broken blocks of consecutive observations) by the equation

$$\theta^T_{n,G}\phi_{n,G} = w_n \ \text{ if } b_{n,G} \neq 0, \ \text{ and } u_n = w_n \ \text{ otherwise,} \tag{4.44}$$

where $\theta_{n,G}$ is a modified stochastic gradient algorithm estimating $\widetilde{\theta}$ defined recursively as follows: Choose $\theta_{0,G} = \cdots = \theta_{n_1+d-1,G}$ such that its component $b_{0,G}$ estimating b_1 is nonzero. Let $\phi_{n,G} = (y_n, \cdots, y_{n-p(d)+1}, u_n, \cdots, u_{n-k-d+2}, w_{n-1}, \cdots, w_{n-h})^T$. For $m_{j-1} + d - 1 < n < n_j + d$, define $\theta_{n,G} = \theta_{m_{j-1}+d-1,G}$. For $n_j + d \le n \le m_j + d - 1$ define

$$\theta_{n,G} = \theta_{n-1,G} + (1/r_{n-d})(y_n - w_{n-d})\phi_{n-d,G}, \tag{4.45a}$$

$$r_{n-d} = r_{n-d-1} + \|\phi_{n-d,G}\|^2, \ r_{n_j-1} = \sum_{i=1}^{j-1}\sum_{t=n_i+d}^{m_i+d-1} \|\phi_{n-d,G}\|^2 +$$

$$\sum_{i=1}^{j}(1 \vee \sum_{\nu=0}^{p(d)-1} y^2_{n_i-\nu} \vee \sum_{\nu=1}^{k+d-2} u^2_{n_i-\nu})\log^2\{\sum_{i=1}^{j}(2 \vee \sum_{\nu=0}^{p(d)-1} y^2_{n_i-\nu} \vee \sum_{\nu=1}^{k+d-2} u^2_{n_i-\nu})\}. \tag{4.45b}$$

Assuming furthermore that $B(z)$ is stable and that $z^p A(z^{-1}), z^{k-1}B(z^{-1})$ and $z^h C(z^{-1})$ are coprime polynomials, [38] shows that the input-output data at stages $t \in J$ have the following excitation property:

$$(\sum_{i=1}^{j} \sum_{t=n_i+p+k+h+d}^{m_i} Z_t U_t^T)/\#_j \text{ converges a.s. to a nonrandom matrix}$$

of full rank $p+k$, where $U_t = (y_{t-1}, \cdots, y_{t-p}, u_{t-d}, \cdots, u_{t-d-k+1})^T$,

$$Z_t = (y_{t-h-1}, \cdots, y_{t-h-k+1}, w_{t-d}, \cdots, w_{t-d-p-k+1})^T, \#_j = \sum_{i=1}^{j}(m_i - n_i). \quad (4.46)$$

In view of (4.46), [38] defines consistent estimates λ_{m_j} of $\lambda = (-a_1, \cdots, -a_p, b_1, \cdots, b_k)^T$ by

$$\lambda_{m_j} = (V_j^T V_j)^{-1} V_j^T \sum_{i=1}^{j} \sum_{t=n_i+p+k+h+d}^{m_i} Z_t y_t, \text{ where } V_j = \sum_{i=1}^{j} \sum_{t=n_i+p+k+h+d}^{m_i} Z_t U_t^T.$$

In fact, [38] uses (4.46) to show that $\lambda_{m_j} - \lambda = o(\#_j^{-1/2+\delta})$ a.s. for every $\delta > 0$. Using adaptive spectral factorization; [38] then estimates $(c_1, \cdots, c_h)$ with an error of $o(\#_j^{-1/2+\delta})$ a.s. This in turn provides strongly consistent estimates $\widetilde{\theta}_{m_j}$ of $\widetilde{\theta}$ such that $\widetilde{\theta}_{m_j} = \widetilde{\theta} + o(\#_j^{-1/2+\delta})$ a.s. for every $\delta > 0$. Pick any $0 < \delta < 1/2$ and let $\mathbf{I}_j$ denote the cube with center $\widetilde{\theta}_{m_j}$ and width $\#_j^{-1/2+\delta}$. Then (4.38) clearly holds.

Suppose that the target value y_n^* at stage n is $\mathcal{F}_{n-d}$-measurable and that $\sum_1^n y_i^{*2} = O(n)$ and $y_n^{*2} = o((\log n)^\gamma)$ a.s. for some $0 < \gamma < 1$. To define the sequence $n_1 < m_1 < n_2 < \cdots$ associated with the modified stochastic gradient algorithm $\theta_{n,G}$, take a nondecreasing sequence of constants $K_n \geq d+h$ such that $K_n \to \infty$, $K_n = O((\log n)^{1/2\wedge(1-\gamma)})$ and $K_{2n} = O(K_n)$. Define inductively the stopping times

$$n_j = \inf\{n > m_{j-1} : \sum_{i=1}^{j-1}(m_i - n_i) \leq K_n \text{ and}$$
$$(\sum_{r=0}^{p(d)-1} y_{n-r}^2) \vee (\sum_{r=1}^{k+d-2} u_{n-r}^2) \leq K_n / \log^2(K_n+2)\},$$

$$m_j = n_j + [K_{n_j} \vee (1 \vee \sum_{r=0}^{p(d)-1} y_{n_j-r}^2 \vee \sum_{r=1}^{k+d-2} u_{n_j-r}^2) \log^2\{\sum_{i=1}^{j}(2 \vee \sum_{r=0}^{p(d)-1} y_{n_i-r}^2 \vee \sum_{r=1}^{k+d-2} u_{n_i-r}^2)\}].$$

Let $J_n = \{t \in J : t \leq n\}$ and let $\#(J_n)$ denote the number of elements of J_n. The adaptive control scheme in [38] defines the input u_n by (4.44) using the modified stochastic gradient scheme $\theta_{n,G}$ and white-noise perturbation w_n at stage $n \in J$. At stage $n \notin J$, it uses the certainty-equivalence equation (4.43) based on the monitored recursive maximum likelihood algorithm $\theta_{n,M}$ to determine the input u_n if $\#(J_n) > K_n^{1/2}$. If $n \notin J$ and $\#(J_n) \leq K_n^{1/2}$ (signalling too few white-noise excitations), [38] uses a stochastic gradient certainty-equivalence rule to determine the input u_n. The regret of the resultant rule is shown in [38] to have the logarithmic order (4.36), and in the case $\log(1 + \sum_{i-1}^{n} y_i^{*2}) = o(\log n)$, to have even the smaller order (4.37). This basically follows from (4.42) since as shown in Lemmas 8 and 9 of [38], $\#(J_n) = O(K_n)$ and $\liminf_{n\to\infty} \#(J_n)/K_n > 0$ a.s.

REFERENCES

[1] P. R. Kumar, *A survey of some results in stochastic adaptive control*, SIAM J. Contr. Optimiz., **23** (1985), pp. 329-380.

[2] D. Feldman, *Contributions to the "two-armed bandit" probl. m*, Ann. Math. Staist., **33** (1962), pp. 847-856.

[3] J. Fabius and W. R. van Zwet, *Some remarks on the two-armed bandit*, Ann. Math. Statist., **41** (1970), pp. 1906-1916.

[4] D. A. Berry, *A Bernoulli two-armed bandit*, Ann. Math. Statist., **43** (1972), pp. 871-897.

[5] J. C. Gittins and D. M. Jones, *A dynamic allocation index for the sequential design of experiments*, in Progress in Statistics (J. Gani et al., Eds.), North Holland, Amsterdam, 1974, pp. 241-266.

[6] J. C. Gittins, *Bandit processes and dynamic allocation indices*, J. Roy. Statist. Soc. Ser. B, **41** (1979), pp. 148-177.

[7] P. Whittle, *Multi-armed bandits and the Gittins index*, J. Roy. Statist. Soc. Ser. B, **42** (1980), pp. 143-149.

[8] F. Chang and T. L. Lai, *Optimal stopping and dynamic allocation*, Adv. Appl. Prob., **19** (1987), pp. 829-853.

[9] T. L. Lai, *Adaptive treatment allocation and the multi-armed bandit problem*, Ann. Statist., **16** (1987), pp. 1091-1114.

[10] H. Chernoff and S. N. Ray, *A Bayes sequential sampling inspection plan*, Ann. Math. Statist., **36** (1965), pp. 1387-1407.

[11] H. Chernoff, *Sequential models for clinical trials*, Proc. Fifth Berkeley Symp. Math. Statist. Prob., **4** (1967), Univ. California Press, pp. 805-812.

[12] H. Robbins, *Some aspects of the sequential design of experiments*, Bull. Amer. Math. Soc., **55** (1952), pp. 527-535.

[13] T. L. Lai and H. Robbins, *Asymptotically efficient adaptive allocation rules*, Adv. Appl. Math., **6** (1985), pp. 4-22.

[14] T. L. Lai, *Asymptotic solutions of bandit problems*, in Stochastic Differential Systems, Stochastic Control Theory and Applications (W. Fleming and P. L. Lions, Eds.), Springer-Verlag, New York-Berlin-Heidelberg, 1988, pp. 275-292.

[15] V. Ananthraman, P. Varaiya, and J. Walrand, *Asymptotically efficient allocation rules for multi-armed bandit problems with multiple plays. Part I: I.I.D. Rewards. Part II: Markovian Rewards*, IEEE Trans. Automat. Contr., **AC-32** (1987), pp. 968-982.

[16] R. Agrawal, M. Hedge and D. Teneketzis, *Asymptotically efficient rules for the multi-armed bandit problem with switching costs*, IEEE Trans. Automat. Contr., **AC-33** (1988), pp. 899-906.

[17] A. Zellner, *An Introduction to Bayesian Inference in Econometrics*, Wiley, New York, 1971.

[18] G. C. Chow, *Analysis and Control of Dynamic Economic Systems*, Wiley, New York, 1975.

[19] E. C. Prescott, *The multiperiod control problem under uncertainty*, Econometrica, **40** (1972), pp. 1043-1058.

[20] M. Aoki, *On some price adjustment schemes*, Ann. Econ. Soc. Measurements, **3** (1974), pp. 95-116.

[21] T. W. Anderson and J. Taylor, *Some experimental results on the statistical properties of least squares estimates in control problems*, Econometrica, **44** (1976), pp. 1289-1302.

[22] T. L. Lai and H. Robbins, *Iterated least squares in multiperiod control*, Adv. Appl. Math., **3** (1982), pp. 50-73.

[23] T. L. Lai and H. Robbins, *Adaptive design and the multiperiod control problem*, in Statistical Decision Theory and Related Topics III, Vol. 2 (S. S. Gupta and J. O. Berger, Eds.) Academic Press, 1982, pp. 103-120.

[24] T. L. Lai and H. Robbins, *Adaptive design and stochastic approximtion*, Ann. Statist., **7** (1979), pp. 1196-1221.

[25] H. Robbins and S. Monro, *A stochastic approximation method*, Ann. Math. Statist., **22** (1951), pp. 400-407.

[26] C. Z. Wei, *Multivariate adaptive stochastic approximation*, Ann. Statist., **15** (1987), pp. 1115-1130.

[27] T. L. Lai, *Asymptotically efficient adaptive control in stochastic regression models*, Adv. Appl. Math., **7** (1986), pp. 23-45.

[28] A. A. Feldbaum, *The theory of dual control I-IV*, Automation and Remote Control, **21** (1961), pp. 874-883 (Part I) and pp. 1033-1039 (Part II), **22** (1962), pp. 1-12 (Part III) and pp. 109-121 (Part IV).

[29] K. J. Åström, *Theory and applications of adaptive control – A survey*, Automatica-J. IFAC, **19** (1983), pp. 471-486.

[30] K. J. Åström and B. Wittenmark, *On self-tuning regulators*, Automatica-J. IFAC, **9** (1973), pp. 195-199.

[31] T. L. Lai and C. Z. Wei, *Asymptotically efficient self-tuning regulators*, SIAM J. Contr. Optimiz., **25** (1987), pp. 466-481.

[32] T. L. Lai and C. Z. Wei, *On the concept of excitation in least squares identification and adaptive control*, Stochastics, **16** (1986), pp. 227-254.

[33] V. Solo, *The convergence of AML*, IEEE Trans. Automat Contr., **AC-21** (1979), pp. 958-962.

[34] G. C. Goodwin, P. J. Ramadge, and P. E. Caines, *Discrete time stochastic adaptive control*, SIAM J. Contr. Optimiz., **19** (1981), pp. 829-853.

[35] G. C. Goodwin and K. S. Sin, *Adaptive Filtering, Prediction and Control*, Prentice-Hall, Englewood Cliffs, 1984.

[36] T. L. Lai and C. Z. Wei, *Least squares estimates in stochastic regression models with applications to identification and control of dynamic systems*, Ann. Statist., **10** (1982), pp. 154-166.

[37] T. L. Lai and C. Z. Wei, *Extended least squares and their applications to adaptive control and prediction in linear systems*, IEEE Trans. Automat. Contr., **AC-31** (1986), pp. 898-906.

[38] T. L. Lai and Z. Ying, *Parallel recursive algorithms in asymptotically efficient adaptive control of linear stochastic systems*, SIAM J. Contr. Optimiz., **29** (1991), in press.

[39] L. Ljung and T. Söderström, *Theory and Practice of Recursive Identification*, MIT Press, Cambridge, 1983.

[40] P. E. Caines and S. Lafortune, *Adaptive control with recursive identification for stochastic linear systems*, IEEE Trans. Automat. Contr., **AC-29** (1984), pp. 312-321.

Stability of Markov Chains on Topological Spaces

with applications to adaptive control and time series analysis

Sean P. Meyn

Coordinated Science Laboratory
1101 W. Springfield ave.
Urbana, Illinois 61801

Work supported in part by NSF initiation grant # ECS 8910088

Part I

Introduction and Background

1 Introduction

In this paper we present recent results from the theory of Markov chains on topological state spaces. We show how certain regularity conditions, either topological or measure-theoretic, may be used to obtain an ergodic decomposition of the state space, analogous to what is well known to hold in the countable state space case. Essentially, the state space may be written as the disjoint union of ergodic classes, together with a transient set.

When these regularity properties are satisfied, we obtain simple criteria for tightness of the distributions of the stochastic process under consideration; a form of stability for the process.

These results are used to develop a more general Lyapunov-Foster test function approach to stability, and to establish rates of convergence of the distributions of a Markov chain to an invariant probability.

Our motivation comes from problems in control theory, signal processing, time series analysis, and other applied areas. Because of the extreme complexity of the stochastic processes encountered in these fields, it is of interest to have available a general theory to aid in the search for a stability proof, and to give some qualitative measure of performance.

The results described in this article will be expanded upon and derived in detail in the text [39], which is currently in preparation.

1.1 Structure of the paper

The paper is divided into five main parts.

Part I consists of introductory material, basic definitions, and motivation for our definitions.

We describe two standard ideas from dynamical systems theory; Lagrange and Lyapunov stability. Under the hypotheses of the paper, the distributions of the stochastic processes under consideration are generated by a dynamical system whose state space consists of probability measures. In this context, Lagrange stability becomes the tightness hypothesis of [3], and Lyapunov stability is equivalent to equicontinuity of the transition function of the process, as formulated in [47,19].

The first condition is called *boundedness in probability*, and a process which satisfies the second stability criterion is called an *e-process*. Part II of the paper investigates the consequences of these two hypotheses, and the relationship between them.

In Section 3 we establish an analogue of the Doeblin Decomposition of [6,21,58,54] for a bounded in probability e-process: The state space is decomposed into the union of a possibly uncountable collection of disjoint ergodic classes, together with a transient set.

To obtain stability, the Lyapunov-Foster test function approach is considered in Section 4. It is shown that the existence of one of the test functions introduced in [10,55,38] implies that an e-process is bounded in probability in an average sense.

In Section 5 we demonstrate that the distributions of a bounded in probability e-process converge in a Cesaro sense for each initial condition. When the recurrent states are aperiodic, Cesaro convergence is strengthened to ordinary convergence, and the stability results of Section 4 are similarly strengthened. Sample path results are also obtained: It is shown that the weak law of large numbers holds for every initial condition. If the occupation probabilities are tight, then the strong law of large numbers is obtained.

In Part III of the paper we consider a different class of processes for which the various stability and recurrence concepts are closely related: these *T-processes* are precisely those which admit a continuous component T, as defined in [54].

For T-processes, we show that boundedness in probability is equivalent to a generalization of positive Harris recurrence. If the underlying distributions are not tight, then the trajectories of the Markov chain $\mathbf{\Phi}$ are either recurrent in a strong sense, or leave every compact subset of the state space.

In Section 7, the class of *petite sets* is introduced, and their properties explored. We show that positive Harris recurrence is essentially equivalent to finiteness of the mean return time $\mathsf{E}[\tau_A]$ to a petite set A. Further, under a mild stability condition we show that $\mathbf{\Phi}$ is a T-process if and only if the state space admits a covering by open petite sets, and that all compact sets are petite if and only the process is an irreducible T-process.

In Section 8, we demonstrate that stochastic stability is intimately connected with the return times to petite sets. Lyapunov-Foster criteria are developed for the various stability hypotheses, and to obtain rates of convergence for the underlying distributions of $\mathbf{\Phi}$.

In Part IV we develop techniques for verifying our conditions for Markov chains generated by a nonlinear stochastic system.

Forward accessibility of an associated control system is seen to imply that $\mathbf{\Phi}$ is a T-process.

Equicontinuity may be established if a linearized system is stable. If the linearized system is exponentially stable, then we obtain precise rates of convergence of the distributions of $\mathbf{\Phi}$ towards an invariant probability.

In Part V these results are illustrated with a simple adaptive system and a bilinear time series model.

Acknowledgements

This paper is the result of collaborations with several people to whom I would like to express my gratitude.

Peter Caines and I did much of the research on nonlinear systems and ergodic theory during the period that I was working towards my doctoral degree at McGill University in Montréal.

Lei Guo from the Academia Sinica in Beijing and I worked on several aspects of probabilistic modeling and stochastic stability while we were postdoctoral fellows at the Australian National University at Canberra, Australia. The examples in this paper are largely based upon this collaboration.

The results for T-processes and Harris recurrence follow mainly from Richard Tweedie's own work, and our collaboration in Australia and the U.S.

I am grateful to these people not just for our academic interactions, but for our friendship and, in particular, the comradeship they provided while touring Hawaii, Florence, Montréal, and Australia in the pursuit of fun, sun, and a drinkable bottle of chianti.

2 Background and General Definitions

2.1 General state space Markov chains

Here we describe the general framework to be considered, and introduce some basic definitions.

We assume that the Markov chain $\mathbf{\Phi}$ evolves on a metric space X, which we assume is locally compact and separable. We let $\mathcal{B}(\mathsf{X})$ denote its Borel σ-field.

The transition probabilities of $\mathbf{\Phi}$ are defined by a *Markov transition function* P, which is a function on $\mathsf{X} \times \mathcal{B}(\mathsf{X})$ satisfying the defining property that

$$\mathsf{P}\{\Phi_{k+1} \in A \mid \Phi_0, \dots, \Phi_k\} = P(\Phi_k, A) \qquad a.s.$$

for any $A \in \mathcal{B}(\mathsf{X})$, $\Phi_0 \in \mathsf{X}$, and $k \geq 0$.

The transition law P acts on measurable functions f by the operation

$$Pf(x) \triangleq \int P(x, dy) f(y), \qquad x \in \mathsf{X},$$

so that letting P^k denote the kth composition of this operator we have

$$\mathsf{E}[f(\Phi_{n+k}) \mid \sigma\{\Phi_0, \dots, \Phi_n\}] = P^k f(\Phi_n) \qquad a.s.$$

For a probability μ on $\mathcal{B}(\mathsf{X})$ we let μP denote the probability which is defined as

$$\mu P\{A\} \triangleq \int \mu(dx) P(x, A) \qquad A \in \mathcal{B}(\mathsf{X})$$

A probability π on $\mathcal{B}(\mathsf{X})$ with the property $\pi P = \pi$ will be called *invariant.*

2.2 Dynamical systems and Markov chains

A Markov transition function P gives rise to a deterministic map from $\mathcal{M}$, the space of probabilities on $\mathcal{B}(\mathsf{X})$, to itself. The stability issue of interest is the behavior of the compositions P^k, $k \geq 0$, when k becomes large. Our hope is that this sequence will be bounded in some sense, or converge to some fixed probability $\pi \in \mathcal{M}$.

Dynamical systems theory is concerned with precisely these same questions. In order to help make reasonable definitions of stability for the Markov chain $\mathbf{\Phi}$, and to motivate other results and definitions, we present here some ideas from this field.

A discrete time (semi) dynamical system is a triple $(T, \mathcal{X}, d)$ where $(\mathcal{X}, d)$ is a metric space, and $T: \mathcal{X} \to \mathcal{X}$ is continuous. A basic concern is the structure of the *orbit*, or *trajectory* $\{T^k x : k \in \mathbb{Z}_+\}$, where $x \in \mathcal{X}$ is an *initial condition*, $T^0 x \triangleq x$, and we define inductively $T^{k+1}x \triangleq T^k(Tx)$ for $k \geq 1$.

Two useful formulations of stability for a dynamical system $(T, \mathcal{X}, d)$ are

(i) *Lagrange stability*: For each $x \in \mathcal{X}$, the orbit starting at x is a precompact subset of $\mathcal{X}$;

(ii) *Stability in the sense of Lyapunov*: For all initial conditions $y, x \in \mathcal{X}$,

$$\lim_{y \to x} \sup_{k \geq 0} d(y_k, x_k) \triangleq \lim_{y \to x} \sup_{k \geq 0} d(T^k y, T^k x) = 0,$$

where d denotes the metric on $\mathcal{X}$.

Observe that the second criterion says nothing about the actual stability or boundedness of the orbits of $(T, \mathcal{X}, d)$, since it is simply continuity of the map T^k, uniformly in $k \geq 0$. An example of a system which is stable in the sense of Lyapunov is the recursion $x_{k+1} = x_k + 1$, $k \geq 0$, where x evolves on $\mathbb{R}^1$. Although distinct trajectories stay close together if their initial conditions are similarly close, we would not consider this system stable in any ordinary sense of the word.

In order to construct a dynamical system from the transition function P, we must specify a metric, and hence also a topology on $\mathcal{M}$. We take the topology of weak convergence: Let C denote the set of all continuous bounded functions $f: \mathsf{X} \to \mathbb{R}$. A sequence of probabilities $\{\mu_k : k \in \mathbb{Z}_+\} \subset \mathcal{M}$ converges *weakly* to $\mu_\infty \in \mathcal{M}$ (denoted $\{\mu_k : k \in \mathbb{Z}_+\} \xrightarrow{\text{weakly}} \mu_\infty$) if

$$\lim_{k \to \infty} \int f \, d\mu_k = \int f \, d\mu_\infty$$

for every $f \in \mathsf{C}$.

Due to our restrictions on the state space X, the topology of weak convergence is induced by a metric d_m on $\mathcal{M}$. This metric may be expressed

$$d_m(\mu, \nu) = \sum_{k=0}^{\infty} \left| \int f_k \, d\mu - \int f_k \, d\nu \right| 2^{-k}, \qquad \mu, \nu \in \mathcal{M} \tag{1}$$

where $\{f_k\}$ is a fixed subset of C_c, the set of continuous functions on X with compact support, such that (i) $\sup_{x \in \mathsf{X}} |f_k(x)| = 1$ for all k, and (ii) the span of the functions $\{f_k\}$ is dense in C_c with respect to uniform convergence on compact subsets of X. This is possible because X is locally compact and separable, and hence by Lindelöf's theorem [24], X may be expressed as the countable union of open, precompact subsets of X.

The question arises: When will P be a *continuous* map on $\mathcal{M}$? If P is continuous, then we must have in particular that if a sequence of point masses $\{\delta_{x_k} : k \in \mathbb{Z}_+\} \subset \mathcal{M}$ converge to some point mass $\delta_{x_\infty} \in \mathcal{M}$, then

$$\delta_{x_k} P \xrightarrow{\text{weakly}} \delta_{x_\infty} P \qquad \text{as } k \to \infty$$

Equivalently, $\lim_{k \to \infty} Pf(x_k) = Pf(x_\infty)$ for all $f \in \mathsf{C}$. That is, if the Markov transition function induces a *continuous* map on $\mathcal{M}$, then Pf must be continuous for any bounded continuous function f.

This is known as the *Feller property*. Conversely, we may show that for any Feller Markov transition function P, the associated operator P on $\mathcal{M}$ is continuous.

We collect these ideas together in a proposition. For related results see [49,27].

Proposition 2.1 *The triple $(P, \mathcal{M}, d_m)$ is a dynamical system if and only if the Markov transition function P has the Feller property.*

Although we do not get a great deal of milage from this observation - there do not exist many results in the dynamical systems theory literature to be picked out and exploited - this idea is valuable in that it motivates the stability concepts to be introduced below.

We now consider Lagrange and Lyapunov stability for the dynamical system $(P, \mathcal{M}, d_m)$.

Lagrange stability of Markov chains

Below is our first basic formulation of stochastic stability. Boundedness in probability is used in [16,33,38], whilst the concept ***boundedness in probability on average*** is introduced in [33].

> The Markov chain $\boldsymbol{\Phi}$ is called
>
> (i) ***bounded in probability*** if for each initial condition $x \in \mathsf{X}$ and each $\varepsilon > 0$, there exists a compact subset $K \subset \mathsf{X}$ such that
>
> $$\liminf_{k\to\infty} \mathsf{P}_x\{\Phi_k \in K\} \geq 1 - \varepsilon.$$
>
> (ii) ***bounded in probability on average*** if for each initial condition $x \in \mathsf{X}$ and each $\varepsilon > 0$, there exists a compact subset $K \subset \mathsf{X}$ such that
>
> $$\liminf_{N\to\infty} \frac{1}{N} \sum_{k=1}^{N} \mathsf{P}_x\{\Phi_k \in K\} \geq 1 - \varepsilon.$$

Boundedness in probability is simply the tightness hypothesis of [3] for the collection of probabilities $\{P^k(x, \cdot) : k \geq 1\}$. For the class of Markov chains considered in the present article, this is a minimal stability assumption which implies the existence of an invariant probability.

Recall from [45] that a set of probabilities $A \subset \mathcal{M}$ is *tight* if and only if it is precompact in the metric space $(\mathcal{M}, d_m)$ defined in the previous section. This proves

Proposition 2.2 *$\boldsymbol{\Phi}$ is bounded in probability if and only if the dynamical system $(P, \mathcal{M}, d_m)$ is Lagrange stable.*

Lyapunov stability of Markov chains

We now present our second principal stability assumption: equicontinuity of the transition function P.

It has been noted in [54] that the Feller property is too weak to relate topological formulations of stability with measure theoretic stability concepts such as Harris recurrence. The *strong Feller property*, that Pf is continuous for any bounded measurable function, does imply significant connections between various stability concepts [13,54], but this assumption is not satisfied in many applications. These considerations led to the development of continuous components in [54,38,37], a condition which retains the strong implications of the strong Feller property and is also, in principle, verifiable for a large class of examples.

Rather than weaken the strong Feller property, here we strengthen the ordinary Feller property to build a new set of tools for studying the stability and asymptotic behavior of the Markov chain $\boldsymbol{\Phi}$.

> The Markov transition function P is called ***equicontinuous*** if for each $f \in \mathsf{C}_c$ the sequence of functions $\{P^k f : k \in \mathbb{Z}_+\}$ is equicontinuous on compact sets.
>
> A Markov chain which possesses an equicontinuous Markov transition function will be called an *e-process*.

Equicontinuity may be compared to *uniform stability* [19] or *regularity* [8].

The following result creates a further link between classical dynamical systems theory, and the theory of Markov chains on topological state spaces.

Proposition 2.3 *The Markov transition function P is equicontinuous if and only if the dynamical system $(P, \mathcal{M}, d_m)$ is stable in the sense of Lyapunov.*

There is also a direct link between equicontinuity and Lagrange stability. The space C is a normed linear space, where we take the norm $\|\cdot\|_c$ to be defined for $f \in \mathsf{C}$ as

$$\|f\|_c \triangleq \sum_{k=0}^{\infty} 2^{-k} (\sup_{x \in K_k} |f(x)|)$$

where $\{K_k\}$ is a sequence of open precompact sets whose union is equal to X. The associated metric d_c generates the topology of uniform convergence on compact subsets of X.

If P is a Feller Markov transition function, then the mapping P on C is continuous, and in this case the triple (P, C, d_c) is a dynamical system.

By Ascoli's theorem, (P, C, d_c) will be Lagrange stable if and only if for each initial condition $f \in \mathsf{C}$, the orbit $\{P^k f : k \in \mathbb{Z}_+\}$ is uniformly bounded, and equicontinuous on compact subsets of X. This fact easily implies

Proposition 2.4 *Suppose that Φ is bounded in probability. Then Φ is an e-process if and only if the dynamical system (P, C, d_c) is Lagrange stable.*

To summarize, our two main stability assumptions are precisely Lagrange stability, and stability in the sense of Lyapunov for the dynamical system $(P, \mathcal{M}, d_m)$. These stability conditions are both satisfied if and only if the dynamical system $(P, \mathcal{M}, d_m)$ and its dual (P, C, d_c) are simultaneously Lagrange stable.

Stability in the sense of Lyapunov is a useful concept when a stationary point for the dynamical system $(T, \mathcal{X}, d)$ exists, which is a state $x \in \mathcal{X}$ such that $Tx = x$. If x is a stationary point and the dynamical system is stable in the sense of Lyapunov, then trajectories which start near x will stay near x, and this turns out to be a very useful notion of stability. For the dynamical system $(P, \mathcal{M}, d_m)$, a stationary point is an invariant probability. These connections suggest that equicontinuity will be a useful tool for studying the limiting behavior of the distributions governing the Markov chain Φ, a belief which will be justified in the results below.

Part II

Markov Chains with Equicontinuous Transition Functions

3 An Ergodic Decomposition

In this section we obtain an ergodic decomposition for a bounded in probability e-process. This result is similar to the well known decomposition theorem for a countable state space Markov chain, although the cardinality of the collection of all ergodic classes will not be countable in general.

We are concerned with sets satisfying the following: A set $M \subset \mathsf{X}$ is called *absorbing* if $P(x, M) = 1$ for $x \in M$. M is called *minimal* if it is closed, absorbing, and does not contain as a proper subset any closed absorbing sets.

A minimal set is called an *ergodic kernel* in [13] where strong Feller Markov chains on a compact state space are considered. Under these conditions there always exists a non-empty, finite collection of minimal sets.

The following result allows us to define a "center" of the state space where the ergodic decomposition takes place. A *kernel* Π on $(\mathsf{X}, \mathcal{B}(\mathsf{X}))$ is a function $\Pi: \mathsf{X} \times \mathcal{B}(\mathsf{X}) \to \mathbb{R}_+$ such that $\Pi(x, \cdot)$ is a finite measure for each $x \in \mathsf{X}$, and $\Pi(\cdot, A)$ is measurable for each $A \in \mathcal{B}(\mathsf{X})$. For $\rho < 1$, $k \in \mathbb{Z}_+$, define the Markov transition functions K_ρ, A_k by

$$K_\rho \triangleq (1-\rho)\sum_{k=0}^{\infty} \rho^k P^k \qquad\qquad A_k \triangleq \frac{1}{k}\sum_{\ell=1}^{k} P^\ell$$

A subprobability is a positive measure with total mass less than or equal to one, and a set of subprobabilities $\{\nu_k\}$ converges *vaguely* to a subprobability ν (denoted $\nu_k \xrightarrow{\text{vaguely}} \nu$) if

$$\lim_{k\to\infty} \int f\, d\nu_k = \int f\, d\nu$$

for every function $f \in \mathsf{C}_c$. If the subprobabilities $\{\nu_k\}$ and ν are in fact probabilities on $\mathcal{B}(\mathsf{X})$, then vague convergence coincides with weak convergence.

Lemma 3.1 *If Φ is an e-process, then there exists a kernel Π such that*

$$K_\rho(x, \cdot) \xrightarrow{\text{vaguely}} \Pi(x, \cdot) \qquad \textit{as } \rho \uparrow 1 \tag{2}$$

$$A_k(x, \cdot) \xrightarrow{\text{vaguely}} \Pi(x, \cdot) \qquad \textit{as } k \to \infty \tag{3}$$

for all $x \in \mathsf{X}$.

For each j, k, $\ell \in \mathbb{Z}_+$ we have

$$P^j \Pi^k P^\ell = \Pi, \tag{4}$$

and hence $\Pi(x, \cdot)$ is an invariant subprobability for all $x \in \mathsf{X}$.

The proof follows from considering a vaguely convergent subsequence of the probabilities $\{A_k(x, \cdot) : k \in \mathbb{Z}_+\}$ or $\{K_\rho(x, \cdot) : \rho < 1\}$.

We assume throughout the remainder of this section that Φ is an e-process, and that Φ is bounded in probability on average. If this is the case then Lemma 3.1 implies that $A_N f$ converges uniformly on compact subsets of X to Πf when f is bounded and continuous. It follows therefore that for each such f, the function Πf is continuous. That is, the Markov transition function Π has the Feller property.

Let μ be any probability on $\mathcal{B}(\mathsf{X})$ with the property that $\int f\, d\mu > 0$ for all positive functions $f \in \mathsf{C}$ which do not vanish identically. That is, $\operatorname{supp}\mu = \mathsf{X}$. Let ν denote the invariant probability $\nu \triangleq \mu\Pi$, and let $S \triangleq \operatorname{supp}\nu$ denote its support. The set S is called the *center* of X [51], and is equal to the *conservative* part of X under the present conditions [9]. For $x \in \mathsf{X}$ we let S_x denote the support of the invariant probability $\Pi(x, \cdot)$.

The main result in this section extends results of [22,20,51,50] to non-compact state spaces.

Theorem 3.1 *If Φ is a bounded in probability on average e-processes then*

(i) *For each $x \in S$,*

$$\Pi(y, \cdot) = \Pi(x, \cdot) \qquad \textit{for all } y \in S_x;$$

(ii) *For each $x \in S$ the set S_x is a minimal set, and every minimal set is of this form.*

It follows from Lemma 3.1, Theorem 3.1 and the definitions that for $f \in \mathsf{C}$,

$$\lim_{N\to\infty} A_N f(\Phi_0) = \Pi f(x)$$

for any initial condition $\Phi_0 \in S_x$. By minimality S_x is absorbing, and for any two states $x, y \in S$, the sets S_x and S_y are either disjoint or identical.

Hence, the sets $\{S_x : x \in S\}$ are analogues of the Harris sets $\{E_i\}$ obtained in the Doeblin decompositions of for instance [6,21,58,54] (See Theorem 6.1 below).

To prove Theorem 3.1 we require a few preliminary results. The following result follows from the representation $\pi = \pi\Pi$, valid for any invariant probability π, and the observation that Π has the Feller property.

Lemma 3.2 *Suppose that the conditions of Theorem 3.1 hold. Then for every invariant probability π we have $\operatorname{supp}\pi \subset S$, and hence in particular $S_x \subset S$ for each $x \in \mathsf{X}$.*

By construction and the Law of Large Numbers for Markov chains for each $g \in \mathsf{C}$,

$$\Pi g = E[g \mid \Sigma_I] \qquad a.e.\ [\nu] \tag{5}$$

where Σ_I denotes the σ-algebra of ν-invariant sets (see [7]). That is, the collection of all Borel sets A with the property that

$$P(x, A) = \mathbf{1}_A(x) \qquad \text{for almost every } x \in \mathsf{X}\ [\nu].$$

This may be used to prove

Lemma 3.3 *Suppose that the conditions of Theorem 3.1 hold. Then*

(i) *For each $f, g \in \mathsf{C}$,*

$$\Pi\big[g\Pi f\big](x) = \big(\Pi g(x)\big)\big(\Pi f(x)\big)$$

for $x \in S$;

(ii) *if $g \in \mathsf{C}$, $\Pi g = g$ on S, and $\alpha\colon \mathsf{IR} \to \mathsf{IR}$ is continuous then*

$$\Pi(\alpha \circ g) = \alpha \circ g \qquad on\ S.$$

The proof of Theorem 3.1 will now be easy:

Proof of Theorem 3.1

Let $f \in \mathsf{C}$ and set $g = \Pi f$. For each fixed $x \in S$ the equality

$$\Pi(g - g(x)) = (g - g(x))$$

holds everywhere on X, and hence by Lemma 3.3 we have for each $x \in S$,

$$\Pi\big[(g - g(x))^2\big](x) = (g - g(x))^2(x) = 0.$$

Thus $g(y) = g(x)$ for *a.e.* $y \in \mathsf{X}$ with respect to the probability $[\Pi(x, \cdot)]$, and by continuity it follows that $g(y) = g(x)$ for all $y \in S_x = \operatorname{supp}\Pi(x, \cdot)$. This proves (i).

To prove (ii) let $x \in S$, and $C \subset \mathsf{X}$ be closed and absorbing. If $y \in C \cap S_x$ then, because C is closed and absorbing, $P^k(y, C^c) = 0$ for all $k \in \mathbb{Z}_+$. Hence $\Pi(y, C^c) = 0$, showing that $S_y \subset C$. Further, by (i) above $S_x = S_y$, which shows that $S_x \subset C$.

Hence either $S_x \cap C = \phi$ or $S_x \subset C$, and this shows that S_x is a minimal set.

Conversely, if S_0 is a minimal set, then it follows from minimality and the construction of π that $S_x = S_0$ for all $x \in S_0$. □

4 Criteria For Stability

Here we show how the second method of Lyapunov may be extended to establish boundedness in probability for an e-process. This test function approach to obtaining stability of a Markov chain was initiated by Foster in [10], and extended by many others including [55,28,38].

A basic hypothesis in each of these works is some form of irreducibility for the Markov chains involved (although see [57,15] for other results). We show in this section that these results may be extended to prove boundedness in probability and ergodicity for an e-process.

4.1 Compact sets as test sets for stability

Since weak and vague convergence coincide when the limiting subprobability is a probability, it follows from Lemma 3.1 that $\mathbf{\Phi}$ is bounded in probability on average if and only if $\Pi(x, \mathsf{X}) = 1$ for all $x \in \mathsf{X}$. Hence the following result shows that compact sets serve as test sets for stability: If a fixed compact set is reachable from all initial conditions, and if $\mathbf{\Phi}$ is reasonably well behaved from initial conditions on that compact set, then $\mathbf{\Phi}$ will be bounded in probability on average.

The first entrance times σ_K and τ_K are defined as

$$\sigma_K = \min(k \geq 0 : \Phi_k \in K) \qquad \tau_K = \min(k \geq 1 : \Phi_k \in K)$$

with the usual conventions if $\mathbf{\Phi}$ does not reach the set K.

Theorem 4.1 *If $\mathbf{\Phi}$ is an e-process then*

(i) $\max_{x\in\mathsf{X}} \Pi(x, \mathsf{X})$ *exists, and is equal to zero or one;*

(ii) *If* $\min_{x\in\mathsf{X}} \Pi(x, \mathsf{X})$ *exists, then it is equal to zero or one;*

(iii) *Suppose that there exists a compact set $K \subset \mathsf{X}$ such that*

$$\mathsf{P}_x\{\tau_K < \infty\} = 1 \qquad x \in \mathsf{X}.$$

Then $\min_{x\in\mathsf{X}} \Pi(x, \mathsf{X})$ *exists, and is attained on K:*

$$\inf_{x\in\mathsf{X}} \Pi(x, \mathsf{X}) = \min_{x\in K} \Pi(x, \mathsf{X});$$

(iv) *If $K \subset \mathsf{X}$ is compact, then*

$$\inf_{x\in K} \Pi(x, \mathsf{X}) \geq \left(\sup_{x\in K} \mathsf{E}_x[\tau_K]\right)^{-1}.$$

We have immediately the

Corollary *Let $\mathbf{\Phi}$ be an e-process, and $K \subset \mathsf{X}$ be compact. If $\mathsf{P}_x\{\tau_K < \infty\} = 1$, $x \in \mathsf{X}$, and $\sup_{x\in K} \mathsf{E}_x[\tau_K] < \infty$, then $\mathbf{\Phi}$ is bounded in probability on average.*

Proof of Theorem 4.1

(i) This result follows from the ergodic theorem for Markov chains [7]. If $\Pi(x,\mathsf{X}) > 0$ for some $x \in \mathsf{X}$, then an invariant probability π exists. In fact, we may take $\pi = \Pi(x,\,\cdot\,)/\Pi(x,\mathsf{X})$. By the ergodic theorem, the probabilities $\{A_N(y,\,\cdot\,) : N \geq 1\}$ are tight for a.e. $y \in \mathsf{X}$ $[\pi]$. This shows that $\Pi(y,\mathsf{X}) = 1$ for a.e. $y \in \mathsf{X}$ $[\pi]$, proving (i) of the theorem.

(ii) Let $\alpha = \inf_{x\in\mathsf{X}} \Pi(x,\mathsf{X})$, and let

$$S_\alpha = \{x \in \mathsf{X} : \Pi(x,\mathsf{X}) = \alpha\}.$$

By the assumptions of (ii), $S_\alpha \neq \phi$. Letting $u(\,\cdot\,) \triangleq \Pi(\,\cdot\,,\mathsf{X})$, we have $Pu = u$. That is, the function $\Pi(\,\cdot\,,\mathsf{X})$ is *harmonic*, and this implies that the set S_α is absorbing. Since u is lower semi-continuous, the set S_α is also a closed subset of X.

Using the fact that S_α is closed and absorbing, we now show that α is equal to zero or one.

Since S_α is closed, it follows by vague convergence that for all $x \in \mathsf{X}$,

$$\liminf_{N\to\infty} A_N(x, S_\alpha^c) \geq \Pi(x, S_\alpha^c).$$

Since S_α is absorbing, this shows that

$$\Pi(x, S_\alpha^c) = 0 \qquad \text{for all } x \in S_\alpha \tag{6}$$

Suppose now that $0 \leq \alpha < 1$. We have remarked in the proof of (i) that

$$\pi\{y \in \mathsf{X} : \Pi(y,\mathsf{X}) = 1\} = 1$$

for any invariant probability π, and hence

$$\Pi(x, S_\alpha) \leq \Pi(x, \{y \in \mathsf{X} : \Pi(y,\mathsf{X}) < 1\}) = 0. \tag{7}$$

Equations (6, 7) show that for any $x \in S_\alpha$,

$$\alpha = \Pi(x,\mathsf{X}) = \Pi(x, S_\alpha) + \Pi(x, S_\alpha^c) = 0,$$

and this proves (ii).

(iii) Recall that $u(x) \triangleq \Pi(x,\mathsf{X})$. Since u is lower semi-continuous we have

$$\inf_{x\in K} u(x) = \min_{x\in K} u(x).$$

That is, the infimum is attained.

Since the function u is harmonic, the process $(u(\Phi_k))$ is a martingale, which converges to a random variable u_∞ satisfying $\mathsf{E}_x[u_\infty] = u(x)$, $x \in \mathsf{X}$.

By standard results from the general theory of Markov chains [41], the assumption that $\mathsf{P}_x\{\tau_K < \infty\} \equiv 1$ implies that

$$\mathsf{P}_x\{\Phi \in K \text{ } i.o.\} = 1, \qquad x \in \mathsf{X}. \tag{8}$$

If $\Phi_k \in K$ for some $k \in \mathbb{Z}_+$, then obviously $u(\Phi_k) \geq \min_{x\in K} u(x)$, which by (8) implies that

$$u_\infty = \lim_{k\to\infty} u(\Phi_k) \geq \min_{x\in K} u(x) \qquad a.s.$$

Taking expectations shows that $u(y) \geq \min_{x\in K} u(x)$ for all $y \in \mathsf{X}$, proving part (iii) of the theorem.

(iv) For the first entrance time τ_K to the compact set K, let θ^{τ_K} denote the τ_K-fold shift on sample space, defined so that

$$\theta^{\tau_K} f(\Phi_k) = f(\Phi_{k+\tau_K})$$

for any measurable function f on X.

Fix $x \in K$, $0 < \rho < 1$, and observe that by conditioning at time τ_K and using the strong Markov property,

$$\begin{aligned} K_\rho(x,K) &= (1-\rho)\mathsf{E}_x\Big[\sum_{k=0}^{\infty} \rho^k \mathbf{1}\{\Phi_k \in K\}\Big] \\ &= (1-\rho)\mathsf{E}_x\Big[1 + \sum_{k=0}^{\infty} \rho^{\tau_K+k}(\theta^{\tau_K}\mathbf{1}\{\Phi_k \in K\})\Big] \\ &= (1-\rho) + (1-\rho)\mathsf{E}_x\Big[\rho^{\tau_K}\mathsf{E}_{\Phi_{\tau_K}}\Big[\sum_{k=0}^{\infty} \rho^k \mathbf{1}\{\Phi_k \in K\}\Big]\Big] \\ &\geq (1-\rho) + \mathsf{E}_x[\rho^{\tau_K}] \inf_{y\in K} K_\rho(y,K) \end{aligned}$$

Taking the infimum over all $x \in K$, we obtain

$$\inf_{y\in K} K_\rho(y,K) \geq (1-\rho) + \inf_{y\in K} \mathsf{E}_y[\rho^{\tau_K}] \inf_{y\in K} K_\rho(y,K) \tag{9}$$

By Jensen's inequality we have the bound $\mathsf{E}[\rho^{\tau_K}] \geq \rho^{\mathsf{E}[\tau_K]}$. Hence letting $M_K = \sup_{x\in K} \mathsf{E}_x[\tau_K]$ it follows from (9) that

$$\inf_{y\in K} K_\rho(y,K) \geq (1-\rho) + \big(\inf_{y\in K} K_\rho(y,K)\big)\rho^{M_K} \geq \frac{1-\rho}{1-\rho^{M_K}}$$

Letting $\rho \uparrow 1$ and using the assumption that K is compact together with vague convergence, we have for each $y \in K$,

$$\Pi(y,K) \geq \limsup_{\rho\uparrow 1} K_\rho(y,K) \geq \lim_{\rho\uparrow 1}\left(\frac{1-\rho}{1-\rho^{M_K}}\right) = \frac{1}{M_K}.$$

□

4.2 A Lyapunov-Foster criterion

Let $\{V_k : k \in \mathbb{Z}_+\}$ denote a family of positive Borel measurable functions $V_k : \mathsf{X}^{k+1} \to \mathbb{R}_+$. By imposing the appropriate drift condition on $V_k = V_k(\Phi_0, \dots, \Phi_k)$, we may obtain bounds on moments of first entrance times to certain subsets of X [38]. This result will be used to obtain boundedness in probability, and hence also the ergodic properties of interest for $\mathbf{\Phi}$.

Let $\mathcal{F}_k = \sigma\{\Phi_0, \dots, \Phi_k\}$, $k \geq 0$. The stochastic Lyapunov-Foster criterion we consider here is as follows.

(V1)	For some $\varepsilon > 0$, a compact set $K \subset \mathsf{X}$, and a positive measurable function $f \geq 1$, $$\mathsf{E}_x[V_{k+1} \mid \mathcal{F}_k] \leq V_k - \varepsilon f(\Phi_k) \qquad a.s.,$$ when $\sigma_K > k$, $k \in \mathbb{Z}_+$.

If $V : \mathsf{X} \to \mathbb{R}_+$ is a measurable function which satisfies the bound

$$PV(x) \leq V(x) - \varepsilon f \qquad x \in K^c,$$

then the stochastic process $V_k = V(\Phi_k)$ satisfies V1.

Condition V1 is often satisfied by verifying the following stronger condition. A function $f: \mathsf{X} \to \mathbb{R}_+$ is called a *moment* if $f(x) \to \infty$ as $x \to \infty$ [2]. Assume that there exists a moment $f: \mathsf{X} \to \mathbb{R}_+$ with the property

$$V_k(x_0, \ldots, x_k) \geq f(x_k) \geq 0 \tag{10}$$

for all $k \in \mathbb{Z}_+$ and all $x_i \in \mathsf{X}$. The criterion we consider for such a $\{V_k\}$ is

(V2)	There exists $\lambda < 1$, $L < \infty$ and an adapted sequence $\{V_k, \mathcal{F}_k\}$ satisfying (10), such that $$\mathsf{E}_x[V_{k+1} \mid \mathcal{F}_k] \leq \lambda V_k + L \qquad a.s.$$ for all $k \in \mathbb{Z}_+$ and all initial conditions $x \in \mathsf{X}$.

The next result generalizes Foster's criterion to the class of processes considered in this paper.

Theorem 4.2 *Let Φ be an e-process, suppose that condition V1 is satisfied, and that*

$$\sup_{x \in K} (\mathsf{E}_x[V_0(\Phi_1)] + f(x)) \leq B_f < \infty.$$

Then the Markov chain Φ is bounded in probability on average, and $\int f\, d\pi \leq B_f/\varepsilon$ for any invariant probability π.

Theorem 4.2 will be strengthened below in the corollary to Theorem 5.1.

Proof of Theorem 4.2

If $\{V_k\}$ satisfies Assumption V1 then by a discrete-time formulation of Dynkin's formula (see Theorem 3.1 (i) of [38]),

$$\mathsf{E}_x[\tau_K] \leq \begin{cases} \varepsilon^{-1} V_0(x) & x \in K^c \\ \varepsilon^{-1} \mathsf{E}_x[V_0(\Phi_1)] + 1 & x \in K \end{cases}$$

Hence the conditions of the corollary to Theorem 4.1 are satisfied, which shows that Φ is bounded in probability on average.

The bound on $\int f\, d\pi$ follows from Corollary 4.2 of [38] (see also [26]). □

5 Ergodic Theorems

In this section we illustrate the close connections between the stability and regularity properties described in Part I, the existence of invariant probabilities, and the convergence of the distributions of a Markov chain. We will also establish versions of the weak and strong Law of Large Numbers.

5.1 Mean ergodic theorems

A state $x \in \mathsf{X}$ is called *recurrent* if $\mathsf{P}_x\{\Phi \in O \ i.o.\} = 1$ for every open set O containing x (see [48]). A recurrent state x is called *aperiodic* if $P^k(x, O) > 0$ for each open set O containing x, and all $k \in \mathbb{Z}_+$ sufficiently large.

When the state space X is discrete, these definitions coincide with the classical formulations of recurrence and aperiodicity.

Our first result justifies our definition of aperiodicity. For a proof the reader is referred to [30].

Lemma 5.1 *Suppose that Φ is a bounded in probability on average e-process. Let $x^* \in \mathsf{X}$ be recurrent and aperiodic, and let $\pi = \Pi(x^*, \cdot)$. Then for each initial condition y lying in the support of π,*

$$P^k(y, \cdot) \xrightarrow{\text{weakly}} \pi \qquad \text{as } k \to \infty \tag{11}$$

An immediate consequence of this result is that every state in $\operatorname{supp} \pi$ is aperiodic if a single state in $\operatorname{supp} \pi$ is aperiodic.

We now extend Lemma 5.1 to arbitrary initial conditions:

Theorem 5.1 *Suppose that Φ is a bounded in probability on average e-process. Then*

(i) *there exists a Markov transition function Π on $(\mathsf{X}, \mathcal{B}(\mathsf{X}))$ such that*

$$A_k(x, \cdot) \xrightarrow{\text{weakly}} \Pi(x, \cdot)$$

for each $x \in \mathsf{X}$.

(ii) *Suppose further that every recurrent state is aperiodic. Then*

$$P^k(x, \cdot) \xrightarrow{\text{weakly}} \Pi(x, \cdot)$$

for $x \in \mathsf{X}$.

Proof The first result follows directly from Lemma 3.1.

The proof of (ii) uses Lemma 5.1 and the same argument that is used to prove Theorem 5.2 below. We omit the details.

□

A version of Theorem 5.1 is given in [19] in the special case where the state space is compact. We stress that compactness is not required here.

A sample path version of Theorem 5.1 will be presented below.

The following corollary follows immediately:

Corollary *Let Φ be an e-process, and suppose that every recurrent state is aperiodic. Then*

(i) *Φ is bounded in probability if and only if Φ is bounded in probability on average;*

(ii) *If condition V1 is satisfied, and $\sup_{x \in K} \mathsf{E}_x[V_0(\Phi_1)] < \infty$, then Φ is bounded in probability.*

As is the case with Markov chains possessing a continuous component (see [38,33] and Part III), under the conditions of Theorem 5.1, a unique invariant probability exists if some given state may be "almost reached" for all initial conditions.

To be precise, call a state $x^* \in \mathsf{X}$ *attracting* if

$$\sum_{k=1}^{\infty} P^k(y, O) > 0$$

for every state $y \in \mathsf{X}$, and every open set O containing x^*. If the state space is discrete, or if an appropriate continuous component exists, then the existence of an attracting state is equivalent to irreducibility of the Markov chain.

If an attracting state exists for an e-process then the decomposition described in Theorem 3.1 is trivial:

Theorem 5.2 *Suppose that Φ is an e-process which is bounded in probability on average. Then*

(i) *A unique invariant probability π exists if and only if an attracting state $x^* \in \mathsf{X}$ exists;*

(ii) *If an aperiodic attracting state $x^* \in \mathsf{X}$ exists, then for each initial state $x \in \mathsf{X}$,*

$$P^k(x, \cdot) \xrightarrow{\text{weakly}} \pi \qquad \text{as } k \to \infty, \tag{12}$$

where π is the unique invariant probability for Φ. Conversely, if (12) holds for all $x \in \mathsf{X}$ then every state in $\operatorname{supp} \pi$ is attracting and aperiodic.

Theorem 5.2 (ii) is a generalization of a result of [8]. Related results may also be found in [19] and its references, where again the state space is assumed compact.

Proof The proof of (i) follows easily from Theorem 3.1, and so we leave the details to the reader. The converse of (ii) is also straightforward.

So, to prove the proposition we assume that the state $x^* \in \mathsf{X}$ is attracting and aperiodic, and show that equation (12) holds for all initial conditions.

Suppose that $\int f \, d\pi = 0$, $|f(x)| \le 1$ for all x, and for fixed $\varepsilon > 0$ define the set

$$O_\varepsilon \triangleq \{x \in \mathsf{X} : \limsup_{k\to\infty} |P^k f| < \varepsilon\}.$$

Because the Markov transition function P is equicontinuous, and because Lemma 5.1 implies that (12) holds for all initial conditions in $\operatorname{supp} \pi$, the set O_ε is an open neighborhood of $\operatorname{supp} \pi$.

Hence $\pi\{O_\varepsilon\} = 1$, and since O_ε is open, it follows from Theorem 5.2 (i) that

$$\lim_{N\to\infty} A_N(x, O_\varepsilon) = 1$$

Fix $x \in \mathsf{X}$, and choose $N_0 \in \mathbb{Z}_+$ such that $P^{N_0}(x, O_\varepsilon) \ge 1 - \varepsilon$. We then have by the definition of O_ε and Fatou's Lemma,

$$\begin{aligned} \limsup_{k\to\infty} |P^{N_0+k} f(x)| &\le P^{N_0}(x, O_\varepsilon^c) + \limsup_{k\to\infty} \int_{O_\varepsilon} P^{N_0}(x, dy) |P^k f(y)| \\ &\le 2\varepsilon \end{aligned}$$

Since ε is arbitrary, this completes the proof. □

We now state a sample path version of Theorem 5.1.

5.2 The law of large numbers

Define the *occupation probabilities* as

$$\tilde{\mu}_N\{A\} \triangleq \frac{1}{N}\sum_{k=1}^{N} \mathbf{1}\{\Phi_k \in A\} \qquad N \in \mathbb{Z}_+, \quad A \in \mathcal{B}(\mathsf{X}). \tag{13}$$

Observe that $\{\tilde{\mu}_k\}$ are not probabilities in the usual sense, but are probability-valued random variables.

Doob's Strong Law of Large Numbers for Markov chains implies that if an invariant probability π exists, then the occupation probabilities converge with probability one for each initial condition lying in a set of full π-measure.

One of the main difficulties in establishing the Law of Large Numbers for arbitrary initial conditions is finding a candidate limit probability $\tilde{\pi}$. To this end, suppose that $\mathbf{\Phi}$ is a bounded in probability on average e-process. Under these conditions, Theorem 5.1 implies that a Markov transition function Π exists satisfying

$$A_N(x,\,\cdot\,) \xrightarrow{\text{weakly}} \Pi(x,\,\cdot\,) \qquad \text{as } N \to \infty \qquad\qquad x \in \mathsf{X}. \tag{14}$$

If this is the case, then the function Πf is harmonic for each bounded measurable function f on X. That is, the stochastic process $\{\Pi f(\Phi_k)\}$ is a convergent martingale. We then make the definition

$$\tilde{\pi}\{A\} \triangleq \lim_{k\to\infty} \Pi(\Phi_k, A) \tag{15}$$

The following result generalizes [4,20] to non-compact state spaces.

Theorem 5.3 *Suppose that $\mathbf{\Phi}$ is an e-process. Then,*

(i) *If $\mathbf{\Phi}$ is bounded in probability on average then for any $f \in \mathsf{C}$,*

$$\int f\,d\tilde{\mu}_k \to \int f\,d\tilde{\pi} \qquad \text{as } k \to \infty$$

in probability, for each initial condition;

(ii) *If for each initial condition of the Markov chain the occupation probabilities are almost surely tight, then*

$$\tilde{\mu}_k \xrightarrow{\text{weakly}} \tilde{\pi} \quad \text{as } k \to \infty \qquad a.s.\ [\mathsf{P}_x], \qquad x \in \mathsf{X}. \tag{16}$$

Theorem 5.3 and a result of [29] provides a test function approach to establishing the Law of Large Numbers for $\mathbf{\Phi}$:

Theorem 5.4 *If condition V2 holds then $\mathbf{\Phi}$ is bounded in probability, and the occupation probabilities are almost surely tight for each initial condition.*

Part III

Markov Chains with Continuous Components

The results above relied on equicontinuity of the transition function for the process, a condition which is somewhat stronger than the Feller property. In this part of the paper we describe a different approach for obtaining similar stability and ergodic results for a Markov process on a topological state space.

For a probability $\mathbf{F}$ on $\mathbb{Z}_+$ we define the Markov transition function K_F as

$$K_F \triangleq \sum_{i=1}^{\infty} F(i)P^i. \tag{17}$$

For two probabilities $\mathbf{F}$ and $\mathbf{G}$ on $\mathbb{Z}_+$ we have $K_F K_G = K_{F\star G}$, where $\mathbf{F}\star\mathbf{G}$ denotes the convolution of $\mathbf{F}$ and $\mathbf{G}$.

We let $\mathbf{F}_\varepsilon \triangleq (1-\varepsilon)^{-1}(1,\varepsilon,\varepsilon^2,\dots)$ for $0<\varepsilon<1$, and $\mathbf{e}_j$ denote the probability on $\mathbb{Z}_+$ which is supported on $j \in \mathbb{Z}_+$. We let K_ε denote K_F in the specific case where $\mathbf{F} = \mathbf{F}_\varepsilon$, and $\mathbf{F}\star\varepsilon$ denote $\mathbf{F}\star\mathbf{F}_\varepsilon$. In this situation we always assume ε is a fixed positive number.

A kernel T is called a *continuous component* of a function $K\colon(\mathsf{X},\mathcal{B}(\mathsf{X}))\to \mathbb{R}_+$ if

(i) For $A\in\mathcal{B}(\mathsf{X})$ the function $T(\,\cdot\,,A)$ is lower semi-continuous;

(ii) For all $x\in\mathsf{X}$ and $A\in\mathcal{B}(\mathsf{X})$, the measure $T(x,\,\cdot\,)$ satisfies

$$K(x,A)\geq T(x,A).$$

The continuous component T is called *non-trivial* at x if $T(x,\mathsf{X})>0$.

We will be interested exclusively in continuous components of the Markov transition function K_F, as defined in (17). A process will be called a *T-process* if, for some $\mathbf{F}$, the Markov transition function K_F admits a continuous component T which is non-trivial for all $x\in\mathsf{X}$.

6 The Doeblin Decomposition

Here we show that the celebrated Doeblin decomposition holds for a T-process. This result is very similar to the decomposition theorem described in Section 3, where the minimal sets $\{S_x : x\in S\}$ are replaced here by *Harris sets* (c.f. [41]). The minimal sets constructed in Theorem 3.1 and the Harris sets in the Decomposition Theorem below are absorbing, and each of these sets supports a unique invariant measure (which for a Harris set may have infinite mass). The key difference is that while the sets S_x satisfy a strong topological form of recurrence, Harris sets possess a stronger measure-theoretic recurrence property.

A Harris set is called *positive* if it supports an invariant probability.

We let L denote the hitting probabilities of the process, defined as

$$L(x,A) = \mathsf{P}_x\{\bigcup_{k=1}^{\infty}\{\Phi_k\in A\}\}.$$

The probability $Q(x,A)$ is defined as

$$Q(x,A)=\mathsf{P}_x\{\mathbf{\Phi}\in A \text{ i.o.}\}.$$

We say that a sample path of $\mathbf{\Phi}$ *converges to infinity* (denoted $\mathbf{\Phi}\to\infty$) if the trajectory visits each compact set only finitely often. A Markov chain $\mathbf{\Phi}$ will be called *non-explosive* if $\mathsf{P}_x\{\mathbf{\Phi}\to\infty\}=0$ for each $x\in\mathsf{X}$.

Theorem 6.1 (Decomposition Theorem) *Suppose that $\mathbf{\Phi}$ is a T-process. Then*

(i) X *can be decomposed into the disjoint union*

$$\mathsf{X} = \sum_{i\in I} H_i + E$$

where the index set I is countable, each H_i is a Harris set;

(ii) *For each compact set $C \subset \mathsf{X}$, $H_i \cap C = \emptyset$ for all but a finite number of $i \in I$;*

(iii) *For each initial condition $x \in \mathsf{X}$,*

$$\mathsf{P}_x\Big\{\{\Phi \to \infty\} \cup \{\Phi \text{ enters } \sum_{i \in I} H_i\}\Big\} = 1,$$

and hence the index set I is non-empty if and only if $P_x\{\Phi \to \infty\} < 1$ for some initial condition $x \in \mathsf{X}$;

(iv) *If Φ is non-explosive then $L(x, \sum_{i\in I} H_i) \equiv 1$. For each $i \in I$, the Harris set H_i is positive if and only if $\limsup_{k\to\infty} P^k(x, C) > 0$ for some $x \in H_i$, and some compact set $C \subset \mathsf{X}$.*

The proof of this result is based upon consideration of the recurrent states in X. If $x \in \mathsf{X}$ is recurrent, then we define

$$H_x \triangleq \{y \in \mathsf{X} : \mathsf{P}_y\{\Phi \in O \ i.o.\} = 1 \quad \text{for any open set } O \text{ containing } x\}$$

If T is a continuous component, then with $\nu \triangleq T(x, \cdot)$ we have

$$\nu\{A\} > 0 \Rightarrow \mathsf{P}_y\{\Phi \in A \ i.o.\} = 1$$

for any $y \in H_x$, which shows that H_x satisfies the defining property of a Harris set. For the remaining details the reader is referred to [54,38].

If an attracting state exists, the Doeblin Decomposition gives the following connections immediately.

Corollary *Suppose that Φ is a T-process, and that an attracting state exists. Then*

(i) *Φ is non-explosive if and only if Φ is Harris recurrent*

(ii) *Φ is bounded in probability if and only if Φ is positive Harris recurrent*

From the Decomposition Theorem we also see that in the reducible case, if Φ is a T-process, then Φ is non-explosive if and only if every trajectory enters some Harris set with probability one. Again, a stronger result is in fact true.

Theorem 6.2 *Suppose that Φ is a T-process. Then Φ is bounded in probability if and only if it is non-explosive, and every Harris set is positive.*

Proof The proof of Theorem 2.2 in [33] can be adapted, given the results of the Decomposition Theorem, and we omit the details. □

7 Petite Sets

7.1 Petite sets and stability conditions

We now define a class of sets which will be critical in linking stability properties of the process with various criteria for those properties to hold.

A set $A \in \mathcal{B}(\mathsf{X})$ and a sub-probability measure φ on $\mathcal{B}(\mathsf{X})$ are called *petite* if for some probability $\mathbf{F}$ on $\mathbb{Z}_+$,

$$K_F(x, \cdot) \geq \varphi(\cdot)$$

for all $x \in A$. If Φ is positive Harris recurrent with invariant probability π, then the class of petite sets extends the class of small sets, defined in [41], which are petite sets with $\mathbf{F} = \mathbf{e}_j$ for some j.

We now use Theorem 6.1 to give a stronger link between the existence of petite compact sets and T-process properties. For a definition of *irreducibility* see [41].

Theorem 7.1 (i) *Suppose that* $\mathsf{P}_x\{\Phi \to \infty\} < 1$ *for one* x. *Then every compact set is petite if and only if* Φ *is an irreducible T-process.*

(ii) *Suppose that* $\mathsf{P}_x\{\Phi \to \infty\} < 1$ *for all* $x \in \mathsf{X}$. *Then* Φ *is a T-process if and only if every compact set admits a finite cover by open petite sets.*

We link the properties of petite sets, T-processes and stability conditions in

Theorem 7.2 *Suppose that every compact subset of* X *is petite.*

(i) *A Harris set* $H \subset \mathsf{X}$ *exists and is unique if and only if* $\mathsf{P}_x\{\Phi \to \infty\} < 1$ *for some* $x \in \mathsf{X}$, *and if this is the case, then for all* $y \in \mathsf{X}$,

$$\mathsf{P}_y\Big\{\{\Phi \to \infty\} \cup \{\Phi \text{ enters } H\}\Big\} = 1; \quad .$$

(ii) Φ *is Harris recurrent if and only if it is non-explosive;*

(iii) Φ *is positive Harris recurrent if and only if it is bounded in probability.* □

It is well-known that forms of stochastic stability are closely related to the return time behavior of the process on sets in the "center" of the space: individual points on countable spaces, compacta on topological spaces, and so on.

By deriving conditions for appropriately finite return times to such sets, criteria for stability have been found in [55,56,41]. We show here that petite sets play this same role, and then proceed to develop criteria for them to have finite return times.

Proposition 7.1 (i) *The chain* Φ *is Harris recurrent if and only if a petite set* A *exists with* $L(x, A) = 1$ *for all* $x \in \mathsf{X}$.

(ii) *The chain* Φ *is positive Harris recurrent if and only if a petite set* A *exists with* $L(x, A) = 1$ *for all* $x \in \mathsf{X}$, *and* $\sup_{x \in A} \mathsf{E}_x[\tau_A] < \infty$.

(iii) *If* Φ *possesses a sub-invariant measure* π, *then every petite set has finite* π*-measure.*

Proof We first prove (iii), which does not depend on any irreducibility hypothesis.

If A is petite we have $K_F(x, B) \geq \varphi\{B\}$ for all $x \in A$, $B \in \mathcal{B}(\mathsf{X})$. Suppose that π is a sub-invariant measure with $\pi\{A\} \neq 0$. It is easy to see that $\varphi \prec \pi$, and since π is σ-finite there exists a set B such that $\pi\{B\} < \infty$ and $\varphi\{B\} > 0$. In this case we have

$$\infty > \pi\{B\} \geq (\pi I_A K_F)\{B\} \geq \pi\{A\}\varphi\{B\}$$

To see (i), note that the chain is irreducible if a petite set A exists with $L(x, A) = 1$ for all $x \in \mathsf{X}$. By (iii), and the results of Sections 3 and 5 of [55], the chain is Harris recurrent. The converse is well-known.

To see (ii), assume that the chain is positive recurrent.

From Proposition 5.13 of [41] there is an increasing sequence of petite sets satisfying the conditions stated.

Conversely, since the chain is irreducible as in the proof of (i), from (iii) and Section 5 of [55], it is straightforward that the chain is positive recurrent as required. □

In this sense petite sets mimic the behavior we would expect of compact sets, and they play very much this role for the chain Φ.

8 Criteria for Stability and Ergodicity: Irreducible Chains

The topological conditions of non-explosiveness and boundedness in probability do not require any irreducibility. We have seen that, when an attracting state exists, they equate to Harris recurrence and positive Harris recurrence under mild conditions.

In this section we review and extend known results which refine these concepts of stability, and which, in particular, relate the moments of hitting times on petite sets to rates of convergence to stable regimes. We can then use the Lyapunov-Foster approach to find criteria for these rates of convergence to hold.

8.1 Criteria for Harris ergodicity

The results of this section require the following extension of the total variation norm.

For any positive measurable function $f \geq 1$ and any signed measure μ on $\mathcal{B}(\mathsf{X})$ we write

$$\|\mu\|_f = \sup_{|g| \leq f} |\mu(g)|$$

Note that the total variation norm $\|\mu\|$ is $\|\mu\|_f$ in the special case where $f \equiv 1$.

Our next result considers criteria for, and consequences of, positive Harris recurrence.

Theorem 8.1 *Let $f \geq 1$ be measurable. Then*

(i) *The following are equivalent:*

 (a) *the chain $\mathbf{\Phi}$ is positive Harris recurrent with invariant probability π, and $\pi(f) < \infty$.*

 (b) *there exists a petite set A such that $L(x, A) \equiv 1$ and*

$$\sup_{y \in A} \mathsf{E}_y\Big[\sum_{k=1}^{\tau_A} f(\Phi_{k-1})\Big] < \infty. \tag{18}$$

(ii) *If either of the equivalent conditions (a) or (b) holds then*

$$\mathsf{E}_x\Big[\sum_{k=1}^{\tau_A} f(\Phi_{k-1})\Big] < \infty \tag{19}$$

for a.e. $x \in \mathsf{X}$ $[\pi]$;

(iii) *If $\mathbf{\Phi}$ is positive Harris recurrent with invariant probability π and period $m \geq 1$ (cf [44]), and if $\pi(f) < \infty$, then for any petite set A satisfying (18), and initial condition x satisfies (19), the distributions converge in f-norm:*

$$\Big\|\frac{1}{m}\sum_{i=1}^{m} P^{i+\ell}(x, \cdot\,) - \pi\Big\|_f \to 0. \tag{20}$$

Proof (i) and (ii): If $\mathbf{\Phi}$ is positive Harris recurrent, then (18) and (19) hold for some petite set by Proposition 5.13 of [41]. Conversely, if (b) holds, then since A is petite, and $f \geq 1$, (18) implies trivially that

$$\sup_{y \in A} \mathsf{E}_y[\tau_A] < \infty$$

and so $\mathbf{\Phi}$ is positive Harris recurrent from Section 5 of [55] and Proposition 7.1 (iii): (18) then gives $\pi(f) < \infty$ from a representation of π given in [38].

(iii): This convergence result follows from the case $R = 1$ of [43]: note that the null set N_g in that result can be taken as the set on which (19) fails, when $R = 1$, since the set N_1 in that proof can be taken as empty in this case.

□

When Φ is aperiodic (that is, the period $m = 1$), and (20) holds for some $f \geq 1$ and all x, then Φ is called *Harris ergodic.*

Theorem 8.2 *Suppose that A is petite, that $f \geq 1$ is measurable, and that (V1) is satisfied with $f + PV_0$ bounded on A.*

Then Φ is positive Harris recurrent, and (20) holds for all x and for the same f, so that in the aperiodic case, Φ is Harris ergodic.

Proof This follows from Dynkin's formula, and Theorem 8.1 (see [38]). Notice that we do not have to assume irreducibility when A is petite, and that the π-null set on which (19) fails is empty under the conditions of the theorem, so Harris ergodicity does hold in the aperiodic case. □

This result is taken from [38]. It extends Theorem 9.1 of [55] from the situation where V is a simple function on X.

8.2 Criteria for geometric recurrence

We now turn to the concepts of *geometric recurrence and geometric ergodicity* which are refinements of the idea of positive recurrence and Harris ergodicity (cf [41,42,56]). Suppose that for some measurable $f \geq 1$, a petite set A, and $r > 1$,

$$\sup_{x \in A} \mathsf{E}_x\Big[\sum_{k=1}^{\tau_A} r^k f(\Phi_{k-1})\Big] < \infty. \tag{21}$$

Then Φ is called *geometrically recurrent.*

Theorem 8.3 *Suppose that the chain Φ satisfies (21) for some A and r, and is hence geometrically recurrent. Then there exists $\rho < 1$, $R < \infty$, and a probability π such that*

$$\Big\|\frac{1}{m}\sum_{i=1}^{m} P^{i+\ell}(x,\,\cdot\,) - \pi\Big\|_f \leq R\mathsf{E}_x\Big[\sum_{k=1}^{\tau_A} r^k f(\Phi_{k-1})\Big]\rho^\ell, \qquad x \in \mathsf{X},\ \ell \in \mathbb{Z}_+, \tag{22}$$

where the right hand side is finite for a.e. $x \in \mathsf{X}$ $[\pi]$, and m denotes the period of the chain [41].

Conversely, if (22) holds for some $r > 1$, $R < \infty$, $\rho < 1$, π a probability, and A petite, with the right hand side finite for a.e. $x \in \mathsf{X}$ $[\pi]$, then Φ is geometrically recurrent.

If the bound (22) holds for all $x \in \mathsf{X}$ with $m = 1$ and the RHS everywhere finite, then Φ is called *geometrically ergodic.*

The proof of this result is based upon consideration of the "split chain" [40,41,1]. This device allows us to shift our analysis from petite sets, to a recurrent atom lying in an enlarged state space. An *atom* α is by definition a subset of the state space for which $P(x,\,\cdot\,) = P(y,\,\cdot\,)$ for all $x, y \in \alpha$. Hence an atom is essentially a single state, and by the strong Markov property, the trajectories of the process between visits to an atom are independent and identically distributed [1]. See [38] for a proof.

The classification in Theorem 8.3 together with an application of Dynkin's formula easily gives

Corollary *Suppose that A is petite, $f \geq 1$ is measurable, and that (V2) is satisfied with $f + V_0$ bounded on A. Then for some $r > 1$ the chain Φ satisfies (21) with the same f, and and the RHS of (22) is finite for all $x \in \mathsf{X}$. Hence Φ is geometrically recurrent.*

In the aperiodic case there exists $R < \infty$ and $\rho < 1$ such that

$$\|P^n(x,\,\cdot\,) - \pi\|_f \leq R(V_0(x) + f(x))\rho^n, \qquad n \in \mathbb{Z}_+,\ x \in \mathsf{X}.$$

Hence in this case Φ is geometrically ergodic.

We remark that the precise rate of convergence established in the corollary was used in an essential way in [37] to obtain extensions to the continuous time case.

Part IV

Criteria For Equicontinuity and Criteria for The Existence of Continuous Components

The results of the previous sections are not of much value unless techniques are available to prove that a given transition function is equicontinuous, or that a non-trivial continuous component exists.

This topic is taken up here, where we consider Markov chains generated by nonlinear stochastic systems. We consider a Markov chain Φ generated by the *Markovian system*,

$$\Phi_{k+1} = F(\Phi_k, w_{k+1}), \qquad k \in \mathbb{Z}_+, \tag{23}$$

where for all k, $\Phi_k \in \mathsf{X}$ = an open subset of IR^n, $w_k \in \mathsf{IR}^p$, and $F : \mathsf{X} \times \mathsf{IR}^p \to \mathsf{X}$ is smooth (C^∞).

We assume that the initial condition Φ_0 and the disturbance process $\mathbf{w}$ satisfy

(N1) $(\Phi_0, \mathbf{w})$ are random variables on the probability space $(\Omega, \mathcal{F}, \mathsf{P}_{\Phi_0})$;

(N2) Φ_0 is independent of $\mathbf{w}$;

(N3) $\mathbf{w}$ is an independent and identically distributed (i.i.d) process.

Our first topic in this part of the paper shows how equicontinuity is related to stability of a linearized system.

9 Equicontinuity and Stability of the Linearized System

Associated with the Markovian system (23) is the random linear system

$$\Delta_{k+1} = [DF(\Phi_k, w_{k+1})]\Delta_k, \qquad k \in \mathbb{Z}_+ \tag{24}$$

where Δ takes values in the set of $n \times n$-matrices, and DF denotes the derivative of F with respect to its first variable.

It follows from the chain rule and induction that

$$\Delta_k = \frac{d}{d\Phi_0}\Phi_k \qquad \text{for all } k \in \mathbb{Z}_+,$$

when $\Delta_0 = I$.

The system (24) is simply the system (23), linearized about the sample path $(\Phi_0, \Phi_1, \ldots)$. If the original system (23) is reasonably stable then we might expect that the linearized system will also be stable.

By *stable* we mean that the system (24) satisfies one of the stability conditions given in the introduction. In fact we will see that we often obtain much more, including some generalization of exponential asymptotic stability to the stochastic framework (See [26] for a development of these forms of stability).

Theorem 9.1 *Suppose that conditions N1 - N3 hold for the Markovian system (23). Then letting Δ_k denote the derivative of Φ_k with respect to Φ_0, $k \in \mathbb{Z}_+$, we have*

(i) *If for some open convex set $N \subset \mathsf{X}$,*

$$\mathsf{E}[\sup_{\Phi_0 \in N} \|\Delta_k(\Phi_0)\|] < \infty \tag{25}$$

then for all $x \in N$,

$$\frac{d}{dx}\mathsf{E}_x[\Phi_k] = \mathsf{E}_x[\Delta_k].$$

(ii) *Suppose that (25) holds for each $y_0 \in \mathsf{X}$, and all sufficiently small neighborhoods N of y_0. Suppose further that for any compact set $C \subset \mathsf{X}$,*

$$\sup_{y \in C} \sup_{k \geq 0} \mathsf{E}_y[\|\Delta_k\|] < \infty.$$

Then $P^k f$ is differentiable for any $f \in \mathsf{C}^\infty \cap \mathsf{C}_c$, the derivative being uniformly bounded in $k \in \mathbb{Z}_+$, and for initial conditions x lying in compact subsets of X. In particular, P is equicontinuous.

It may seem that the technical assumption (25) will be difficult to verify in practice. However, to take one large class of examples, consider the case where the i.i.d. process $\mathbf{w}$ is uniformly bounded. It follows from the smoothness condition on F that $\sup_{\Phi_0 \in N} \|\Delta_k(\Phi_0)\|$ is almost surely finite for any compact subset $N \subset \mathsf{X}$, which shows that in this case (25) is trivially satisfied.

The following result provides another large class of models for which (25) is satisfied. Observe that the conditions imposed on $\mathbf{w}$ in Proposition 9.1 are satisfied for any i.i.d. Gaussian process. The proof is straightforward.

Proposition 9.1 *For the Markov chain (23), suppose that conditions N1 - N3 hold, that F is a rational function of its arguments, and that for some $\varepsilon_0 > 0$,*

$$\mathsf{E}[\exp(\varepsilon_0 |w_1|)] < \infty.$$

Then letting Δ_k denote the derivative of Φ_k with respect to Φ_0, we have for any compact set $C \subset \mathsf{X}$, and any $k \geq 0$,

$$\mathsf{E}[\sup_{\Phi_0 \in C} \|\Delta_k\|] < \infty.$$

Hence under these conditions,

$$\frac{d}{dx}\mathsf{E}_x[\Phi_k] = \mathsf{E}_x[\Delta_k].$$

Proof of Theorem 9.1

The first result is a consequence of the Dominated Convergence Theorem.

To prove the second result, let $f \in \mathsf{C}_c \cap \mathsf{C}^\infty$. Then

$$\left|\frac{d}{dx}\mathsf{E}[f(\Phi_k)]\right| \leq \|f'\|_\infty \mathsf{E}_x[\|\Delta_k\|]$$

which by the assumptions of (ii), implies that the sequence of functions $\{P^k f : k \in \mathbb{Z}_+\}$ is equicontinuous on compact subsets of X. Since $\mathsf{C}^\infty \cap \mathsf{C}_c$ is dense in C_c, this completes the proof. □

10 Rates of convergence

Here we demonstrate how the rate of convergence of P^kV to $\int V\,d\pi$ is determined by the rate at which the derivative process $\mathbf{\Delta}$ decays to zero.

Theorem 10.1 *Suppose that* $\mathsf{X} \subset \mathrm{IR}^n$ *is an open convex set, and let* $V\colon \mathsf{X} \to \mathrm{IR}$ *be a fixed continuous function.*

Suppose that there exists a convex function $B\colon \mathsf{X} \to \mathrm{IR}_+$, *and a sequence* $\{b_k : k \geq 1\}$ *such that*

$$|\frac{d}{dx}P^kV(x)| \leq b_kB(x), \qquad x \in \mathsf{X}, \quad k \in \mathbb{Z}_+. \tag{26}$$

Then

(i) *for all* $x, y \in \mathsf{X}$,

$$|\mathsf{E}_x[V(\Phi_k)] - \mathsf{E}_y[V(\Phi_k)]| \leq \frac{b_k}{2}(B(x) + B(y))|x - y| \tag{27}$$

(ii) *Suppose further that an invariant probability* π *exists for which*

$$\int |x|(B(x) + 1)\,\pi(dx) < \infty.$$

Then for all $x \in \mathsf{X}$, $k \in \mathbb{Z}_+$,

$$|\mathsf{E}_x[V(\Phi_k)] - \mathsf{E}_\pi[V(\Phi_0)]| \leq \frac{b_k}{2}\int\Big[(B(x) + B(y))|x - y|\Big]\,\pi(dy) \tag{28}$$

In the paper [11] it is shown that when $\Phi_0 \sim \pi$, so that $\mathbf{\Phi}$ is a strictly stationary process,

$$\lim_{k\to\infty}\frac{1}{k}\log\|\Delta_k\| = \lim_{k\to\infty}\frac{1}{k}\mathsf{E}[\log\|\Delta_k\|] = \bar{\lambda}, \tag{29}$$

where the *top Lyapunov exponent* $\bar{\lambda}$ may be positively or negatively infinite. It is also assumed that the process $\mathbf{\Phi}$ is metrically transitive, a condition which holds when the support of the invariant probability π is a minimal set (see Section 3).

If the top Lyapunov exponent is negative then by conditioning on Φ_0, we may show that for almost every initial condition $\Phi_0 = x \in \mathsf{X}$ $[\pi]$, almost every sample path will decay to zero at least as fast as $\exp(\bar{\lambda}k)$. This together with Theorem 10.1 strongly suggests that the distributions of $\mathbf{\Phi}$ will converge at the same rate to an invariant probability.

Proof of Theorem 10.1

Let $v_k(t) = P^kV\left((1-t)x + ty\right)$, $0 \leq t \leq 1$. Then

$$\begin{aligned}\frac{dv_k}{dt} &= \Big(\frac{d}{dx}P^kV\left((1-t)x + ty\right)\Big)(y - x)\\ &\leq b_k((1-t)B(x) + tB(y))|x - y|\end{aligned}$$

Integrating gives the bound

$$v_k(1) \leq v_k(0) + \int_0^1 |\frac{dv_k}{dt}|\,dt \leq b_k/2(B(x) + B(y))|x - y|$$

By symmetry (between y and x) and the definition of v_k, this proves part (i). The second result follows from the first by integrating with respect to π. ∎

11 Criteria for Continuous Components

To show how petite sets may be constructed we first consider the Markovian system (23) with $\mathsf{X} = \mathrm{IR}$. We assume that the distribution of w_k possesses a lower semi continuous density p_w, which we have already assumed to be independent of k.

Although $\mathbf{w}$ will be assumed i.i.d. throughout this section, we will also consider the disturbance w_k as a deterministic control variable. The open set $\mathcal{O}_w \triangleq \{x : p_w(x) > 0\}$ is then the set of possible values that the "control" w_k may take on. In this setting, $\mathbf{\Phi}$ is a deterministic state process whose trajectories are determined by its inital condition, and the control sequence $\mathbf{w}$.

The process $\mathbf{\Phi}$ generated by the Markovian system (23) will be a T-process if the associated control system is such that the sequence $\mathbf{w}$ may drive the state $\mathbf{\Phi}$ to a sufficiently large subset of the state space. By the implicit function theorem it may be seen that this condition will be satisfied if the derivative of the state with respect to the control sequence $\mathbf{w}$ has full rank. Details of these results are developed in [25,17,18,33,29]. Here we give a brief sketch of the main ideas.

Assume that for some $(x^*, w^*) \in \mathsf{X} \times \mathcal{O}_w$,

$$\frac{\partial F}{\partial w}(x^*, w^*) \neq 0 \tag{30}$$

We will construct an open petite set containing x^* under this condition.

By the implicit function theorem, there exist open sets B_1 and B_2, containing x^* and w^* respectively, and a smooth function G mapping an open subset of IR^2 into IR such that

$$G \circ F(x, w) = w, \qquad (x, w) \in B_1 \times B_2. \tag{31}$$

By reducing the size of B_2, if necessary, we may assume that

$$p_w(t) \geq \varepsilon > 0, \qquad t \in B_2 \tag{32}$$

for some $\varepsilon > 0$ sufficiently small.

It follows that for any $x \in \mathsf{X}$, $A \in \mathcal{B}(\mathsf{X})$,

$$P(x, A) = \int_{\mathrm{IR}} \mathbf{1}\{F(x,t) \in A\} p_w(t)\, dt \geq \varepsilon \int_{B_2} \mathbf{1}\{F(x,t) \in A\}\, dt. \tag{33}$$

Using the change of variables

$$y = F(x,t); \quad t = G(x,y); \quad dy = \Big|\frac{\partial F}{\partial t}(x,t)\Big| dt; \quad dt = \Big|\frac{\partial G}{\partial y}(x,y)\Big| dy,$$

we obtain from (33),

$$P(x, A) \geq \varepsilon \int \mathbf{1}\{G(x,y) \in B_2\} \mathbf{1}\{y \in A\} |\frac{\partial G}{\partial y}(x,y)|\, dy \tag{34}$$

$x \in B_1$, $A \in \mathcal{B}(\mathsf{X})$.

Hence $P(x, A)$ dominates a kernel on $\mathsf{X} \times \mathcal{B}(\mathsf{X})$ with lower semi continuous density

$$q(x,y) \triangleq \varepsilon \mathbf{1}\{x \in B_1\} \mathbf{1}\{G(x,y) \in B_2\} |\frac{\partial G}{\partial y}(x,y)| \tag{35}$$

By construction we see that $q(x^*, y^*) > 0$, where $y^* = F(x^*, w^*)$. By lower semi continuity, we may find $\delta > 0$, and open sets $\mathcal{O}$ and $\mathcal{V}$ containing x^* and y^* respectively for which $q(x,y) \geq \delta$ for all $(x,y) \in \mathcal{O} \times \mathcal{V}$.

By (34) and (35) we conclude that

$$P(x,A) \geq c\mu_{\mathcal{V}}\{A\}, \qquad x \in \mathcal{O}$$

where $\mu_{\mathcal{V}}$ denotes the uniform distribution on $\mathcal{V}$, and $c = \delta$ times the Lebesgue measure of $\mathcal{V}$.

Hence, under the rank condition (30), an open petite set exists which contains the state x^*.

We now describe how this construction may be generalized to multidimensional models. Assume that X is an open subset of $\mathbb{R}^n$, and that the distribution of w_k possesses a density p_w on $\mathbb{R}^p$ which is lower semi continuous.

The *extended transition map* $S_x^k : \mathbb{R}^{kp} \to \mathbf{X}$ of the Markovian system (23) is defined inductively for $k \in \mathbb{Z}_+$, $x \in \mathbf{X}$, and $\mathbf{z} = (z_1, \dots, z_k)^\mathsf{T} \in \mathbb{R}^{pk}$ by

$$\begin{aligned} S_x^k(\mathbf{z}) &= F\left(S_x^{k-1}(z_1, \cdots, z_{k-1}), z_k\right), \quad k \geq 1, \\ S_x^0 &= x. \end{aligned}$$

The extended transition map is so named because for all $k \in \mathbb{Z}_+$,

$$\Phi_k = S_x^k(w_1, \ \dots \ , w_k) \quad \text{when } \Phi_0 = x.$$

Given two measures ν and μ on $\mathcal{B}(\mathsf{X})$ we say that ν is *absolutely continuous* with respect to μ (denoted $\nu \prec \mu$) if $\nu\{A\} = 0$ whenever $\mu\{A\} = 0$. The measures ν and μ are called *equivalent* (denoted $\nu \approx \mu$) if $\nu \prec \mu$ and $\mu \prec \nu$. We let $\mathbf{1}_A\mu$ denote the measure defined for $B \in \mathcal{B}(\mathsf{X})$ by $(\mathbf{1}_A\mu)\{B\} \triangleq \mu\{A \cap B\}$.

The following definitions of stochastic controllability are taken from [33]:

(i) The Markovian system (23) is called *locally stochastically controllable* if for each initial condition $x \in \mathbf{X}$ there exists $T = T(x) \in \mathbb{Z}_+$ and an open set $\mathcal{O}_x \subset \mathbf{X}$ such that $P^T(x, \cdot) \approx \mathbf{1}_{\mathcal{O}_x}\mu^{\mathrm{Leb}}$.

(ii) The Markovian system (23) is called *weakly stochastically controllable* if for each initial condition $x \in \mathbf{X}$ there exists $T = T(x) \in \mathbb{Z}_+$ and an open set $\mathcal{O}_x \subset \mathbf{X}$ such that $P^T(x, \cdot) \succ \mathbf{1}_{\mathcal{O}_x}\mu^{\mathrm{Leb}}$.

If (23) is locally stochastically controllable then the probability $P^T(x, \cdot)$ possesses a density which is strictly positive on an open set $\mathcal{O}_x$ and zero elsewhere. Similarly, if (23) is weakly stochastically controllable then the Radon-Nikodym derivative of the probability $P^T(x, \cdot)$ (with respect to Lebesgue measure) is strictly positive on $\mathcal{O}_x$.

One consequence of these definitions may be roughly described as follows: If (23) is weakly or locally stochastically controllable, and if starting at a point $y \in \mathbf{X}$ it is possible to reach a point $z \in \mathcal{O}_y$ at time T, then at time T all points in some neighborhood of z are reachable from y. In fact, under assumptions A1-A3 and the smoothness condition made on F, it is easily verified that with the disturbance sequence w viewed as an input, the forward accessibility condition of [18] is equivalent to weak stochastic controllability. The terminology may also be motivated by the fact that if $F : \mathbf{X} \times \mathbb{R}^p \to \mathbf{X}$ is linear then the notions of local stochastic controllability, weak stochastic controllability, and controllability in the usual sense are equivalent.

For $y \in \mathbf{X}$ and a sequence $\{z_k : z_k \in \mathbb{R}^p, \, k \in \mathbb{Z}_+\}$ let $\{\mathrm{A}_k, \, \mathrm{B}_k : k \in \mathbb{Z}_+\}$ denote the matrices

$$\begin{aligned} \mathrm{A}_k = \mathrm{A}_k(y, z_1, \dots, z_{k+1}) &\triangleq \left[\frac{\partial F}{\partial x}\right]_{(S_y^k, z_{k+1})} \\ \mathrm{B}_k = \mathrm{B}_k(y, z_1, \dots, z_{k+1}) &\triangleq \left[\frac{\partial F}{\partial z}\right]_{(S_y^k, z_{k+1})}, \end{aligned}$$

and let $C_y^k = C_y^k(z_1, \ldots, z_k)$ denote the *generalized controllability matrix* (along the sequence z_1^k)

$$C_y^k \triangleq [A_{k-1} \cdots A_1 B_0 | A_{k-1} \cdots A_2 B_1 | \ \cdots \ | A_{k-1} B_{k-2} | B_{k-1}] \tag{36}$$

We remark that if F is of the form

$$F(y, z) = Ay + Bz$$

then the generalized controllability matrix becomes the familiar controllability matrix

$$\left[A^{T-1}B | A^{T-2}B | \cdots | AB | B\right].$$

Note that all quantities in the matrix (36) are deterministic.

Here we give necessary and sufficient conditions for local and weak stochastic controllability in terms of the generalized controllability matrix defined above. Alternative conditions for weak stochastic controllability involving the dimension of a certain Lie algebra and substantially stronger conditions on the function F may be found in [18].

Let $\mathcal{O}_w^T$ denote the T-fold Cartesian product of $\mathcal{O}_w$ (recall that $\mathcal{O}_w$ is the open set which supports p_w).

Theorem 11.1 *Suppose that Φ is of the form (23) and that conditions A1 - A3 hold. Then*

(i) *The Markovian system (23) is locally stochastically controllable if and only if for all initial conditions $x \in \mathrm{X}$ there exists $T \geq 1$ such that*

$$\operatorname{rank} C_x^T(\lambda) = n \ \textit{for all} \ \lambda \in \mathcal{O}_w^T \setminus Z_x \tag{37}$$

where $Z_x \cap \mathcal{O}_w^T$ has zero Lebesgue measure.

(ii) *The Markovian system (23) is weakly stochastically controllable if and only if for all initial conditions $x \in \mathbf{X}$, there exists $T \geq 1$ and $\lambda \in \mathcal{O}_w^T$ such that*

$$\operatorname{rank} C_x^T(\lambda) = n. \tag{38}$$

(iii) *If $\operatorname{rank} C_x^T(\lambda) = n$ for some $\lambda \in \mathcal{O}_w^T$, then there exists $c > 0$, and open sets $\mathcal{U}_x^\lambda$, $\mathcal{V}_x^\lambda$ containing x and $S_x^T(\lambda)$ respectively such that*

$$P^T(y, A) \geq c\mu^{\text{Leb}}\{A \cap \mathcal{V}_x^\lambda\} \tag{39}$$

for all $A \in \mathcal{B}(\mathrm{X})$ and $y \in \mathcal{U}_x^\lambda$.

Theorem 11.1 shows that if Φ is weakly stochastically controllable, then the state space may be covered by open petite sets. Applying Theorem 7.1, we see that a weakly stochastically controllable process is also a T-process, showing that these processes admit a Doeblin Decomposition and that the remaining results described in Part III of the paper are valid.

Because of the additional structure that weakly stochastically controllable processes possess, additional results may be obtained. For instance, if F is invertable in the sense of [18] or if Φ is locally stochastically controllable, then each invariant probability possesses a lower semi continuous density with respect to Lebesgue measure. This additional structure also unites the two theories presented in this article: Propositions 3.3 and 3.5 of [33] imply that, when Φ is weakly stochastically controllable and bounded in probability, the probabilities $\{P^k(x, \cdot)\}$ may be written as a convex combination of probabilities which converge cyclically and uniformly for x in compact subsets of X. This fact may be used to prove that Φ is an e-process under these conditions.

We now turn to the examples.

Part V

Examples

12 Adaptive Control

Consider the single input single output random parameter system model

$$y_{k+1} = \theta_k y_k + u_k + v_{k+1} \qquad k \in \mathbb{Z}_+ \tag{40}$$

where the parameter process $\boldsymbol{\theta}$ is the output of the $AR1$ model

$$\theta_{k+1} = \alpha\theta_k + e_{k+1} \qquad k \in \mathbb{Z}_+ \tag{41}$$

and $\alpha \in (-1,1)$. We adopt the conditions imposed in [31,52]. In particular we assume that $\mathbf{v}$ and $\mathbf{e}$ are mutually independent Gaussian white noise processes with variance σ_v^2, σ_e^2 respectively. We assume that $\sigma_v^2 > 0$, to avoid trivialities. We also assume that $\sigma_e < 1$, a necessary condition for L_2-boundedness of the output process $\mathbf{y}$.

The parameter process $\boldsymbol{\theta}$ is assumed to be unknown, but is estimated by the Kalman filter based algorithm

$$\begin{aligned}
\widehat{\theta}_{k+1} &= \alpha\widehat{\theta}_k - \alpha V_k y_{k+1} y_k (V_k y_k^2 + \sigma_v^2)^{-1} \\
V_{k+1} &= \sigma_e^2 + \alpha^2\sigma_v^2 V_k (V_k y_k^2 + \sigma_v^2)^{-1}, \quad k \geq 1,
\end{aligned}$$

Our goal is to choose a control $u_k \in \mathcal{Y}_k = \sigma\{y_0, \ldots, y_k\}$ which makes the mean square output error $\mathrm{E}[y_k^2]$ as small as possible. Applying the certainty equivalence control law $u_k = -\widehat{\theta}_k y_k$ and letting $\widetilde{\theta}_k \triangleq \theta_k - \widehat{\theta}_k$, the closed loop system becomes

$$\widetilde{\theta}_{k+1} = \alpha\widetilde{\theta}_k - \alpha V_k y_{k+1} y_k (V_k y_k^2 + \sigma_v^2)^{-1} + e_k \tag{42}$$

$$y_k = \widetilde{\theta}_k y_k + v_k \tag{43}$$

$$V_{k+1} = \sigma_e^2 + \alpha^2\sigma_v^2 V_k (V_k y_k^2 + \sigma_v^2)^{-1}, \quad k \geq 1. \tag{44}$$

It is well known that

$$\widetilde{\theta}_k = \theta_k - \mathrm{E}[\theta_k \mid \mathcal{Y}_k] \text{ and } V_k = \mathrm{E}[\widetilde{\theta}_k^2 \mid \mathcal{Y}_k]$$

under the appropriate conditions on the initial conditions of the system (see [5,31,52]). We will use these facts to obtain a new, elementary proof of L_2-boundedness for the closed loop system.

From (42) and (44) we have for any initial condition which guarantees that the Kalman filter generates the conditional mean and covariance of θ_k,

$$\begin{aligned}
\mathsf{E}[V_{k+1} y_{k+1}^2 \mid \mathcal{Y}_k] &= V_{k+1}\mathsf{E}[y_{k+1}^2 \mid \mathcal{Y}_k] \\
&= V_{k+1}(V_k y_k^2 + \sigma_v^2) \\
&= \sigma_e^2 V_k y_k^2 + \sigma_e^2\sigma_v^2 + \alpha^2\sigma_v^2 V_k
\end{aligned}$$

This shows that

$$\limsup_{k\to\infty} \mathsf{E}[V_k y_k^2] \leq (1-\sigma_e^2)^{-1}(\sigma_e^2\sigma_v^2 + \alpha^2\sigma_v^2 \limsup_{k\to\infty} \mathsf{E}[V_k])$$

which, from the bound $\sigma_e^2 \leq V_k \leq \sigma_e^2/(1-\alpha^2)$, implies that $\mathsf{E}[y_k^2]$ is uniformly bounded.

The drawback with this result is that it cannot be generalized to the more general case where the statistics of (v, e) are only partially known, and it only gives stability for a special class of initial conditions. The theory presented in this paper gives a more general stability proof, and stronger stability properties for the system.

It is evident that the state process $\mathbf{\Phi} \triangleq (\mathbf{y}, \boldsymbol{\theta}, \mathbf{V})^{\mathsf{T}}$ is a Markov chain of the form (23) with state space $\mathsf{X} \triangleq \mathbb{R}^2 \times [\sigma_e^2, \sigma_e^2/(1-\alpha^2)]$.

It is shown in [14] that the state process $\mathbf{\Phi}$ is weakly stochastically controllable, and that 0 is an attracting state for the closed loop system. Our next task is to find a moment satisfying (V2) so that we may apply Theorem 8.3.

Let $y{:}\,\mathsf{X} \to \mathbb{R}$, $\tilde{\theta}{:}\,\mathsf{X} \to \mathbb{R}$, $V{:}\,\mathsf{X} \to \mathbb{R}$ denote the coordinate variables on X so that

$$y_k = y(\Phi_k) \quad \tilde{\theta}_k = \tilde{\theta}(\Phi_k) \quad V_k = V(\Phi_k) \qquad k \in \mathbb{Z}_+,$$

and define the test function U on X by

$$U(y, \tilde{\theta}, P) = \tilde{\theta}^4 + \varepsilon_0 \tilde{\theta}^2 y^2 + \varepsilon_0^2 y^2 \tag{45}$$

The stability proofs in [32,14] may be adapted to prove that the process $\{U(\Phi_k)\}$ satisfies condition (V2). This immediately gives

Theorem 12.1 *The Markov chain $\mathbf{\Phi}$ is geometrically ergodic, and for all initial conditions $x \in \mathsf{X}$,*

$$\lim_{k\to\infty} \mathrm{E}_x[y_k^2] = \int y^2 \, d\pi < \infty$$

at a geometric rate, and

$$\lim_{N\to\infty} \frac{1}{N} \sum_{k=1}^{N} y_k^2 = \int y^2 \, d\pi < \infty \qquad a.s.\ [\mathrm{P}_x]$$

13 Doubly Stochastic Time Series

Consider the time series model

$$y_{k+1} = \theta_k^{\mathsf{T}} \varphi_k + v_{k+1} \tag{46}$$

where $\mathbf{v}$ is an independent and identically distributed (i.i.d.) process on $\mathbb{R}$ and $\binom{\theta}{\varphi}$ is a stochastic proces taking values in $\mathbb{R}^{2m}$, both defined on a probability space $(\Omega, \mathcal{F}, \mathsf{P})$.

This model typically appears in adaptive control problems. In this setting φ is a regression vector consisting of inputs and outputs of a system to be controlled, and $\boldsymbol{\theta}$ represents either a partially observed time varying parameter, or a parameter estimation error sequence [12,5].

The process $\mathbf{y}$ defined by (46) is called a *doubly stochastic autoregressive process of order* m, or (DSAR(m)) [53]. It is shown in [53] that this class of models contains a broad range of popular nonlinear time series models. A survey of recent results may be found in the introduction of [46], and some recent breakthroughs may be found in [23].

One instance of (46) is the scalar case where $\boldsymbol{\theta}$ itself is the output of a stable linear state space model and $\phi_k = y_k$. We will concentrate on this special case due to lack of space, although broad extensions are possible. Some possible extensions will be described below.

Assume that the joint process $(\boldsymbol{\theta}, \mathbf{y})$ may be expressed

$$\theta_{k+1} = \alpha\theta_k + e_{k+1}, \qquad |\alpha| < 1 \tag{47}$$

$$y_{k+1} = \theta_k y_k + v_{k+1} \tag{48}$$

These equations are similar to those considered in the adaptive control problem described in the previous section. It is surprising that this apparently simpler model is much more difficult to analyze than the adaptive system (42) - (44).

We assume that the model (47, 48) satisfies the following conditions:

(E1) $\mathrm{w} \triangleq (\mathrm{e}, \mathrm{v})^{\mathsf{T}}$ is i.i.d., is independent of (θ_0, y_0), and **e** and **v** are mutually independent.

(E2) For some $p > 2$,

$$\mathsf{E}\Big[|v_1|^p\Big] < \infty, \qquad \gamma^p \triangleq \mathsf{E}\Big[\exp\Big\{\frac{p}{1-|\alpha|}|e_1| - p\Big\}\Big] < 1$$

Let $\mathsf{X} = \mathsf{IR}^2$, and define the Markov transition function P on $(\mathsf{X}, \mathcal{B}(\mathsf{X}))$ for $x \in \mathsf{X}$ and a measurable function $f : \mathsf{X} \to \mathsf{IR}$ by

$$Pf(\theta, y) = \int f(\alpha\theta + e_1, \theta y + v_1)\, d\mu_w(e_1, v_1)$$

Then P is the Markov transition function for the Markov chain $\mathbf{\Phi} \triangleq \binom{\theta}{y}$.

The following result is taken from [30]:

Proposition 13.1 *Suppose that Assumptions E1 and E2 hold for the system (47, 48). Then* $\frac{d}{dx}\mathsf{E}_x[\Phi_k]$ *exists for all k, and we have for some constant $C_0 < \infty$,*

$$\|\frac{d}{dx}\mathsf{E}_x[\Phi_k]\| \le kC_0 \exp(\frac{2|\theta_0|}{1-|\alpha|})(1 + |y_0|)(\gamma^k + \alpha^k).$$

Hence $\mathbf{\Phi}$ is an e-process.

L_p boundedness of the process is established in [36], and under slightly stronger conditions it is shown that $\mathbf{\Phi}$ is a T-process. From L_p boundedness and Proposition 13.1 we will prove

Proposition 13.2 *For the Markov chain (47, 48), suppose that conditions E1 and E2 hold with $p = 4$. Then*

(i) *there exists a unique invariant probability π, and for all initial conditions $\Phi_0 = x \in \mathsf{X}$,*

$$\begin{array}{rcll} \tilde{\mu}_k & \xrightarrow{\text{weakly}} & \pi & a.s.\ [\mathsf{P}_x] \\ P^k(x, \cdot\,) & \xrightarrow{\text{weakly}} & \pi & as\ k \to \infty; \end{array}$$

(ii) *For all initial conditions $\Phi_0 = x \in \mathsf{X}$,*

$$\lim_{k\to\infty} \mathsf{E}_x[y_k^2] = \int y^2\, d\pi;$$

(iii) *There exists a fixed positive function $B\colon \mathsf{X} \to \mathsf{IR}_+$ such that for any function $f \in \mathsf{C} \cap \mathsf{C}^\infty$ with a uniformly bounded derivative,*

$$|\mathsf{E}_x[f(\Phi_k)] - \int f\, d\pi| \le k\|f'\|_\infty B(x)(\gamma^k + \alpha^k)$$

for $k \in \mathbb{Z}_+$, $x \in \mathsf{X}$.

Proof The third result follows from L_p boundedness (which implies that an invariant probability exists) together with Proposition 13.1, and Theorem 10.1. The required moments on the invariant probability exist by the calculations in [35].

It follows from (iii) that a unique invariant probability π exists, and that the distributions of $\mathbf{\Phi}$ converge weakly to π.

In [35] a stochastic process $\{V_k\}$ satisfying (V2) is constructed for the process (47, 48). Hence we may apply Theorem 5.3 and Theorem 5.4 to conclude that the occupation probabilities converge weakly to π.

To prove (ii) we use L_p boundedness, which implies that the function y^2 is uniformly integrable with respect to the probabilities $(P^k(x, \cdot\,))$ for each $x \in \mathsf{X}$.

□

Colored noise

So far we have assumed that the disturbance process $\mathbf{w} = \binom{\mathbf{e}}{\mathbf{v}}$ is i.i.d. Let us consider the colored noise case, where $\mathbf{v}$ is the output of the stable linear system

$$X_{k+1} = AX_k + Bz_{k+1} \tag{49}$$

$$v_{k+1} = CX_k, \tag{50}$$

the state X evolves on IR^M and $\mathbf{z}$ evolves on IR^N, the joint process $\binom{\mathbf{e}}{\mathbf{z}}$ is i.i.d., and the initial condition $(X_0, \theta_0, y_0)^\top \in \mathrm{IR}^{M+2}$ is a constant, or a random variable which is independent of $\binom{\mathbf{e}}{\mathbf{z}}$. The matrices A, B, and C are of suitable dimension, and are constant. Since A is assumed stable, there exists an initial distribution which makes $\mathbf{v}$ a strictly stationary process with rational spectral density. However, we do not restrict the initial condition of (49).

Under these assumptions, the joint process $\mathbf{\Phi} = \begin{pmatrix} X \\ \theta \\ y \end{pmatrix}$ is a Feller Markov chain evolving on $\mathsf{X} = \mathrm{IR}^{M+2}$.

The derivatives of $\boldsymbol{\theta}$ and X with respect to Φ_0 decay geometrically, so to establish equicontinuity it is enough to bound the derivative of $\mathbf{y}$, which has the form

$$\frac{dy_{k+1}}{d\Phi_0} = \theta_k \frac{dy_k}{d\Phi_0} + y_k \frac{d\theta_k}{d\Phi_0} + \frac{dv_{k+1}}{d\Phi_0}$$

If the "core" of this model $z_{k+1} = \theta_k z_k$ is stable, then because the terms $\frac{d\theta_k}{d\Phi_0}$ and $\frac{dv_{k+1}}{d\Phi_0}$ decay to zero at a geometric rate, $\frac{dy_{k+1}}{d\Phi_0}$ will also decay to zero at a geometric rate.

This shows that, as in the white noise case, $\mathbf{\Phi}$ is an e-process if it is suitably stable.

Multidimensional models

Consider the multidimensional case (46) in which the process φ has the form

$$\varphi_{k+1} = \begin{bmatrix} & \theta_k^\top \\ & 0 \\ \mathrm{I} & \vdots \\ & 0 \end{bmatrix} \varphi_k + v_{k+1} \tag{51}$$

This model has been shown to be bounded in probability using the same methods as in the scalar case. See [35] for the case where $\boldsymbol{\theta}$ is an autoregressive process, or [34] for the case where φ represents the partial state of a control system, and $\boldsymbol{\theta}$ is a parameter estimation error sequence.

The derivative process may be computed as

$$\frac{d\varphi_{k+1}}{d\Phi_0} = \begin{bmatrix} & \theta_k^\top \\ & 0 \\ \mathrm{I} & \vdots \\ & 0 \end{bmatrix} \frac{d\varphi_k}{d\Phi_0} + \begin{bmatrix} \varphi_k^\top \frac{d\theta_k}{d\Phi_0} \\ 0 \end{bmatrix} \tag{52}$$

where we have assumed for simplicity that $\mathbf{v}$ is i.i.d.

This system is stable if, roughly speaking, the original system is stable, and the derivative $\frac{d\theta_k}{d\Phi_0}$ is bounded. For example, in one class of adaptive control problems, $\boldsymbol{\theta}$ denotes a parameter estimation error sequence and has the form

$$\theta_{k+1} = (1-\sigma)\theta_k - \mu \frac{\varphi_k y_{k+1}}{1+|\varphi_k|^2} + \sigma\bar{\theta}$$

where μ and σ are small positive constants, $\bar{\theta}$ is constant, and $\varphi_k^\top = (y_k, \dots, y_{k-n+1})$. In this case

$$\|\frac{d\theta_{k+1}}{d\Phi_0}\| \leq (1-\sigma)\|\frac{d\theta_k}{d\Phi_0}\| + \mu C(\|\frac{d\varphi_k}{d\Phi_0}\| + \|\frac{dy_{k+1}}{d\Phi_0}\|)$$

which suggests that the derivative process will be stable if $\mathbf{\Phi}$ is suitably stable and μ is sufficiently small.

References

[1] ATHREYA, K. B., AND NEY, P. A new approach to the limit theory of recurrent Markov chains. *Transactions of the American Mathematical Society 245* (1978), 493–501.

[2] BENĚS, V. E. Finite regular invariant measures for Feller processes. *J. Applied Prob. 5* (1968), 203.

[3] BILLINGSLEY, P. *Convergence of Probability Measures.* John Wiley & Sons, New York, NY, 1968.

[4] BREIMAN, L. The strong law of large numbers for a class of Markov chains. *Annual Mathematics Statist. 31* (1960), 801–803.

[5] CAINES, P. E. *Linear stochastic systems.* John Wiley & Sons, New York, NY, 1988.

[6] DOEBLIN, W. Eléments d'une théorie générale des chaînes simples constantes de Markov. *Annales Scientifiques de l'Ecole Normale Supérieure 57,* III (1940), 61–111. Paris.

[7] DOOB, J. L. *Stochastic Processes.* John Wiley & Sons, New York, NY, 1953.

[8] FELLER, W. *An Introduction to Probability Theory and its Applications,* vol. 2. John Wiley & Sons, New York, NY, 1971.

[9] FOGUEL, S. The ergodic theory of positive operators on continuous functions. *Annual Scuola Norm. Sup. Pisa 27* (1973), 19–51.

[10] FOSTER, F. G. On the stochastic matrices associated with certain queueing processes. *Annals of Mathematical Statistics 24* (1953), 355–360.

[11] FURSTENBERG, H., AND KESTEN, H. Products of random matrices. *The Annals of Mathematical Statistics 31* (1960), 457–469.

[12] GOODWIN, G. C., AND SIN, K. S. *Adaptive Filtering, Prediction and Control.* Prentice-Hall, Englewood Cliffs, NJ, 1984.

[13] GRIGORESCU, S. Ergodic decomposition for continuous Markov chains. *Rev. Roum. Mathematics Pures Et Appl. XXI,* 6 (1976), 683–698.

[14] GUO, L., AND MEYN, S. P. Adaptive control for time-varying systems: A combination of Martingale and Markov chain techniques. *International Journal of Adaptive Control and Signal Processing 3* (1989), 1–14.

[15] HAJEK, B. Hitting-time and occupation-time bounds implied by drift analysis with applications. *Advances in Applied Probability 14* (1982), 502–506.

[16] HAS'MINSKII, R. Z. *Stochastic stability of differential equations.* Sijthoff & Noordhoff Alphen an den Rijn, The Netherlands, Rockville, MD USA, 1980.

[17] ICHIHARA, K., AND KUNITA, H. A classification of the second order degenerate elliptic operators and its probabilistic characterization. *Z. Wahrscheinlichkeitstheorie verw. Gebiete 30* (1974), 235–254.

[18] JAKUBCZYK, B., AND SONTAG, E. D. Controllability of nonlinear discrete time systems: A lie-algebraic approach. To be published in *SIAM Journal on Control Optimization,* April, 1990.

[19] JAMISON, B. Asymptotic behavior of successive iterates of continuous functions under a Markov operator. *J. Mathematics Anal. Appl. 9* (1964), 203–214. MR29 #6295.

[20] JAMISON, B. Ergodic decomposition induced by certain Markov operators. *Trans. American Mathematics Society 117* (1965), 451–468.

[21] JAMISON, B. A result in Doeblin's theory of Markov chains implied by Suslin's conjecture. *Zeitschrift fur Wahrscheinlichkeitstheorie und verwandte Gebiete 24* (1972), 287–293.

[22] JAMISON, B., AND SINE, R. Sample path convergence of stable Markov processes. *Z. Wahrscheinlichkeitstheorie 28* (1974), 173–177.

[23] KARLSEN, H. A. Existence of moments in a stationary stochastic difference equation. *Advances in Applied Probability 22*, 1 (March 1990), 129–146.

[24] KELLEY, J. L. *General Topology.* Van Nostrand, Princeton, NJ, 1955.

[25] KLIEMANN, W. Recurrence and invariant measures for degenerate diffusions. *The Annals of Probability 15*, 2 (1987), 690–707.

[26] KUSHNER, H. J. *Stochastic stability and control.* Academic Press, New York, NY, 1967.

[27] KUSHNER, H. J. Stochastic stability. In *Lecture Notes in Mathematics.* Springer-Verlag, New York, NY, 1972.

[28] LAMPERTI, J. Criteria for the recurrence or transience of stochastic processes I. *J. Mathematics Anal. Appl. 1* (1960), 314–330.

[29] MEYN, S. P. Ergodic theorems for discrete time stochastic systems using a stochastic Lyapunov function. *SIAM Journal on Control and Optimization 27*, 6 (November 1989), 1409–1439.

[30] MEYN, S. P. A stability theory for Feller Markov chains with applications to adaptive control and time series analysis. Technical report, University of Illinois, Urbana-Champaign, IL, 1991.

[31] MEYN, S. P., AND CAINES, P. E. A new approach to stochastic adaptive control. *IEEE Transactions on Automatic Control AC-32*, 3 (1987), 220–226.

[32] MEYN, S. P., AND CAINES, P. E. Stochastic controllability and stochastic Lyapunov functions with applications to adaptive and nonlinear systems. Presented at the *Fourth Bad Honnef Conference on Differetial Systems*, June 1988. Germany.

[33] MEYN, S. P., AND CAINES, P. E. Asymptotic behavior of stochastic systems processing Markovian realizations. To appear in SIAM J. Control and Opt., March 1991.

[34] MEYN, S. P., AND GUO, L. Stability, convergence, and performance of an adaptive control algorithm applied to a randomly varying system. In *Proceedings of the 28th IEEE Conference on Decision and Control* (Tampa, FL, December 1989), pp. 2108–2113.

[35] MEYN, S. P., AND GUO, L. Adaptive control of time varying stochastic systems. In *Proceedings of the 11th IFAC World Congress* (August 13-17 1990), V. Utkin and O. Jaaksoo, Eds., vol. 3, pp. 198–202. Tallinn, Estonia, USSR.

[36] MEYN, S. P., AND GUO, L. Geometric ergodicity of a bilinear time series model. To appear in *Journal of Time Series Analysis*, 1990.

[37] MEYN, S. P., AND TWEEDIE, R. L. Criteria for stability of Markovian processes II: Continuous time processes. Submitted to *Advances Applied Prob.*

[38] MEYN, S. P., AND TWEEDIE, R. L. Criteria for stability of Markovian processes I: discrete time chains. Submitted to *Advances Applied Prob.*, March 1990.

[39] MEYN, S. P., AND TWEEDIE, R. L. Markov chains and stochastic stability. To appear in *Control and Communication in Engineering, Springer-Verlag*, 1991.

[40] NUMMELIN, E. A splitting technique for Harris recurrent Markov chains. *Zeitschrift fur Wahrscheinlichkeitstheorie und verwandte Gebiete 43* (1978), 309–318.

[41] NUMMELIN, E. *General Irreducible Markov Chains and Non-Negative Operators.* Cambridge University Press, Cambridge, MA, 1984.

[42] NUMMELIN, E., AND TUOMINEN, P. Geometric ergodicity of Harris recurrent Markov chains with applications to renewal theory. *Stochastic Processes and their Applications 12* (1982), 187–202.

[43] NUMMELIN, E., AND TWEEDIE, R. L. Geometric ergodicity and r-positivity for general Markov chains. *The Annals of Probability 6*, 3 (1978), 404–420.

[44] OREY, S. *Limit Theorems for Markov Chain Transition Probabilities.* Van Nostrand Reinhold Mathematical Studies 34, London, 1971.

[45] PARTHASARATHY, T. *Probability measures on metric spaces.* Academic Press, New York, NY, 1967.

[46] POURAHMADI, M. On stationarity of the solution of a doubly stochastic model. *J. Time Series Anal. 7*, 2 (1986), 123–131.

[47] ROSENBLATT, M. Equicontinuous Markov operators. *Teor. Verojatnost. i Primenen 9* (1964), 205–222. MR 30 #1549.

[48] ROSENBLATT, M. Recurrent points and transition functions acting on continuous functions. *Z. Wahrscheinlichkeitstheorie verw. Gebiete 30* (1974), 173–183.

[49] SAPERSTONE, S. H. *Semidynamical Systems in Infinite Dimensional Spaces.* Springer-Verlag, New York, NY, 1981.

[50] SINE, R. Convergence theorems for weakly almost periodic Markov operators. *Israel Journal of Mathematics 19*, 3 (1974), 246–255.

[51] SINE, R. On local uniform mean convergence for Markov operators. *Pacific Journal of Mathematics 60*, 2 (1975), 247–252.

[52] SOLO, V. Stochastic adaptive control and Martingale limit theory. *IEEE Transactions on Automatic Control 35*, 1 (January 1990), 66–70.

[53] TJØSTHEIM, D. Some doubly stochastic time series models. *J. Time Series Anal. 7*, 1 (1986), 51–72.

[54] TUOMINEN, P., AND TWEEDIE, R. L. Markov chains with continuous components. In *Proceedings London Mathematics Society* (1979), vol. 3.

[55] TWEEDIE, R. L. Criteria for classifying general Markov chains. *Advances in Applied Probability 8* (1976), 737–771.

[56] TWEEDIE, R. L. Criteria for rates of convergence of Markov chains with application to queueing and storage theory. In *Probability, Statistics and Analysis*, J. F. C. Kingman and G. E. H. Reuter, Eds., 79 ed. London Mathematics Society Lecture Note Series, Cambridge University Press, Cambridge, MA, 1983.

[57] TWEEDIE, R. L. Invariant measures for Markov chains with no irreducibility assumptions. *J. Applied Probability* (1988), 275–285.

[58] WINKLER, W. Doeblin's and Harris's theory of Markov processes. *Z. Wahrscheinlichkeitstheorie verw. Gebiete 31* (1975), 79–88.

Lecture Notes in Control and Information Sciences

Edited by M. Thoma and A. Wyner

Vol. 117: K.J. Hunt
Stochastic Optimal Control Theory with Application in Self-Tuning Control
X, 308 pages, 1989.

Vol. 118: L. Dai
Singular Control Systems
IX, 332 pages, 1989

Vol. 119: T. Başar, P. Bernhard
Differential Games and Applications
VII, 201 pages, 1989

Vol. 120: L. Trave, A. Titli, A. M. Tarras
Large Scale Systems: Decentralization, Structure Constraints and Fixed Modes
XIV, 384 pages, 1989

Vol. 121: A. Blaquière (Editor)
Modeling and Control of Systems in Engineering, Quantum Mechanics, Economics and Biosciences
Proceedings of the Bellman Continuum Workshop 1988, June 13–14, Sophia Antipolis, France
XXVI, 519 pages, 1989

Vol. 122: J. Descusse, M. Fliess, A. Isidori, D. Leborgne (Eds.)
New Trends in Nonlinear Control Theory
Proceedings of an International Conference on Nonlinear Systems, Nantes, France, June 13–17, 1988
VIII, 528 pages, 1989

Vol. 123: C. W. de Silva, A. G. J. MacFarlane
Knowledge-Based Control with Application to Robots
X, 196 pages, 1989

Vol. 124: A. A. Bahnasawi, M. S. Mahmoud
Control of Partially-Known Dynamical Systems
XI, 228 pages, 1989

Vol. 125: J. Simon (Ed.)
Control of Boundaries and Stabilization
Proceedings of the IFIP WG 7.2 Conference Clermont Ferrand, France, June 20–23, 1988
IX, 266 pages, 1989

Vol. 126: N. Christopeit, K. Helmes M. Kohlmann (Eds.)
Stochastic Differential Systems
Proceedings of the 4th Bad Honnef Conference June 20–24, 1988
IX, 342 pages, 1989

Vol. 127: C. Heij
Deterministic Identification of Dynamical Systems
VI, 292 pages, 1989

Vol. 128: G. Einarsson, T. Ericson, I. Ingemarsson, R. Johannesson, K. Zigangirov, C.-E. Sundberg
Topics in Coding Theory
VII, 176 pages, 1989

Vol. 129: W. A. Porter, S. C. Kak (Eds.)
Advances in Communications and Signal Processing
VI, 376 pages, 1989.

Vol. 130: W. A. Porter, S. C. Kak, J. L. Aravena (Eds.)
Advances in Computing and Control
VI, 367 pages, 1989

Vol. 131: S. M. Joshi
Control of Large Flexible Space Structures
IX, 196 pages, 1989.

Vol. 132: W.-Y. Ng
Interactive Multi-Objective Programming as a Framework for Computer-Aided Control System Design
XV, 182 pages, 1989.

Vol. 133: R. P. Leland
Stochastic Models for Laser Propagation in Atmospheric Turbulence
VII, 145 pages, 1989.

Vol. 134: X. J. Zhang
Auxiliary Signal Design in Fault Detection and Diagnosis
XII, 213 pages, 1989

Vol. 135: H. Nijmeijer, J. M. Schumacher (Eds.)
Three Decades of Mathematical System Theory
A Collection of Surveys at the Occasion of the 50th Birthday of Jan C. Willems
VI, 562 pages, 1989

Vol. 136: J. Zabczyk (Ed.)
Stochastic Systems and Optimization
Proceedings of the 6th IFIP WG 7.1 Working Conference, Warsaw, Poland, September 12–16, 1988
VI, 374 pages. 1989

Lecture Notes in Control and Information Sciences

Edited by M. Thoma and A. Wyner

Vol. 137: S. L. Shah, G. Dumont (Eds.)
Adaptive Control Strategies for Industrial Use
Proceedings of a Workshop
Kananaskis, Canada, 1988
VI, 360 pages. 1989

Vol. 138: D. C. McFarlane, K. Glover
Robust Controller Design Using Normalized Coprime Factor Plant Descriptions
X, 206 pages. 1990

Vol. 139: V. Hayward, O. Khatib (Eds.)
Experimental Robotics I
The First International Symposium
Montreal, June 19-21, 1989
XVII, 613 pages. 1990

Vol. 140: Z. Gajic, D. Petkovski, X. Shen
Singularly Perturbed and Weakly Coupled Linear Control Systems
A Recursive Approach
VII, 202 pages. 1990

Vol. 141: S. Gutman
Root Clustering in Parameter Space
VIII, 153 pages. 1990

Vol. 142: A. N. Gündeş, C. A. Desoer
Algebraic Theory of Linear Feedback Systems with Full and Decentralized Compensators
V, 176 pages. 1990

Vol. 143: H.-J. Sebastian, K. Tammer (Eds.)
System Modelling and Optimization
Proceedings of the 14th IFIP Conference
Leipzig, GDR, July 3-7, 1989
X, 960 pages. 1990

Vol. 144: A. Bensoussan, J. L. Lions (Eds.)
Analysis and Optimization of Systems
Proceedings of the 9th International Conference, Antibes, June 12-15, 1990
XII, 992 pages. 1990

Vol. 145: M. B. Subrahmanyam
Optimal Control with a Worst-Case Performance Criterion and Applications
X, 133 pages. 1990

Vol. 146: D. Mustafa, K. Glover
Minimum Entropy H_∞ Control
IX, 144 pages. 1990

Vol. 147: J. P. Zolésio (Ed.)
Stabilization of Flexible Structures
Third Working Conference,
Montpellier, France, January 1989
V, 327 pages, 1990

Vol. 148: In preparation

Vol. 149: K.H. Hoffmann, W. Krabs (Eds.)
Optimal Control of Partial Differential Equations
Proceedings of the IFIP WG 7.2 International Conference, Irsee, April 9-12, 1990
VI, 245 pages. 1991

Vol. 150: L.C.G.J.M. Habets
Robust Stabilization in the Gap-topology
IX, 126 pages. 1991

Vol. 151: J. M. Skowronski, H. Flashner, R. S. Guttalu (Eds.)
Mechanics and Control
Proceedings of the 3rd Workshop on Control Mechanics in Honor of the 65th Birthday of George Leitmann
January 22-24, 1990, University of Southern California
IV, 497 pages. 1991

Vol. 152: J. D. Aplevich
Implicit Linear Systems
XI, 176 pages. 1991

Vol. 153: O. Hájek
Control Theory in the Plane
X, 272 pages. 1991

Vol. 154: A. Kurzhanski, I. Lasiecka (Eds.)
Modelling and Inverse Problems of Control for Distributed Parameter Systems
Proceedings of IFIP (W.G. 7.2)-IIASA Conference
Laxenburg, Austria, July 24-28, 1989
VII, 179 pages. 1991

Vol. 155: M. Bouvet, G. Bienvenu (Eds.)
High-Resolution Methods in Underwater Acoustics
V, 249 pages. 1991